Stem Cell-Based Tissue Repair

Stem Cell-Based Tissue Repair

Edited by

Raphael Gorodetsky
Sharett Institute of Oncology, Hadassah Hebrew-University, Jerusalem, Israel

Richard Schäfer
Harvard Stem Cell Institute, Harvard University, Cambridge MA, USA

RSC Publishing

ISBN: 978-1-84973-001-3

A catalogue record for this book is available from the British Library

© The Royal Society of Chemistry 2011

The RSC is not responsible for individual opinions expressed in this work.

Published by The Royal Society of Chemistry,
Thomas Graham House, Science Park, Milton Road,
Cambridge CB4 0WF, UK

Registered Charity Number 207890

For further information see our web site at www.rsc.org

Printed and bound in Great Britain by CPI Antony Rowe, Chippenham and Eastbourne

Preface

Regenerative medicine is a new medical science aimed to overcome the biological limitations in the restoration of damaged or missing organs in humans. In general, severely damaged or amputated organs in mammals would not regenerate spontaneously, as happens in organisms lower in the phylogenetic tree. This young field of medical science is still struggling to solve basic biological barriers and obstacles associated with the limited postnatal tissue regeneration potential.

Based on the need to design and replace damaged tissues, a new approach termed "tissue engineering" was conceptualized with the aim of incorporating and merging material science technologies with transplantation of cellular biological components. New biodegradable and biocompatible materials have been designed, attempting to create scaffolds which do not induce toxic, immunogenic or inflammatory responses. The simpler proposed technologies were based on implants of the cell-friendly matrices alone. The cells from the damaged tissue were expected to be induced to migrate, proliferate and fill the empty scaffolds to restore the function of the repaired organ.

Nevertheless, the anticipation to yield a functional tissue just by implants of bio-mimetic materials has been too optimistic. It became clear that such bio-technologies should incorporate the major players in the regenerative process in the form of implanted functional cells. Unfortunately, merely injecting isolated lineage-specific differentiated cells into the failing organs could not induce a significant repair.

The only routinely used cell based therapy that is applied is bone marrow transplantation, where progenitors of the hematopoietic system are injected. But even in such commonly used treatments, the delivered cells need the support of their specific homing niche in the form of the bone marrow stroma to survive and regenerate the hematopoietic system.

With the recognition of the cellular component as a critical factor in regenerative technologies, the introduction of stem cells of different types and

Stem Cell-Based Tissue Repair
Edited by Raphael Gorodetsky and Richard Schäfer
© The Royal Society of Chemistry 2011
Published by the Royal Society of Chemistry, www.rsc.org

origins drew much of the attention in this field. The fact that such cells can be induced to differentiate to a wide spectrum of cell types within the same germ layer, or in the case of very immature cells, to trans-differentiate to even a wider range of cells types, seemed to render them the holy grail for regenerative medicine. Moreover, recently it was proposed that not only cell differentiation but also their cross talk with resident cell types of the host and immuno-modulation may be a major factor in successful stem/progenitor based tissue regeneration.

The research and accumulated knowledge on stem cells derivation, identification and specific markers has soared in recent years. Nevertheless, the goal of finding ways to introduce the cells into the organs to regain lost functions has so far yielded only very limited results. The field still waits for major break-throughs to get closer to the clinics.

This book aims to present a wide range of approaches on stem cells based regenerative medicine. Issues related to isolation, derivation, identification and differentiation of progenitor cells and assembly of matrices to carry the cells are reviewed. These include know-how on new approaches of targeting and transport the adequate cells to their destination and enable their integration in the target damaged organs.

The contributors to this book include experts in a wide range of areas associated with regenerative medicine. Different points of view and possible approaches for tissue regeneration, as well as relevant considerations of how to incorporate the wide range of stem cells sources in different regenerative treatment regimens are presented. The book also emphasizes the remaining unsolved issues to be addressed by future research.

The editors wish to express their gratitude to the authors of the different chapters for their important contribution and for their success to highlight diverse issues associated with stem cells based regenerative medicine. Though this book attempts to present an up-to-date overview of this field, considering its very dynamic nature, the editors would appreciate any comments or sug-gestions from the readership. Such input could be incorporated in the next editions.

Raphael Gorodetsky and Richard Schaefer

Contents

Stem Cell-Based Tissue Repair
Edited by Raphael Gorodetsky and Richard Schäfer
© The Royal Society of Chemistry 2011
Published by the Royal Society of Chemistry, www.rsc.org

Promises and Limitations in the Application of Cell Therapy for Tissue Regeneration

RAPHAEL GORODETSKY

Laboratory of Biotechnology and Radiobiology, Sharett Institute of Oncology, Hadassah Hebrew University Medical Center, Jerusalem, Israel

1.1 Factors Affecting Morphogenesis and Normal Cellular Organization in Tissues and Organs

Differentiated tissues are composed of different types of specialized cells organized into complex structures that form functional organs. Mammals higher in the phylogenetic tree are estimated to have at least 250 different cell types, which develop from early common embryonic stem cells and get organized in the different organs by a complex homing process which is so far only partially understood.[1–5] In spite of the vast interest in this issue, very limited information is available with regard to the full cascades which regulate the migration of progenitor cells and integration in damaged tissues.[5–14]

From early embryogenesis onward, the different cell types in complex organs get organized in a manner dictated by the fine balance between their rate of proliferation and self renewal on one hand, and cell death or apoptosis on the other hand. Signaling by a large number of membrane cell receptors is a crucial factor in the organization of tissues and organs from single cells, but the exact role of only a small portion of such signals has so far been investigated and deciphered. With current knowledge it is clear that the behavior of cells in

Stem Cell-Based Tissue Repair
Edited by Raphael Gorodetsky and Richard Schäfer
© The Royal Society of Chemistry 2011
Published by the Royal Society of Chemistry, www.rsc.org

normal tissues and tumors and their targeting and homing is dictated by interactions with the extracellular matrix (ECM) and chemokines.[15–18] There are also numerous direct messages transmitted by physical contact and various physical interactions, membrane potential, chemical messages, signaling agents that diffuse out from adjacent cells *via* cell–cell junctions, signals that are delivered systemically by the circulation and lymph pathways, or controlled stimulation by nerve ends through synapses.[15–21] Not only the cells, but the whole organs do not operate autonomously and the function of one organ may be immediately affected by malfunctioning of others—even in cases where this interaction is not straightforward. For instance, lack of nerve input to tissues such as muscle will result in the degeneration of the muscles and massive cell loss in this tissue though it may be otherwise intact.[22–25]

1.2 Endogenous Cell-based Repair of Damaged Tissues

Since in many disorders and injuries the repair of the damage and tissue regeneration is attributed to cells that reside in the damaged tissues, it is tempting to try to seek external delivery of reinforcement in the form of implanted differentiated or multipotent cells to help the organs overcome the complex healing process. The different options of cell-based therapies with different cell sources, matrix scaffolds and growth factors are summarized in Figure 1.1.

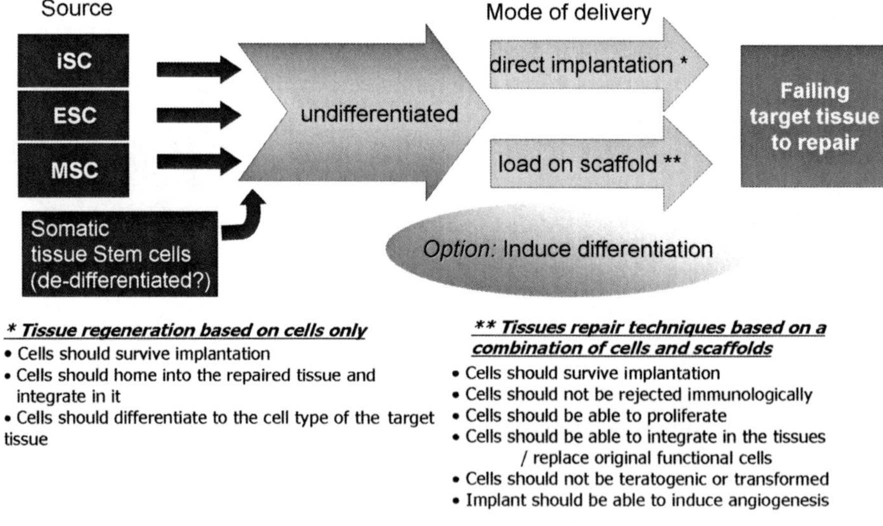

Figure 1.1 Schematic presentation of the optional potential combinations of different cellular components, differentiation procedures and delivery for regenerative medicine. Key: ESC = embryonic stem cells; iSC = inducted stem cells; MSC = mesenchymal stem cells.

It should be noted that, in general, though the adult animals higher in the phylogenetic tree have the ability to somehow overcome tissue damage and repair trauma, this ability is limited to moderate injuries and in most cases such repair is also associated with the formation of non-functional scar tissue that replaces severely damaged tissue and may even interfere with the function of the healed organ. With a few exceptions,[26] the ability to replace damaged organs is lost from the early stage of prenatal development and after birth, with a gradual decrease in the natural ability to fully repair severe damage and regenerate tissues in damaged organs.[12,27–37] Therefore, it was suggested that organs can maintain functionality and repair spontaneously limited damage using a reservoir of an adequate stock of progenitor cells within the tissue.[38–45] Such cells should exist as progenitors in soft and hard tissues with fast cell turnover. For instance in the gut, where the crypt cells divide regularly to repair and replace worn-out damaged gut tissue, and in bones, where the osteoblasts are constantly going through natural turnover which is regulated, among other factors, by physical load and growth hormones. In some cases it has been shown that the adult progenitor cells may have also limited trans-differentiation potential, which renders them partially multipotent stem cells.[46,47] In most cases, the healing and cell replacement in damaged tissues is not performed by the differentiated functional cells, which have lost their proliferative potential along with their specialization. It is disputed as to what extent progenitor cells within tissues (often referred as pericytes) are involved in the repair of the damaged tissue by providing an adequate cell reservoir to replace functional cells in the tissues where they reside.[46,48–53]

1.3 Cellular Implants for Regenerative Therapies of Damaged Organs and Tissues

The discovery of progenitors and stem cells of all kinds raised the expectation that cell-based therapy will be able to solve major problems associated with tissue regeneration, especially where autologous cell implants are involved (Figure 1.2). With these high expectations, numerous studies have been published in this field of which only a few showed significant prospects of feasibility.

The simple approach of progenitor cells delivery in suspension for the induction of organ regeneration seems highly appealing. Nevertheless, in most cases where this approach has been investigated, it was not found to provide an easy and straightforward solution.[54–58] Even allogeneic cells were examined, though it is expected that such cells will not survive long in the regenerated tissue due to rejection.[59–64] As to autologous cells, though they are not expected to be rejected by the immune system, both experimental and early preliminary clinical works showed that in the best case only a small fraction of cells could become incorporated in the damaged tissues after local or systemic delivery as a cell suspension; this was also the case when progenitors that had been induced to differentiate specifically to the cells of the target organs were implanted.[65–68] The fact that similar results were obtained in many cases with the implant of

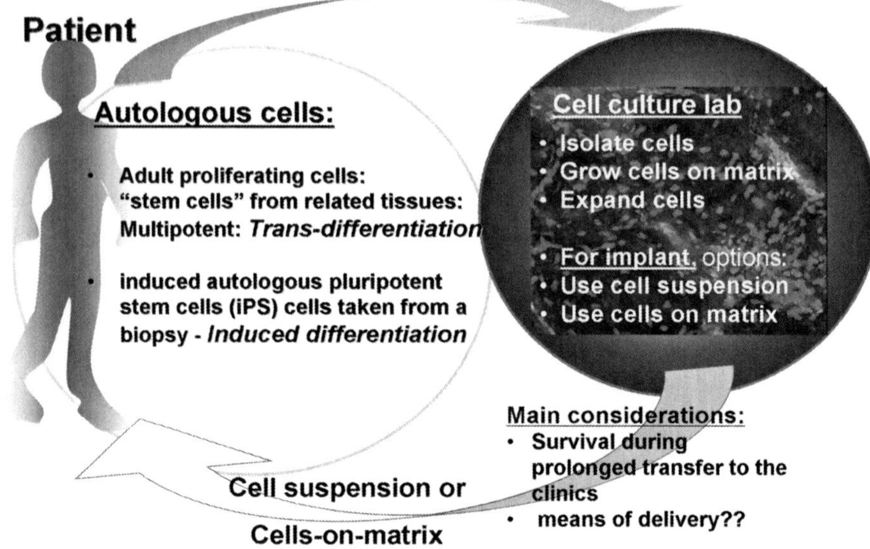

Figure 1.2   The basics of concepts on the use of self (autologous) progenitor cells for tissue engineering using autologous and induced stem cells.

cells from syngeneic or allogeneic sources[60,62,64,69] hints that the cells were probably not incorporated in the repaired tissue and that their claimed fringe effect on the regeneration of the damaged tissues was indirect.

1.4   Potential Practical Application of Stem Cells for Tissue Regeneration

1.4.1   Definition of Stem Cells

Stem cells have no clear-cut definition. In general this terminology refers to cells that are less differentiated or are part of a reservoir of unspecialized precursor which can divide and differentiate to form adult specialized cells. All these qualities can be summarized in two major properties—self-renewal by multiple cell division and the ability to differentiate into the target specialized cells. The highest degree of 'stemness' is associated with the ability of such cells to turn into a wide range of phenotypes which are found in the three main germ layers. This quality, which is also termed pluripotency, is normally associated with the early embryonic stage. With the progress in embryonic development, most cells irreversibly lose their plasticity to trans-differentiate while irreversibly shutting down the expression of most genes by an epigenetic process of non-reversible gene inactivation.[70–77] Nevertheless, some small populations of cells with the ability to form a few cell types, mostly of the same germ layer, still remain in adults and may contribute to the maintenance of the adult organs and their ability to overcome damage. Those are called multipotent stem cells. The other feature of stem cells is associated with their ability to proliferate and have multiple divisions. This ability is believed to be mainly associated with the ability of stem cells to maintain and regenerate their longer chromosomal telomeres, which are normally shortened in specialized adult cells along cell divisions and with aging.[78–82]

1.4.2 Somatic ('Adult') Multipotent Stem Cells

The first clonogenic stem cells were first identified in bone marrow, where they could replace radiation-irradiated bone marrow.[83] This is also the organ from which hematopoietic stem cells have been routinely and successfully used for the last few decades in bone marrow transplantation (termed bone marrow hematopoietic stem cell transplantation—HSCT) to treat many disorders associated with a malfunctioning or malignant hematopoietic system. Such stem cells may derive from different sources besides bone marrow, including mobilized blood (growth factors induced acceleration of hematopoietic cell generation) or cord blood. But in addition to the vast number of hematopoietic stem cells, a fraction of mesodermal cells can be isolated from the stroma of bone marrow. They were initially termed stromal cells which can be induced to differentiate to various cell phenotype of the mesenchymal germ layer. Therefore, these cells have been referred to in the last decade as mesenchymal stem cells. Among other major sources that could provide such multipotent cells are the adipose derived progenitor, which can be available in high numbers from simple liposuction procedures.[84–91] The easily accessible raw material renders it a major candidate for future regenerative medicine.

A claim for the existence of a naturally occurring subpopulation within adult bone marrow derived stem cell progenitors that are multipotent (multipotent adult progenitor cells—MAPC), which can produce cell types of the three germ layers was proposed and gained a lot of attention.[92] Some of the data behind these findings were eventually found to be problematic and a relevant publication was later withdrawn due to a claim that part of the data were flawed. This could serve as a good example as how too high expectations in this field can distort the scientific integrity of the bench researchers who are keen to deliver in spite of objective unsolved difficulties.

1.4.3 'Embryonic' Pluripotent Stem Cells

Researchers discovered ways to derive embryonic stem cells from early mouse embryos in 1981.[93,94] The detailed study of the biology of mouse stem cells led to the discovery of a method to derive stem cells from human embryos and grow them in the laboratory.[95] The embryos used in these studies were created for reproductive purposes through *in vitro* fertilization procedures. The isolated cells could be grown with feeder cells in special medium conditions. They produced non-differentiated cell lines that could be induced to differentiate to cells from different germlines.[96–102] Their other main feature is their tendency to form spontaneously teratomas—tumors of a mixture of different partially differentiated embryonic tissues.[103–107] This turned out to be the main obstacle to their use apart from the difficulties in their derivation and expansion. Their tendency to regain the donor human leukocyte antigen (HLA) phenotype after differentiation and thereby to be rejected post-implantation is another major concern in the attempt to use such cells for tissue regeneration.[108–111]

1.4.4 'Induced' Pluripotent Stem Cells

In 2006 a breakthrough was made with the identification of conditions that would allow some specialized adult cells to be 'reprogrammed' genetically to assume a stem cell-like state and reverse their epigenetic barriers. This new type of stem cells was termed 'induced pluripotent stem cells' (iPS cells).[71–77,112,113] By practicing this approach any adult cell could be 'reversed' to embryonic-like pluripotent cells by the introduction of a selected set of regulatory genes, either by viral vectors or other non-viral transfection. The advantage of this approach is that it could provide individualized therapy with no need to develop numerous cell lines and without the risk of rejection due to the self origin of the donor cell. It was suggested that iPS cells have a lower proliferative potential than embryonic stem cells[114] and in that perspective they may behave more like other adult progenitors. It was also proposed that they can form teratomas or be hazardous due to the use of viral vectors.[115] These issues pose a major difficulty in their future application.

1.5 Regenerative Medicine

1.5.1 The Combination of Artificial Scaffolds and Cells

Regenerative medicine was initiated a few decades ago under the title of 'tissue engineering'. This field has been led by researchers many of whom originated from engineering, technical and material science.[116–122]

The basis of tissue engineering is to try to replace non-functioning tissues and organs by their reconstruction, or to repair damaged organs by artificial cellular implants. The rational is that the cells that handle normal wear may have only limited ability to handle major damage. Therefore, it was proposed to aid natural spontaneously occurring cellular maintenance–repair by external intervention with major biological cellular components that can replace missing functional cells in the damaged tissues.[123–127] Essentially, this approach is based on the introduction of the tissue derived cells or stem cells, which could be delivered either directly into damaged organs or through the circulation, counting on the possible homing of the cells to replace the non-functional cells and thereby help repair or even reconstruct failing organs (Figure 1.1).

This approach is based on the expectation that complex biological systems could operate according to the principles behind sophisticated mechanics and material science. It was anticipated that components such as cells and scaffolds

(Figure 1.1) could efficiently replace the naturally organized cells in the tissues and organs.[128–137]

Indeed, tissue engineering has captured the imagination of professionals, as well as non-professionals, with a logarithmic increase of works on this direction. Nevertheless, even the early works in this field indicated that there are enormous drawbacks in trying to use cells from different organs to help repair and enhance the healing of organs, not mentioning their re-engineering.[138] Most techniques that have been proposed are based on *in vitro* prepared cell-bearing scaffolds, synthetic or biological polymer materials— either with the differentiated cells of the damaged tissue or pluripotent stem cells with the addition of a cocktail of growth factors. Nevertheless, long experimental records over the years indicate that tissues cannot so easily be engineered by this simplified approach as implantable functional tissue replacement.

1.5.2 Main Biological Difficulties Associated with Tissue Engineering

Matrices for tissue engineering need to be cell friendly to allow cells to load and attach, to secrete extracellular matrices, to proliferate and to gain functionality. For *in vitro* studies this may be sufficient, but for *in vivo* applications, the problem is that most artificial matrices are conceived by the body as foreign bodies and are prone to induce inflammation and rejection. In this process the cells are also prone to be lost within the inflamed area. A matrix containing foreign proteins may also be immunogenic. Among the more practical approaches to overcome this problem is the attempt to reduce rejection by implants coated with relatively inert polymers or made of human derived or modified proteins[139–146] and macromolecules that are expected to be accepted by the recipient's immune system.[104,147–149]

The main problem in tissue engineering that, even with optimal choice of cells and scaffolds, in most tissues the integration of artificial cellular constructs into the damaged tissues requires immediate formation of a vascular network to nourish the cells. In organ transplantation, the first concern is to connect the implanted organ to the host circulation immediately. Each of the implanted living cells needs to be located at least 100–150 µm from an active flowing blood capillary to survive. This cannot be achieved with most currently proposed artificial 3D tissue engineering methods. Even in cases where angiogenesis is eventually induced by the implant, the process does not occur immediately. Consequently, the cells impregnated within the matrix are doomed to die from severe hypoxia before such angiogenesis occurs.

1.5.3 Regulatory Burden in Practicing Tissue Regeneration Technologies

A major issue to be considered when practicing cellular tissue engineering technologies is the regulatory burden (Figure 1.3). Organ or tissue transplants

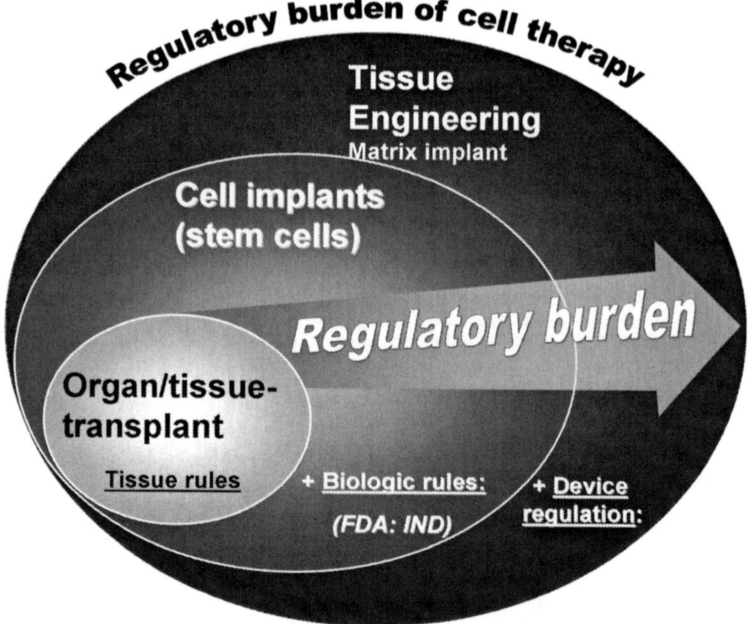

Figure 1.3 A presentation of the accumulated regulatory burden that accompanies tissue regeneration technologies. While the transplant of biological tissues and organs has the least regulatory burden, the implants of cell or stem cells have more restrictions and regulations. The combination of cells loaded on artificial matrices as practised in tissue engineering technologies has to confront the higher regulatory burden. Key: FDA: IND = US Food and Drug Administration Investigational New Drug.

are regarded as surgical procedures and are the easiest from a regulatory point of view as the technical validity of the operation is proven. The additional core regulatory burden relates to the non-cellular matrix or scaffold implantation. The regulatory burden of using this component is still not the highest, since only the matrices need to be tested to confirm they are biocompatible and not toxic or teratogenic. Here again, if surgery is involved, the surgical procedure requires an additional approval. When cellular constructs are proposed, the regulatory burden gets even harder due to a further need to confirm the biocompatibility of the scaffold, to approve the cell source, growth conditions for the cell loading on the matrix in terms of the safety of preparation and proving the added efficacy of the component of the cell implant. All this renders cellular tissue engineering the hardest to obtain approval for.

1.6   Examples of Tissue Engineering Approaches

So far most attempts to introduce tissue engineering products to clinical practice have been directed at skin. Current experience with 'bioengineered skin' produced by different commercial entities is relatively extensive;[134,150–164] elaborate techniques have been used to ensure that the artificial matrix would be 'cell friendly', non-immunogenic and biodegradable. Notwithstanding this, where foreign allogeneic cells were used, they were prone to be ultimately rejected due to immunologic reaction. But even using a biocompatible matrix embedded with non-rejected cells in 3D is problematic because such an implant lacks the necessary vascularization that may allow cell survival. It should be noted that even grafted natural full depth skin in the same individual may not take unless vascularization is maintained by using flap techniques that allow minimal blood perfusion into the area (as commonly practiced by plastic surgeons). Therefore, a more realistic approach by some R&D companies in this field is to admit that the main benefit from the cellular skin implant, in which the cells do not survive long, derives from the secretion of growth factors to enhance the regeneration of the damaged skin.

The difficulties with skin replacements attest to the problems with general practical tissue engineering in other fields where grafted cellular materials with appropriate cells such as autologous stem cells and extracellular matrices are used.

Of major interest are the attempts to re-engineer organs where adequate vascularization is most critical such as the heart and specifically the myocardium.[97,99,102,105,143] In normal circumstances, the native cardiac muscle would not survive more than a few minutes without a blood supply even if the blood vessels are reconnected and reperfusion occurs. Therefore, there is conceptual problem in attempts to build artificially cardiac muscles and implant them to replace damaged myocardium without connecting the implant immediately into a functioning intense vascularization. Moreover, even if this problem could also be overcome, a major risk of arrhythmia due to non-synchronized pacing of the implant with the damaged heart may occur.

Other non-tissue-engineering approaches have been proposed for enhancing the function of failing hearts and other tissues. These include the direct injection of stem cells of different origin into the damaged tissue or the circulation, with the hope that enough cells out of the vast number of the injected cells will survive and integrate in the damaged tissue to rescue the myocardium by inducing angiogenesis or even enhance repair of the damaged muscles. The results so far are controversial.[97,102,105] Interesting experiments of de-cellularization of heart tissue and replacement of the cells in the naked native scaffold with stem cells that can form and replace the original myocytes and blood vessels coating have been recently reported.[165–169] But this approach is far from being feasible as a potential treatment to replace whole heart transplants.

The exception in tissue regeneration is tissues such as cartilage in which the cells are normally trapped as small islands and, in normal circumstances, are set to survive with minimum vascularization. Such tissues may be best candidates for cell-based bulk tissue engineering.[170–172] Another major drawback with regeneration of this tissue is that, with the lack of natural process of tissue regeneration, it is hard to integrate the implant as an integral part of the cartilage to replace damaged or worn-out tissue.

One of the approaches for more successful delivery of appropriate cells in conditions that will allow them to stay viable and even integrate in the implanted tissues is their transplantation attached to biocompatible small cell carriers. A carrier loaded with a high titer of cells *in vitro* could deliver cells to the target site, where they would download and degrade the carrier while rebuilding the damaged tissue. If the carrier is both biocompatible and bio-degradable at an adequately slow rate, it could download the cells more efficiently. An example of this approach is the use of dense fibrin microbeads made of pure condensed fibrin to which adult stem cells could attach and further proliferate, and slowly download into the vascularized tissue to be repaired after their implantation.[31,142,173]

References

1. M. Hristov and C. Weber, Progenitor cell trafficking in the vascular wall, *J. Thromb. Haemost.*, 2009, **7**(Suppl 1), 31–34.
2. K. L. Kroeze, W. J. Jurgens, B. Z. Doulabi, F. J. van Milligen, R. J. Scheper and S. Gibbs, Chemokine-mediated migration of skin-derived stem cells, predominant role for CCL5/RANTES, *J. Invest. Dermatol.*, 2009, **129**, 1569–1581.
3. D. J. Laird, U. H. von Andrian and A. J. Wagers, Stem cell trafficking in tissue development, growth and disease, *Cell*, 2008, **132**, 612–630.
4. M. Mione, D. Baldessari, G. Deflorian. G. Nappo and C. Santoriello, How neuronal migration contributes to the morphogenesis of the CNS: insights from the zebrafish, *Dev. Neurosci.*, 2008, **30**, 65–81.
5. C. E. Burns and L. I. Zon, Homing sweet homing: odyssey of hematopoietic stem cells, *Immunity*, 2006, **25**, 859–862.
6. M. Hristov, W. Erl and P. C. Weber, Endothelial progenitor cells: mobilization, differentiation and homing, *Arterioscler. Thromb. Vasc. Biol.*, 2003, **23**, 1185–1189.
7. G. Chamberlain, J. Fox, B. Ashton and J. Middleton, Concise review: mesenchymal stem cells, their phenotype, differentiation capacity, immunological features and potential for homing, *Stem Cells*, 2007, **25**, 2739–2749.
8. M. Hristov, A. Zernecke, E. A. Liehn and C. Weber, Regulation of endothelial progenitor cell homing after arterial injury, *Thromb. Haemost.*, 2007, **98**, 274–277.
9. O. C. Velazquez, Angiogenesis and vasculogenesis: inducing the growth of new blood vessels and wound healing by stimulation of bone marrow-derived progenitor cell mobilization and homing, *J. Vasc. Surg.*, 2007, **45**(Suppl A), A39–47.
10. S. Khaldoyanidi, Directing stem cell homing, *Cell Stem Cell*, 2008, **2**, 198–200.
11. J. M. Karp and G. S. Leng Teo, Mesenchymal stem cell homing: the devil is in the details, *Cell Stem Cell*, 2009, **4**, 206–216.

12. S. C. Wheatley, C. M. Isacke and P. H. Crossley, Restricted expression of the hyaluronan receptor, CD44, during postimplantation mouse embryogenesis suggests key roles in tissue formation and patterning, *Development*, 1993, **119**, 295–306.
13. S. Corti, F. Locatelli, D. Papadimitriou, C. Donadoni, R. Del Bo, F. Fortunato, S. Strazzer, S. Salani, N. Bresolin and G. P. Comi, Multipotentiality, homing properties and pyramidal neurogenesis of CNS-derived LeX(ssea-1)+/CXCR4+ stem cells, *FASEB J.*, 2005, **19**, 1860–1862.
14. G. A. Colvin, J. F. Lambert, M. S. Dooner, J. Cerny and P. J. Quesenberry, Murine allogeneic *in vivo* stem cell homing(,), *J. Cell. Physiol.*, 2007, **211**, 386–391.
15. J. E. Meredith Jr, B. Fazeli and M. A. Schwartz, The extracellular matrix as a cell survival factor, *Mol. Biol. Cell*, 1993, **4**, 953–961.
16. K. M. Yamada and K. Clark, Cell biology, survival in three dimensions, *Nature*, 2002, **419**, 790–791.
17. L. A. Liotta and E. C. Kohn, The microenvironment of the tumour–host interface, *Nature*, 2001, **411**, 375–379.
18. J. L. Lee, M. J. Wang, P. R. Sudhir and J. Y. Chen, CD44 engagement promotes matrix-derived survival through the CD44-SRC-integrin axis in lipid rafts, *Mol. Cell. Biol.*, 2008, **28**, 5710–5723.
19. C. C. Chen and L. F. Lau, Functions and mechanisms of action of CCN matricellular proteins, *Int. J. Biochem. Cell Biol.*, 2009, **41**, 771–783.
20. M. Ebadi, R. M. Bashir, M. L. Heidrick, F. M. Hamada, H. E. Refaey, A. Hamed, G. Helal, M. D. Baxi, D. R. Cerutis and N. K. Lassi, Neurotrophins and their receptors in nerve injury and repair, *Neurochem. Int.*, 1997, **30**, 347–374.
21. T. Gordon, T. M. Brushart and K. M. Chan, Augmenting nerve regeneration with electrical stimulation, *Neurol. Res.*, 2008, **30**, 1012–1022.
22. S. Rochkind, S. Geuna and A. Shainberg, Chapter 25: Phototherapy in peripheral nerve injury, effects on muscle preservation and nerve regeneration, *Int. Rev. Neurobiol.*, 2009, **87**, 445–464.
23. I. Biros and S. Forrest, Spinal muscular atrophy, untangling the knot?, *J. Med. Genet.*, 1999, **36**, 1–8.
24. A. E. Emery, The nosology of the spinal muscular atrophies, *J. Med. Genet.*, 1971, **8**, 481–495.
25. M. R. Lunn and C. H. Wang, Spinal muscular atrophy, *Lancet*, 2008, **371**, 2120–2133.
26. J. S. Price, S. Allen, C. Faucheux, T. Althnaian and J. G. Mount, Deer antlers: a zoological curiosity or the key to understanding organ regeneration in mammals?, *J. Anat.*, 2005, **207**, 603–618.
27. S. P. Bruder, K. H. Kraus, V. M. Goldberg and S. Kadiyala, The effect of implants loaded with autologous mesenchymal stem cells on the healing of canine segmental bone defects, *J. Bone Joint Surg. Am.*, 1998, **80**, 985–996.
28. X. Guo, C. Wang, Y. Zhang, R. Xia, M. Hu, C. Duan, Q. Zhao, L. Dong. J. Lu and Y. Qing Song, Repair of large articular cartilage defects with

implants of autologous mesenchymal stem cells seeded into b-tricalcium phosphate in a sheep model, *Tissue Eng.*, 2004, **10**, 1818–1829.

29. D. Hannouche, H. Terai, J. R. Fuchs, S. Terada, S. Zand, B. A. Nasseri, H. Petite, L. Sedel and J. P. Vacanti, Engineering of implantable cartilaginous structures from bone marrow-derived mesenchymal stem cells, *Tissue Eng.*, 2007, **13**, 87–99.

30. R. Rivkin, A. Ben-Ari, I. Kassis, L. Zangi, E. Gaberman, L. Levdansky, G. Marx and R. Gorodetsky, High-yield isolation, expansion and differentiation of murine bone marrow-derived mesenchymal stem cells using fibrin microbeads (FMB), *Cloning Stem Cells*, 2007, **9**, 157–175.

31. L. Zangi, R. Rivkin, I. Kassis, L. Levdansky, G. Marx and R. Gorodetsky, High-yield isolation, expansion and differentiation of rat bone marrow-derived mesenchymal stem cells with fibrin microbeads, *Tissue Eng.*, 2006, **12**, 2343–2354.

32. R. Shainer, E. Gaberman, L. Levdansky and R. Gorodetsky, Efficient isolation and chondrogenic differentiation of adult mesenchymal stem cells with fibrin microbeads and micronized collagen sponges, *Regen. Med.*, 2010, **5**, 255–265.

33. J. M. Lane, E. Tomin and M. P. Bostrom, Biosynthetic bone grafting, *Clin. Orthop. Relat. Res.*, 1999, **367**(Suppl), S107–117.

34. L. Jackson, D. Jones, P. Scotting and V. Sottile, Adult mesenchymal stem cells: differentiation potential and therapeutic applications, *J. Postgrad. Med.*, 2007, **53**, 121–127.

35. J. Ringe, C. Kaps, G.-R. Burmester and M. Sittinger, Stem cells for regenerative medicine: advances in the engineering of tissues and organs, *Naturwissenschaften*, 2002, **89**, 338–351.

36. H. Desai, Ageing and wounds. Part 1: foetal and postnatal healing, *J. Wound Care*, 1997, **6**, 192–196.

37. D. A. Cox, S. Kunz, N. Cerletti, G. K. McMaster and R. R. Burk, Wound healing in aged animals: effects of locally applied transforming growth factor beta 2 in different model systems, *EXS*, 1992, **61**, 287–295.

38. P. Prang, D. Del Turco and J. P. Kapfhammer, Regeneration of entorhinal fibers in mouse slice cultures is age dependent and can be stimulated by NT-4, GDNF, and modulators of G-proteins and protein kinase C, *Exp. Neurol.*, 2001, **169**, 135–147.

39. D. M. Lewis and H. Schmalbruch, Effects of age on aneural regeneration of soleus muscle in rat, *J. Physiol.*, 1995, **488**, 483–492.

40. J. Westermann, K. U. Willfuhr and R. Pabst, Influence of donor and host age on the regeneration and blood flow of splenic transplants, *J. Pediatr. Surg.*, 1988, **23**, 835–838.

41. R. S. Tare, J. C. Babister, J. Kanczler and R. O. Oreffo, Skeletal stem cells: phenotype, biology and environmental niches informing tissue regeneration, *Mol. Cell. Endocrinol.*, 2008, **288**, 11–21.

42. C. C. Hughes, Endothelial–stromal interactions in angiogenesis, *Curr. Opin. Hematol.*, 2008, **15**, 204–209.

43. B. Eyden, The myofibroblast: phenotypic characterization as a prerequisite to understanding its functions in translational medicine, *J. Cell. Mol. Med.*, 2008, **12**, 22–37.

44. M. S. Davidoff, R. Middendorff, D. Muller and A. F. Holstein, The neuroendocrine Leydig cells and their stem cell progenitors, the pericytes, *Adv. Anat. Embryol. Cell Biol.*, 2009, **205**, 1–107.

45. S. Bhagavati, Stem cell based therapy for skeletal muscle diseases, *Curr. Stem Cell Res. Ther.*, 2008, **3**, 219–228.

46. G. T. Huang, S. Gronthos and S. Shi, Mesenchymal stem cells derived from dental tissues *vs.* those from other sources, their biology and role in regenerative medicine, *J. Dent. Res.*, 2009, **88**, 792–806.

47. G. T. Huang, Pulp and dentin tissue engineering and regeneration: current progress, *Regen. Med.*, 2009, **4**, 697–707.

48. A. S. Breathnach, Development and differentiation of dermal cells in man, *J. Invest. Dermatol.*, 1978, **71**, 2–8.

49. A. C. Nag, Study of non-muscle cells of the adult mammalian heart: a fine structural analysis and distribution, *Cytobios*, 1980, **28**, 41–61.

50. L. Leibnitz and B. Bar, A blood capillaries-bridging cell type in adult mammalian brains, *J. Hirnforsch.*, 1988, **29**, 367–375.

51. H. Takahashi-Iwanaga, The three-dimensional cytoarchitecture of the interstitial tissue in the rat kidney, *Cell Tissue Res.*, 1991, **264**, 269–281.

52. L. Diaz-Flores, R. Gutierrez, A. Lopez-Alonso, R. Gonzalez and H. Varela, Pericytes as a supplementary source of osteoblasts in periosteal osteogenesis, *Clin. Orthop. Relat. Res.*, 1992, **275**, 280–286.

53. K. J. Mitchell, A. Pannerec, B. Cadot, A. Parlakian, V. Besson, E. R. Gomes, G. Marazzi and D. A. Sassoon, Identification and characterization of a non-satellite cell muscle resident progenitor during postnatal development, *Nat. Cell Biol.*, 2010, **12**, 257–266.

54. K. C. Wollert, Cell therapy for acute myocardial infarction, *Curr. Opin. Pharmacol.*, 2008, **8**, 202–210.

55. H. M. Wei, P. Wong, L. F. Hsu and W. Shim, Human bone marrow-derived adult stem cells for post-myocardial infarction cardiac repair, current status and future directions, *Singapore Med. J.*, 2009, **50**, 935–942.

56. U. Noth, L. Rackwitz, A. F. Steinert and R. S. Tuan, Cell delivery therapeutics for musculoskeletal regeneration, *Adv. Drug Deliv. Rev.*, 2010, **62**, 765–783.

57. J. K. Leach and D. J. Mooney, Bone engineering by controlled delivery of osteoinductive molecules and cells, *Expert Opin. Biol. Ther.*, 2004, **4**, 1015–1027.

58. L. K. Branski, G. G. Gauglitz, D. N. Herndon and M. G. Jeschke, A review of gene and stem cell therapy in cutaneous wound healing, *Burns*, 2009, **35**, 171–180.

59. S. Ishikane, S. Ohnishi, K. Yamahara, M. Sada, K. Harada, K. Mishima, K. Iwasaki, M. Fujiwara, S. Kitamura, N. Nagaya and T. Ikeda, Allogeneic injection of fetal membrane-derived mesenchymal stem cells

induces therapeutic angiogenesis in a rat model of hind limb ischemia, *Stem Cells*, 2008, **26**, 2625–2633.

60. W. van't Hof, N. Mal, Y. Huang, M. Zhang, Z. Popovic, F. Forudi, R. Deans and M. S. Penn, Direct delivery of syngeneic and allogeneic large-scale expanded multipotent adult progenitor cells improves cardiac function after myocardial infarct, *Cytotherapy*, 2007, **9**, 477–487.

61. L. C. Amado, A. P. Saliaris, K. H. Schuleri, M. St John, J. S. Xie, S. Cattaneo, D. J. Durand, T. Fitton, J. Q. Kuang, G. Stewart, S. Lehrke, W. W. Baumgartner, B. J. Martin, A. W. Heldman and J. M. Hare, Cardiac repair with intramyocardial injection of allogeneic mesenchymal stem cells after myocardial infarction, *Proc. Natl. Acad. Sci. USA.*, 2005, **102**, 11474–11479.

62. K. H. Schuleri, L. C. Amado, A. J. Boyle, M. Centola, A. P. Saliaris, M. R. Gutman, K. E. Hatzistergos, B. N. Oskouei, J. M. Zimmet, R. G. Young, A. W. Heldman, A. C. Lardo and J. M. Hare, Early improvement in cardiac tissue perfusion due to mesenchymal stem cells, *Am. J. Physiol. Heart Circ. Physiol.*, 2008, **294**, H2002–2011.

63. C. A. Rasmussen, A. L. Gibson, S. J. Schlosser, M. J. Schurr and B. L. Allen-Hoffmann, Chimeric composite skin substitutes for delivery of autologous keratinocytes to promote tissue regeneration, *Ann. Surg.*, 2010, **251**, 368–376.

64. P. R. Adha, K. H. Chua, A. L. Mazlyzam, K. C. Low, B. S. Aminuddin and B. H. Ruszymah, Usage of allogeneic single layered tissue engineered skin enhance wound treatment in sheep, *Med. J. Malaysia*, 2008, **63**(Suppl A), 30–31.

65. C. M. Rosario, B. D. Yandava, B. Kosaras, D. Zurakowski, R. L. Sidman and E. Y. Snyder, Differentiation of engrafted multipotent neural progenitors towards replacement of missing granule neurons in meander tail cerebellum may help determine the locus of mutant gene action, *Development*, 1997, **124**, 4213–4224.

66. T. S. Li, M. Takahashi, M. Ohshima, S. L. Qin, M. Kubo, K. Muramatsu and K. Hamano, Myocardial repair achieved by the intramyocardial implantation of adult cardiomyocytes in combination with bone marrow cells, *Cell Transplant.*, 2008, **17**, 695–703.

67. A. Leo and D. Grande, Mesenchymal stem cells in tissue engineering, *Cells Tissues Organs*, 2006, **183**, 112–122.

68. A. Ben-Ari, R. Rivkin, M. Frishman, E. Gaberman, L. Levdansky and R. Gorodetsky, Isolation and implantation of bone marrow-derived mesenchymal stem cells with fibrin micro beads to repair a critical-size bone defect in mice, *Tissue Eng. Part A*, 2009, **15**, 2537–2546.

69. D. Wolf, A. Reinhard, A. Seckinger, L. Gross, H. A. Katus and A. Hansen, Regenerative capacity of intravenous autologous, allogeneic and human mesenchymal stem cells in the infarcted pig myocardium-complicated by myocardial tumor formation, *Scand. Cardiovasc. J.*, 2009, **43**, 39–45.

70. T. Tada, M. Tada, K. Hilton, S. C. Barton, T. Sado, N. Takagi and M. A. Surani, Epigenotype switching of imprintable loci in embryonic germ cells, *Dev. Genes Evol.*, 1998, **207**, 551–561.
71. M. F. Chan, R. van Amerongen, T. Nijjar, E. Cuppen, P. A. Jones and P. W. Laird, Reduced rates of gene loss, gene silencing, and gene mutation in Dnmt1-deficient embryonic stem cells, *Mol. Cell. Biol.*, 2001, **21**, 7587–7600.
72. N. Rodic, M. Oka, T. Hamazaki, M. R. Murawski, M. Jorgensen, D. M. Maatouk, J. L. Resnick, E. Li and N. Terada, DNA methylation is required for silencing of ant4, an adenine nucleotide translocase selectively expressed in mouse embryonic stem cells and germ cells, *Stem Cells*, 2005, **23**, 1314–1323.
73. P. J. Rugg-Gunn, A. C. Ferguson-Smith and R. A. Pedersen, Human embryonic stem cells as a model for studying epigenetic regulation during early development, *Cell Cycle*, 2005, **4**, 1323–1326.
74. N. Liu, M. Lu, X. Tian and Z. Han, Molecular mechanisms involved in self-renewal and pluripotency of embryonic stem cells, *J. Cell. Physiol.*, 2007, **211**, 279–286.
75. S. Atkinson and L. Armstrong, Epigenetics in embryonic stem cells, regulation of pluripotency and differentiation, *Cell Tissue Res.*, 2008, **331**, 23–29.
76. D. J. Lees-Murdock and C. P. Walsh, DNA methylation reprogramming in the germ line, *Adv. Exp. Med. Biol.*, 2008, **626**, 1–15.
77. P. Aranda, X. Agirre, E. Ballestar, E. J. Andreu, J. Roman-Gomez, I. Prieto, J. L. Martin-Subero, J. C. Cigudosa, R. Siebert, M. Esteller and F. Prosper, Epigenetic signatures associated with different levels of differentiation potential in human stem cells, *PLoS One*, 2009, **4**, e7809.
78. M. A. Blasco, Telomere length, stem cells and aging, *Nat. Chem. Biol.*, 2007, **3**, 640–649.
79. L. Harrington, Does the reservoir for self-renewal stem from the ends?, *Oncogene*, 2004, **23**, 7283–7289.
80. S. E. Holt, W. E. Wright and J. W. Shay, Multiple pathways for the regulation of telomerase activity, *Eur. J. Cancer*, 1997, **33**, 761–766.
81. F. Ishikawa, Regulation mechanisms of mammalian telomerase. A review, *Biochemistry (Mosc.)*, 1997, **62**, 1332–1337.
82. M. Mimeault and S. K. Batra, Recent insights into the molecular mechanisms involved in aging and the malignant transformation of adult stem/progenitor cells and their therapeutic implications, *Ageing Res. Rev.*, 2009, **8**, 94–112.
83. J. E. Till and E. A. McCulloch, Repair processes in irradiated mouse hematopoietic tissue, *Ann. N. Y. Acad. Sci.*, 1964, **114**, 115–125.
84. J. Gimble and F. Guilak, Adipose-derived adult stem cells, isolation, characterization, and differentiation potential, *Cytotherapy*, 2003, **5**, 362–369.
85. H. Mizuno and H. Hyakusoku, Mesengenic potential and future clinical perspective of human processed lipoaspirate cells, *J. Nippon Med. Sch.*, 2003, **70**, 300–306.

86. K. C. Hicok, T. V. Du Laney, Y. S. Zhou, Y. D. Halvorsen, D. C. Hitt, L. F. Cooper and J. M. Gimble, Human adipose-derived adult stem cells produce osteoid *in vivo*, *Tissue Eng.*, 2004, **10**, 371–380.
87. Y. S. Choi, S. N. Park and H. Suh, Adipose tissue engineering using mesenchymal stem cells attached to injectable PLGA spheres, *Biomaterials*, 2005, **26**, 5855–5863.
88. M. J. Seo, S. Y. Suh, Y. C. Bae and J. S. Jung, Differentiation of human adipose stromal cells into hepatic lineage *in vitro* and *in vivo*, *Biochem. Biophys. Res. Commun.*, 2005, **328**, 258–264.
89. Y. Xu, P. Malladi, D. R. Wagner and M. T. Longaker, Adipose-derived mesenchymal cells as a potential cell source for skeletal regeneration, *Curr. Opin. Mol. Ther.*, 2005, **7**, 300–305.
90. A. M. Parker and A. J. Katz, Adipose-derived stem cells for the regeneration of damaged tissues, *Expert Opin. Biol. Ther.*, 2006, **6**, 567–578.
91. Y. Yamada, X. D. Wang, S. Yokoyama, N. Fukuda and N. Takakura, Cardiac progenitor cells in brown adipose tissue repaired damaged myocardium, *Biochem. Biophys. Res. Commun.*, 2006, **342**, 662–670.
92. Y. Jiang, B. N. Jahagirdar, R. L. Reinhardt, R. E. Schwartz, C. D. Keene, X. R. Ortiz-Gonzalez, M. Reyes, T. Lenvik, T. Lund, M. Blackstad, J. Du, S. Aldrich, A. Lisberg, W. C. Low, D. A. Largaespada and C. M. Verfaillie, Pluripotency of mesenchymal stem cells derived from adult marrow, *Nature*, 2002, **418**, 41–49.
93. M. Monk, A stem-line model for cellular and chromosomal differentiation in early mouse-development, *Differentiation*, 1981, **19**, 71–76.
94. M. J. Evans and M. H. Kaufman, Establishment in culture of pluripotential cells from mouse embryos, *Nature*, 1981, **292**, 154–156.
95. J. A. Thomson, J. Itskovitz-Eldor, S. S. Shapiro, M. A. Waknitz, J. J. Swiergiel, V. S. Marshall and J. M. Jones, Embryonic stem cell lines derived from human blastocysts, *Science*, 1998, **282**, 1145–1147.
96. A. E. Bishop and H. J. Rippon, Stem cells—potential for repairing damaged lungs and growing human lungs for transplant, *Expert Opin. Biol. Ther.*, 2006, **6**, 751–758.
97. P. Gallo and G. Condorelli, Human embryonic stem cell-derived cardiomyocytes, inducing strategies, *Regen. Med.*, 2006, **1**, 183–194.
98. J. G. Grudeva-Popova, Cellular therapy–the possible future of regenerative medicine, *Folia Med. (Plovdiv.)*, 2005, **47**, 5–10.
99. I. Kehat and L. Gepstein, Human embryonic stem cells for myocardial regeneration, *Heart Fail. Rev.*, 2003, **8**, 229–236.
100. H. J. Rippon, S. Lane, M. Qin, N. S. Ismail, M. R. Wilson, M. Takata and A. E. Bishop, Embryonic stem cells as a source of pulmonary epithelium *in vitro* and *in vivo*, *Proc. Am. Thorac. Soc.*, 2008, **5**, 717–722.
101. I. L. Weissman, Translating stem and progenitor cell biology to the clinic, barriers and opportunities, *Science*, 2000, **287**, 1442–1446.
102. Y. F. Xiao, Cardiac application of embryonic stem cells, *Sheng Li Xue Bao*, 2003, **55**, 493–504.

103. H. Hentze, P. L. Soong, S. T. Wang, B. W. Phillips, T. C. Putti and N. R. Dunn, Teratoma formation by human embryonic stem cells: evaluation of essential parameters for future safety studies, *Stem Cell Res.*, 2009, **2**, 198–210.

104. J. G. Lees, S. A. Lim, T. Croll, G. Williams, S. Lui, J. Cooper-White, L. R. McQuade, B. Mathiyalagan and B. E. Tuch, Transplantation of 3D scaffolds seeded with human embryonic stem cells, biological features of surrogate tissue and teratoma-forming potential, *Regen. Med.*, 2007, **2**, 289–300.

105. J. Nussbaum, E. Minami, M. A. Laflamme, J. A. Virag, C. B. Ware, A. Masino, V. Muskheli, L. Pabon, H. Reinecke and C. E. Murry, Transplantation of undifferentiated murine embryonic stem cells in the heart, teratoma formation and immune response, *FASEB J.*, 2007, **21**, 1345–1357.

106. M. Dihne, C. Bernreuther, C. Hagel, K. O. Wesche and M. Schachner, Embryonic stem cell-derived neuronally committed precursor cells with reduced teratoma formation after transplantation into the lesioned adult mouse brain, *Stem Cells*, 2006, **24**, 1458–1466.

107. T. Fujikawa, S. H. Oh, L. Pi, H. M. Hatch, T. Shupe and B. E. Petersen, Teratoma formation leads to failure of treatment for type I diabetes using embryonic stem cell-derived insulin-producing cells, *Am. J. Pathol.*, 2005, **166**, 1781–1791.

108. G. W. Basak, S. Yasukawa, A. Alfaro, S. Halligan, A. S. Srivastava, W. P. Min, B. Minev and E. Carrier, Human embryonic stem cells hemangioblast express HLA-antigens, *J. Transl. Med.*, 2009, **7**, 27.

109. Z. Master and B. Williams-Jones, The global HLA banking of embryonic stem cells requires further scientific justification, *Am. J. Bioeth.*, 2007, **7**, 45–46, discussion W44–46.

110. C. J. Taylor, E. M. Bolton, S. Pocock, L. D. Sharples, R. A. Pedersen and J. A. Bradley, Banking on human embryonic stem cells, estimating the number of donor cell lines needed for HLA matching, *Lancet*, 2005, **366**, 2019–2025.

111. M. Drukker, G. Katz, A. Urbach, M. Schuldiner, G. Markel, J. Itskovitz-Eldor, B. Reubinoff, O. Mandelboim and N. Benvenisty, Characterization of the expression of MHC proteins in human embryonic stem cells, *Proc. Natl. Acad. Sci. U. S. A.*, 2002, **99**, 9864–9869.

112. A. Hotta and J. Ellis, Retroviral vector silencing during iPS cell induction, an epigenetic beacon that signals distinct pluripotent states, *J. Cell. Biochem.*, 2008, **105**, 940–948.

113. K. Okita, T. Ichisaka and S. Yamanaka, Generation of germline-competent induced pluripotent stem cells, *Nature*, 2007, **448**, 313–317.

114. Q. Feng, S. J. Lu, I. Klimanskaya, I. Gomes, D. Kim, Y. Chung, G. R. Honig, K. S. Kim and R. Lanza, Hemangioblastic derivatives from human induced pluripotent stem cells exhibit limited expansion and early senescence, *Stem Cells*, 2010, **28**, 704–712.

115. K. Miura, Y. Okada, T. Aoi, A. Okada, K. Takahashi, K. Okita, M. Nakagawa, M. Koyanagi, K. Tanabe, M. Ohnuki, D. Ogawa, E. Ikeda, H. Okano and S. Yamanaka, Variation in the safety of induced pluripotent stem cell lines, *Nat. Biotechnol.*, 2009, **27**, 743–745.
116. M. J. Levesque, J. F. Cornhill and R. M. Nerem, Vascular casting. A new method for the study of the arterial endothelium, *Atherosclerosis*, 1979, **34**, 457–467.
117. R. M. Nerem, Tissue engineering:, the hope, the hype and the future, *Tissue Eng.*, 2006, **12**, 1143–1150.
118. R. M. Nerem, Tissue engineering: confronting the transplantation crisis, *Proc. Inst. Mech. Eng. H*, 2000, **214**, 95–99.
119. T. Ziegler and R. M. Nerem, Tissue engineering a blood vessel: regulation of vascular biology by mechanical stresses, *J. Cell. Biochem.*, 1994, **56**, 204–209.
120. L. G. Cima, J. P. Vacanti, C. Vacanti, D. Ingber, D. Mooney and R. Langer, Tissue engineering by cell transplantation using degradable polymer substrates, *J. Biomech. Eng.*, 1991, **113**, 143–151.
121. R. Langer and J. P. Vacanti, Tissue engineering, *Science*, 1993, **260**, 920–926.
122. C. A. Vacanti and J. P. Vacanti, Bone and cartilage reconstruction with tissue engineering approaches, *Otolaryngol. Clin. North. Am.*, 1994, **27**, 263–276.
123. K. Naka, M. Ohmura and A. Hirao, Regulation of the self-renewal ability of tissue stem cells by tumor-related genes, *Cancer Biomark.*, 2007, **3**, 193–201.
124. D. Metcalf, Concise review: hematopoietic stem cells and tissue stem cells, current concepts and unanswered questions, *Stem Cells*, 2007, **25**, 2390–2395.
125. W. J. Huh, X. O. Pan, I. U. Mysorekar and J. C. Mills, Location, allocation, relocation: isolating adult tissue stem cells in three dimensions, *Curr. Opin. Biotechnol.*, 2006, **17**, 511–517.
126. S. S. Tholpady, R. Llull, R. C. Ogle, J. P. Rubin, J. W. Futrell and A. J. Katz, Adipose tissue: stem cells and beyond, *Clin. Plast. Surg.*, 2006, **33**, 55–62, vi.
127. M. Loeffler and I. Roeder, Tissue stem cells: definition, plasticity, heterogeneity, self-organization and models–a conceptual approach, *Cells Tissues Organs*, 2002, **171**, 8–26.
128. B. M. Abdallah and M. Kassem, Human mesenchymal stem cells: from basic biology to clinical applications, *Gene Ther.*, 2007, **15**, 109–116.
129. L. Bonassar and C. Vacanti, Tissue engineering: the first decade and beyond, *J. Cell. Biochem. Suppl.*, 1998, **30–31**, 297–303.
130. J. Young, J. Teumer, P. Kemp and N. Parenteau, *Approaches to transplanting engineered cells and tissues*, in *Principles of Tissue Engineering*, R. Lanza, R. Langer and W. Chick, (ed.), R.G. Landes Company, Austin, TX, 1997, pp. 297–307.
131. D. Michaeli and M. McPherson, Immunologic study of artificial skin used in the treatment of thermal injuries, *J. Burn Care Rehabil.*, 1990, **11**, 21–26.

132. A. T. Truong, A. Kowal-Vern, B. A. Latenser, D. E. Wiley and R. J. Walter, Comparison of dermal substitutes in wound healing utilizing a nude mouse model, *J. Burns Wounds*, 2005, **4**, e4.
133. C. Pham, J. Greenwood, H. Cleland, P. Woodruff and G. Maddern, Bioengineered skin substitutes for the management of burns: a systematic review, *Burns*, 2007, **33**, 946–957.
134. M. Ehrenreich and Z. Ruszczak, Update on tissue-engineered biological dressings, *Tissue Eng.*, 2006, **12**, 2407–2424.
135. G. D. Gentzkow, S. D. Iwasaki, K. S. Hershon, M. Mengel, J. J. Prendergast, J. J. Ricotta, D. P. Steed and S. Lipkin, Use of dermagraft, a cultured human dermis, to treat diabetic foot ulcers, *Diabetes Care*, 1996, **19**, 350–354.
136. L. Griffith, Emerging design principles in biomaterials and scaffolds for tissue engineering, *Ann. N.Y. Acad. Sci.*, 2002, **961**, 83–95.
137. Z. Lokmic, F. Stillaert, W. A. Morrison, E. W. Thompson and G. M. Mitchell, An arteriovenous loop in a protected space generates a permanent, highly vascular, tissue-engineered construct, *FASEB J.*, 2007, **21**, 511–522.
138. Y. Onuki, U. Bhardwaj, F. Papadimitrakopoulos and D. J. Burgess, A review of the biocompatibility of implantable devices: current challenges to overcome foreign body response, *J. Diabetes Sci. Technol.*, 2008, **2**, 1003–1015.
139. O. Gurevich, A. Vexler, G. Marx, T. Prigozhina, L. Levdansky, S. Slavin, I. Shimeliovich and R. Gorodetsky, Fibrin microbeads for isolating and growing bone marrow-derived progenitor cells capable of forming bone tissue, *Tissue Eng.*, 2002, **8**, 661–672.
140. V. Falanga, S. Iwamoto, M. Chartier, T. Yufit, J. Butmarc, N. Kouttab, D. Shrayer and P. Carson, Autologous bone marrow derived cultured mesenchymal stem cells delivered in a fibrin spray accelerate healing in murine and human cutaneous wounds, *Tissue Eng.*, 2007, **13**, 1299–1312.
141. L. F. Brown, N. Lanir, J. McDonagh, K. Tognazzi, A. Dvorak and H. Dvorak, Fibroblast migration in fibrin gel matrices, *Am. J. Pathol.*, 1993, **142**, 273–283.
142. R. Gorodetsky, A. Vexler, L. Levdansky and G. Marx, Fibrin microbeads (FMB) as biodegradable carriers for culturing cells and for accelerating wound healing, *Methods Mol. Biol.*, 2007, **238**, 11–24.
143. G. Zhang, Q. Hu, E. A. Braunlin, L. J. Suggs and J. Zhang, Enhancing efficacy of stem cell transplantation to the heart with a PEGylated fibrin biomatrix, *Tissue Eng. Part A*, 2008, **14**, 1025–1036.
144. W. D. Thompson, C. M. Stirk, W. T. Melvin and E. B. Smith, Plasmin, fibrin degradation and angiogenesis, *Nat. Med.*, 1996, **2**, 493.
145. R. Gorodetsky, A. Vexler, J. An, X. Mou and G. Marx, Haptotactic and growth stimulatory effects of fibrin(ogen) and thrombin on cultured fibroblasts, *J. Lab. Clin. Med.*, 1998, **131**, 269–280.
146. C. A. Acevedo, C. Weinstein-Oppenheimer, D. I. Brown, H. Huebner, R. Buchholz and M. E. Young, A mathematical model for the design of

fibrin microcapsules with skin cells, *Bioprocess. Biosyst. Eng.*, 2009, **32**, 341–351.

147. J. Y. Lee, K. H. Kim, S. Y. Shin, I. C. Rhyu, Y. M. Lee, Y. J. Park, C. P. Chung and S. J. Lee, Enhanced bone formation by transforming growth factor-beta1-releasing collagen/chitosan microgranules, *J. Biomed. Mater. Res. A*, 2006, **76**, 530–539.

148. G. E. Friedlaender, C. R. Perry, J. D. Cole, S. D. Cook, G. Cierny, G. F. Muschler, G. A. Zych, J. H. Calhoun, A. J. LaForte and S. Yin, Osteogenic protein-1 (bone morphogenetic protein-7) in the treatment of tibial nonunions, *J. Bone Joint Surg. Am.*, 2001, **83–A**(Suppl 1), S151–158.

149. T. Aigner and J. Stove, Collagens—major component of the physiological cartilage matrix, major target of cartilage degeneration, major tool in cartilage repair, *Adv. Drug Deliv. Rev.*, 2003, **55**, 1569–1593.

150. J. P. Vacanti, R. Langer, J. Upton and J. J. Marler, Transplantation of cells in matrices for tissue regeneration, *Adv. Drug Deliv. Rev.*, 1998, **33**, 165–182.

151. R. E. Horch, A. M. Munster and B. M. Achauer, *Cultured Human Keratinocytes and Tissue Engineered Skin Substitutes*, Thieme, Stuttgart, 2001.

152. J. Mansbridge, Skin substitutes to enhance wound healing, *Expert Opin. Investig. Drugs*, 1998, **7**, 803–809.

153. S. Gohari, C. Gambla, M. Healey, G. Spaulding, K. B. Gordon, J. Swan, B. Cook, D. P. West and J. C. Lapiere, Evaluation of tissue-engineered skin (human skin substitute) and secondary intention healing in the treatment of full thickness wounds after Mohs micrographic or excisional surgery, *Dermatol. Surg.*, 2002, **28**, 1107–1114, discussion 1114.

154. K. G. Donohue, P. Carson, M. Iriondo, L. Zhou, L. Saap, K. Gibson and V. Falanga, Safety and efficacy of a bilayered skin construct in full-thickness surgical wounds, *J. Dermatol.*, 2005, **32**, 626–631.

155. M. Griffiths, N. Ojeh, R. Livingstone, R. Price and H. Navsaria, Survival of Apligraf in acute human wounds, *Tissue Eng.*, 2004, **10**, 1180–1195.

156. E. Epstein, Evidence for living cells: DNA fragments are not enough, *Arch. Dermatol.*, 2003, **139**, 541, author reply, 541.

157. S. K. De, E. D. Reis and M. D. Kerstein, Wound treatment with human skin equivalent, *J. Am. Podiatr. Med. Assoc.*, 2002, **92**, 19–23.

158. D. M. Briscoe, V. R. Dharnidharka, C. Isaacs, G. Downing, S. Prosky, P. Shaw, N. L. Parenteau and J. Hardin-Young, The allogeneic response to cultured human skin equivalent in the hu-PBL-SCID mouse model of skin rejection, *Transplantation*, 1999, **67**, 1590–1599.

159. Y. M. Bello, A. F. Falabella and W. H. Eaglstein, Tissue-engineered skin. Current status in wound healing, *Am. J. Clin. Dermatol.*, 2001, **2**, 305–313.

160. A. Langer and W. Rogowski, Systematic review of economic evaluations of human cell-derived wound care products for the treatment of venous leg and diabetic foot ulcers, *BMC Health Serv. Res.*, 2009, **9**, 115.

161. W. A. Marston, Dermagraft: a bioengineered human dermal equivalent for the treatment of chronic non-healing diabetic foot ulcer, *Expert Rev. Med. Devices*, 2004, **1**, 21–31.

162. G. K. Naughton, W. R. Tolbert and T. M. Grillot, Emerging developments in tissue engineering and cell technology, *Tissue Eng.*, 1995, **1**, 211–219.

163. T. P. Economou, M. D. Rosenquist, R. W. Lewis 2nd and G. P. Kealey, An experimental study to determine the effects of Dermagraft on skin graft viability in the presence of bacterial wound contamination, *J. Burn Care Rehabil.*, 1995, **16**, 27–30.

164. J. Still, P. Glat, P. Silverstein, J. Griswold and D. Mozingo, The use of a collagen sponge/living cell composite material to treat donor sites in burn patients, *Burns*, 2003, **29**, 837–841.

165. Y. Eitan, U. Sarig, N. Dahan and M. Machluf, Acellular cardiac extracellular matrix as a scaffold for tissue engineering: *in vitro* cell support, remodeling and biocompatibility, *Tissue Eng. Part C Methods*, 2010, **16**, 671–683.

166. M. Yang, C. Z. Chen, X. N. Wang, Y. B. Zhu and Y. J. Gu, Favorable effects of the detergent and enzyme extraction method for preparing decellularized bovine pericardium scaffold for tissue engineered heart valves, *J. Biomed. Mater. Res. B Appl. Biomater.*, 2009, **91**, 354–361.

167. P. M. Dohmen and W. Konertz, Decellularization of xenogenic biologic tissue, *Ann. Thorac. Surg.*, 2008, **85**, 2163, author reply, 2163–2164.

168. R. L. Knight, H. E. Wilcox, S. A. Korossis, J. Fisher and E. Ingham, The use of acellular matrices for the tissue engineering of cardiac valves, *Proc. Inst. Mech. Eng. H*, 2008, **222**, 129–143.

169. H. J. Wei, S. C. Chen, Y. Chang, S. M. Hwang, W. W. Lin, P. H. Lai, H. K. Chiang, L. F. Hsu, H. H. Yang and H. W. Sung, Porous acellular bovine pericardia seeded with mesenchymal stem cells as a patch to repair a myocardial defect in a syngeneic rat model, *Biomaterials*, 2006, **27**, 5409–5419.

170. M. Weidenbecher, H. M. Tucker, A. Awadallah and J. E. Dennis, Fabrication of a neotrachea using engineered cartilage, *Laryngoscope*, 2008, **118**, 593–598.

171. G. Lisignoli, S. Cristino, A. Piacentini, S. Toneguzzi, F. Grassi, C. Cavallo, N. Zini, L. Solimando, N. Mario Maraldi and A. Facchini, Cellular and molecular events during chondrogenesis of human mesenchymal stromal cells grown in a three-dimensional hyaluronan based scaffold, *Biomaterials*, 2005, **26**, 5677–5686.

172. D. Magne, C. Vinatier, M. Julien, P. Weiss and J. Guicheux, Mesenchymal stem cell therapy to rebuild cartilage, *Trends Mol. Med.*, 2005, **11**, 519–526.

173. R. Rivkin, A. Ben-Ari, I. Kassis, L. Zangi, E. Gaberman, L. Levdansky, G. Marx and R. Gorodetsky, High-yield isolation, expansion, and differentiation of murine bone marrow-derived mesenchymal stem cells using fibrin microbeads (FMB), *Cloning Stem Cells*, 2007, **9**, 157–175.

Adult Non-hematopoietic Stem Cells and Progenitor Cells (aNHSCs)

R. SCHÄFER[a, b] AND L. DAHÉRON[a]

[a] Harvard Stem Cell Institute, [b] Department of Stem Cell and Regenerative Biology, Harvard University, 7 Divinity Avenue, Cambridge, MA 02138, USA

2.1 Introduction and General Remarks on the Nature of aNHSCs

Stem cells are primitive cells which are capable of self-renewal and differentiation into more specialized cells. Self-renewal is characterized by cell division retaining the stem cell fate in the daughter cells whereas, according to the current knowledge, the ability for self-renewal is lost during differentiation. The model of asymmetric cell division, represented by two sub-models (divisional asymmetry and environmental asymmetry), gives a reasonable idea of the fate of the stem cells in their local environment—called the 'stem cell niche'. During divisional asymmetry, stem cell fate determinants are asymmetrically localized and retained to one of the two daughter cells by the process of cell division, whilst the other daughter cell differentiates. Environmental asymmetry is mainly characterized by the influence of the microenvironment: one daughter cell remains in the stem cell niche leading to a retained stem cell fate whilst the other daughter cell leaves the stem cell niche and, exposed to a differentiation-promoting microenvironment, receives the signals to differentiate.[1]

Stem Cell-Based Tissue Repair
Edited by Raphael Gorodetsky and Richard Schäfer
© The Royal Society of Chemistry 2011
Published by the Royal Society of Chemistry, www.rsc.org

Referring to the ability to differentiate or 'differentiation potential', a distinction is drawn between totipotent, pluripotent and multipotent stem cells.

- Totipotent stem cells, exemplified by cells from the embryo until the eight-cell stadium, are capable of generating a complete organism.
- Pluripotent stem cells like embryonic stem (ES) cells or induced pluripotent stem (iPS) cells are self-renewing cells. They can give rise to all tissues of the body, plus many of the cells that support the embryo development, but they are unable to produce a new individual on their own.[2]
- Multipotent stem cells can only differentiate into a limited variety of cell types, with rare evidence of real 'trans-differentiation' (differentiation into cells of a germ layer other than the own).

Moreover, their self-renewing capacity is controversially discussed and lower than in ES cells.[3] Progenitor cells are less clearly defined; they are currently regarded as cells in a transient state, destined eventually to become one or more differentiated types[4] (Figure 2.1).

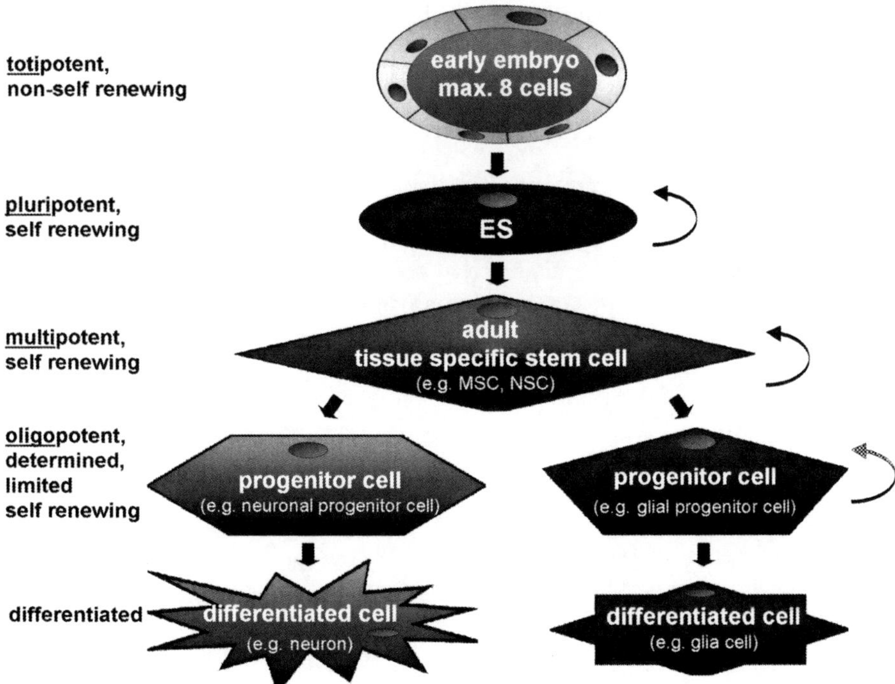

Figure 2.1   Scheme of hierarchical model of stem cell fate and development. Stem cells can be classified for their 'stemness' by their plasticity. This is the range of their differentiation potential from totipotent, pluripotent, multipotent and oligopotent to generate fully differentiated cells.

As outlined above, stem cells are characterized by their ability to differentiate along multiple lineages and their capacity to self-renew. Self-renewal is a common feature of stem cells and is highly regulated by the microenvironment or niche. The differentiation potential of stem cells becomes more and more restricted as the cells go through embryonic development to the adult body. At later embryonic stages and in the adult, stem cells are multipotent. Their potential is generally restricted to one tissue type.

In an attempt to better understand the mechanisms regulating the self-renewal and lineage differentiation of stem cells, two research groups compared the transcriptional profiling of embryonic stem cells, neuronal and hematopoietic stem cells.[5,6] Both reports generated a list of about 250 common genes that were overexpressed in these three types of stem cells. Interestingly, most of these genes were not specific for stem cells.

Ramalho-Santos *et al.*[5] identified 216 stem cell enriched genes, among them several genes known to be implicated in self-renewal:

- Janus kinases/signal transducers and activators of transcription (JAK/ STAT);
- transforming growth factor β (TGFβ);
- Notch signaling pathways;
- genes involved in chromatin remodeling (DNA helicases, DNA methylases and histone deacetylases); and
- genes involved in post-transcriptional regulation (RNA helicases).

The list from Ivanova *et al.*[6] included 283 genes; some of these had previously been shown to be important for the regulation of stem cells such as transcription factors Edr1 and Tcf3.

Although the aim of these two groups was to identify a core set of genes defining 'stemness', only 15 genes were common between the two lists. This relatively small number of common genes might be due to differences in the methodology used by these two groups (stem cell isolation as well as algorithm used to analyze the changes in gene expression). But it could also suggest that there is no common stem cell signature and that each stem cell has a unique signature tailored to its niche. The transcriptional profiling of additional cell types (*e.g.* intestinal epithelial progenitors[7] and skin stem cells[8]) has been revealed and compared to the data available for embryonic stem cells, hematopoietic stem cells and neuronal stem cells. Few members of the TGFβ, JAK/ STAT and Notch signaling pathways have emerged from this comparison consistent with their predicted roles in stem cell self-renewal.

Although stem cells share the fundamental property of self-renewal, a number of reports have demonstrated that the molecular mechanisms of self-renewal are very different in various stem cells. In fact, some factors act in opposite manners in different stem cell types. For instance, the activation of the leukemia inhibitory factor (LIF) signaling pathway and bone morphogenetic protein (BMP) pathway is critical for mouse embryonic stem cell self renewal,[9] while inhibition of the BMP pathway in combination with basic fibroblast growth factor (bFGF) has been shown to maintain human embryonic stem cell self-renewal.[10] BMP signaling has also been revealed to inhibit intestinal stem cell self-renewal.[11] The Wnt signaling pathway has been shown to regulate self-renewal for a variety of stem cells including embryonic stem cells, hematopoietic stem cells, neuronal stem cells and skin stem cells. In contrast, its activation can induce muscle stem cell to differentiate.[12] The Notch signaling pathway plays an important role in the maintenance of hematopoietic stem cells,[13] but a transient activation of Notch was shown to induce rapid glial differentiation of neural crest stem cells.[14] Together, these reports support the idea that stem cell self-renewal is governed by different signaling pathways which are activated by the neighboring cells in their specific niche.

The other property shared by stem cells is their ability to give rise to multiple differentiated cell types. Although this property is shared, each stem cell has a unique set of cells into which they can differentiate. Therefore, the molecular signature or 'stemness' might be distinctive for each stem cell type, reflecting their unique potential. In 2005, Rick Young's group unveiled the core transcriptional regulatory circuitry of human embryonic stem cells.[15] By combining chromatin immunoprecipitation (Chip) with DNA microarrays (Chip–Chip analysis), they identified a wide range of targets for the main pluripotent transcription factors—Oct4, Sox2 and Nanog. Interestingly, these three factors co-localize on the promoters of many genes in human (h)ES cells, suggesting an intertwined network controlling the state of hES cells. These target genes can be divided into two sets: one that is transcriptionally active in hES cells and one that is repressed in hES cells. About half of the genes bound by Oct4, Sox2 and Nanog are activated. These include genes known to be crucial for self-renewal and pluripotent state including Oct4, Nanog and Sox2 themselves, as well as components of the TGFβ and Wnt signaling pathways. The three pluripotent transcription factors also bind to a set of genes that are repressed in hES cells. This list of genes includes many developmental transcription factors known to be essential for lineage commitment and cell differentiation. This work demonstrated that the three transcription factors (Oct4, Sox2 and Nanog) are master regulators of ES cells and that their complex regulatory network is essential to maintain the balance between self-renewal and differentiation.

In 2006, Shinya Yamanaka's group showed that mouse fibroblasts could be converted into pluripotent cells (called induced pluripotent stem cells) by introducing four transcription factors (Oct4, Sox2, Klf4 and c-Myc).[16] This groundbreaking discovery supports the concept that a small set of core genes is critical for establishing the stem cell fate. Although Nanog was not among the four factors in Yamanaka's protocol, another group led by Jamie Thomson showed that fibroblasts could be reprogrammed with a different set of genes including Nanog (Oct4, Sox2, Nanog and Lin28).[17]

This raised the question of whether direct cell reprogramming of one differentiated cell into another differentiated cell (without reversion to a pluripotent state) can be achieved by overexpression of the right combination of transcription factors. Several attempts to convert one cell type onto another by overexpressing few key genes have been reported in the last few years.

For instance, the overexpression of myogenic differentiation (MyoD) into fibroblasts induced the conversion of these cells into muscle-like cells. Introduction of interleukin (IL)-2 and granulocyte macrophage-colony stimulating factor (GM-CSF) receptor turned myeloid cells into lymphoid progenitors. Fibroblasts were switched to macrophage-like cells by overexpression of PU.1 and CCAAT/enhancer-binding protein alpha (Cebpa). B cells could be converted into macrophages or common lymphoid progenitors by overexpressing Cebpa or repressing paired box (PAX)5, respectively.

One caveat from all these reports is the limited data on the function of these converted cells. In 2008, Doug Melton's group used an *in vivo* approach to reprogram pancreatic exocrine cells into cells that closely resemble beta-cells.[18] They introduced three transcription factors—Neurogenin (NGN)3, pancreatic and duodenal homeobox (Pdx)1 and Mafa—into exocrine cells using an adenoviral system. Ten days following transduction, 20% of these exocrine cells turned into cells with beta-cell characteristics, *i.e.* small spindle shaped cells with the capacity to secrete insulin. They expressed the typical markers of beta-cells and synthesized vascular endothelial growth factor (VEGF) that induced angiogenesis. Moreover, they showed an improvement of fasting blood glucose level in diabetic mice injected with the three transcription factors compared to a green fluorescent protein (GFP) virus control, suggesting that these induced beta-cells are functional. Interestingly, although these three transcription factors were sufficient to reprogrammed exocrine cells into beta-cells, they were unable to turn fibroblasts or skeletal muscle into beta-cells suggesting that interchanging the fate of closely related cell types is much easier than unrelated cells.

Nonetheless, a recent *Nature* paper showed that the cell fate restriction can also be crossed between cells from different germ layer. Vierbuchen *et al.*[19] directly reprogrammed fibroblasts into neurons using a cocktail of only three transcription factors. By overexpressing Ascl1, Brn2 and Myt1l in embryonic or adult fibroblasts, they generated induced neuronal (iN) cells with an efficiency close to 20%. These iN cells could produce action potentials and form functional synapses; however, they were limited to one subtype of neurons—excitatory glutamatergic neurons. This finding strongly supports the notion that few master regulator genes can establish cell fate. It is reasonable to speculate that, in the next few years, the set of genes that regulate the fate of many stem cell types will be uncovered. This approach will offer a new technology to derive specific stem cell types *in vitro*.

As outlined in Chapter 1, adult hematopoietic stem cells (HSCs) are well-characterized cells which have been firmly integrated in the treatment of hematological malignancies for decades. Moreover, modern hematology, as we know it today, would not exist without the successful clinical-scale isolation, storage and application of HSCs.

How different is the current situation for stem/progenitor cell entities with non-hematopoietic properties gathered in the following simplifying name 'non-hematopoietic stem/progenitor cells' (NHSCs). Since Julius Friedrich Cohnheim, a German pathologist and pupil of Rudolf Virchow, speculated in the 19th century about the role of bone marrow derived 'fibroblasts' in wound healing,[20] our knowledge of cells of fibroblastoid morphology in the adult bone marrow has increased steadily. After Alexander Friedenstein's reports in the 1970s on fibroblastoid cells obtained from the bone marrow showing properties usually assigned to stem cells such as colony formation and *in vitro* as well *in vivo* differentiation potential,[21,22] investigations on postnatal non-hematopoietic cells with stem cell/progenitor like properties increased in a remarkably dynamic way leading in the following decades to the identification of tissue-specific adult stem cells.

Common features of these cells are their ability to self-renew and to differentiate into multiple lineages. Referring to the hierarchical model of stem cell fate and development described above (Figure 2.1), aNHSCs may be classified as ontogenetically downstream of ES cells. This is supported by multiple evidence: ES cells or iPS cells can be differentiated into all germ layers as shown by embryoid body differentiation and *in vivo* teratoma formation (pluripotency)[23–25] whereas aNHSCs show a reduced differentiation potential (multipotency).[26–28] An interesting complementary hypothesis is drawn by Jonathan Slack who argues that tissue-specific stem cells are in a similar state of developmental commitment than the embryonic rudiment that produced them. Some few cells permanently retain their embryonic qualities because they are located in the stem cell niche. Some of these are preserved as tissue-specific stem cells due to signals provided by adjacent mesenchymal cells. The rest differentiates to the final tissue maintained by the sustained production of differentiated cells from the local stem cell pool.[4] In other words and in a mechanistic way, aNHSCs can be regarded as the regenerative cellular reserve of a damaged or aging organ.

Although the distinction between adult stem cells and adult progenitor cells is not simple and appears to be elusive, some principal differences can be pointed out: Progenitor cells, continuously replaced by stem cells, show a limited lifespan whereas stem cells show a prolonged lifespan. Progenitor cells divide quickly whereas true stem cells are regarded to divide comparatively slowly. The growth kinetic of the latter is a major characteristic of the status of 'quiescence' and correlates to their long lifespan.[29] However, alterations of the microenvironment of the stem cell niche (*e.g.* during regeneration) may initiate the generation of fast dividing stem cells in order to expand the pool of stem cells (symmetric cell division).[29,30]

Currently, the question where aNHSCs are localized in the adult organism is controversially discussed. Some researchers hypothesize that the bone marrow may serve as a storage compartment of 'multifunctional' stem cells such as mesenchymal stem cells (MSCs). Upon demand such as damage or cell turnover in the respective organ, these cells may be released and home to the organ where they undergo differentiation into the required lineage. This broad differentiation potential is also described by the term 'plasticity'. According to this model, plasticity is regarded as essential, whereas self-renewal and hierarchy are optional[31] and major mechanisms of tissue regeneration by aNHSCs are *de novo* generation of tissue cells by (trans)differentiation of aNHSCs or fusion of aNHSCs with local terminally differentiated tissue cells.[32,33]

In fact, there is evidence that the bone marrow may represent a unique niche for stem cells participating in the formation and regeneration of tissues other than bone marrow: In a recent study it was shown that the pancreatic mesenchyme originates partly from cells derived from the CD45 + component of bone marrow.[34] As the expression of CD45 is assigned to cells of the hematopoietic lineage but also to freshly isolated MSCs,[35] further characterization of these cells is needed.

However, currently there is only rare evidence that (trans)differentiation or cell fusion of primary non-resident aNHSCs may be major players in the fate of aNHSCs.[4,26] Therefore, one may opine that the majority of aNHSCs are cells present in every postnatal organ capable of differentiating into cells of the lineage of their 'home organ' and hereby contributing significantly to the cell turnover and regeneration of their 'home organ'.[36–38]

In the following chapter, aNHSCs from different organs as central nervous system (CNS), skeletal muscle, heart, vasculature, intestine, pancreas and liver are briefly characterized followed by a more comprehensive overview on mesenchymal stem/stromal cells (MSCs). Chapters 11 and 12 refer specifically to the role of MSCs in the regeneration of bone and cartilage.

2.2 Characterization of aNHSCs

2.2.1 Adult Neural Stem Cells (aNSCs)

Multipotent aNSCs capable of self-renewal and generation of multiple neural lineages including astrocytes, oligodendrocytes and functional neurons have been derived from the adult CNS.[39] However, the existence of a tri-potent aNSC with capacity to generate neurons, astrocytes and oligodendrocytes in the adult brain remains to be demonstrated at the clonal level *in vivo*.

Adult neural stem/progenitor cells represent populations of multipotent neural cells mainly located in two specialized niches of the adult mammalian brain—the subventricular zone (SVZ) of the lateral ventricle wall and the subgranular zone (SGZ) of the dentate gyrus. These cells, derived from radial glia cells, are regarded to maintain neurogenesis and gliogenesis throughout adult life.

Type B NSCs in the SVZ share many characteristics with astrocytes. These are in intimate contact with all other SVZ cell types, including the rapidly dividing transit-amplifying type C cells and the lineage-committed migratory neuronal type A cells.[40] However, injuries and pathological stimuli such as stroke may activate the neurogenesis program outside of the SVZ or SGZ.[41] Moreover, there is evidence that distinct cells in the adult CNS can serve as aNSCs mediating neurogenesis under normal conditions or after dramatic injuries, similar to adult neurogenesis in the olfactory epithelium.[39] Adult NSCs can be characterized as follows:

- Nestin + ;
- RC2 + ;

- sex determining region Y-box (SOX)2+;
- brain lipid binding protein (BLBP)+;
- glutamate aspartate transporter (GLAST)+;
- PAX6+; CD44+; and
- musashi+.

Culturing aNSCs *in vitro* requires the addition of growth factors such as fibroblast growth factor (FGF)2 and/or epidermal growth factor (EGF).

2.2.2 Satellite Cells (SatCs) and Myoblasts

The cell entities representing aNHSCs in skeletal muscle tissue are SatCs. SatCs play a pivotal role in skeletal muscle regeneration and they contribute to the postnatal growth of muscle fibers.[42] SatCs reside in an anatomically defined niche constituting a distinct membrane-enclosed compartment underneath the basal lamina within the muscle fiber.[38] To date, the distinct progenitor cell type which gives rise to SatCs remains to be identified. SatCs can be detected in healthy adult muscle as quiescent cells and represent 2.5–6% of all nuclei of a given muscle fiber.[42]

After activation (*e.g.* by muscle injury), they are able to generate large numbers of new myofibers. The activation of SatCs can be mediated by hepatocyte growth factor (HGF), FGF, insulin-like growth factor (IGF) and nitric oxide (NO).[42] These factors can initiate the asymmetric division of SatCs generating Pax7+, MyoD1+, and myogenic regulatory factor (Myf)5+ differentiating cells and Pax7+ MyoD- cells returning to quiescence in order to maintain a pool of progenitors.[42] After activation, the proliferating cells are classified as myoblasts further differentiating to Pax7- MyoD+ myocytes (Table 2.1).

2.2.3 Cardiac Progenitor Cells (CPCs)

Intensive investigations in cardiac cell biology resulted in a shift from the paradigm of the heart as a terminally differentiated postmitotic organ to an organ with limited self-renewing properties.[43,44] This perception is mainly based on the identification of resident adult cardiac progenitor cells (CPCs) which, generating myocytes, endothelial cells, smooth muscle cells and even coronary arteries, regulate the homeostasis of the heart in postnatal development and adulthood.[36,45,46] Moreover, recently distinct subpopulations of cardiomyocytes with different morphological and functional properties have been described.[47,48]

CPCs have been shown to contribute to the regeneration of the infarcted myocardium[36] and the activation of CPCs [*e.g.* by cytokines such as hepatocyte growth factor (HGF) or insulin-like growth factor (IGF)-1] leads to a regeneration of even scarred post-infarct myocardium[49] as well as to a reversion of the senescent cardiac phenotype and prolongation of lifespan.[50] To date,

Table 2.1 Characterization of SatCs and myoblasts.

Markers	Cell phenotypes		
	Quiescent SatCs	*Activated SatCs*	*Myoblasts*
Pax3	+		
CXCR4	+		
Pax7	+	+	
Syndecan 3 and 4	+	+	
VCAM-1	+	+	
c-met	+	+	
forkhead box (Fox)k1	+	+	
CD34[a]	+	+	
β1 integrin	+	+	
Myf5	+	+	+
caveolin-1	+	+	+
α7 integrin	+	+	+
CD56	+	+	+
M-cadherin[b]	+	+	+
MyoD		+	+

[a]Not in humans.
[b]Inconstant in humans.

several molecules have been identified to regulate the cardiac homeostasis: The serine–threonine kinases Akt and Pim-1 as well as the proteins atrial natriuretic peptide (ANP) and nucleostemin play a pivotal role in cardiomyocyte survival,[51–54] and Akt is involved in the survival and vascular differentiation of CPCs.[46] Moreover, under hypoxic conditions, stromal-cell derived factor 1 (SDF-1) promotes the differentiation of CPCs.[46]

CPCs possess the HGF–c-Met system and the IGF-1–IGF-1 receptor (R) system. The HGF–c-Met system modulates CPC migration whereas the IGF-1–IGF-1R modulates primarily CPC division and survival.[55,56] Islet (ISL)1 + CPCs can be isolated from the fetal heart as well as from the postnatal heart[57,58] and generated from ES cells and iPS cells.[59] Moreover, the Wnt/β-catenin pathway plays an important role in cardiac development including renewal and differentiation of ISL1 + cardiovascular progenitor cells.[60]

CPCs can be characterized as follows:

- ISL1 + or CD117 (c-kit) + ;[i]
- CD34-;
- CD133-;
- CD45-;
- VEGFR2-;
- CD31-; and
- Sca-1 + (mouse).

[i]There is evidence that the expression of CD117 in the human heart may not exclusively be related to CPCs but also to mast cells.[61]

2.2.4 Endothelial Progenitor Cells (EPCs)

Blood- or vasculature-derived endothelial progenitor cells (EPCs) have been shown to generate vasculature by forming long-lasting vessels.[62,63] Human EPCs can be isolated from the peripheral blood of adult individuals, cultured *in vitro* and committed into an endothelial lineage.[64] For several years, the isolation and successful large-scale *ex vivo*-expansion of this cell entity was difficult to perform. A significant progress in this problem may be achieved using a protocol describing the large-scale expansion of EPCs using 10% pooled human platelet lysate.[63]

The precise definition of EPC phenotype and function is still controversially discussed[65] and extensive phenotyping as well as functional analyses may be necessary to distinguish clearly hematopoietic from endothelial lineage cells, since monocytes may mimic EPC phenotype.[62,66,67]

EPCs can be characterized as follows:

- CD14+;
- CD31+;
- CD105+;
- CD144+;
- CD146+;
- c-kit+ (in early culture);
- VEGFR-2+;
- von Willebrand factor+;
- eNOS+; and
- ac-LDL/lectin+.

2.2.5 Gastrointestinal Stem Cells, Pancreatic Stem Cells and Hepatic Stem Cells

In the past decades, aNHSCs of the small intestine and the colon have been extensively characterized.[29] However, the identification and characterization of gastric, esophageal, pancreatic and hepatic aNHSCs remain incomplete today.

The niche of intestinal aNHSCs was identified in epithelial regions closely located to regions of high cell turnover,[29] *i.e.* in the colon at the base of the crypts and in the small intestine above the Paneth cell zone.[68] There is evidence that, in the intestine, quiescent and activated aNHSCs may co-exist.[29] Moreover, the local mesenchyme seems to play an important role in the regulation of the activity of local aNHSCs, as there is evidence of cross-talk between these cell entitites.[69] In the stomach, the oxyntic glands were shown to contain multipotent stem cells[70] and, in the esophagus, aNHSCs were detected in the basal layer of the stratified squamous epithelium.[71] In the adult liver, oval cells, anatomically associated to the biliary system, display a bipotent differentiation potential which enables them to generate hepatic and biliary cells.[72] Under physiological conditions, the maintenance of the beta-cell pool in the adult pancreas is provided by dividing beta-cells.[73] However, under experimental pathological conditions (duct ligation), endogenous progenitor cells which were able to generate beta-cells could be identified in the ductal epithelium[74] and, recently, self-renewing differentiated BMI1 + cells were identified in the acini of the adult pancreas.[75] Gastrointestinal stem cells can be characterized as follows:

- Esophagus: Hoechst dye + , CD34 + ;
- Stomach: double cortin CaM kinase-like (DCLK)1 + ; and
- Intestine: leucine-rich orphan G protein-coupled receptor (LGR)5 + , BMI1 + , DCLK1 + .

Hepatic stem cells (oval cells) can be characterized as follows:

- Prominin1 + ; and
- FOXL1 + .

Pancreatic stem cells can be characterized as follows:

- NGN3 + ;
- PDX1 + ;
- and BMI1 + .

2.2.6 Mesenchymal Stem/Stromal Cells (MSCs)

Adult mesenchymal stem/stromal cells (MSCs) are defined as multipotent cells which can differentiate into mesenchymal and non mesenchymal lineages. Although extensive investigations on MSCs have been performed during the last decades, knowledge about these cells remains incomplete. Numerous subpopulations of cells with more or less MSC-like properties [*e.g.* 'multipotent adult progenitor cells' (MAPCs) and 'marrow isolated adult multilineage inducible' (MIAMI) cells[76]] are grouped under the roof of 'MSCs' increasing the complexity of the field. Therefore, comprehensive characterization of these cells is one major goal of today's MSC research.

MSCs have been isolated from:

- various postnatal organs and tissues, *e.g.* from brain, spleen, liver, kidney, lung, muscle, thymus, pancreas, adipose tissue, blood vessels and umbilical cord blood[77,78] and
- various species, *e.g.* humans, rats, mice, cats, dogs, rabbits, pigs, baboons.[79]

Their distribution throughout the organism seems to be related to their prevalence in a perivascular niche.[77] There is evidence that MSCs are not, or only in very low numbers, present in adult human peripheral blood[77] (exempt pathological status or mobilization procedures[80]—refer to Chapter 18). The most common source of MSCs is the bone marrow. Here, the frequency of MSCs is up to 0.01% of the mononuclear cells.[81–83] In bone marrow they interact with HSCs supporting them in their niche.[84–86]

MSCs can be isolated from bone marrow in comparatively high numbers. Although a number of different methods has been described for isolating MSCs (including antibody- or aptamer- based negative or positive enrichment/depletion techniques and culture-based selection methods[87,88]), the most frequently applied technique is the selection of MSCs by plastic adherence using the affinity of MSCs to plastic surfaces.[84] However, with this technique, contamination with other adherent cells without MSC-like properties is unavoidable.

In disease models such as myocardial infarction, cerebral ischemic stroke, pulmonary fibrosis, nephropathy and osteogenesis imperfecta, systematically administered MSC engraft preferentially to the site of injury.[84] These observations combined with their differentiation potential led to the mechanistic assumption that MSCs may represent ideal candidates for tissue regeneration. Actually, the transplantation of MSCs resulted in clinical improvement in various diseases such as osteogenesis imperfecta, lung injury, kidney disease, diabetes, myocardial infarction and neurological disorders like cerebral ischemia and other diseases of the central nervous system including neurodegenerative and inflammatory disorders.[26,89–91] Interestingly, the therapeutical effects of MSCs seem not necessarily to require the *in vivo* differentiation of the MSCs. Moreover, only a minority of the MSCs show long-term survival after transplantation.[92] Therefore, it is reasonable to assume that a paracrine activity of MSCs (*e.g.* secretion of growth factors and other cytokines) leads to tissue regeneration.[26,84,93,94] Due to the gap between the low survival and *in vivo* trans-differentiation rates and the evident clinical effects, the paracrine activity of MSCs could possibly be regarded as the most relevant mechanism for tissue repair compared with differentiation.[26] Referring to autologous regeneration, it is likely that MSCs act as an internal cellular resource in order to repair damaged tissue (compare with the other entities of aNHSCs). This hypothesis is supported by the following observations:

- MSCs of fetal origin can be detected in maternal tissues for a long time after pregnancy, participating in tissue regeneration.[79]
- MSCs residing in bone marrow can be mobilized after tissue injury. These cells had been shown to participate in the regeneration of myocardial infarction, skeletal muscle damage or cerebral stroke.[76]

Some interesting molecules have been described which may be useful in identifying MSC (sub)populations (CD271, GD2, CD49a, W7C5, W8B2, C15, CDCP1, CD340, CD349, SSEA4).[26,83,95–97] However, a distinct exclusively MSC-restricted marker has yet to be identified. Furthermore, the cellular shape by itself is not sufficient to determine the cells because the morphology of MSCs may vary from a spindle shape to a broad trapezoid shape depending on culture conditions and passages.[79] Due to the heterogeneity of MSC-like cell populations and biased sample preparations,[87,98] it is necessary to standardize the characterization of MSCs. For this purpose, a useful approach is the application of the minimal criteria for defining MSCs published by the International Society for Cellular Therapy (ISCT).[99] These defining criteria are based on the phenotypical and functional characteristics of cultured MSCs:

2.2.6.1 Surface Antigen Pattern

The following antigen pattern is characteristic for human MSCs only. MSCs isolated from other species, particularly mouse, may show a different pattern which varies between strains.[100] Moreover, the surface antigen pattern fails to reflect the developmental potential of the MSCs.[98]

The typical surface antigen pattern of cultured, non stimulated and/or not differentiated MSC (positive antigen expression is defined $\geq 95\%$ positive counts, negative antigen expression is defined $\leq 2\%$ positive counts by flow cytometry) comprises:

- CD73 +, CD90 +, CD105 +; and
- CD14- or CD11b-, CD34-, CD45-, CD79α- or CD19-, HLA-DR-

2.2.6.2 Multipotent Differentiation Potential

MSCs should show the ability to differentiate into adipogenic, osteogenic and chondrogenic lineage (tri-lineage differentiation potential) after treatment with the respective differentiation media.[99]

The *in vitro* differentiation is usually determined by specific staining techniques: MSCs cultured in adipogenic differentiation medium show lipid vacuoles which can be stained with oil red O. Osteogenesis is shown, for example, by staining for alkaline phosphatase and calcium. Glycosaminoglycans in pellets of MSCs undergoing chondrogenic differentiation can be stained with toluidine blue.[84]

The ISCT-definition of MSCs requires the ability of plastic adherence. Accounting for reports on non-adherent MSC subpopulations,[101,102] plastic adherence may not be an essential aspect of MSC characterization.

MSC preparations from different species (human, mouse, rat) and sources (bone marrow, adipose tissue or umbilical cord blood) may vary in their surface epitope pattern, differentiation or proliferation capacity.[78,100,103]

It should be pointed out, that these issues of characterization are the result of pure *in vitro* investigations. This implicates that the characterization and comparability of cell populations in respect to their MSC-like properties *in vitro* is feasible. However, there is evidence that the transferability of the *in vitro* characteristics into the *in vivo* situation is limited. The expression of CD45 and CD29 on MSCs, for example, is completely different depending on the environment surrounding the MSCs. Freshly isolated MSCs express high CD45 but do not express CD29 on their surface, and, under culture conditions, CD45 is quickly lost whereas CD29 is upregulated.[35] Moreover, the antigen expression can be influenced, for example, by stimulation with interferon-γ (upregulation of HLA-DR)[99] or malignant transformation (loss of CD90),[104] and antigens that are usually detected on human MSCs such as CD166 can also be detected on other cell types such as malignant melanoma.[105]

2.2.6.3 Immunomodulatory Potential

There is impressive proof for the immunomodulatory properties of MSCs as shown by *in vivo* observations: MSCs can improve the outcome after allogeneic transplantations by promotion of hematopoietic engraftment and amelioration of graft *versus* host disease[106,107] as well as the amelioration of inflammatory response. Moreover, MSCs can improve the clinical course in experimental allergic encephalomyelitis.[108] Usually, immunomodulation by MSC is effective as immunosuppression. This is the result of interactions of MSCs with T- and B-cells: T-cell proliferation and function is suppressed, B-cell proliferation, differentiation and chemotaxis are inhibited.[106] Moreover, MSCs have been shown to suppress allo-specific antibody production *in vitro*.[109]

References

1. A. Wilson and A. Trumpp, Bone-marrow haematopoietic-stem-cell niches., *Nat. Rev. Immunol.*, 2006, **6**, 93–106.
2. C. M. Verfaillie, M. F. Pera and P. M. Lansdorp, Stem cells: hype and reality. Hematology., *Am. Soc. Hematol. Educ. Program.*, 2002, **1**, 369–391.
3. K. Ksiazek, A comprehensive review on mesenchymal stem cell growth and senescence., *Rejuvenation. Res.*, 2009, **12**, 105–116.
4. J. M. Slack, Origin of stem cells in organogenesis, *Science*, 2008, **322**, 1498–1501.
5. M. Ramalho-Santos, S. Yoon, Y. Matsuzaki, R. C. Mulligan and D. A. Melton, "Stemness": transcriptional profiling of embryonic and adult stem cells., *Science*, 2002, **298**, 597–600.
6. N. B. Ivanova, J. T. Dimos, C. Schaniel, J. A. Hackney, K. A. Moore and I. R. Lemischka, A stem cell molecular signature., *Science*, 2002, **298**, 601–604.
7. T. S. Stappenbeck, J. C. Mills and J. I. Gordon, Molecular features of adult mouse small intestinal epithelial progenitors., *Proc. Natl. Acad. Sci. U. S. A.*, 2003, **100**, 1004–1009.
8. T. Tumbar, G. Guasch, V. Greco, C. Blanpain, W. E. Lowry, M. Rendl and E. Fuchs, Defining the epithelial stem cell niche in skin., *Science.*, 2004, **303**, 359–363.
9. Q. L. Ying, J. Nichols, I. Chambers and A. Smith, BMP induction of Id proteins suppresses differentiation and sustains embryonic stem cell self-renewal in collaboration with STAT3., *Cell.*, 2003, **115**, 281–292.
10. R. H. Xu, R. M. Peck, D. S. Li, X. Feng, T. Ludwig and J. A. Thomson, Basic FGF and suppression of BMP signaling sustain undifferentiated proliferation of human ES cells., *Nat. Methods.*, 2005, **2**, 185–190.

11. X. C. He, J. Zhang, W. G. Tong, O. Tawfik, J. Ross, D. H. Scoville, Q. Tian, X. Zeng, X. He, L. M. Wiedemann, Y. Mishina and L. Li, BMP signaling inhibits intestinal stem cell self-renewal through suppression of Wnt-beta-catenin signaling., *Nat. Genet.*, 2004, **36**, 1117–1121.

12. A. S. Brack, I. M. Conboy, M. J. Conboy, J. Shen and T. A. Rando, A temporal switch from notch to Wnt signaling in muscle stem cells is necessary for normal adult myogenesis., *Cell Stem Cell.*, 2008, **2**, 50–59.

13. S. Stier, T. Cheng, D. Dombkowski, N. Carlesso and D. T. Scadden, Notch1 activation increases hematopoietic stem cell self-renewal in vivo and favors lymphoid over myeloid lineage outcome., *Blood*, 2002, **99**, 2369–2378.

14. S. J. Morrison, S. E. Perez, Z. Qiao, J. M. Verdi, C. Hicks, G. Weinmaster and D. J. Anderson, Transient Notch activation initiates an irreversible switch from neurogenesis to gliogenesis by neural crest stem cells., *Cell.*, 2000, **101**, 499–510.

15. L. A. Boyer, T. I. Lee, M. F. Cole, S. E. Johnstone, S. S. Levine, J. P. Zucker, M. G. Guenther, R. M. Kumar, H. L. Murray, R. G. Jenner, D. K. Gifford, D. A. Melton, R. Jaenisch and R. A. Young, Core transcriptional regulatory circuitry in human embryonic stem cells., *Cell.*, 2005, **122**, 947–956.

16. K. Takahashi and S. Yamanaka, Induction of pluripotent stem cells from mouse embryonic and adult fibroblast cultures by defined factors., *Cell.*, 2006, **126**, 663–676.

17. J. Yu, M. A. Vodyanik, K. Smuga-Otto, J. Antosiewicz-Bourget, J. L. Frane, S. Tian, J. Nie, G. A. Jonsdottir, V. Ruotti, R. Stewart, I. I. Slukvin and J. A. Thomson, Induced pluripotent stem cell lines derived from human somatic cells., *Science*, 2007, **318**, 1917–1920.

18. Q. Zhou, J. Brown, A. Kanarek, J. Rajagopal and D. A. Melton, In vivo reprogramming of adult pancreatic exocrine cells to beta-cells., *Nature*, 2008, **455**, 627–632.

19. T. Vierbuchen, A. Ostermeier, Z. P. Pang, Y. Kokubu, T. C. Südhof and M. Wernig, Direct conversion of fibroblasts to functional neurons by defined factors., *Nature*, 2010, **463**, 1035–1041.

20. J. F. Cohnheim, Ueber Entzündung und Eiterung., *Archiv für Pathologische Anatomie und Physiologie und für klinische Medicin*, 1867, **40**, 1–79.

21. A. J. Friedenstein, R. K. Chailakhjan and K. S. Lalykina, The development of fibroblast colonies in monolayer cultures of guinea-pig bone marrow and spleen cells., *Cell Tissue Kinet.*, 1970, **3**, 393–403.

22. A. J. Friedenstein, Precursor cells of mechanocytes., *Int. Rev. Cytol.*, 1976, **47**, 327–359.

23. K. Takahashi, K. Tanabe, M. Ohnuki, M. Narita, T. Ichisaka, K. Tomoda and S. Yamanaka, Induction of pluripotent stem cells from adult human fibroblasts by defined factors., *Cell.*, 2007, **131**, 861–872.

24. K. Osafune, L. Caron, M. Borowiak, R. J. Martinez, C. S. Fitz-Gerald, Y. Sato, C. A. Cowan, K. R. Chien and D. A. Melton, Marked differences

in differentiation propensity among human embryonic stem cell lines., *Nat. Biotechnol.*, 2008, **26**, 313–315.

25. I. H. Park, N. Arora, H. Huo, N. Maherali, T. Ahfeldt, A. Shimamura, M. W. Lensch, C. Cowan, K. Hochedlinger and G. Q. Daley, Disease-specific induced pluripotent stem cells., *Cell.*, 2008, **134**, 877–886.

26. D. G. Phinney and D. J. Prockop, Concise review: mesenchymal stem/ multipotent stromal cells: the state of transdifferentiation and modes of tissue repair--current views., *Stem Cells.*, 2007, **25**, 2896–2902.

27. R. A. Rose, H. Jiang, X. Wang, S. Helke, J. N. Tsoporis, N. Gong, S. C. Keating, T. G. Parker, P. H. Backx and A. Keating, Bone marrow-derived mesenchymal stromal cells express cardiac-specific markers, retain the stromal phenotype, and do not become functional cardiomyocytes in vitro., *Stem Cells.*, 2008, **26**, 2884–2892.

28. R. A. Rose, A. Keating and P. H. Backx, Do mesenchymal stromal cells transdifferentiate into functional cardiomyocytes?, *Circ. Res.*, 2008, **103**, e120.

29. M. Quante and T. C. Wang, Stem cells in gastroenterology and hepa-tology., *Nat. Rev. Gastroenterol. Hepatol.*, 2009, **6**, 724–737.

30. E. Fuchs, The tortoise and the hair: slow-cycling cells in the stem cell race., *Cell.*, 2009, **137**, 811–819.

31. D. Zipori, The stem state: plasticity is essential, whereas self-renewal and hierarchy are optional., *Stem Cells.*, 2005, **23**, 719–726.

32. J. M. Weimann, C. A. Charlton, T. R. Brazelton, R. C. Hackman and H. M. Blau, Contribution of transplanted bone marrow cells to Purkinje neurons in human adult brains., *Proc. Natl. Acad. Sci. U. S. A.*, 2003, **100**, 2088–2093.

33. E. Mezey, et al. Transplanted bone marrow generates new neurons in human brains., *Proc. Natl. Acad. Sci. U. S. A.*, 2003, **100**, 1364–1369.

34. V. Sordi, R. Melzi, A. Mercalli, R. Formicola, C. Doglioni, F. Tiboni, G. Ferrari, R. Nano, K. Chwalek, E. E. Lammert, E. Bonifacio, D. Borg and L. Piemonti, Mesenchymal cells appearing in pancreatic tissue culture are bone marrow-derived stem cells with the capacity to improve trans-planted islet function., *Stem Cells.*, 2010, **28**, 140–151.

35. K. T. Guo, R. Schaefer, A. Paul, A. Gerber, G. Ziemer and H. P. Wendel, A new technique for the isolation and surface immobilization of mesenchymal stem cells from whole bone marrow using high-specific DNA aptamers., *Stem Cells.*, 2006, **24**, 2220–2231.

36. C. Bearzi, M. Rota, T. Hosoda, J. Tillmanns, A. Nascimbene, A. De Angelis, S. Yasuzawa-Amano, I. Trofimova, R. W. Siggins, N. Lecapitaine, S. Cascapera, A. P. Beltrami, D. A. D'Alessandro, E. Zias, F. Quaini, K. Urbanek, R. E. Michler, R. Bolli, J. Kajstura, A. Leri and P. Anversa, Human cardiac stem cells., *Proc. Natl. Acad. Sci. U. S. A.*, 2007, **104**, 14068–14073.

37. K. Urbanek, D. Torella, F. Sheikh, A. De Angelis, D. Nurzynska, F. Silvestri, C. A. Beltrami, R. Bussani, A. P. Beltrami, F. Quaini, R. Bolli, A. Leri, J. Kajstura and P. Anversa, Myocardial regeneration by

activation of multipotent cardiac stem cells in ischemic heart failure., *Proc. Natl. Acad. Sci. U. S. A.*, 2005, **102**, 8692–8697.

38. B. D. Cosgrove, A. Sacco, P. M. Gilbert and H. M. Blau, A home away from home: challenges and opportunities in engineering in vitro muscle satellite cell niches., *Differentiation.*, 2009, **78**, 185–194.

39. X. Duan, E. Kang, C. Y. Liu, G. L. Ming and H. Song, Development of neural stem cell in the adult brain., *Curr. Opin. Neurobiol.*, 2008, **18**, 108–115.

40. L. Conti and E. Cattaneo, Neural stem cell systems: physiological players or in vitro entities?, *Nat. Rev. Neurosci.*, 2010, **11**, 176–187.

41. G. L. Ming and H. Song, Adult neurogenesis in the mammalian central nervous system., *Annu. Rev. Neurosci.*, 2005, **28**, 223–250.

42. F. S. Tedesco, A. Dellavalle, J. az-Manera, G. Messina and G. Cossu, Repairing skeletal muscle: regenerative potential of skeletal muscle stem cells., *J. Clin. Invest.*, 2010, **120**, 11–19.

43. P. Anversa, A. Lerin, M. Rota, T. Hosoda, C. Bearzi, K. Urbanek, J. Kajstura and R. Bolli, Concise review: stem cells, myocardial regeneration, and methodological artifacts., *Stem Cells.*, 2007, **25**, 589–601.

44. P. Anversa, J. Kajstura and A. Leri, If I can stop one heart from breaking., *Circulation.*, 2007, **115**, 829–832.

45. A. P. Beltrami, K. Urbanek, J. Kajstura, S. M. Yan, N. Finato, R. Bussani, B. Nadal-Ginard, F. Silvestri, A. Leri, C. A. Beltrami and P. Anversa, Evidence that human cardiac myocytes divide after myocardial infarction., *N. Engl. J. Med.*, 2001, **344**, 1750–1757.

46. J. Tillmanns, M. Rota, T. Hosoda, Y. Misao, G. Esposito, A. Gonzales, S. Vitale, C. Parolin, S. Yasuzawa-Amano, J. Muraski, A. De Angelis, N. Lecapitaine, R. W. Siggins, M. Loredo, C. Bearzi, R. Bolli, K. Urbanek, A. Leri, J. Kajstura and P. Anversa, Formation of large coronary arteries by cardiac progenitor cells., *Proc. Natl. Acad. Sci. U. S. A.*, 2008, **105**, 1668–1673.

47. X. Chen, R. M. Wilson, H. Kubo, R. M. Berretta, D. M. Harris, X. Zhang, N. Jaleel, S. M. MacDonnel, C. Bearzi, J. Tillmanns, I. Trofimova, T. Hosoda, F. Mosna, L. Cribbs, A. Leri, J. Kajstura, P. Anversa and S. R. Houser, Adolescent feline heart contains a population of small, proliferative ventricular myocytes with immature physiological properties., *Circ. Res.*, 2007, **100**, 536–544.

48. M. Rota, T. Hosoda, A. De Angelis, M. L. Arcarese, G. Esposito, R. Rizzi, J. Tillmanns, D. Tugal, E. Musso, O. Rimoldi, C. Bearzi, K. Urbanek, P. Anversa, A. Leri and J. Kajstura, The young mouse heart is composed of myocytes heterogeneous in age and function., *Circ. Res.*, 2007, **101**, 387–399.

49. M. Rota, M. E. Padin-Iruegas, Y. Misao, A. De Angelis, S. Maestroni, J. Ferreira-Martins, E. Fiumana, R. Rastaldo, M. L. Arcarese, T. S. Mitchell, A. Boni, R. Bolli, K. Urbanek, T. Hosoda, P. Anversa, A. Leri and J. Kajstura, Local activation or implantation of cardiac progenitor

cells rescues scarred infarcted myocardium improving cardiac function., *Circ. Res.*, 2008, **103**, 107–116.

50. A. Gonzalez, M. Rota, D. Nurzynska, Y. Misao, J. Tillmans, C. Ojaimi, M. E. Padin-Iruegas, P. Müller, G. Esposito, C. Bearzi, S. Vitale, B. Dawn, S. K. Sanganalmath, M. Baker, T. H. Hintze, R. Bolli, K. Urbanek, T. Hosoda, P. Anversa, J. Kajstura and A. Leri, Activation of cardiac progenitor cells reverses the failing heart senescent phenotype and prolongs lifespan., *Circ. Res.*, 2008, **102**, 597–606.

51. J. A. Muraski, M. Rota, Y. Misao, J. Fransioli, C. Cottage, N. Gude, G. Esposito, F. Delucchi, M. Arcarese, R. Alvarez, S. Siddiqi, G. N. Emmanuel, W. Wu, K. Fischer, J. J. Martindale, C. C. Glembotski, A. Leri, J. Kajstura, N. Magnuson, A. Berns, R. M. Beretta, S. R. Houser, E. M. Schaefer, P. Anversa and M. A. Sussman, Pim-1 regulates cardiomyocyte survival downstream of Akt., *Nat. Med.*, 2007, **13**, 1467–1475.

52. Y. Tsujita, J. Muraski, I. Shiraishi, T. Kato, J. Kajstura, P. Anversa and M. A. Sussman, Nuclear targeting of Akt antagonizes aspects of cardiomyocyte hypertrophy., *Proc. Natl. Acad. Sci. U. S. A.*, 2006, **103**, 11946–11951.

53. T. Kato, J. Muraski, Y. Chen, Y. Tsujita, J. Wall, C. C. Glembotski, E. Schaefer, M. Beckerle and M. A. Sussman, Atrial natriuretic peptide promotes cardiomyocyte survival by cGMP-dependent nuclear accumulation of zyxin and Akt., *J. Clin. Invest.*, 2005, **115**, 2716–2730.

54. S. Siddiqi, N. Gude, T. Hosoda, J. Muraski, M. Rubio, G. Emmanuel, J. Fransioli, S. Vitale, C. Parolin, D. D'Amario, E. Schaefer, J. Kajstura, A. Leri, P. Anversa and M. A. Sussman, Myocardial induction of nucleostemin in response to postnatal growth and pathological challenge., *Circ. Res.*, 2008, **103**, 89–97.

55. A. Linke, P. Müller, D. Nurzynska, C. Casarsa, D. Torella, A. Nascimbene, C. Castaldo, S. Cascapera, M. Böhm, F. Quaini, K. Urbanek, A. Leri, T. H. Hintze, J. Kajstura and P. Anversa, Stem cells in the dog heart are self-renewing, clonogenic, and multipotent and regenerate infarcted myocardium, improving cardiac function., *Proc. Natl. Acad. Sci. U. S. A.*, 2005, **102**, 8966–8971.

56. K. Urbanek, M. Rota, S. Cascapera, C. Bearzi, A. Nascimbene, A. De Angelis, T. Hosoda, S. Chimenti, M. Baker, F. Limana, D. Nurzynska, D. Torella, F. Rotatori, R. Rastaldo, E. Musso, F. Quaini, A. Leri, J. Kajstura and P. Anversa, Cardiac stem cells possess growth factor-receptor systems that after activation regenerate the infarcted myocardium, improving ventricular function and long-term survival., *Circ. Res.*, 2005, **97**, 663–673.

57. L. Bu, X. Jiang, S. Martin-Puig, L. Caron, S. Zhu, Y. Shao, D. J. Roberts, P. L. Huang, I. J. Domian and K. R. Chien, Human ISL1 heart progenitors generate diverse multipotent cardiovascular cell lineages., *Nature.*, 2009, **460**, 113–117.

58. K. L. Laugwitz, A. Moretti, J. Lam, P. Gruber, Y. Chen, S. Woodward, L. Z. Lin, C. L. Cai, M. M. Lu, M. Reth, O. Platoshyn, J. X. Yuan, S. Evans and K. R. Chien, Postnatal isl1 + cardioblasts enter fully differentiated cardiomyocyte lineages., *Nature.*, 2005, **433**, 647–653.
59. A. Moretti, M. Bellin, C. B. Jung, T. M. Thies, Y. Takashima, A. Bernshausen, M. Schiemann, S. Fischer, S. Moosmang, A. G. Smith, J. T. Lam and K. L. Laugwitz, Mouse and human induced pluripotent stem cells as a source for multipotent Isl1 + cardiovascular progenitors., *FASEB J.*, 2009, **24**, 700–711.
60. Y. Qyang, S. Martin-Puig, M. Chiravuri, S. Chen, H. Xu, L. Bu, X. Jiang, L. Lin, A. Granger, A. Moretti, L. Caron, X. Wu, J. Clarke, M. M. Taketo, K. L. Laugwitz, R. T. Moon, P. Gruber, S. M. Evans, S. Ding and K. R. Chien, The renewal and differentiation of Isl1 + cardiovascular progenitors are controlled by a Wnt/beta-catenin pathway., *Cell Stem Cell.*, 2007, **1**, 165–179.
61. Y. Zhou, P. Pan, L. Yao, M. Su, P. He, N. Niu, M. A. McNutt and J. Gu, CD117-positive cells of the heart: progenitor cells or mast cells?, *J. Histochem. Cytochem.*, 2010, **58**, 309–316.
62. M. C. Yoder, L. E. Mead, D. Prater, T. R. Krier, K. N. Mroueh, F. Li, R. Krasich, C. J. Temm, J. T. Prchal and D. A. Ingram, Redefining endothelial progenitor cells via clonal analysis and hematopoietic stem/progenitor cell principals., *Blood.*, 2007, **109**, 1801–1809.
63. A. Reinisch, C. Bartmann, E. Rohde, K. Schallmoser, V. Bjelic-Radisic, G. Lanzer, W. Linkesch and D. Strunk, Humanized system to propagate cord blood-derived multipotent mesenchymal stromal cells for clinical application., *Regen. Med.*, 2007, **2**, 371–382.
64. T. Asahara, Cell therapy and gene therapy using endothelial progenitor cells for vascular regeneration., *Handbook of Exp. Pharmacol.*, 2007, **180**, 181–194.
65. K. K. Hirschi, D. A. Ingram and M. C. Yoder, Assessing identity, phenotype, and fate of endothelial progenitor cells., *Arterioscler. Thromb. Vasc. Biol.*, 2008, **28**, 1584–1595.
66. E. Rohde, C. Malischnik, D. Thaler, T. Maierhofer, W. Linkesch, G. Lanzer, C. Guelly and D. Strunk, Blood monocytes mimic endothelial progenitor cells., *Stem Cells.*, 2006, **24**, 357–367.
67. E. Rohde, C. Bartmann, K. Schallmoser, A. Reinisch, G. Lanzer, W. Linkesch, C. Guelly and D. Strunk, Immune cells mimic the morphology of endothelial progenitor colonies in vitro., *Stem Cells.*, 2007, **25**, 1746–1752.
68. T. G. Fellous, S. A. McDonald, J. Burkert, A. Humphries, S. Islam, N. M. De-Alwis, L. Gutierrez-Gonzalez, P. J. Tadrous, G. Elia, H. M. Kocher, S. Bhattacharya, L. Mears, M. El-Bahrawy, D. M. Turnbull, R. W. Taylor, L. C. Greaves, P. F. Chinnery, C. P. Day, N. A. Wright and M. R. Alison, A methodological approach to tracing cell lineage in human epithelial tissues., *Stem Cells.*, 2009, **27**, 1410–1420.
69. D. H. Scoville, T. Sato, X. C. He and L. Li, Current view: intestinal stem cells and signaling., *Gastroenterology.*, 2008, **134**, 849–864.

70. S. A. McDonald, L. C. Greaves, L. Gutierrez-Gonzalez, M. Rodriguez-Justo, M. Deheragoda, S. J. Leedham, R. W. Taylor, C. Y. Lee, S. L. Preston, M. Lovell, T. Hunt, G. Elia, D. Oukrif, R. Harrison, M. R. Novelli, I. Mitchell, D. L. Stoker, D. M. Turnbull, J. A. Jankowski and N. A. Wright, Mechanisms of field cancerization in the human stomach: the expansion and spread of mutated gastric stem cells., *Gastroenterology.*, 2008, **134**, 500–510.

71. J. P. Seery and F. M. Watt, Asymmetric stem-cell divisions define the architecture of human oesophageal epithelium., *Curr. Biol.*, 2000, **10**, 1447–1450.

72. P. A. Lysy, D. Campard, F. Smets, M. Najimi and E. M. Sokal, Stem cells for liver tissue repair: current knowledge and perspectives., *World J. Gastroenterol.*, 2008, **14**, 864–875.

73. Y. Dor, J. Brown, O. I. Martinez and D. A. Melton, Adult pancreatic beta-cells are formed by self-duplication rather than stem-cell differentiation., *Nature.*, 2004, **429**, 41–46.

74. X. Xu, J. D'Hoker, G. Stangé, S. Bonné, N. De Leu, X. Xiao, M. Van de Casteele, G. Mellitzer, Z. Ling, D. Pipeleers, L. Bouwens, R. Scharfmann, G. Gradwohl and H. Heimberg, Beta cells can be generated from endogenous progenitors in injured adult mouse pancreas., *Cell.*, 2008, **132**, 197–207.

75. E. Sangiorgi and M. R. Capecchi, Bmi1 lineage tracing identifies a self-renewing pancreatic acinar cell subpopulation capable of maintaining pancreatic organ homeostasis., *Proc. Natl. Acad. Sci. U. S. A.*, 2009, **106**, 7101–7106.

76. M. Z. Ratajczak, E. K. Zuba-Surma, B. Machalinski and M. Kucia, Bone-marrow-derived stem cells - our key to longevity?, *J. Appl. Genet.*, 2007, **48**, 307–319.

77. M. L. da Silva, P. C. Chagastelles and N. B. Nardi, Mesenchymal stem cells reside in virtually all post-natal organs and tissues., *J. Cell Sci.*, 2006, **119**, 2204–2213.

78. S. Kern, H. Eichler, J. Stoeve, H. Kluter and K. Bieback, Comparative Analysis of Mesenchymal Stem Cells from Bone Marrow, Umbilical Cord Blood or Adipose Tissue., *Stem Cells*, 2006, **24**, 1294–301.

79. E. H. Javazon, K. J. Beggs and A. W. Flake, Mesenchymal stem cells: paradoxes of passaging., *Exp. Hematol.*, 2004, **32**, 414–425.

80. M. Z. Ratajczak, M. Kucia, R. Reca, M. Majka, A. Janowska-Wieczorek and J. Ratajczak, Stem cell plasticity revisited: CXCR4-positive cells expressing mRNA for early muscle, liver and neural cells 'hide out' in the bone marrow., *Leukemia.*, 2004, **18**, 29–40.

81. M. F. Pittenger, A. M. Mackay, S. C. Beck, R. K. Jaiswal, R. Douglas, J. D. Mosca, M. A. Moorman, D. W. Simonetti, S. Craig and D. R. Marshak, Multilineage potential of adult human mesenchymal stem cells., *Science*, 1999, **284**, 143–147.

82. T. Tondreau, L. Lagneaux, M. Dejeneffe, A. Delforge, M. Massy, C. Mortier and D. Bron, Isolation of BM mesenchymal stem cells by

plastic adhesion or negative selection: phenotype, proliferation kinetics and differentiation potential., *Cytotherapy.*, 2004, **6**, 372–379.

83. N. Quirici, D. Soligo, P. Bossolasco, F. Servida, C. Lumini and G. L. Deliliers, Isolation of bone marrow mesenchymal stem cells by anti-nerve growth factor receptor antibodies., *Exp. Hematol.*, 2002, **30**, 783–791.

84. D. Baksh, J. E. Davies and P. W. Zandstra, Soluble factor cross-talk between human bone marrow-derived hematopoietic and mesenchymal cells enhances in vitro CFU-F and CFU-O growth and reveals heterogeneity in the mesenchymal progenitor cell compartment., *Blood.*, 2005, **106**, 3012–3019.

85. K. Dorshkind, A. Johnson, L. Collins, G. M. Keller and R. A. Phillips, Generation of purified stromal cell cultures that support lymphoid and myeloid precursors., *J. Immunol. Methods.*, 1986, **89**, 37–47.

86. A. Johnson and K. Dorshkind, Stromal cells in myeloid and lymphoid long-term bone marrow cultures can support multiple hemopoietic lineages and modulate their production of hemopoietic growth factors., *Blood.*, 1986, **68**, 1348–1354.

87. R. Schäfer, M. Dominici, I. Müller, F. Dazzi, K. Bieback, K. Godthardt, K. Le Blanc, R. Meisel, R. Pochampally, R. Richter, T. Skutella, G. Steinhoff, M. Mitterberger, H. P. Wendel, J. Wiskirchen, R. Handgretinger and H. Northoff, Progress in characterization, preparation and clinical applications of non-hematopoietic stem cells, 29-30 September 2006, Tubingen, Germany., *Cytotherapy.*, 2007, **9**, 397–405.

88. K. T. Guo, R. Schäfer, A. Paul, G. Ziemer and H. P. Wendel, Aptamer-based strategies for stem cell research., *Mini. Rev. Med. Chem.*, 2007, **7**, 701–705.

89. M. Dezawa, Insights into autotransplantation: the unexpected discovery of specific induction systems in bone marrow stromal cells., *Cell Mol. Life Sci.*, 2006, **63**, 2764–2772.

90. A. Uccelli, E. Zappia, F. Benvenuto, F. Frassoni and G. Mancardi, Stem cells in inflammatory demyelinating disorders: a dual role for immuno-suppression and neuroprotection., *Expert. Opin. Biol. Ther.*, 2006, **6**, 17–22.

91. T. Honma, O. Honmou, S. Iihoshi, K. Harada, K. Houkin, H. Hamada and J. D. Kocsis, Intravenous infusion of immortalized human mesenchymal stem cells protects against injury in a cerebral ischemia model in adult rat., *Exp. Neurol.*, 2006, **199**, 56–66.

92. Y. Iso, J. L. Spees, C. Serrano, B. Bakondi, R. Pochampally, Y. H. Song, B. E. Sobel, P. Delafontaine and D. J. Prockop, Multipotent human stromal cells improve cardiac function after myocardial infarction in mice without long-term engraftment., *Biochem. Biophys. Res. Commun.*, 2007, **354**, 700–706.

93. M. Gnecchi, H. He, N. Noiseaux, O. D. Liang, L. Zhang, F. Morello, H. Mu, L. G. Melo, R. E. Pratt, J. S. Ingwall and V. J. Dzau, Evidence supporting paracrine hypothesis for Akt-modified mesenchymal stem cell-mediated cardiac protection and functional improvement., *FASEB J.*, 2006, **20**, 661–669.

94. Y. Dai, M. Xu, Y. Wang, Z. Pasha, T. Li and M. Ashraf, HIF-1 alpha induced-VEGF overexpression in bone marrow stem cells protects cardiomyocytes against ischemia., *J. Mol. Cell Cardiol.*, 2007, **42**, 1036–1044.

95. C. Martinez, T. J. Hofmann, R. Marino, M. Dominici and E. M. Horwitz, Human bone marrow mesenchymal stromal cells express the neural ganglioside GD2: a novel surface marker for the identification of MSCs., *Blood.*, 2007, **109**, 4245–4248.

96. C. Giesert, A. Marxer, D. R. Sutherland, A. C. Schuh, L. Kanz and H. J. Bühring, Antibody W7C5 defines a CD109 epitope expressed on CD34 + and CD34- hematopoietic and mesenchymal stem cell subsets., *Ann. N. Y. Acad. Sci.*, 2003, **996**, 227–230.

97. H. J. Bühring, S. Kuçi, T. Conze, G. Rathke, K. Bartoloviæ, F. Grüne-bach, M. Scherl-Mostageer, T. H. Brümmendorf, N. Schweifer and R. Lammers, CDCP1 identifies a broad spectrum of normal and malignant stem/progenitor cell subsets of hematopoietic and nonhematopoietic origin., *Stem Cells.*, 2004, **22**, 334–343.

98. D. G. Phinney, Building a consensus regarding the nature and origin of mesenchymal stem cells., *J. Cell Biochem. Suppl.*, 2002, **38**, 7–12.

99. M. Dominici, K. Le Blanc, I. Müller, I. Slaper-Cortenbach, F. Marini, D. Krause, R. Deans, A. Keating, D. J. Prockop and E. Horwitz, Minimal criteria for defining multipotent mesenchymal stromal cells. The International Society for Cellular Therapy position statement. Cytotherapy, 2006, **8**, 315–317.

100. A. Peister, J. A. Mellad, B. L. Larson, B. M. Hall, L. F. Gibson and D. J. Prockop, Adult stem cells from bone marrow (MSCs) isolated from different strains of inbred mice vary in surface epitopes, rates of proliferation, and differentiation potential., *Blood*, 2004, **103**, 1662–1668.

101. M. Dominici, C. Pritchard, J. E. Garlits, T. J. Hofmann, D. A. Persons and E. M. Horwitz, Hematopoietic cells and osteoblasts are derived from a common marrow progenitor after bone marrow transplantation., *Proc. Natl. Acad. Sci. U. S. A.*, 2004, **101**, 11761–11766.

102. D. Baksh, P. W. Zandstra and J. E. Davies, A non-contact suspension culture approach to the culture of osteogenic cells derived from a CD49e(low) subpopulation of human bone marrow-derived cells., *Biotechnol. Bioeng.*, 2007, **98**, 1195–1208.

103. R. Schäfer, R. Kehlbach, J. Wiskirchen, R. Bantleon, J. Pintaske, B. R. Brehm, A. Gerber, H. Wolburg, C. D. Claussen and H. Northoff, Transferrin Receptor Upregulation: In Vitro Labeling of Rat Mesenchymal Stem Cells with Superparamagnetic Iron Oxide., *Radiology*, 2007, **244**, 514–523.

104. M. Miura, Y. Miura, H. M. Padilla-Nash, A. A. Molinolo, B. Fu, V. Patel, B. M. Seo, W. Sonoyama, J. J. Zheng, C. C. Baker, W. Chen, T. Ried and S. Shi, Accumulated chromosomal instability in murine bone marrow mesenchymal stem cells leads to malignant transformation., *Stem Cells.*, 2006, **24**, 1095–1103.

105. W. M. Klein, B. P. Wu, S. Zhao, H. Wu, A. J. Klein-Szanto and S. R. Tahan, Increased expression of stem cell markers in malignant melanoma., *Mod. Pathol.*, 2007, **20**, 102–107.
106. G. Chamberlain, J. Fox, B. Ashton and J. Middleton, Concise review: mesenchymal stem cells: their phenotype, differentiation capacity, immunological features, and potential for homing., *Stem Cells.*, 2007, **25**, 2739–2749.
107. K. Le Blanc, I. Rasmusson, B. Sundberg, C. Götherström, M. Hassan, M. Uzunel and O. Ringdén, Treatment of severe acute graft-versus-host disease with third party haploidentical mesenchymal stem cells., *Lancet.*, 2004, **363**, 1439–1441.
108. E. Zappia, S. Casazza, E. Pedemonte, F. Benvenuto, I. Bonanni, E. Gerdoni, D. Giunti, A. Ceravolo, F. Cazzanti, F. Frassoni, G. Mancredi and A. Uccelli, Mesenchymal stem cells ameliorate experimental auto-immune encephalomyelitis inducing T-cell anergy., *Blood.*, 2005, **106**, 1755–1761.
109. P. Comoli, F. Ginevri, R. Maccario, M. A. Avanzini, M. Marconi, A. Groff, A. Cometa, M. Cioni, L. Porretti, W. Barberi, F. Frassoni and F. Locatelli, Human Mesenchymal Stem Cells Inhibit Antibody Production Induced In Vitro by Allostimulation., *Nephrol. Dial. Transplant.*, 2008, **23**, 1196–202.

Screening Approaches for Stem Cells

DAVID G. BUSCHKE,[a, c] DEREK J. HEI,[b] KEVIN W. ELICEIRI[a, c] AND BRENDA M. OGLE[a, c, d]

[a] Department of Biomedical Engineering; [b] School of Medicine and Public Health; [c] Laboratory for Optical and Computational Instrumentation; [d] Material Sciences Program, University of Wisconsin at Madison, Madison, Wisconsin 53706, USA

3.1 Dynamics of Stem Cells

The two primary biological properties of stem cells are self-renewal and the capacity to differentiate into multiple cell types (plasticity). These unique properties distinguish stem cells from lineage-restricted somatic cells and position stem cells to serve as predictive models of development,[1,2] pathogenesis[3,4] and drug toxicity,[1,2] and ultimately to enhance regenerative therapies. For example, embryonic stem cells (ESCs), which originate from the inner cell mass, are frequently used to test the effects of toxicants such as second-hand smoke on pre-implantation development.[2] Alternatively ESCs can be genetically altered to replicate human genetic defects. In particular, an *in vitro* model of Down's Syndrome has been accomplished by incorporating a single human chromosome 21 into murine ESCs.[4] Model systems such as these are valuable tools for basic research and useful alternatives to animal models. However, if stem cells (here defined as embryonic stem cells, induced pluripotent stem cells, adult hematopoietic stem cells or adult mesenchymal stem cells) are to be effective in this capacity, or in the context of regenerative therapies, it is

Stem Cell-Based Tissue Repair
Edited by Raphael Gorodetsky and Richard Schäfer
© The Royal Society of Chemistry 2011
Published by the Royal Society of Chemistry, www.rsc.org

imperative that the maintenance and loss of plasticity be tightly regulated and closely monitored. In this chapter we review the types of signals and cues that could potentially be exploited for *in vitro* and *in vivo* stem cell model characterization and discuss current and emerging screening technologies that can be used and adapted to detect these parameters.

Stem cell plasticity is dependent on endogenous properties as well as exogenous cues provided by the local microenvironment (Figure 3.1). *In vivo*, the local microenvironment or niche is comprised of components and interactions including growth factors, cell–cell contacts and cell–matrix interactions. Significant efforts have been made to replicate the *in vivo* niche in tissue culture environments in an attempt to either maintain multipotency or to induce specific somatic cell differentiation. Toward this end, a whole host of chemical,[5–13] mechanical[14–23] and electrical stimuli[24–27] have been thrust upon stem cells.

Perhaps the most common chemical stimulus secreted by stem cells and nearby niche cells *in vivo* are growth factors. Growth factors are tightly regulated with development, normal tissue maintenance and repair, and so play a crucial role in regulating stem cell fate.[28–33] In culture, several approaches have been applied to control the spatial and temporal exposure of stem cells to growth factors. One approach is the miniaturization of cell culture platforms *via* microtechnology to tightly control the inputs (including addition of soluble factors such as growth factors and hormones), the placement of cells and the system outputs (image analysis or use of inline sensors) in a high-throughput manner.[34]

Recent studies show that microtechnology and microfluidic platforms can more closely mimic the scale and dynamics of living systems than their macroscale counterparts. For example, microfluidic devices constructed to generate gradients of fibroblast growth factor and platelet-derived growth factor stimulate growth and differentiation of neural stem cells in a tightly controlled, concentration-dependent manner.[35] Microfluidic systems can also be modified to include three-dimensional matrices (*i.e.* synthetic polymers, natural extracellular matrix proteins and/or peptides) to preferentially deliver or sequester growth factors or their corresponding cellular binding motif. With Xenopus embryonic spinal neurons, for example, microfluidic control of brain-derived neurotrophic factor in conjunction with a laminin gradient stimulates concentration-dependent neuronal growth cone responses.[36] Continued advances in micro- and nanotechnology will spur development of well-defined *in vitro* niches for tighter regulation of chemical stimuli and, consequently, stem cell fate.

The *in vivo* niche is also subject to mechanical forces:

- those exerted at the macroscale as a result of tissue function (*i.e.* lung expansion, heart contraction, blood flow); and
- those exerted by the cell on the local environment (*i.e.* extracellular matrix remodeling).

Mechanical forces have been shown to influence stem cell fate and so attempts to control cell fate in culture have included the incorporation of culture

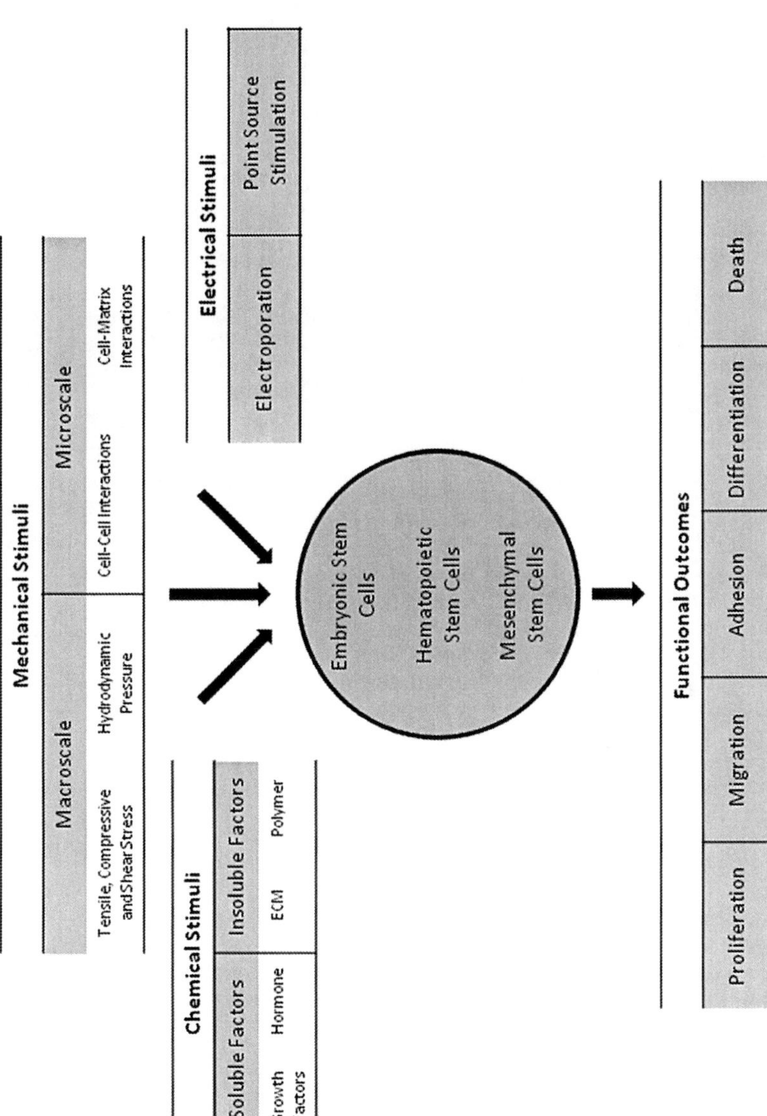

Figure 3.1 Environmental stimuli capable of changing stem cell state. Environmental stimuli can be divided into at least three categories including: chemical, mechanical and electrical stimuli. Such stimuli are known to alter the fate of (stem) cells *in vivo*, and have thus been extensively applied in vitro to maintain pluripotency or multipotency, or induce differentiation. Investigators measure phenotypic changes and functional outcomes before and after applying a stimulus, to begin to define the state of a stem cell in terms of quantifiable, phenotypic markers.

substrates of varying stiffness or porosity[37,38] and/or exposure of cultured cells to tensile, compressive or shear stress.[16,22,39–42] Mesenchymal stem cells, for example, will assume a myogenic phenotype when exposed to collagen-coated gels that mimic the elasticity of muscle, but will assume an osteogenic phenotype when grown on stiffer gels mimicking precalcified bone.[37] Matrix sensing that drives lineage specification has been shown to involve cytoskeletal motors, non-muscle myosin II isoforms A–C. Embryonic stem cells will undergo cardiovascular differentiation (*i.e.* contracting foci and vasculogenesis) when exposed to 10% tensile strain.[22] Mechanical strain is postulated to elevate intracellular reactive oxygen species capable of stimulating ERK1, 2 and JNK-dependent signaling pathways involved in differentiation.

Electrical potential, created by an unequal distribution of ions between the extra- and intracellular compartments, is crucial for neural transmission as well as initiation and maintenance of cardiac contraction in the developing organism.[43,44] Recent reports suggest that electrical potential and subsequent electrical activation might also impact the postnatal fate of stem cells. For example, embryonic stem cells stimulated with an electroporator under several voltage conditions (0–20 V) showed potential to differentiate into various types of neurons *in vitro* and contributed to injured spinal cord as neuronal cells *in vivo*.[45] Similarly, point source electrical stimulation (< 1 V) of embryonic stem cells induced enhanced levels of cardiomyocyte differentiation compared with unstimulated controls.[24] The mechanism by which electrical stimulation induces cardiac differentiation is poorly understood, but the techniques thus far applied might be refined to increase understanding of molecular pathways in addition to increasing the numbers of stem cell-derived cardiomyocytes for therapeutic interventions.

3.2   Characterization is Critical

Creation of *in vitro* culture environments to maintain multipotency/pluripotency has been largely successful. Testaments to this success are stem cell banks, which routinely generate and distribute multipotent/pluripotent cells. For example, standard methods have been developed for propagating and freezing human embryonic stem cells (hESCs) with intent to maintain high quality and functionality of the resulting cells (National Stem Cell Bank, Madison, WI, USA).[46–49] Current propagation methods include feeder-free culture methods that rely on defined media components and few or no animal-derived components. During and after propagation, many are screened to verify identity and purity. Toward this end, tests are performed to confirm that the cells have a normal karyotype (G-Band) and possess characteristics that are indicative of human embryonic stem cells (*e.g.* expression of hESC-specific markers by flow cytometry). In addition, many are tested for teratoma formation to demonstrate pluripotent function (Table 3.1).

Human ESCs are typically frozen prior to distribution. Freezing is yet another stimulus found to alter stem cell behavior, in some cases inducing

Table 3.1 Summary of testing for hESCs (National Stem Cell Bank, Madison, WI, USA).

Test	*Description*	*Specification*
Identity	Short tandem repeat (STR) testing Human leukocyte antigen (HLA) testing	Matches parent cell line
Bacteria/fungi	Bacteria, fungi testing according to 21 CFR 610.12 Direct transfer method with bacteriastasis and fungistasis testing	No contamination detected
Mycoplasma	Direct culture in broth and agar and indirect test using indicator culture (Vero)/DNA stain; meets requirements for FDA's Points to Consider (PTC)	No contamination detected
Post-thaw recovery	Test for the ability to recover viable hESC colonies without excessive differentiation	Viable colonies recovered
Karyotype	G-band analysis on 20 metaphase spreads	Normal karyotype
Growth characteristics	Report doubling time	Report value
HESC marker expression	Flow cytometry for SSEA-1, SSEA-3, SSEA-4, TRA-1-60, TRA-1-81, Oct-3/4 Comparative genome hybridization—NimbleGen 380K array	>70% for each marker
Copy number changes	Cross-check all copy number variants with the normal copy Number Variant Database Gene expression profiling using DNA microarray analysis	Report data only
Gene expression	Analyze for expression of core hESC genes as well as markers of differentiated cell types Testing for murine, bovine and porcine pathogens Testing for human pathogens by polymerase chain reaction (PCR)	Report data only
Adventitious agents	*In vitro* and *in vivo* testing for adventitious agents Retrovirus assays (cell culture and electron microscopy)	No contamination detected
Pluripotency	Teratoma formation in SCID mice	Teratoma formation with detection of cell types representative of endoderm, mesoderm and ectoderm

apoptosis[50] and in other cases triggering differentiation.[51] Fortunately, several groups have made significant progress in improving cell yields utilizing Rho-associated kinase (ROCK) inhibitor Y-27632 to suppress the apoptosis typically encountered during both cryopreservation and the early stages of post-thaw culture.[52–54] Additional efforts for improving cryopreservation have included optimization of freezing profiles using programmable controlled rate freezers[55] and improvement of the dissociation method used to prepare cells for cryopreservation.[56] Those cells frozen utilizing modified techniques such as these are subject to the same testing regime as described above for propagation. This model for cell banking reflects the current practices recommended for cell lines that will be used to produce cell therapeutics for human clinical applications as indicated in guidance documents published by the US Food and Drug Administration (FDA) and the International Conference on Harmonization (ICH).[57,58]

Creation of *in vitro* culture environments to induce differentiation has been less successful than those described above to maintain pluripotency. In most cases success has been tempered by the inability of multiple laboratories to replicate results. This difficulty does not reflect carelessness on the part of researchers, but instead an inability to easily identify and track heterogeneity within populations of stem cells. The list of potential contributions to heterogeneity is long and onerous. For embryonic stem cells, for example,[59] contributions include:

- **Differences due to origin of cells lines**—genomic diversity, stage of pre-implantation embryo at derivation, divergent derivation procedures, use of feeder layers during culture, varying densities, culture substrates, media, additives and freezing methods;
- **Differences arising over time in culture**—genetic changes (loss or gain of specific sequences or chromosomes), general or specific epigenetic changes (DNA methylation, histone acetylation, microRNAs); and
- **Differences due to mosaicism during culture**—partial or terminal differentiation of subpopulations within cultures, variation among epigenetic and genetic changes.

Heterogeneity is even more problematic for adult-derived stem cells due to dramatic differences in donor age and so stem cell composition. For example, we conducted a study wherein adult, bone marrow-derived mesenchymal stem cells (CD105+, CD73+, CD90+) isolated *via* fluorescence-activated cell sorting were seeded on various extracellular matrix (ECM) proteins and allowed to grow for 14 days (akin to protocols used to generate tissue-engineered constructs).[60] Not surprisingly, ECM proteins were found to drive differentiation of stem cells. Surprisingly, individual ECM proteins could stimulate multilineage differentiation (*i.e.* type I collagen initiates differentiation of cardiomyocytes, osteoblasts and adipocytes). In addition, the kinetics of differentiation varied dramatically between stem cell donors despite identical isolation procedures and culture conditions. Clearly not all stimuli capable of driving stem cell differentiation have been identified. In addition, the mechanisms responsible for stimulated differentiation with known stimuli have not been identified.

The complexity of discerning the stimuli and corresponding mechanism of action is compounded by the many different types of stem cells and the varying conditions by which stem cells are obtained and subsequently cultured. Thus the extraordinary heterogeneity of stem cells, both adult and embryonic and plentiful means to stimulate differentiation by largely unknown mechanisms, should strike a cautionary note for both basic research and clinical therapy. Perhaps the most striking example is provided by a recent study in which adult bone marrow-derived stem cells were injected into the myocardium of rats with intent to regenerate myocardium.[61] Instead extensive intramyocardial calcification was observed. These results stress the importance of comprehensive screening and continued monitoring of stem cells or stem cells seeded within matrices prior to *in vitro* study or *in vivo* application.

3.3 What is Stem Cell Screening?

Stem cell screening in the context of this chapter is defined as the process by which a stem cell is distinguished from a stem cell of another type or from a differentiated cell (Figure 3.2). Stem cell screening can be high-throughput or low-throughput, high-content or low-content, and can employ one or a combination of several analysis techniques (see below). The primary goals of stem cell screening (Figure 3.2) are:

1. To characterize stem cells or differentiated progeny based on existing biomarkers.
2. To determine the mechanism whereby the biomarker is altered.

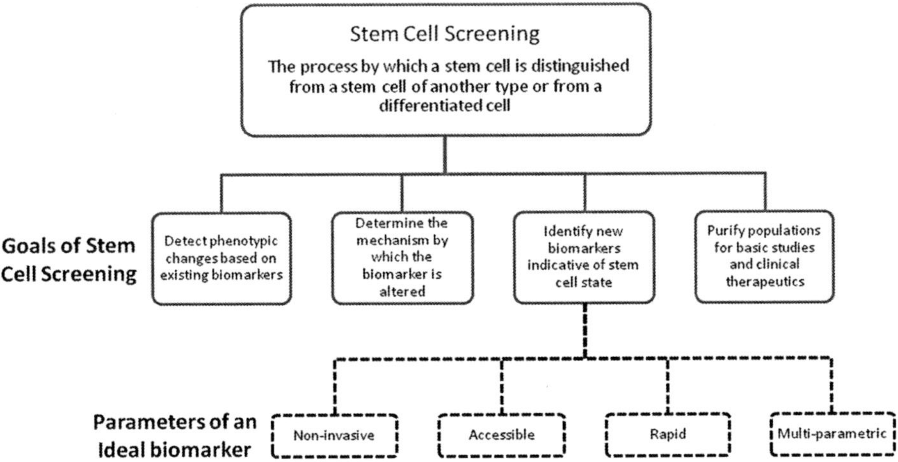

Figure 3.2 Stem cells screening process.

3. To identify new biomarkers indicative of stem cell state.
4. To purify populations for basic studies and clinical therapeutics.

To meet these goals, one might define a quintessential stem cell screening modality as one that would meet the following specifications (Figure 3.2):

1. The modality would precisely detect the presence of a biomarker(s) that is stable and easily distinguished from the presence of said biomarker(s) in other cell types.
2. Detection of the biomarker would be noninvasive such that the cell(s) could be maintained in a healthy, unaltered state during and after screening.
3. The modality would be robust and accessible to most research and clinical laboratories to ensure consistent, standardized characterization.
4. The modality would rapidly screen thousands of events and determine the state of multiple parameters simultaneously.

3.4 Stem Cell Screening

3.4.1 Current Approaches

Current approaches to stem cell screening have been derived principally from the fields of cell and molecular biology. Stem cells were first characterized by their ability to undergo functional, multilineage differentiation. Later, protein expression profiles were defined for each type of stem cell to avoid the cumbersome functional assays initially employed (Table 3.2).

Adult hematopoietic stem cells (HSCs) were the first multipotent cell type to be robustly characterized. HSCs are found primarily in the bone marrow and give rise to all cells of hematopoietic lineage including lymphocytes, macrophages, neutrophils and basophils. Early on, definitive proof of the HSC phenotype required production of a long-lasting clone able to give rise to all of the hematopoietic lineages *in vivo*. Long-term culture-initiating cell assays or colony forming assays were used to confirm multilineage differentiation of HSCs. These assays are time consuming and cumbersome and so to avoid this difficulty, genetic and proteomic profiling were performed to define an expression panel specific to HSCs—principally $CD34^+$, $CD133^+$, $CD150^+$ and $CD45^-$. Similarly, adult mesenchymal stem cells (MSCs) found in many tissue types including bone marrow, fat, cord blood and dental pulp were first characterized by their ability to undergo multilineage differentiation upon stimulation *in vitro*. Today, an expression panel indicative of MSCs has been defined, principally $CD73^+$, $CD90^+$, $CD105^+$ and $CD45^-$. Finally, embryonic stem cells (ESCs), derived from the inner cell mass of early stage embryos, were first confirmed by their ability to form teratomas *in vivo*. Today, routine confirmation of ESCs is focused largely on expression of Oct-4, Nanog, SSEA-3, SSEA-4, TRA-1-60 and TRA-1-81.

Table 3.2 Phenotypic, morphological and functional biomarkers of stem cells.

Stem cell	Human protein expression	Mouse protein expression	Colony morphology	Proliferative capacity	Differentiation capacity
Embryonic	Oct4$^+$ Telomerase TRA-1-60$^+$ TRA-1-81$^+$	Oct4$^+$	Grown on MEF feeder layers (irradiated), conditioned media, and recently TeSR.	Capable of hundreds of population doublings.	Pluripotent. Forms mature derivatives of all three embryonic germ layers.
	Alkaline phosphatase SSEA-3$^+$ SSEA-4$^+$ SSEA-1$^-$ Nanog$^+$	Alkaline phosphatase SSEA-3$^-$ SSEA-4$^-$ SSEA-1$^+$ FoxD3	Requires mechanical removal of differentiated cells determined by morphology	Doubling time 24–48 hours.	Pluripotency determined by teratoma formation in SCID mice, spontaneous differentiation into all three germ layers *in vitro*, or injection of ES cells from inner cell mass of one blastocyst into another blastocyst to create a chimeric mouse.
	GTCM-2$^+$ Sox 2$^+$ Rex1$^+$		Clonogenic		
	FoxD3		Long-term culture-initializing cell (LTC-IC).		
Hematopoeitic	CD34$^+$	c-kit$^+$		Expansion *in vitro* is limited.	
	Lin$^-$ CD133$^+$	Lin$^-$ CD90$^-$	Cobble-stone area forming cell (maintain viability under stromal cells for 5–7 weeks, progenitors die earlier).		Can differentiate into all mature blood cell types from lymphoid, myeloid and erythroid lineages.
	CXCR4$^+$	Sca$^-$		Can be expanded with multiple cytokines.	

Table 3.2 (*continued*)

Stem cell	Human protein expression	Mouse protein expression	Colony morphology	Proliferative capacity	Differentiation capacity
	Aldehyde dehydrogenase(hi) CD150$^+$	Aldehyde dehydrogenase(hi) CD150$^+$	CFU-S (forms large colonies in spleen by day 12.	Difficult to recreate *in vivo* niche.	Capable of restoring hematopoiesis in a lethally irradiated mouse.
	CD48$^-$ CD244$^-$ Sca-1$^+$	CD48$^-$ CD244$^-$ CD38$^+$	Human implantation into NOD/SCID mice—detect human CD45 (mature cells).		
Mesenchymal	CD73$^+$ CD90$^+$ CD105$^+$ CD45$^-$ CD34$^-$ CD14$^-$	CD90$^+$ Sca-1$^+$ CD135$^+$ CD45$^-$ CD34$^-$ CD11b$^-$	Fusiform, fibroblast-like cells. Colony forming unit fibroblasts [CFU-f].	Limited. *In vitro* replicative senescence.	Under *in vitro* conditions, capable of forming osteocytes, adipocytes and chondrocytes.
	CD29$^+$ CD166$^+$ CD44$^+$ Stro-1$^+$	CD31(low) CD44$^+$ Vcam-1$^-$	Adherent to tissue culture polystyrene.	Maximum population doublings (PD) dependent on age of donor (24–40 PD).	

Protein expression panels for each stem cell type are typically confirmed *via* analysis of gene-specific RNA or protein content *via* traditional cell and molecular biology approaches. Approaches such as real-time polymerase chain reaction (RT-PCR) and Northern blotting are used to detect RNA expression while *in situ* immunofluorescent labeling and Western blotting are used to detect protein expression (Table 3.3). Unfortunately each of these methods is low-throughput, time-consuming and invasive, and this prohibits long-term study of a particular population. To avert these limitations, methodologies have been developed which are capable of rapid characterization with no deleterious effects on the cells. An overview of these methods is presented below together with a discussion of their applicability to stem cells.

3.4.2 Recent Advances: High-throughput and High-content Screening

Advances in instrumentation technology in the last 15 years have improved the speed and efficiency with which to detect and track key stem cell bio-markers.

First, techniques have been developed to 'print' nucleic acid and amino acid sequences on large arrays containing hundreds to hundreds of thousands of sequences which can be tested for compatibility with cellular products. Used routinely in many laboratories today, these arrays serve as a powerful probing mechanism to:[62–74]

- rapidly detect known expression profiles of stem cells and their progeny;
- determine a more comprehensive expression profile of stem cells; and
- determine changes in gene and protein expression throughout various stages in a differentiation scheme.

Downregulated genes after differentiation typically consist of those that play an important role in maintaining potency or 'stemness', while upregulated genes are typically characteristic of a mature, differentiated phenotype.

Secondly, intact proteins can also be bound in array format for high-throughput analysis of protein function. Protein arrays, coupled with advanced dielectrophoresis (2-DE) and mass spectrometry allow for high-throughput, quantitative detection of protein expression. Differential protein expression analysis is of most interest to stem cell researchers in determining how a cell will change its proteome based on an applied stimulus. An extensive body of research has been established on differential protein expression to begin to build a comprehensive database of the stem cell proteome (reviewed in ref. 59). Combined, these evolving technologies have been and will continue to be invaluable tools for the elucidation of molecular machinery of stem cells. Unfortunately, these techniques still require destruction of cells for analysis. Sample destruction is especially problematic for stem cells which, given their dynamic nature, would benefit from long-term tracking.

Table 3.3 Current stem cell screening modalities.

Biomarker	Method	Resolution	Sacrificial	Throughput	Primary advantage
RNA/DNA (bulk)	Southern/Northern blot	Low	Y	Binary	High specificity, detects NA sequences based on defined primers
	Polymerase chain reaction (PCR)	Low	Y	Semi-quantitative	Highly sensitive (compared to blots)
	Quantitative PCR	Low	Y	Quantitative	Highly sensitive (compared to blots)
	Microarray	High	Y	Differential expression	Differential expression analysis of very large arrays of genes
Protein (bulk)	ICC/IHC	Low	Y	Semi-quantitative	Protein localization in cells/tissues
	Western blot	Low	Y	Semi-quantitative	High specificity
	Protein array	High	Y	Differential expression	Differential expression analysis for thousands of proteins
					More sensitive measure of phenotype than microarray

	Mass spectrometry	High	Y	Amino acid sequence	Highly sensitive measure of mass : charge ratio of proteins Compare results with protein databases
Cytometry (single cells)	Single cell flow cytometry	High	Y	Quantitative	Some instruments capable of analyzing 15 parameters per cell
	Conventional cell sorters	High	N	Quantitative	Sort and recover cells based on user defined criteria
	Imaging flow cytometry	Enhanced	Y	Quantitative + morphology	Capable of capturing image of cells in the context of flow
	Confocal flow cytometry	Enhanced	N	Quantitative + morphology	Maintain cells *in vitro* while capturing five images per second
	Non-linear excitation flow cytometry	Enhanced	N	Quantitative + morphology	Maintenance of 3D construct and detection of intrinsic fluorescence

Surprisingly few high-throughput analytical tools allow for the preservation of stem cells after analysis. As one example, advances in robotics, chemistry and biology has led to the creation of a high-throughput screening (HTS) technique that allows a scientist to quickly run a series of drug and biochemical tests in a single well of a multi-well plate. Using automated approaches with robotics and software high-throughput screening can quickly assess a large number of wells for different responses to drug candidates. This technique, while lacking single cell resolution, has become standard practice for drug research. Current cell-based, high-throughput screening methods have evolved from simple non-imaged light scattering or fluorescence measurements to cell imaging systems where cells in individual wells of a multi-well plate are imaged using wide-field fluorescence microscopy techniques. Another technology that allows for the conservation of stem cells during and after analysis is high-content live imaging. The basic idea behind such instruments is to maintain cells in sterile *in vitro* culture systems, while incorporating automated imaging and fluidic handling features to analyze large quantities of cells both spatially and temporally. Several groups have taken advantage of this technology to track and analyze stem cells in an automated, highly controlled environment.[75–79]

Perhaps the most commonly used high throughput analysis technique of intact cells is flow cytometry. Flow cytometry is widely used amongst stem cell researchers as a multiparametric means to verify the presence or absence of biomarkers. The most common approach to determine the features of cells usually involves preparation with extrinsic factors such as fluorescence-coupled antibodies or cytoplasmic stains to gain such information. For example, depending on the type of instrument available for use, a single population of MSCs could be analyzed for the presence of CD73, CD90 and CD105 and absence of CD45 (the defining biomarkers for MSCs) in a matter of minutes. Not only do flow cytometers have the capacity to detect said populations of cells simultaneously, but also have the ability to sort based on those characteristics. This powerful property allows for recovery of particular subpopulations and continued culture or manipulation of said cells after analysis (see below for a further discussion of sorting by this approach).

Despite the great utility of flow cytometry, limitations still exist. Here, an in-depth description of flow cytometry and its traditional uses sets the groundwork for future advances. Flow cytometry is a relatively mature, high-throughput biological and clinical instrument that was first established in 1969.[80] Cell preparation for analysis by flow cytometry consists of the same type of protocols necessary for immunohistochemistry and fluorescence imaging.[81] Today, a myriad of measurements can be rapidly made on flow cytometers including cell count, viability, size and granularity, as well as DNA, RNA and protein expression.

All flow cytometers consist of a fluidic delivery component that focuses a sample stream usually consisting of single cells onto a light interrogation point within a flow cell. Principles of laminar flow allow for a buffered solution, commonly known as sheath fluid, to physically focus cells into a confined stream. This so-called hydrodynamic focusing effect is most commonly implemented with sheath fluid in commercial systems, but has recently been accomplished in micro-flow cytometry applications with the use of electric fields[82–84] and air streams.[85,86] Light excitation in commercial systems is typically supplied by mercury lamps, along with argon and krypton ion lasers. The incorporation of optical excitation filters increases the sensitivity of the instrument by delivering light at a wavelength that caters to the excitation spectrum of the fluorophores being used. Once the fluorophore has been excited, photons are emitted from it at a higher wavelength and directed to specific detectors—typically photomultiplier tubes (PMTs)—where they are amplified and digitized to an electric pulse to be analyzed by the software. Many of these instruments include the capability to detect other cellular parameters based exclusively on light scatter. Slightly deflected light, known as forward scatter, gives a measure of cell size, while light that is deflected perpendicularly, or side-scatter, measures the amount of granularity inside a cell. Some modern instruments incorporate the ability to analyze 15 colors simultaneously, though making sense of such data is a current limitation. Recently, there has also been a push to 'miniaturize' flow cytometry to increase availability to researchers and reduce material use and cost.

One of the great advantages of flow cytometry for analysis of stem cells is its unrivaled ability to sort cells based on the previously mentioned cellular characteristics. The prototype fluorescence activated cell sorter (FACS) was built at Stanford University in 1972, just a few years after the first flow cytometer was built.[87,88] FACS machines accomplish sorting first by creating uniformly sized droplets leaving the sample injection nozzle by way of an ultrasonic nozzle vibrator. While in the nozzle (flow cell), each individual cell passes through the laser beam and is detected as a positive or negative event based on its expression of an associated fluorescent probe. If the cell is detected as a positive event, the droplet that is formed housing that cell is given a positive charge, whereas if no cell is present or a cell is deemed to be a negative event, a negative charge is given to the droplet. Those droplets then pass through an electric field which deflects the droplets based on their charge, effectively sorting cells based on fluorescence levels.

Magnetic activated cell sorting (MACS) is another popular cell separation technique based on its ability to sort large populations of cells very quickly.[89] The principle behind MACS is conjugating a magnetic microbead to a biomarker of interest and drawing positive cells out of the stream by way of a magnet. One drawback is that cells can only be positively selected based on a single parameter (*e.g.* CD105); further selection requires sequential sorts. FACS, on the other hand, rapidly produces highly enriched populations of stem cells based on its ability to select for several parameters simultaneously.

One major limitation of conventional cell sorters is the requirement to prepare single cell populations in suspension, often removing cells from their native environment. As previously discussed, stem cells require proper spatial and temporal cues from their microenvironment to either maintain their undifferentiated phenotype, or progress down their respective differentiation pathway. Stem cell bioengineers, in attempting to mimic the *in vivo* milieu, often grow stem cells in three-dimensional (3D) constructs to recreate such an environment. Removing cells from these constructs can alter their differentiation status[90–92] and can therefore unintentionally alter experimental parameters.

Large particle flow cytometers could alleviate this limitation as sample preparation requires minimal disruption of the stem cell culture environment. Large particle flow cytometers were first engineered in the late 1980s for analyzing and sorting samples of aquatic phytoplankton[93,94] as well as pancreatic islets.[95] The first instrument was developed *de novo* and was capable of analyzing a broad range of phytoplankton particles (0.5–500 µm in diameter and over 2000 µm in length), while the second instrument made modifications to a PARTEC PAS II cell sorter to distinguish islets from exocrine tissue to increase cellular yield for transplantation.

Scaling up flow cytometry instruments to accommodate large particles requires major adjustments to the fluidic and optical components. The fluidic components must be carefully designed to minimize clogging, fouling and shearing of sample particles. Electrical and optical components must be adjusted to accommodate longer delay pulses for events, as well as detect fluorescence measurements that are representative of an entire particle. The Complex Object Parametric Analyzer and Sorter (COPAS), built by Union Biometrica, incorporated the ability to analyze up to five parameters at once on particles up to 1500 µm in diameter. This instrument is capable of probing large cellular entities such as pancreatic islets,[96] stem-cell clusters, *Drosophila melanogaster* embryos and larvae[97,98] and *Caenorhabditis elegans*.[99] The primary limitation of the instrument is the inability to probe deep within multicellular entities. This can be problematic for stem cell research since differences in viability, pluripotency and differentiation status typically differ relative to position within a tissue or cell aggregate.

One technology with potential to address the challenge of analysis of 3D structures with close affiliation to high-content screening approaches is image-based flow cytometry. Image-based flow cytometry combines the quantitative benefits of a traditional flow cytometer with the qualitative benefits of an image, such that a real-time readout of a population of cells includes intensity variations and a simultaneous image of the cell morphology. Recently, the ability to dynamically capture images of cells in flow has been utilized for a more rigorous analysis of cell morphology and phenotype.[100,101] Amnis Corporation developed the ImageStream, a multispectral imaging flow cytometer capable of capturing six images on a single cell at a rate of 100 cells per second with a detection system based on a charge-coupled device (CCD). Custom software allows users to click on any given event on a dot plot and observe localized fluorescence on up to four colors, as well as bright and dark field views. Though currently lacking in throughput (two to three orders of magnitude lower than conventional flow), imaging flow cytometry could find an ideal niche in studying stem cells because of its ability to probe large populations of cells at the single cell level, in addition to visualizing localized fluorescence expression and providing morphological information. In its current incarnations the system is based on widefield microscopy and hence the optical sectioning penetration is insufficient for imaging deep into large bodies; however, such an 'imaging-in-flow' approach could be coupled with a superior optical sectioning method such as confocal[102] or multiphoton laser scanning microscopy[103] as discussed later.

All cytometry systems that utilize flow, including the more advanced versions described above, rely on the application of extrinsic labels such as fluorescent molecules bound to antibodies or peptides. Binding of these entities to the cell surface can alter (stem) cell state.[104,105] One potentially powerful approach to avoid application of extrinsic fluorescence is to instead utilize intrinsic fluorescence signals. Well-characterized contributors to cellular autofluorescence are metabolic intermediates such as reduced nicotinamide adenine dinucleotide (NADH) and flavin adenine dinucleotide (FAD). NADH plays a key role as a carrier of electrons and is involved in many important metabolic pathways including glycolysis.[106] NADH has two forms in cells—free and protein bound. Most bound NADH is found in the mitochondria while free NADH exists in both the cytoplasm and the mitochondria.[107–109] NADH fluorescence intensity changes have been used to study cell metabolic activity *in vivo* for many years.[110–115] In addition, recent studies have used multiphoton laser-scanning microscopy (MPLSM) to characterize NADH and the intrinsic metabolite FAD in cancer cells, and to characterize the metastatic potential. For example, the metabolic state of carcinoma cells, as detected by endogenous fluorescent metabolic intermediates, has been correlated with the identification and metastatic potential of cancers in both animal and human models.[116] Human ESCs undergo a number of changes in mitochondrial characteristics as they differentiate, including an increase in mitochondrial mass and ATP production, suggestive of metabolic differences.[117] Given the known cellular differences of cells with different developmental potentials, it is likely that imaging endogenous fluorophores (*e.g.* NADH and FAD) in stem cells will provide biologically meaningful information that could be utilized to distinguish cells in different states of maturation.

The utility of endogenous fluorescence as a 'fingerprint' for identifying cells in a given state of differentiation requires appropriate technologies for visualizing those endogenous signals. MPLSM is uniquely suited to detect endogenous fluorophores, particularly in 3D structures due to the broad 'tunability' of its excitation sources, as well as improved deep sectioning, viability and signal to noise compared with traditional optical approaches. MPLSM is a nonlinear optical sectioning technique that allows thick biological sections to be imaged noninvasively *via* absorption of two or more low-energy photons in the near infrared range. Sufficient energy for two photon excitation is only present at the plane of focus such that, unlike other fluorescence microscopy approaches, no out-of-plane signal interference and photobleaching occurs. For this reason, in conjunction with the fact that the longer wavelengths of light used are more immune to scattering and less phototoxic, the effective imaging depth can greatly exceed conventional confocal microscopy.[118–120] Effective imaging depth is especially important in the context of the embryoid body (EB), the common multicellular intermediate between human embryonic stem cells or induced pluripotent stem cells and mature cells. The typical EB size range is approximately 100–500 µm and so fluorescence signals that may be generated at the center are difficult to detect with current confocal microscopy or flow cytometry systems, but are readily attained with MPLSM. These features, when utilized either individually or, especially, in combination, provide significant tools to obtain detailed multidimensional data from either exogenous or endogenous fluorophores associated with stem cells.

Given the unique properties of MPLSM and its potential for stem cell imaging, we hypothesized that a novel multiphoton fluorescence excitation flow instrument (MPFC) could be developed to accurately probe cells deep in the interior of multicell aggregates or tissue constructs in an enhanced-throughput manner. Furthermore, if this system is used to excite endogenous fluorophores of cells as intrinsic biomarkers, the application of exogenous fluorescent labels is thereby avoided. Others have considered the possibility of a multiphoton flow instrument (or multiphoton flow cytometry)[121–123] for analysis of cells and cellular aggregates in turbid and non-uniform flow conditions such as may be encountered in blood vessels *in vivo*. Our intent was to devise a fluid-controlled *in vitro* system such that cellular and multicell entities might not only be analyzed, but also sorted based on endogenous fluorescent properties. Cells sorted with this system would in principle be viable and unperturbed, unlike the output of current flow cytometry sorting systems and so could be used directly for clinical application. We developed an MPFC system composed of:

- a flow cell through which large particles stream past a light interrogation point;
- an optics system with simultaneous transmitted and two-photon excitation capability; and
- data acquisition software to provide real-time qualitative and quantitative readout data.

We found the system was capable of noninvasively probing the interior of large entities in a high throughput manner (Figure 3.3A,B). In addition, we have shown reliable detection of intrinsic fluorescence—most likely representing NADH, a key metabolic intermediate (Figure 3.3C–E). Future studies will continue to explore the potential of intrinsic fluorescence and help to show and validate the potential of the intrinsic fluorophores as biomarkers for flow cytometry. The development of the MPFC required the merging of advanced imaging modalities with high-throughput techniques to gain new insights into stem cell behavior. As described below, this approach is likely to be repeated in future efforts to better screen stem cells.

3.4.3 Future Stem Cell Screening Approaches

The future of stem cell research both in terms of understanding the basic mechanistic underpinnings and clinical potential (including transplantation) requires a thorough understanding of stem cells as both a biological and biomedical research model. Stem cells are unlike established tissue culture cell lines in that they are typically a heterogeneous population of cells in diverse stages of differentiation. In addition they are often in the form of complex 3D aggregates. These features present a considerable challenge to the development of traditional automated screens based on simple optical properties.

Whether for basic research or clinical research, imaging has always been and will continue to be a definitive tool in understanding the lifecycle, activity and behavior of biological entities. This has been particularly true in cell biology where advanced imaging approaches such as fluorescence and confocal microscopy have enabled new avenues of research on cellular behavior and fate, and when coupled with genetic and biochemical tools, allow for unprecedented ability to understand biological phenomena from the cellular level. These methods had to first mature in their respective field and also allow the stem cell community to increase their knowledge of the basic types of cells and fundamental aspects (*e.g.* how to effectively culture them) before these advanced methods could be systemically applied to stem cells.

With the many recent breakthroughs in the stem cell research community (including culture and differentiation factors) and the corresponding imaging technologies that allow for multiparametic measurements to be made deep within live samples in high resolution, there is a great opportunity for the application of state-of-the-art imaging approaches coupled in some cases to high-throughput or high-content techniques as routine tools for stem cell characterization. In particular there are microscopy techniques that are currently used in static and sometimes in high-throughput which have great application for use in high-content imaging flow cytometry.

3.4.4 Fluorescence Microscopy Techniques that could be Applied to High-throughput and High-content Screening

Advanced optical techniques can yield information on the physiological state of a cell or the molecular environment of a particular reporter probe. The challenge for rapid screening of live cells is to integrate all these techniques into a robust automated system to identify specific changes in cell architecture or physiology. Some of these techniques are outlined below.

3.4.4.1 *Spectral Imaging*

Fluorescence signals may be derived from probes such as fluorescent protein reporters or endogenous fluorophores such as collagen or reduced nicotinamide adenine dinucleotide.[124] Fluorescence spectroscopy studies using specimens in cuvettes has yielded a wealth of information about how subtle fluorescence spectral shifts may be used to report on molecular environments of fluorophores.[125] The availability of spectral analysis on a pixel-by-pixel basis in an image could extend these analyses to molecules in identified cellular or sub-cellular environments. The current generation of cell-based high-through-put screening systems has very limited spectral discrimination, having only three or four spectral channels. Although this is adequate for identifying fluorophores with well-separated emission spectra, difficulties arise when spectra overlap, or when it is necessary to measure the spectral shifts of certain indicator probes.

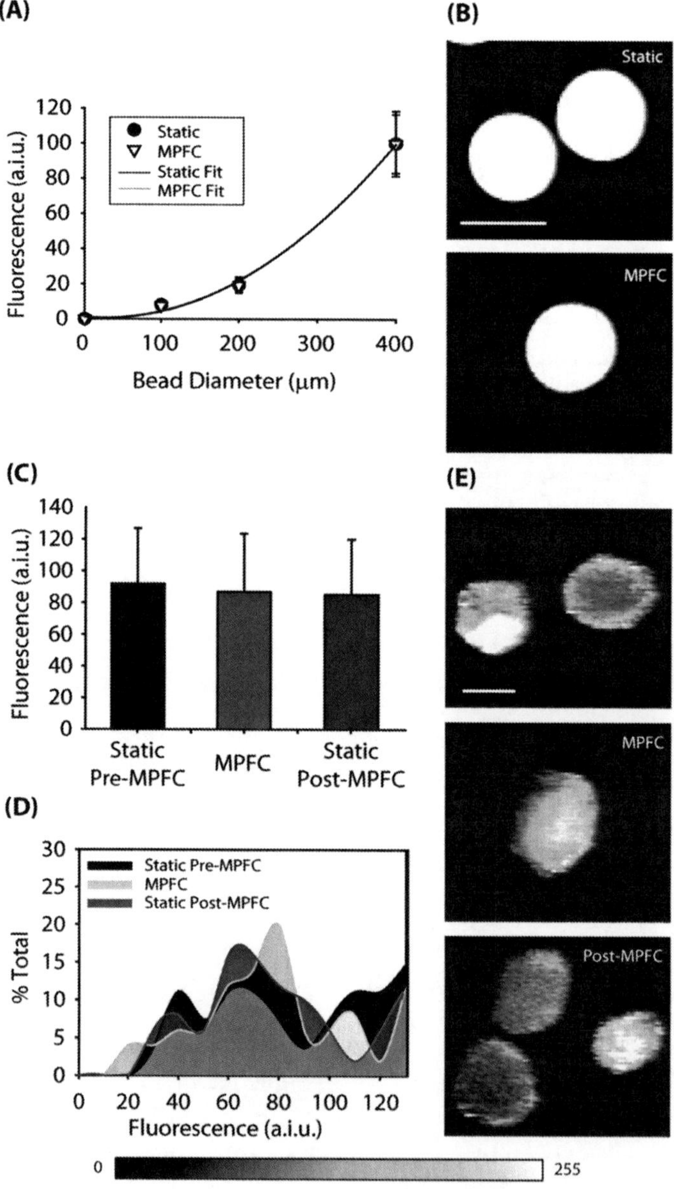

3.4.4.2 Lifetime Imaging

A fluorescence signal contains more information than just intensity and color. The lifetime of the excited state, which gives rise to the fluorescence signal, is diagnostic of the fluorophore and also of its microenvironment.[126] Factors such as ionic strength, hydrophobicity, oxygen concentration, binding to macro-molecules and the proximity of molecules that can deplete the excited state by resonance energy transfer can all modify the lifetime of a fluorophore. Measurements of lifetimes can therefore be used as indicators of these parameters. Fluorescence lifetime measurements are generally absolute, being independent of the concentration of the fluorophore. Furthermore, lifetime properties may be particularly useful in identifying fluorophores with significantly overlapping spectral properties. Fluorescence lifetime imaging has been recently described and the potential of this technique has been demonstrated.[127–129]

Figure 3.3 Multiphoton Flow Cytometry (MPFC) for stem cell screening (adapted with permission). To determine whether the MPFC system was capable of generating information related to the interior portion of large entities by deep optical assessment, we mapped the mean fluorescence intensity as a function of bead size. If detection of fluorescence in the interior of the microsphere were possible, one would expect an exponential trend between the size of the fluorescent sphere and the emitted fluorescent signal, by capturing a true, representative cross-sectional area, and not simply surface fluorescence. (A) *Relationship between bead diameter and normalized fluorescence intensity.* Static measurements of mean bead intensity were made prior to measuring mean bead intensity using the MPFC system. Mean fluorescence intensity was plotted as a function of bead diameter. Best fit regression analysis was applied to the datasets and both static and MPFC conditions yielded second-order exponential relationships; R^2 values of 0.99 and 0.99, respectively. (B) *Fluorescence intensity image of a 400 µm bead under static conditions and using MPFC.* Static images were captured at the location corresponding to the maximum total intensity, thus the "center" of the bead. Mean fluorescence intensity did not vary between static and MPFC acquisition modes for each bead size ($P = 0.58$, $P = 0.72$, $P = 0.74$). a.i.u., arbitrary intensity units. Scale bar, 200 µm. One of the primary benefits of using MPFC as a stem cell screening tool, is the ability to reliably detect autofluorescence corresponding to metabolites including NADH. Knowing that metabolic profiles change with stem cell states, MPFC can be used to assess stem cell state in a non-invasive manner. Here, the Ti-Sapphire laser of an MPFC system was tuned to 780 nm to excite the intrinsic fluorescent metabolite, NADH. (C) *Mean fluorescence intensity of EBs before, during and after analysis with the MPFC system.* Mean fluorescence intensity of EBs did not vary between acquisition conditions ($P > 0.5$) indicating a high level of accuracy of the MPFC system, even for very low level fluorescence emission. (D) *NADH fluorescence intensity distributions before, during, and after analysis.* (E) *Representative images depict NADH intensity acquired pre-MPFC, MPFC, and post-MPFC.* Localized changes in NADH expression could be observed and detected for all three conditions. Color bar represents quantified intensity levels from 0 (black) to 256 (white), scale bar, 200 µm.

3.4.4.3   *Fluorescence Resonant Energy Transfer (FRET)*

FRET is a technique that can detect the close proximity of two fluorescently labeled molecules.[130] Furthermore, this technique may be readily used *in vivo*. If one of the fluorophores (the donor) has a fluorescence emission spectrum that overlaps the excitation spectrum of the second fluorophore (the acceptor), then fluorescence from the donor can be partially quenched if the acceptor molecule is closer than the so-called Förster distance (around 10 nm). Energy is transferred to the acceptor molecule, which increases in fluorescence; the effect varying as the sixth power of the distance separating the two molecules. The technique is capable of detecting when a ligand binds to a receptor and so can be a potent way of visualizing the activation of a signal transduction pathway. However, it is often difficult to obtain reliable FRET estimations from measurements of fluorescence intensity changes, particularly within intracellular domains. Selective bleaching or compartmentalization of the two fluorophores can give rise to changes in relative fluorescence that are indistinguishable from those resulting from a FRET interaction. However, lifetime imaging can potentially circumvent these problems. When a FRET interaction occurs, the donor excited state lifetime increases. This is an absolute change that is independent of intensity. Using lifetime measurements to estimate FRET can potentially yield the percentage of the donor species involved in the FRET interaction (ratio of the shorter to the longer lifetime components) and the proximity of the two interacting molecules (degree of lifetime shortening).

3.4.4.4   *Optical Sectioning*

Optical sectioning fluorescence microscopy has become the method of choice for imaging living specimens as it offers high signal-to-background and the ability to spectrally discriminate between multiple fluorophores. Recently developed techniques such as confocal[102,131] or multiphoton[118,132] imaging allow optical sections to be made of intact specimens. These may be collected as stacks of images at different focal depths to obtain 3D structural data. Stacks of images may be collected at regular time intervals in order to reveal the dynamics of 3D structures in living tissue.[133] Optical sectioning could allow for cells in a high-throughput screen to be grown in a collagen gel, possibly a more natural environment for some cells. Three-dimensional imaging could be used to measure the morphometric parameters of a cell. The association of cells with an extra cellular matrix could be also characterized.

3.4.4.5   *Multiphoton (MP) Imaging*

As mentioned above, multiphoton (MP) imaging has major advantages over other optical imaging approaches for its ability to noninvasively collect high resolution images deep within live samples. An additional advantage of multiphoton imaging is that the nonlinear laser sources employed can be used for other closely related methods on the same optical system.

3.4.4.6 Second Harmonic Imaging

Nonlinear optical effects other than multiphoton fluorescence excitation can occur at the very high photon densities attained at the focus of the scanning excitation beam in a MP microscope. Molecular assemblies with high-order structure (*e.g.* collagen matrices) can generate a second harmonic (SH) signal at half the wavelength of the excitation.[134] Unlike the case in MP imaging, the SH signal has a narrow spectral line-width (determined by the excitation source) and a zero lifetime. These characteristics allow SH signals to be distinguished from MP signals in a laser-scanning microscope with high peak intensity and ultrafast pulse excitation, even if there are fluorescence signals which overlap the SH signal. These characteristics make SH imaging a very useful adjunct to MP imaging when observing cells that are imbedded in an extracellular matrix.

3.4.4.7 Fluorescence Correlation Spectroscopy (FCS)

FCS provides measurements of the mobility of a fluorescent particle. The method uses time-correlated single photon counting techniques to measure the distribution of sojourn times of fluorescent particles within the focal volume of a focused beam of excitation by autocorrelation. The use of multiphoton excitation restricts the excitation (and hence signal) to the focal volume, thereby providing a significant increase in the signal-to-noise ratio of the measurements. The technique can provide a powerful way of measuring protein-protein interactions.[135] FCS capabilities can be incorporated into a laser-scanning microscope and can use the timing digital electronics of a time-domain lifetime measurement system. This configuration allows a specimen to be scanned and then the scanning beam parked at a selected point of interest (*e.g.* in the cytoplasm of an identified cell) where an FCS can be made.

3.4.4.8 Coherent Anti-Stokes Raman Spectroscopy (CARS)

CARS is another nonlinear optical phenomenon that has great advantage for its unique capability to combine high-resolution morphology information with corresponding chemical specificity.[136] Like Raman spectroscopy, CARS can detect vibrational signatures of molecules in order to identify structure without the need of any labels, as the vibrational signatures are intrinsic. Due to the coherent and nonlinear nature of CARS, these signatures can be detected deep in biological samples with good sensitivity and requiring only moderate laser powers for excitation. Several groups[137,138] have begun to investigate the use of CARS for stem cells and, while it has not been used systematically for stem cell screening, it has great potential.

3.4.5 Image Analysis Software for High-throughput and High-content Screening

Advanced microscopy techniques are of little use in high-throughput screening (HTS) or high-content screening (HCS) without corresponding algorithms that can parse image data and extract the relevant parameters for a particular screen. The development of such software can be challenging as it is difficult to develop automatic image analysis systems that approach the performance of a trained operator. However, humans are not so good at the 'high-throughput' aspects of HTS and HCS.

Another problem that faces the developers of rapid cell analysis algorithms is the great diversity of parametric measurements that are required for different screens. For example, one assay may require measurements of cell locomotion, another is the degree of neurite branching, or a third is changes in levels of cytosolic free calcium. Also, several different parameters may have to be measured in order to obtain some meaningful metric of cell state for a particular assay.

Another great challenge in high-throughput studies from a computational point of view is not only the visualization but also practically how to store, track and disseminate the datasets. This has been complicated by the lack of standards for data sharing even for simple microscopy studies and the myriad of proprietary formats used in microscopy and high throughput imaging. The flow cytometry community has been quite active in this area and were leaders in developing a file format standard for flow cytometry, the International Cytometer Standard (ICS), which has been adapted and deployed widely.[139,140] Unfortunately due to the multiparametic nature of some imaging flow cytometry studies or multimodal approaches that have been deployed, some of the efforts of the ICS have not been utilized. There have been recent efforts in the optical microscopy community by the Open Microscopy Environment consortium[141–143] to provide tools for data sharing that could be applied to imaging flow cytometry and other high-content screening approaches. Not only can these tools be compatible with the ICS, but also have the advantages of providing additional data infrastructure such as a database and tools for converting file formats.

3.5 Summary

Screening of stem cells is a tedious, time-intensive and often costly endeavor, but is essential to achieving repeatable experimental results and effective clinical outcomes. Past efforts to screen stem cells have primarily focused on monitoring the spatial (*i.e.* levels of protein expression in specific cell compartments) and temporal (*i.e.* how these levels change over time) aspects of stem cell state. Stem cell screening approaches in the future are likely to take advantage of additional dimensionality features such as chemical (*i.e.* metabolic dynamics) and physical (*i.e.* interaction with and remodeling of extracellular matrices) aspects of stem cell fate and behavior. These qualities are especially appealing as they can be determined with minimal cellular and environment disruption using recently developed advanced optical microscopy approaches. For example, multiphoton based lifetime imaging can be used to determine the fluorescence lifetime of FAD and NADH in stem cells deep within intact environments, effectively allowing for metabolic mapping of the free and bound states of NADH and redox ratio determination (from FAD/NADH measurements). Similarly, second harmonic generation can be used to track structural alignment of extracellular matrix proteins housing stem cells. Merging of these advanced imaging techniques and others to high-throughput and high-content modalities including flow cytometry will yield a spectrum of novel screening techniques. This will likely result in several benefits including:

- advancing screening methods beyond monitoring cell state to tracking cell function;
- identifying and exploiting additional intrinsic biomarkers; and
- ultimately enhancing basic science and clinical outcomes of stem cell research through improved accessibility and standardization.

Acknowledgements

Supported by grants from the UW Stem Cell and Regenerative Medicine Cluster Enhancement Grant, the UW IEDR program, and the Coulter Foundation.

References

1. S. Lin, V. Tran and P. Talbot, Comparison of toxicity of smoke from traditional and harm-reduction cigarettes using mouse embryonic stem cells as a novel model for preimplantation development, *Hum. Reprod.*, 2009, **24**, 386–397.
2. S. J. Flora and A. Mehta, Monoisoamyl dimercaptosuccinic acid abrogates arsenic-induced developmental toxicity in human embryonic stem cell-derived embryoid bodies: comparison with in vivo studies, *Biochem. Pharmacol.*, 2009, **78**, 1340–1349.
3. Z. Wang, G. Xu, Y. Wu, Y. Guan, L. Cui, X. Lei, J. Zhang, L. Mou, B. Sun and Q. Dai, Neuregulin-1 enhances differentiation of cardiomyocytes from embryonic stem cells, *Med. Biol. Eng. Comput.*, 2009, **47**, 41–48.
4. M. Kadota, R. Nishigaki, C. C. Wang, T. Toda, Y. Shirayoshi, T. Inoue, T. Gojobori, K. Ikeo, M. S. Rogers and M. Oshimura, Proteomic signatures and aberrations of mouse embryonic stem cells containing a single human chromosome 21 in neuronal differentiation: an in vitro model of Down syndrome, *Neuroscience*, 2004, **129**, 325–335.
5. K. Kim, S. H. Lee, J. Ha Kim, Y. Choi and N. Kim, NFATc1 induces osteoclast fusion via up-regulation of Atp6v0d2 and the dendritic cell-specific transmembrane protein (DC-STAMP), *Mol. Endocrinol.*, 2008, **22**, 176–185.

6. Z. Zhang, X. Wang and S. Wang, Isolation and characterization of mesenchymal stem cells derived from bone marrow of patients with Parkinson's disease, *In Vitro Cell Dev. Biol. Anim.*, 2008, **44**, 169–177.

7. Y. Narita, A. Yamawaki, H. Kagami, M. Ueda and Y. Ueda, Effects of transforming growth factor-beta 1 and ascorbic acid on differentiation of human bone-marrow-derived mesenchymal stem cells into smooth muscle cell lineage, *Cell Tissue Res.*, 2008, **333**, 449–459.

8. H. Park, J. S. Temenoff, Y. Tabata, A. I. Caplan, R. M. Raphael, J. A. Jansen and A. G. Mikos, Effect of dual growth factor delivery on chondrogenic differentiation of rabbit marrow mesenchymal stem cells encapsulated in injectable hydrogel composites, *J. Biomed. Mater. Res. A*, 2009, **88**, 889–897.

9. L. A. McMahon, P. J. Prendergast and V. A. Campbell, A comparison of the involvement of p38, ERK1/2 and PI3K in growth factor-induced chondrogenic differentiation of mesenchymal stem cells, *Biochem. Biophys. Res. Commun.*, 2008, **368**, 990–995.

10. C. H. Lee, E. K. Moioli and J. J. Mao, Fibroblastic differentiation of human mesenchymal stem cells using connective tissue growth factor, *Conf. Proc. IEEE Eng. Med. Biol. Soc.*, 2006, **1**, 775–778.

11. Q. Wang, B. Sun, D. Wang, Y. Ji, Q. Kong, G. Wang, J. Wang, W. Zhao, L. Jin and H. Li, Murine bone marrow mesenchymal stem cells cause mature dendritic cells to promote T-cell tolerance, *Scand. J. Immunol.*, 2008, **68**, 607–615.

12. X. Q. Kang, W. J. Zang, L. J. Bao, D. L. Li, T. S. Song, X. L. Xu and X. J. Yu, Fibroblast growth factor-4 and hepatocyte growth factor induce differentiation of human umbilical cord blood-derived mesenchymal stem cells into hepatocytes, *World J. Gastroenterol.*, 2005, **11**, 7461–7465.

13. G. Forte, M. Minieri, P. Cossa, D. Antenucci, M. Sala, V. Gnocchi, R. Fiaccavento, F. Carotenuto, P. De Vito, P. M. Baldini, M. Prat and P. Di Nardo, Hepatocyte growth factor effects on mesenchymal stem cells: proliferation, migration, and differentiation, *Stem Cells*, 2006, **24**, 23–33.

14. Y. K. Luu, E. Capilla, C. J. Rosen, V. Gilsanz, J. E. Pessin, S. Judex and C. T. Rubin, Mechanical stimulation of mesenchymal stem cell proliferation and differentiation promotes osteogenesis while preventing dietary-induced obesity, *J. Bone Miner. Res.*, 2009, **24**, 50–61.

15. F. van Eijk, D. B. Saris, L. B. Creemers, J. Riesle, W. J. Willems, C. A. van Blitterswijk, A. J. Verbout and W. J. Dhert, The effect of timing of mechanical stimulation on proliferation and differentiation of goat bone marrow stem cells cultured on braided PLGA scaffolds, *Tissue Eng. Part A*, 2008, **14**, 1425–1433.

16. M. C. Qi, J. Hu, S. J. Zou, H. Q. Chen, H. X. Zhou and L. C. Han, Mechanical strain induces osteogenic differentiation: Cbfa1 and Ets-1 expression in stretched rat mesenchymal stem cells, *Int. J. Oral Maxillofac. Surg.*, 2008, **37**, 453–458.

17. S. Saha, L. Ji, J. J. de Pablo and S. P. Palecek, TGFbeta/Activin/Nodal pathway in inhibition of human embryonic stem cell differentiation by mechanical strain, *Biophys. J.*, 2008, **94**, 4123–4133.
18. W. Y. Sim, S. W. Park, S. H. Park, B. H. Min, S. R. Park and S. S. Yang, A pneumatic micro cell chip for the differentiation of human mesenchymal stem cells under mechanical stimulation, *Lab Chip*, 2007, **7**, 1775–1782.
19. D. F. Ward, Jr., R. M. Salasznyk, R. F. Klees, J. Backiel, P. Agius, K. Bennett, A. Boskey and G. E. Plopper, Mechanical strain enhances extracellular matrix-induced gene focusing and promotes osteogenic differentiation of human mesenchymal stem cells through an extra-cellular-related kinase-dependent pathway, *Stem Cells Dev.*, 2007, **16**, 467–480.
20. I. C. Lee, J. H. Wang, Y. T. Lee and T. H. Young, The differentiation of mesenchymal stem cells by mechanical stress or/and co-culture system, *Biochem. Biophys. Res. Commun.*, 2007, **352**, 147–152.
21. Y. Jing, L. Li, Y. Li, M. Chen, W. Wu, H. Chen and X. Liu, [The effect of mechanical strain on proliferation and osteogenic differentiation of bone marrow mesenchymal stem cells from rats], *Sheng Wu Yi Xue Gong Cheng Xue Za Zhi*, 2006, **23**, 542–545.
22. M. Schmelter, B. Ateghang, S. Helmig, M. Wartenberg and H. Sauer, Embryonic stem cells utilize reactive oxygen species as transducers of mechanical strain-induced cardiovascular differentiation, *FASEB J.*, 2006, **20**, 1182–1184.
23. S. Saha, L. Ji, J. J. de Pablo and S. P. Palecek, Inhibition of human embryonic stem cell differentiation by mechanical strain, *J. Cell Physiol.*, 2006, **206**, 126–137.
24. M. Q. Chen, X. Xie, R. Hollis Whittington, G. T. Kovacs, J. C. Wu and L. Giovangrandi, Cardiac differentiation of embryonic stem cells with point-source electrical stimulation, *Conf. Proc. IEEE Eng. Med. Biol. Soc.*, 2008, **2008**, 1729–1732.
25. R. J. Lang, J. M. Haynes, J. Kelly, J. Johnson, J. Greenhalgh, C. O'Brien, E. M. Mulholland, L. Baker, M. Munsie and C. W. Pouton, Electrical and neurotransmitter activity of mature neurons derived from mouse embryonic stem cells by Sox-1 lineage selection and directed differentia-tion, *Eur. J. Neurosci.*, 2004, **20**, 3209–3221.
26. H. Sauer, G. Rahimi, J. Hescheler and M. Wartenberg, Effects of elec-trical fields on cardiomyocyte differentiation of embryonic stem cells, *J. Cell. Biochem.*, 1999, **75**, 710–723.
27. A. W. Macadar, L. M. Chalupa and D. B. Lindsley, Differentiation of brain stem loci which affect hippocampal and neocortical electrical activity, *Exp. Neurol.*, 1974, **43**, 499–514.
28. P. R. Mangan, L. E. Harrington, D. B. O'Quinn, W. S. Helms, D. C. Bullard, C. O. Elson, R. D. Hatton, S. M. Wahl, T. R. Schoeb and C. T. Weaver, Transforming growth factor-beta induces development of the T(H)17 lineage, *Nature*, 2006, **441**, 231–234.

29. C. Schmidt, F. Bladt, S. Goedecke, V. Brinkmann, W. Zschiesche, M. Sharpe, E. Gherardi and C. Birchmeier, Scatter factor/hepatocyte growth factor is essential for liver development, *Nature*, 1995, **373**, 699–702.

30. A. M. Ritter, G. R. Lewin, N. E. Kremer and L. M. Mendell, Requirement for nerve growth factor in the development of myelinated nociceptors in vivo, *Nature*, 1991, **350**, 500–502.

31. M. C. Raff, L. E. Lillien, W. D. Richardson, J. F. Burne and M. D. Noble, Platelet-derived growth factor from astrocytes drives the clock that times oligodendrocyte development in culture, *Nature*, 1988, **333**, 562–565.

32. J. W. Pollard, A. Bartocci, R. Arceci, A. Orlofsky, M. B. Ladner and E. R. Stanley, Apparent role of the macrophage growth factor, CSF-1, in placental development, *Nature*, 1987, **330**, 484–486.

33. C. E. Murry and G. Keller, Differentiation of embryonic stem cells to clinically relevant populations: lessons from embryonic development, *Cell*, 2008, **132**, 661–680.

34. D. van Noort, S. M. Ong, C. Zhang, S. Zhang, T. Arooz and H. Yu, Stem cells in microfluidics, *Biotechnol. Prog.*, 2009, **25**, 52–60.

35. B. G. Chung, L. A. Flanagan, S. W. Rhee, P. H. Schwartz, A. P. Lee, E. S. Monuki and N. L. Jeon, Human neural stem cell growth and differentiation in a gradient-generating microfluidic device, *Lab Chip*, 2005, **5**, 401–406.

36. C. Joanne Wang, X. Li, B. Lin, S. Shim, G. L. Ming and A. Levchenko, A microfluidics-based turning assay reveals complex growth cone responses to integrated gradients of substrate-bound ECM molecules and diffusible guidance cues, *Lab Chip*, 2008, **8**, 227–237.

37. A. J. Engler, S. Sen, H. L. Sweeney and D. E. Discher, Matrix elasticity directs stem cell lineage specification, *Cell*, 2006, **126**, 677–689.

38. K. Saha, A. J. Keung, E. F. Irwin, Y. Li, L. Little, D. V. Schaffer and K. E. Healy, Substrate modulus directs neural stem cell behavior, *Biophys. J.*, 2008, **95**, 4426–4438.

39. H. Zhao, H. Zhou, X. Wang, J. Dong, Y. Yang and X. Zhang, [Effect of mechanical strain on differentiation of mesenchymal stem cells into osteoblasts], *Sheng Wu Yi Xue Gong Cheng Xue Za Zhi*, 2009, **26**, 518–522.

40. J. P. Winer, P. A. Janmey, M. E. McCormick and M. Funaki, Bone marrow-derived human mesenchymal stem cells become quiescent on soft substrates but remain responsive to chemical or mechanical stimuli, *Tissue Eng. Part A*, 2009, **15**, 147–154.

41. Z. Li, L. Kupcsik, S. J. Yao, M. Alini and M. J. Stoddart, Mechanical Load Modulates Chondrogenesis of Human Mesenchymal Stem Cells through the TGF-beta Pathway, *J. Cell. Mol. Med.*, 2009.

42. B. Sen, Z. Xie, N. Case, M. Ma, C. Rubin and J. Rubin, Mechanical strain inhibits adipogenesis in mesenchymal stem cells by stimulating a durable beta-catenin signal, *Endocrinology*, 2008, **149**, 6065–6075.

43. A. E. West, W. G. Chen, M. B. Dalva, R. E. Dolmetsch, J. M. Kornhauser, A. J. Shaywitz, M. A. Takasu, X. Tao and M. E. Greenberg, Calcium

regulation of neuronal gene expression, *Proc. Natl. Acad. Sci. U S A*, 2001, **98**, 11024–11031.

44. M. Zhang, D. Methot, V. Poppa, Y. Fujio, K. Walsh and C. E. Murry, Cardiomyocyte grafting for cardiac repair: graft cell death and anti-death strategies, *J. Mol. Cell. Cardiol.*, 2001, **33**, 907–921.

45. M. Yamada, K. Tanemura, S. Okada, A. Iwanami, M. Nakamura, H. Mizuno, M. Ozawa, R. Ohyama-Goto, N. Kitamura, M. Kawano, K. Tan-Takeuchi, C. Ohtsuka, A. Miyawaki, A. Takashima, M. Ogawa, Y. Toyama, H. Okano and T. Kondo, Electrical stimulation modulates fate determination of differentiating embryonic stem cells, *Stem Cells*, 2007, **25**, 562–570.

46. M. M. Mitalipova, R. R. Rao, D. M. Hoyer, J. A. Johnson, L. F. Meisner, K. L. Jones, S. Dalton and S. L. Stice, Preserving the genetic integrity of human embryonic stem cells, *Nat. Biotechnol.*, 2005, **23**, 19–20.

47. J. S. Draper, K. Smith, P. Gokhale, H. D. Moore, E. Maltby, J. Johnson, L. Meisner, T. P. Zwaka, J. A. Thomson and P. W. Andrews, Recurrent gain of chromosomes 17q and 12 in cultured human embryonic stem cells, *Nat. Biotechnol.*, 2004, **22**, 53–54.

48. C. Allegrucci, Y. Z. Wu, A. Thurston, C. N. Denning, H. Priddle, C. L. Mummery, D. Ward-van Oostwaard, P. W. Andrews, M. Stojkovic, N. Smith, T. Parkin, M. E. Jones, G. Warren, L. Yu, R. M. Brena, C. Plass and L. E. Young, Restriction landmark genome scanning identifies culture-induced DNA methylation instability in the human embryonic stem cell epigenome, *Hum. Mol. Genet.*, 2007, **16**, 1253–1268.

49. J. S. Draper, H. D. Moore, L. N. Ruban, P. J. Gokhale and P. W. Andrews, Culture and characterization of human embryonic stem cells, *Stem. Cells. Dev.*, 2004, **13**, 325–336.

50. B. C. Heng, C. P. Ye, H. Liu, W. S. Toh, A. J. Rufaihah, Z. Yang, B. H. Bay, Z. Ge, H. W. Ouyang, E. H. Lee and T. Cao, Loss of viability during freeze-thaw of intact and adherent human embryonic stem cells with conventional slow-cooling protocols is predominantly due to apoptosis rather than cellular necrosis, *J. Biomed. Sci.*, 2006, **13**, 433–445.

51. Katkov II, M. S. Kim, R. Bajpai, Y. S. Altman, M. Mercola, J. F. Loring, A. V. Terskikh, E. Y. Snyder and F. Levine, Cryopreservation by slow cooling with DMSO diminished production of Oct-4 pluripotency marker in human embryonic stem cells, *Cryobiology*, 2006, **53**, 194–205.

52. S. Mollamohammadi, A. Taei, M. Pakzad, M. Totonchi, A. Seifinejad, N. Masoudi and H. Baharvand, A simple and efficient cryopreservation method for feeder-free dissociated human induced pluripotent stem cells and human embryonic stem cells, *Hum. Reprod.*, 2009.

53. X. Li, R. Krawetz, S. Liu, G. Meng and D. E. Rancourt, ROCK inhibitor improves survival of cryopreserved serum/feeder-free single human embryonic stem cells, *Hum. Reprod.*, 2009, **24**, 580–589.

54. R. Martin-Ibanez, C. Unger, A. Stromberg, D. Baker, J. M. Canals and O. Hovatta, Novel cryopreservation method for dissociated human

embryonic stem cells in the presence of a ROCK inhibitor, *Hum. Reprod.*, 2008, **23**, 2744–2754.

55. P. F. Yang, T. C. Hua, J. Wu, Z. H. Chang, H. C. Tsung and Y. L. Cao, Cryopreservation of human embryonic stem cells: a protocol by programmed cooling, *Cryo. Letters*, 2006, **27**, 361–368.

56. B. C. Heng, H. Liu, Z. Ge and T. Cao, Mechanical dissociation of human embryonic stem cell colonies by manual scraping after collagenase treatment is much more detrimental to cellular viability than is trypsinization with gentle pipetting, *Biotechnol. Appl. Biochem.*, 2007, **47**, 33–37.

57. M. S. Lao, Cell bank validation: Comparing points to consider and ICH guidelines (vol 13, pg 48, 1999), *Biopharm-the Appl. Tech. of Biopharm. Dev.*, 1999, **12**, 12–12.

58. J. Huang, Y. Zhang, A. Bersenev, W. T. O'Brien, W. Tong, S. G. Emerson and P. S. Klein, Pivotal role for glycogen synthase kinase-3 in hematopoietic stem cell homeostasis in mice, *J. Clin. Invest.*, 2009, **119**, 3519–3529.

59. H. Baharvand, A. Fathi, D. van Hoof and G. H. Salekdeh, Concise review: trends in stem cell proteomics, *Stem Cells*, 2007, **25**, 1888–1903.

60. J. A. Santiago, R. Pogemiller and B. M. Ogle, Heterogeneous differentiation of human mesenchymal stem cells in response to extended culture in extracellular matrices, *Tissue Eng. Part A*, 2009, **15**, 3911–3922.

61. W. S. Shim, S. Jiang, P. Wong, J. Tan, Y. L. Chua, Y. S. Tan, Y. K. Sin, C. H. Lim, T. Chua, M. Teh, T. C. Liu and E. Sim, Ex vivo differentiation of human adult bone marrow stem cells into cardiomyocyte-like cells, *Biochem. Biophys. Res. Commun.*, 2004, **324**, 481–488.

62. D. L. Kelly and A. Rizzino, DNA microarray analyses of genes regulated during the differentiation of embryonic stem cells, *Mol. Reprod. Dev.*, 2000, **56**, 113–123.

63. T. S. Tanaka, S. A. Jaradat, M. K. Lim, G. J. Kargul, X. H. Wang, M. J. Grahovac, S. Pantano, Y. Sano, Y. Piao, R. Nagaraja, H. Doi, W. H. Wood, K. G. Becker and M. S. H. Ko, Genome-wide expression profiling of mid-gestation placenta and embryo using a 15,000 mouse developmental cDNA microarray, *Proc. Nat. Acad. Sci. of the U.S.A.*, 2000, **97**, 9127–9132.

64. T. S. Tanaka, T. Kunath, W. L. Kimber, S. A. Jaradat, C. A. Stagg, M. Usuda, T. Yokota, H. Niwa, J. Rossant and M. S. H. Ko, Gene expression profiling of embryo-derived stem cells reveals candidate genes associated with pluripotency and lineage specificity, *Genome Research*, 2002, **12**, 1921–1928.

65. N. B. Ivanova, J. T. Dimos, C. Schaniel, J. A. Hackney, K. A. Moore and I. R. Lemischka, A stem cell molecular signature, *Science*, 2002, **298**, 601–604.

66. R. L. Phillips, R. E. Ernst, B. Brunk, N. Ivanova, M. A. Mahan, J. K. Deanehan, K. A. Moore, G. C. Overton and I. R. Lemischka, The genetic program of hematopoietic stem cells, *Science*, 2000, **288**, 1635–1640.

67. M. Ramalho-Santos, S. Yoon, Y. Matsuzaki, R. C. Mulligan and D. A. Melton, "Stemness": Transcriptional profiling of embryonic and adult stem cells, *Science*, 2002, **298**, 597–600.

68. A. G. Smith, Embryo-derived stem cells: of mice and men, *Annu. Rev. Cell. Dev. Biol.*, 2001, **17**, 435–462.

69. N. Fortunel, J. Hatzfeld, S. Kisselev, M. N. Monier, K. Ducos, A. Cardoso, P. Batard and A. Hatzfeld, Release from quiescence of primitive human hematopoietic stem/progenitor cells by blocking their cell-surface TGF-beta type II receptor in a short-term in vitro assay, *Stem Cells*, 2000, **18**, 102–111.

70. N. O. Fortunel, A. Hatzfeld and J. A. Hatzfeld, Transforming growth factor-beta: pleiotropic role in the regulation of hematopoiesis, *Blood*, 2000, **96**, 2022–2036.

71. C. E. Burns and L. I. Zon, Portrait of a stem cell, *Dev. Cell*, 2002, **3**, 612–613.

72. N. Sato, I. M. Sanjuan, M. Heke, M. Uchida, F. Naef and A. H. Brivanlou, Molecular signature of human embryonic stem cells and its comparison with the mouse, *Dev. Biol.*, 2003, **260**, 404–413.

73. M. J. Abeyta, A. T. Clark, R. T. Rodriguez, M. S. Bodnar, R. A. R. Pera and M. T. Firpo, Unique gene expression signatures of independently-derived human embryonic stem cell lines, *Human Mol. Genetics*, 2004, **13**, 601–608.

74. B. Bhattacharya, T. Miura, R. Brandenberger, J. Mejido, Y. Q. Luo, A. X. Yang, B. H. Joshi, I. Ginis, R. S. Thies, M. Amit, I. Lyons, B. G. Condie, J. Itskovitz-Eldor, M. S. Rao and R. K. Puri, Gene expression in human embryonic stem cell lines: unique molecular signature, *Blood*, 2004, **103**, 2956–2964.

75. C. Bauwens, T. Yin, S. Dang, R. Peerani and P. W. Zandstra, Development of a perfusion fed bioreactor for embryonic stem cell-derived cardiomyocyte generation: oxygen-mediated enhancement of cardiomyocyte output, *Biotechnol. Bioeng.*, 2005, **90**, 452–461.

76. A. Canela, E. Vera, P. Klatt and M. A. Blasco, High-throughput telomere length quantification by FISH and its application to human population studies, *Proc. Natl. Acad. Sci. U. S. A.*, 2007, **104**, 5300–5305.

77. R. Peerani, B. M. Rao, C. Bauwens, T. Yin, G. A. Wood, A. Nagy, E. Kumacheva and P. W. Zandstra, Niche-mediated control of human embryonic stem cell self-renewal and differentiation, *EMBO J.*, 2007, **26**, 4744–4755.

78. P. P. Szotek, H. L. Chang, K. Brennand, A. Fujino, R. Pieretti-Vanmarcke, C. Lo Celso, D. Dombkowski, F. Preffer, K. S. Cohen, J. Teixeira and P. K. Donahoe, Normal ovarian surface epithelial label-retaining cells exhibit stem/progenitor cell characteristics, *Proc. Natl. Acad. Sci. U. S. A.*, 2008, **105**, 12469–12473.

79. Y. M. Tay, W. L. Tam, Y. S. Ang, P. M. Gaughwin, H. Yang, W. Wang, R. Liu, J. George, H. H. Ng, R. J. Perera, T. Lufkin, I. Rigoutsos, A. M. Thomson and B. Lim, MicroRNA-134 modulates the differentiation of

mouse embryonic stem cells, where it causes post-transcriptional attenuation of Nanog and LRH1, *Stem Cells*, 2008, **26**, 17–29.

80. M. A. Van Dilla, T. T. Trujillo, P. F. Mullaney and J. R. Coulter, Cell microfluorometry: a method for rapid fluorescence measurement, *Science*, 1969, **163**, 1213–1214.

81. J. Schwaber and E. P. Cohen, Human X Mouse Somatic-Cell Hybrid Clone Secreting Immunoglobulins of Both Parental Types, *Nature*, 1973, **244**, 444–447.

82. X. Xuan and D. Li, Focused electrophoretic motion and selected electrokinetic dispensing of particles and cells in cross-microchannels, *Electrophoresis*, 2005, **26**, 3552–3560.

83. C. Lancaster, A. Kokoris, M. Nabavi, J. Clemmens, P. Maloney, J. Capadanno, J. Gerdes and C. F. Battrell, Rare cancer cell analyzer for whole blood applications: Microcytometer cell counting and sorting subcircuits, *Methods*, 2005, **37**, 120–127.

84. C. H. Lin, G. B. Lee, L. M. Fu and B. H. Hwey, Vertical focusing device utilizing dielectrophoretic force and its application on microflow cytometer, *J. Microelectromech. Sys.*, 2004, **13**, 923–932.

85. G. Goddard and G. Kaduchak, Ultrasonic particle concentration in a line-driven cylindrical tube, *J. Acous. Soc. Amer.*, 2005, **117**, 3440–3447.

86. G. Goddard, J. C. Martin, M. Naivar, P. M. Goodwin, S. W. Graves, R. Habbersett, J. P. Nolan and J. H. Jett, Single particle high resolution spectral analysis flow cytometry, *Cytometry A*, 2006, **69**, 842–851.

87. W. A. Bonner, H. R. Hulett, R. G. Sweet and L. A. Herzenberg, Fluorescence activated cell sorting, *Rev. Sci. Instrum.*, 1972, **43**, 404–409.

88. H. R. Hulett, W. A. Bonner, R. G. Sweet and L. A. Herzenberg, Development and application of a rapid cell sorter, *Clin. Chem.*, 1973, **19**, 813–816.

89. P. L. Kronick, G. L. Campbell and K. Joseph, Magnetic microspheres prepared by redox polymerization used in a cell separation based on gangliosides, *Science*, 1978, **200**, 1074–1076.

90. B. W. Phillips, R. Horne, T. S. Lay, W. L. Rust, T. T. Teck and J. M. Crook, Attachment and growth of human embryonic stem cells on microcarriers, *J. Biotechnol.*, 2008, **138**, 24–32.

91. D. T. Leong, W. K. Nah, A. Gupta, D. W. Hutmacher and M. A. Woodruff, The osteogenic differentiation of adipose tissue-derived precursor cells in a 3D scaffold/matrix environment, *Curr. Drug Discov. Technol.*, 2008, **5**, 319–327.

92. R. L. Carpenedo, A. M. Bratt-Leal, R. A. Marklein, S. A. Seaman, N. J. Bowen, J. F. McDonald and T. C. McDevitt, Homogeneous and organized differentiation within embryoid bodies induced by microsphere-mediated delivery of small molecules, *Biomaterials*, 2009, **30**, 2507–2515.

93. J. C. Peeters, G. B. Dubelaar, J. Ringelberg and J. W. Visser, Optical plankton analyser: a flow cytometer for plankton analysis, *I: Design considerations, Cytometry*, 1989, **10**, 522–528.

94. G. B. Dubelaar, A. C. Groenewegen, W. Stokdijk, G. J. van den Engh and J. W. Visser, Optical plankton analyser: a flow cytometer for plankton analysis, II: Specifications, *Cytometry*, 1989, **10**, 529–539.

95. D. W. Gray, W. Gohde, N. Carter, T. Heiden and P. J. Morris, Separation of pancreatic islets by fluorescence-activated sorting, *Diabetes*, 1989, **38**(Suppl 1), 133–135.

96. L. A. Fernandez, E. W. Hatch, B. Armann, J. S. Odorico, D. A. Hullett, H. W. Sollinger and M. S. Hanson, Validation of large particle flow cytometry for the analysis and sorting of intact pancreatic islets, *Transplantation*, 2005, **80**, 729–737.

97. E. E. Furlong, D. Profitt and M. P. Scott, Automated sorting of live transgenic embryos, *Nat. Biotechnol.*, 2001, **19**, 153–156.

98. F. Catteruccia, J. P. Benton and A. Crisanti, An Anopheles transgenic sexing strain for vector control, *Nat. Biotechnol.*, 2005, **23**, 1414–1417.

99. R. L. Sprando, N. Olejnik, H. N. Cinar and M. Ferguson, A method to rank order water soluble compounds according to their toxicity using Caenorhabditis elegans, a Complex Object Parametric Analyzer and Sorter, and axenic liquid media, *Food. Chem. Toxicol.*, 2009, **47**, 722–728.

100. T. C. George, B. E. Hall, C. A. Zimmerman, K. Frost, M. Seo, W. E. Ortyn, D. Basiji and P. Morrissey, Distinguishing modes of cell death using imagestream (TM) multispectral imaging cytometry, *Cytometry Part A*, 2004, **59A**, 118–118.

101. T. C. George, B. E. Hall, R. J. Finch, C. A. Zimmerman, D. Basiji and P. J. Morrissey, Distinguishing modes of cell death using the ImageStream (TM), a novel imaging cytometer, *FASEB J.*, 2003, **17**, C148–C148.

102. J. G. White, W. B. Amos and M. Fordham, An evaluation of confocal versus conventional imaging of biological structures by fluorescence light microscopy, *J. Cell Biol.*, 1987, **105**, 41–48.

103. W. Denk, J. H. Strickler and W. W. Webb, Two-photon laser scanning fluorescence microscopy, *Science (New York, N.Y*, 1990, **248**, 73–76.

104. K. Kajiwara, M. Kamamoto, S. Ogata and M. Tanihara, A synthetic peptide corresponding to residues 301-320 of human Wnt-1 promotes PC12 cell adhesion and hippocampal neural stem cell differentiation, *Peptides*, 2008, **29**, 1479–1485.

105. K. Elknerova, Z. Lacinova, J. Soucek, I. Marinov and P. Stockbauer, Growth inhibitory effect of the antibody to hematopoietic stem cell antigen CD34 in leukemic cell lines, *Neoplasma*, 2007, **54**, 311–320.

106. J. M. Berg, J. L. Tymoczko and L. Stryer, *Biochemistry*, W.H. Freeman, New York, 2002.

107. M. Wakita, G. Nishimura and M. Tamura, Some characteristics of the fluorescence lifetime of reduced pyridine nucleotides in isolated mitochondria, isolated hepatocytes, and perfused rat liver in situ, *J. Biochem.*, 1995, **118**, 1151–1160.

108. K. Blinova, S. Carroll, S. Bose, A. V. Smirnov, J. J. Harvey, J. R. Knutson and R. S. Balaban, Distribution of mitochondrial NADH

fluorescence lifetimes: steady-state kinetics of matrix NADH interactions, *Biochemistry*, 2005, **44**, 2585–2594.

109. P. Belenky, K. L. Bogan and C. Brenner, NAD+ metabolism in health and disease, *Trends Biochem. Sci.*, 2007, **32**, 12–19.

110. B. Chance, V. Legallais and B. Schoener, Metabolically linked changes in fluorescence emission spectra of cortex of rat brain, kidney and adrenal gland, *Nature*, 1962, **195**, 1073–1075.

111. D. J. Pappajohn, R. Penneys and B. Chance, NADH spectrofluorometry of rat skin, *J. Appl. Physiol.*, 1972, **33**, 684–687.

112. B. Zhang, R. Z. Wang, Z. G. Lian, Y. Song and Y. Yao, Experimental study on plasticity of proliferated neural stem cells in adult rats after cerebral infarction, *Chin. Med. Sci. J.*, 2006, **21**, 184–188.

113. Q. Zhang, D. W. Piston and R. H. Goodman, Regulation of corepressor function by nuclear NADH, *Science*, 2002, **295**, 1895–1897.

114. N. Ramanujam, M. F. Mitchell, A. Mahadevan-Jansen, S. L. Thomsen, G. Staerkel, A. Malpica, T. Wright, N. Atkinson and R. Richards-Kortum, Cervical precancer detection using a multivariate statistical algorithm based on laser-induced fluorescence spectra at multiple excitation wavelengths, *Photochem. Photobiol.*, 1996, **64**, 720–735.

115. J. R. Lakowicz, H. Szmacinski, K. Nowaczyk and M. L. Johnson, Fluorescence lifetime imaging of free and protein-bound NADH, *Proc. Natl. Acad. Sci. U. S. A.*, 1992, **89**, 1271–1275.

116. M. C. Skala, J. M. Squirrell, K. M. Vrotsos, J. C. Eickhoff, A. Gendron-Fitzpatrick, K. W. Eliceiri and N. Ramanujam, Multiphoton microscopy of endogenous fluorescence differentiates normal, precancerous, and cancerous squamous epithelial tissues, *Cancer Res.*, 2005, **65**, 1180–1186.

117. Y. M. Cho, S. Kwon, Y. K. Pak, H. W. Seol, Y. M. Choi, J. Park do, K. S. Park and H. K. Lee, Dynamic changes in mitochondrial biogenesis and antioxidant enzymes during the spontaneous differentiation of human embryonic stem cells, *Biochem. Biophys. Res. Commun.*, 2006, **348**, 1472–1478.

118. W. Denk, J. H. Strickler and W. W. Webb, Two-photon laser scanning fluorescence microscopy, *Science*, 1990, **248**, 73–76.

119. V. E. Centonze and J. G. White, Multiphoton excitation provides optical sections from deeper within scattering specimens than confocal imaging, *Biophys. J.*, 1998, **75**, 2015–2024.

120. J. M. Squirrell, D. L. Wokosin, J. G. White and B. D. Bavister, Long-term two-photon fluorescence imaging of mammalian embryos without compromising viability, *Nat. Biotechnol.*, 1999, **17**, 763–767.

121. P. E. Hanninen, J. T. Soini and E. Soini, Photon-burst analysis in two-photon fluorescence excitation flow cytometry, *Cytometry*, 1999, **36**, 183–188.

122. A. Diaspro, Two-photon excitation. A new potential perspective in flow cytometry, *Minerva Biotechnol.*, 1999, **11**, 87–92.

123. C. F. Zhong, E. R. Tkaczyk, T. Thomas, J. Y. Ye, A. Myc, A. U. Bielinska, Z. Cao, I. Majoros, B. Keszler, J. R. Baker and T. B. Norris, Quantitative

two-photon flow cytometry--in vitro and in vivo, *J. Biomed. Opt.*, 2008, **13**, 034008.

124. N. Ramanujam, Fluorescence spectroscopy of neoplastic and non-neoplastic tissues, *Neoplasia*, 2000, **2**, 89–117.

125. J. R. Lakowicz, *Principals of Fluorescence Spectroscopy*, Academic Press, New York, 1999.

126. P. I. Bastiaens and A. Squire, Fluorescence lifetime imaging microscopy: spatial resolution of biochemical processes in the cell, *Trends in Cell Biol.*, 1999, **9**, 48–52.

127. J. R. Lakowicz, H. Szmacinski, K. Nowaczyk, K. W. Berndt and M. Johnson, Fluorescence lifetime imaging, *Anal. Biochem.*, 1992, **202**, 316–330.

128. X. F. Wang, A. Periasamy and B. Herman, Fluorescence lifetime imaging microscopy (FLIM): instrumentation and applications, *Crit. Rev. Anal. Chem*, 1992, **23**, 365–369.

129. T. W. J. Gadella, T. M. Jovin and R. M. Clegg, Fluorescence lifetime imaging microscopy (FLIM): Spatial resolution of microstructures on the nanosecond time scale., *Biophys. Chem*, 1993, **48**, 221–239.

130. G. W. Gordon, G. Berry, X. H. Liang, B. Levine and B. Herman, Quantitative fluorescence resonance energy transfer measurements using fluorescence microscopy, *Biophys. J.*, 1998, **74**, 2702–2713.

131. J. E. Pawley, *Handbook of confocal microscopy*, Plenum Press, New York, 1996.

132. J. G. White, J. M. Squirrell and K. W. Eliceiri, Applying multiphoton imaging to the study of membrane dynamics in living cells, *Traffic*, 2001, **2**, 775–780.

133. C. Thomas, P. DeVries, J. Hardin and J. White, Four-dimensional imaging: computer visualization of 3D movements in living specimens, *Science (New York, N.Y*, 1996, **273**, 603–607.

134. P. J. Campagnola and L. M. Loew, Second-harmonic imaging microscopy for visualizing biomolecular arrays in cells, tissues and organisms, *Nat. Biotechnol.*, 2003, **21**, 1356–1360.

135. J. D. Muller, Y. Chen and E. Gratton, Fluorescence correlation spectroscopy, *Methods Enzymol.*, 2003, **361**, 69–92.

136. J. X. Cheng, Y. K. Jia, G. Zheng and X. S. Xie, Laser-scanning coherent anti-Stokes Raman scattering microscopy and applications to cell biology, *Biophys. J.*, 2002, **83**, 502–509.

137. M. Tommila, A. Jokilammi, P. Terho, T. Wilson, R. Penttinen and E. Ekholm, Hydroxyapatite coating of cellulose sponges attracts bone-marrow-derived stem cells in rat subcutaneous tissue, *J. R. Soc. Interface*, 2009.

138. S. O. Konorov, C. H. Glover, J. M. Piret, J. Bryan, H. G. Schulze, M. W. Blades and R. F. Turner, In situ analysis of living embryonic stem cells by coherent anti-stokes Raman microscopy, *Anal. Chem.*, 2007, **79**, 7221–7225.

139. Y. Qian, O. Tchuvatkina, J. Spidlen, P. Wilkinson, M. Gasparetto, A. R. Jones, F. J. Manion, R. H. Scheuermann, R. P. Sekaly and R. R.

Brinkman, FuGEFlow: data model and markup language for flow cytometry, *BMC Bioinformatics*, 2009, **10**, 184.

140. J. A. Lee, J. Spidlen, K. Boyce, J. Cai, N. Crosbie, M. Dalphin, J. Furlong, M. Gasparetto, M. Goldberg, E. M. Goralczyk, B. Hyun, K. Jansen, T. Kollmann, M. Kong, R. Leif, S. McWeeney, T. D. Moloshok, W. Moore, G. Nolan, J. Nolan, J. Nikolich-Zugich, D. Parrish, B. Purcell, Y. Qian, B. Selvaraj, C. Smith, O. Tchuvatkina, A. Wertheimer, P. Wilkinson, C. Wilson, J. Wood, R. Zigon, R. H. Scheuermann and R. R. Brinkman, MIFlowCyt: the minimum information about a Flow Cytometry Experiment, *Cytometry A*, 2008, **73**, 926–930.
141. J. Moore, C. Allan, J. M. Burel, B. Loranger, D. MacDonald, J. Monk and J. R. Swedlow, Open tools for storage and management of quantitative image data, *Methods Cell. Biol.*, 2008, **85**, 555–570.
142. I. G. Goldberg, C. Allan, J. M. Burel, D. Creager, A. Falconi, H. Hochheiser, J. Johnston, J. Mellen, P. K. Sorger and J. R. Swedlow, The Open Microscopy Environment (OME) Data Model and XML file: open tools for informatics and quantitative analysis in biological imaging, *Genome Biol.*, 2005, **6**, R47.
143. J. R. Swedlow, I. Goldberg, E. Brauner and P. K. Sorger, Informatics and quantitative analysis in biological imaging, *Science*, 2003, **300**, 100–102.

Hematopoietic Stem Cells and their Role in Regenerative Medicine

EITAN FIBACH

Department of Hematology, Hadassah – Hebrew University Medical Centre, Jerusalem, Israel

4.1 The Human Hematopoietic Stem Cell

Mature blood cells have a limited lifespan and therefore require a continuous supply of new cells in order to maintain a constant number of functional cells and to cope with changing physiological conditions. The production of new blood cells (hematopoiesis) involves proliferation and differentiation of less mature precursors. The building blocks of this process are relatively few hematopoietic stem cells (HSC) which are endowed with the potentials of self-renewal, proliferation and differentiation.

Most HSC are in a dormant (G_0) phase of the cell cycle.[1] When they are stimulated to proliferate, HSC undergo a cell division process that can be either symmetrical—giving rise to either two daughter stem cells or two more differentiated daughter cells (progenitors), or asymmetrical—giving rise to one stem cell and one progenitor cell. In the process of self-renewal, at least one of the daughter cells is an identical copy without any newly differentiated features—namely, it sustains its stem cell properties.[1] The balance between the symmetrical and asymmetrical self-renewal cell divisions of HSC is not entirely clear, but the ability of a small number of HSC to generate a larger

---

Stem Cell-Based Tissue Repair
Edited by Raphael Gorodetsky and Richard Schäfer
© The Royal Society of Chemistry 2011
Published by the Royal Society of Chemistry, www.rsc.org

number of progeny HSC following bone marrow transplantation indicates that, under conditions of extensive expansion, symmetrical cell divisions must occur.

HSC are pluripotent in their capacity to give rise to all blood cell lineages. As HSC differentiate, they produce progeny with progressively limited potency yielding, eventually, unipotent committed progenitors which give rise to a specific blood cell lineage. In parallel, these cells also lose their self-renewal capacity; each cell division terminates in daughter cells at an advanced stage of differentiation. The balance between self-renewal and differentiation of HSC is critical for the maintenance of an HSC pool—and thus for sustaining hematopoiesis—throughout the life of the organism. There is much interest in the environmental and molecular requirements for HSC self-renewal since understanding these processes may pave the way to improved therapeutic modalities in cases of insufficiency (*e.g.* aplastic anemia) and for restraining the development of leukemia, as well as devising procedures for *ex vivo* expansion of HSC that can be used therapeutically.

During embryonic development, HSC first emerge in the hematopoietic island in the yolk sac and then, as they move into the embryo, in the aorta–gonad–mesonephros (AGM) region, fetal liver and fetal spleen. They are present in the placenta and the umbilical cord blood, and postnatally in the red marrow—first in most skeletal bones and later, as some undergo lipid transformation into yellow marrow, in more specific bones. Under some pathological conditions, stem cells can lodge and sustain hematopoiesis in extramedullary sites (*e.g.* spleen, liver, lymph nodes). HSC do circulate in the peripheral blood, especially following treatment with cytokines such as the granulocyte colony-stimulating factor (G-CSF) which induces their mobilization from the bone marrow. Neonatal cord blood, adult bone marrow and mobilized peripheral blood serve as sources of HSC for transplantation.

HSC have no morphological distinctive features; they resemble small lymphocytes. They are round, non-adherent cells with a rounded nucleus and low cytoplasm-to-nucleus ratio. HSC are identified by flow cytometry based on their small size (low forward light scatter), and low staining (side population) with vital dyes such as rhodamine 123 or Hoechst 33342. They are defined by their positive and negative expression of antigens: they are positive for CD34, CD38, CD90, CD133, CD105, CD45 and CD117 (the receptor for stem cell factor), but are negative for lineage differentiation markers. However, HSC are not homogeneous. For example, some HSC are CD34$^-$/CD38$^-$ and some lack CD117 on the cell surface; and among the CD133$^+$ cells some are CD34$^+$ while others are CD34$^-$.

Purification of HSC can be accomplished based on their antigenic expression using cell sorting flow cytometry or immunomagnetic bead separation technology. In the latter technology, cells labeled with magnetic bead conjugated monoclonal antibodies to specific surface antibodies (CD34 or CD133) are separated from non-labeled cells in a magnetic field. Purified populations of HSC serve for research as well as for transplantation.

4.2 Clinical Uses of HSC

Among all studied stem cells (embryonic or adult), HSC are the only ones so far that are used routinely in well-established clinical practice. HSC transplantation is most often performed for people with hematological diseases or certain types of cancer. HSC transplantation was pioneered in the 1970s when it was shown that bone marrow cells infused intravenously could repopulate the bone marrow and provide a new source of essential blood and immune system cells.

Most recipients of HSC transplants are leukemia patients or others who would benefit from treatment with high doses of chemotherapy or total body irradiation. Other recipients include children with an inborn defect in the HSC such as severe combined immunodeficiency or congenital neutropenia, and children or adults who lost their HSC for various acquired defects (*e.g.* patients with aplastic anemia). For certain diseases (*e.g.* leukemia and specific immunodeficiencies), transplant is the only proven treatment for long-term survival; in other diseases, they are employed only when frontline therapies have failed or when the disease is very aggressive.

Diseases for which HSC transplants are a standard treatment include:

- leukemia;
- lymphomas;
- multiple myeloma;
- other cancers not originating in the blood system (neuroblastoma, retinoblastoma);
- myelodysplastic syndromes;
- inherited erythrocyte abnormalities; and
- various disorders of blood cell proliferation (anemias, inherited platelet abnormalities, myeloproliferative disorders, inherited immune system disorders).

Another use of allogeneic bone marrow transplants is in the treatment of hereditary blood disorders (*e.g.* different types of inherited anemia) and inborn errors of metabolism. The blood disorders include beta-thalassemia, sickle-cell anemia, Blackfan–Diamond syndrome, globoid cell leukodystrophy, severe combined immunodeficiency, X-linked lymphoproliferative syndrome and Wiskott–Aldrich syndrome. Inborn errors of metabolism that are treated with bone marrow transplants include Hunter's syndrome, Hurler's syndrome, Lesch Nyhan syndrome and osteopetrosis.

HSC transplantation can be either autologous or allogeneic. In autologous transplantation, the recipient's own previously harvested stem cells are re-infused. Allogeneic transplantation uses stem cells from a suitable donor identified by human leukocyte antigen (HLA) typing. If a family relative is not identified, then a search can be performed for an HLA-matched, unrelated donor. However, the chances of finding a suitably matched unrelated donor are approximately 30% for Caucasians and significantly less for other ethnic groups. In addition, the search process can take from 3–6 months and is often very expensive. Graft-versus-host disease (GVHD) is the major complication in the post-transplant period.

Chemotherapy of cancer patients aimed at rapidly dividing cancer cells inevitably hits the hematopoietic cells. An autologous stem cell transplant may replace the cells destroyed by chemotherapy. In this procedure HSC are harvested and stored in liquid nitrogen, while the patient undergoes intensive chemotherapy or radiotherapy to destroy the cancer cells. After the drugs have washed out of a patient's body, the stored HSC are transfused back. Because patients get their own cells back, there is no risk of immune mismatch or graft-versus-host disease. The major problem in using autologous HSC transplantation for cancer therapy has been the occurrence of cancer cells in the harvest; upon re-infusion back into the patient, these cells may cause a relapse of the disease. Various purging protocols are being tested in order to overcome this problem.

4.3 *Ex Vivo* Expansion of HSC

HSC transplantation is hampered by the high cell dose required for engraftment. This poses a significant problem in particular for umbilical cord blood transplantation because the number of HSC derived from a single umbilical cord blood is low. Clinical results have clearly shown that the time of recovery of the neutrophil and platelet counts after myeloablative chemotherapy is closely correlated with the number of transplanted HSC.[2] Therefore, umbilical cord blood transplantation is limited to children.[3] To increase the number of HSC, several strategies are currently being explored including:

- *ex vivo* expansion of HSC;
- co-transplantation of HSC and mesenchymal stem cells;
- combined transplantation of several cord blood units; and
- intra-bone marrow injection of the transplanted cells.

4.4 Expanding HSC in Cytokine Cocktails

To date more than 30 hematopoietic factors that contribute to the proliferation and differentiation of HSC have been cloned and identified.[3] Many of these factors were reported to increase the number of HSC *in vitro*. Among them are erythropoietin, granulocyte colony stimulating factor, granulocyte–macrophage colony-stimulating factor, stem cell factor, thrombopoietin, FLt-3/FLt-2 ligand, fibroblast growth factor, interleukin (IL)-1, IL-3, IL-6/sIL-6R and IL-11. For example, Piacibello *et al.* showed that the combination of FLt-3/FLt-2 ligand and thrombopoietin could generate a two million-fold increase in granulocyte–macrophage colony forming cells and a two hundred-fold increase in long-term culture-initiating cells.[4] However, some researchers disputed the reliance of the *in vitro* functional assays such as the number of CD34+ cells, colony-forming cells, cobblestone area-forming cells and long-term culture-initiating cells employed for evaluation of expansion in most studies to reflect the true activities of human HSC.[5] *In vivo* xeno-transplantations of human HSC in fetal sheep and in immunodeficient (*e.g.* NOD/SCID) mice have been employed to evaluate the transplantability of expanded cells. Thus, Gammaitoni *et al.* showed that adult normal bone marrow and mobilized peripheral blood CD34+ cells cultured in the presence of the FLt-3/FLt-2 ligand, thrombopoietin, stem cell factor, IL-6, or IL-3 retained the ability to repopulate NOD/SCID mice after serial transplants.[5] However, other studies have questioned the long-term engraftment potential of expanded cells.[5,6]

The expansion of HSC is limited by cell differentiation and apoptosis. When HSC derived from human umbilical cord blood were cultured in stroma-free, liquid medium supplemented with cytokines, their expansion was limited. Moreover, they lost their potential to function properly following transplantation.[7] We have defined components in the culture milieu that induce HSC differentiation and apoptosis, and devised means to inhibit their activity.

Clinical observations suggest that copper plays a role in stem cell differentiation. Copper is known to generate free radicals and oxidative stress. We investigated the role of copper and oxidative stress on HSC differentiation and expansion.[7-11] We found that free radicals in HSC were increased by copper and were decreased by the copper-binding compound (chelator), tetraethylenepentamine. We showed that this chelator reduced the free copper content of HSC, thereby reducing their oxidative stress and consequently inhibiting differentiation and promoting their *ex vivo* expansion.[11]

Other inducers of cell differentiation are retinoids such as all-*trans* retinoic acid (ATRA), which are present in serum added to the culture. The activity of ATRA involves binding to nuclear receptors which can be inhibited by a pan-antagonist (AGN 194310).[12]

Another compound that was studied for its affect on HSC is nicotinamide, the amide form of niacin (vitamin B3). This compound acts both as a scavenger of free radicals and as an inhibitor of NAD(+)-dependent enzymes endowed with mono- and poly-ADP-ribosyltransferase activities as well as Sir2 proteins, NAD(+)-dependent enzymes endowed with both histone deacetylase and ribosylase activities.

TEPA, AGN 194310 and nicotinamide stimulate self-renewal of a subset of early HSC and thereby promote their long-term expansion *in vitro* and their ability to engraft and repopulate the hematopoietic system in NOD/SCID mice following transplantation. These compounds are being tested for *ex vivo* manipulation of HSC for clinical applications. Following a phase I clinical trial using TEPA for *ex vivo* expansion of cord blood HSC for transplantation, a phase II study is on its way.[13]

4.5 Expansion of HSC by Coculturing with Stroma Feeder Cells

In light of the importance of cellular network support for HSC, initial research on their expansion has focused, with little success, on constructing the hematopoietic microenvironment *in vitro*. Currently, there is an increasing interest in coculturing HSC with various non-hematopoietic cells. These cells include a murine neonatal neural cell line,[14] microvascular endothelial cells,[15] human yolk sac-derived endothelial cells,[16] umbilical vein endothelial cells,[17] osteoblasts[18] and extracellular matrix.[19] Mesenchymal stem cells (MSC) can differentiate into mature mesenchymal cells that support hematopoiesis. MSC derived from human umbilical cord, human placenta and lung were reported to support expansion of cord blood HSC in the presence of cytokines.[20] To improve the hematopoietic supporting ability, human HSC have been immortalized or engineered to overexpress various cytokines or signaling molecules.[21]

In another approach, Kawano *et al.* developed a human stromal cell line that overexpressed the human telomerase catalytic subunit (hTERT) gene (so-called 'hTERT stromal cells').[22] Human HSC cocultured with hTERT stromal cells were expanded more than a thousand fold in the presence of SCF, TPO, and Flk-2/Flt-3 ligand in seven weeks. In addition, the engraftment of SCID-repopulating cells cocultured with hTERT-stromal cells for four weeks was significantly increased.

Ex vivo manipulation of HSC is still an unsolved objective. Success in this field will provide the means for expansion of HSC, as well as more differentiated progenitors for transplantation of donor's cells and for gene therapy using autologous cells. In light of the recent discoveries of the plasticity of HSC to give rise to non-hematopoietic cells, *in vitro* manipulated HSC may offer new horizons for regenerative medicine.

References

1. A. Wilson, E. Laurenti and A. Trumpp, Balancing dormant and self-renewing hematopoietic stem cells, *Curr. Opin. Genet. Dev.*, 2009, **19**, 461–468.
2. Y. Cohena and A. Nagler, Hematopoietic stem-cell transplantation using umbilical-cord blood, *Leuk. Lymphoma*, 2003, **44**, 1287–1299.
3. E. Gluckman, Hematopoietic stem-cell transplants using umbilical-cord blood, *N. Engl. J. Med.*, 2001, **344**, 1860–1861.
4. W. Piacibello, F. Sanavio, L. Garetto, A. Severino, D. Bergandi, J. Ferrario, F. Fagioli, M. Berger and M. Aglietta, Extensive amplification and self-renewal of human primitive hematopoietic stem cells from cord blood, *Blood*, 1997, **89**, 2644–2653.
5. L. Gammaitoni, S. Bruno, F. Sanavio, M. Gunetti, O. Kollet, G. Cavalloni, M. Falda, F. Fagioli, T. Lapidot, M. Aglietta and W. Piacibello, *Ex vivo* expansion of human adult stem cells capable of primary and secondary hemopoietic reconstitution, *Exp. Hematol.*, 2003, **31**, 261–270.
6. E. F. Srour, R. Abonour, K. Cornetta and C. M. Traycoff, *Ex vivo* expansion of hematopoietic stem and progenitor cells, are we there yet?, *J. Hematother.*, 1999, **8**, 93–102.

7. T. Peled, E. Landau, E. Prus, A. J. Treves, A. Nagler and E. Fibach, Cellular copper content modulates differentiation and self-renewal in cultures of cord blood-derived CD34 + cells, *Br. J. Haematol.*, 2002, **116**, 655–661.
8. T. Peled, E. Landau, J. Mandel, E. Glukhman, N. R. Goudsmid, A. Nagler and E. Fibach, Linear polyamine copper chelator tetraethylenepentamine augments long-term *ex vivo* expansion of cord blood-derived CD34 + cells and increases their engraftment potential in NOD/SCID mice, *Exp. Hematol.*, 2004, **32**, 547–555.
9. T. Peled, J. Mandel, R. N. Goudsmid, C. Landor, N. Hasson, D. Harati, M. Austin, A. Hasson, E. Fibach, E. J. Shpall and A. Nagler, Pre-clinical development of cord blood-derived progenitor cell graft expanded *ex vivo* with cytokines and the polyamine copper chelator tetraethylenepentamine, *Cytotherapy*, 2004, **6**, 344–355.
10. E. Prus and E. Fibach, The effect of the copper chelator tetraethylenepentamine on reactive oxygen species generation by human hematopoietic progenitor cells, *Stem Cells Dev*, 2007, **16**, 1053–1056.
11. E. Prus, T. Peled and E. Fibach, The effect of tetraethylenepentamine, a synthetic copper chelating polyamine, on expression of CD34 and CD38 antigens on normal and leukemic hematopoietic cells, *Leuk. Lymphoma*, 2004, **45**, 583–589.
12. E. Prus, R. A. Chandraratna and E. Fibach, Retinoic acid receptor antagonist inhibits CD38 antigen expression on human hematopoietic cells *in vitro*, *Leuk. Lymphoma*, 2004, **45**, 1025–1035.
13. M. de Lima, J. McMannis, A. Gee, K. Komanduri, D. Couriel, B. S. Andersson, C. Hosing, I. Khouri, R. Jones, R. Champlin, S. Karandish, T. Sadeghi, T. Peled, F. Grynspan, Y. Daniely, A. Nagler and E. J. Shpall, Transplantation of ex vivo expanded cord blood cells using the copper chelator tetraethylenepentamine: a phase I/II clinical trial, *Bone Marrow Transplant.*, 2008, **41**, 771–778.
14. D. Bratosin, L. Mitrofan, C. Palii, J. Estaquier and J. Montreuil, Multipotent neural precursors express neural and hematopoietic factors, and enhance *ex vivo* expansion of cord blood CD34 + cells, colony forming units and NOD/SCID-repopulating cells in contact and noncontact cultures, *Cytometry A*, 2005, **66**, 78–84.
15. E. Rosler, J. Brandt, J. Chute and R. Hoffman, Cocultivation of umbilical cord blood cells with endothelial cells leads to extensive amplification of competent CD34 + CD38- cells, *Exp. Hematol.*, 2000, **28**, 841–852.
16. L. Hu, L. Cheng, J. Wang, H. Zhao, H. Duan and G. Lu, Effects of human yolk sac endothelial cells on supporting expansion of hematopoietic stem/progenitor cells from cord blood, *Cell Biol. Int*, 2006, **30**, 879–884.
17. S. Yildirim, A. M. Boehmler, L. Kanz and R. Mohle, Expansion of cord blood CD34 + hematopoietic progenitor cells in coculture with autologous umbilical vein endothelial cells (HUVEC) is superior to cytokine-supplemented liquid culture, *Bone Marrow Transplant.*, 2005, **36**, 71–79.

18. N. Ahmed, M. A. Khokher and H. T. Hassan, Cytokine-induced expansion of human CD34 + stem/progenitor and CD34 + CD41 + early megakaryocytic marrow cells cultured on normal osteoblasts, *Stem Cells*, 1999, **17**, 92–99.

19. S. V. Madihally, A. W. Flake and H. W. Matthew, Maintenance of CD34 expression during proliferation of CD34 + cord blood cells on glycosaminoglycan surfaces, *Stem Cells*, 1999, **17**, 295–305.

20. Y. K. Jang, D. H. Jung, M. H. Jung, D. H. Kim, K. H. Yoo, K. W. Sung, H. H. Koo, W. Oh, Y. S. Yang and S. E. Yang, Mesenchymal stem cells feeder layer from human umbilical cord blood for ex vivo expanded growth and proliferation of hematopoietic progenitor cells, *Ann. Hematol.*, 2006, **85**, 212–225.

21. K. Nishioka, Y. Fujimori, T. Hashimoto-Tamaoki, S. Kai, H. Qiu, N. Kobayashi, N. Tanaka, K. A. Westerman, P. Leboulch and H. Hara, Immortalization of bone marrow-derived human mesenchymal stem cells by removable simian virus 40T antigen gene: analysis of the ability to support expansion of cord blood hematopoietic progenitor cells, *Int. J. Oncol.*, 2003, **23**, 925–932.

22. Y. Kawano, M. Kobune, M. Yamaguchi, K. Nakamura, Y. Ito, K. Sasaki, S. Takahashi, T. Nakamura, H. Chiba, T. Sato, T. Matsunaga, H. Azuma, K. Ikebuchi, H. Ikeda, J. Kato, Y. Niitsu and H. Hamada, *Ex vivo* expansion of human umbilical cord hematopoietic progenitor cells using a coculture system with human telomerase catalytic subunit (hTERT)-transfected human stromal cells, *Blood*, 2003, **101**, 532–540.

Cord and Cord Blood: Valuable Resources with Potential for Liver Therapy

SABA HABIBOLLAH,[a] MARCIN JURGA,[b] NICO FORRAZ[b] AND COLIN MCGUCKIN*,[b]

[a] Institute of Genetics, Newcastle University, Newcastle-upon-Tyne, UK; [b] Cell Therapy Research Institute, CTI-Lyon, Lyon, France

5.1 Introduction

Liver cirrhosis and/or liver malignancies have been nominated as the fifth leading cause of death worldwide. In 2006 the World Health Organization (WHO) reported that 20 million people around the globe suffer from some form or other of severe liver illness. The ultimate fate of end-stage liver disorders is hepatic dysfunction and eventually organ failure. Unfortunately the only curative mode of management for liver failure is liver transplantation, which is subject to many limitations. Novel alternatives such as artificial and bio-artificial support devices only aid in temporary replacement of some liver function until an organ is available for transplantation. These newer modalities also have drawbacks or remain experimental, and still demand further controlled trials to allow proof of concept and safety before transferring them to the bedside.

Regenerative medicine and stem cell therapy have recently shown promise in the management of various human diseases. Recent reports of stem cell plasticity and its multipotentiality has raised hopes of stem cell therapy offering

Stem Cell-Based Tissue Repair
Edited by Raphael Gorodetsky and Richard Schäfer
© The Royal Society of Chemistry 2011
Published by the Royal Society of Chemistry, www.rsc.org

exciting therapeutic possibilities for patients with chronic liver disease. Although there exists a choice of stem cells that have been reported to be capable of self-renewal and differentiation to hepatobiliary cell lineages both *in vitro* and *in vivo* (including rodent and human embryonic stem cells, bone marrow hematopoietic stem cells, mesenchymal stem cells, umbilical cord blood stem cells, fetal liver progenitor cells, adult liver progenitor cells), it may be argued that, with a global population of six billion people and a global birth rate in access of 130 million per year, placenta and the umbilical cord possibly provide the most readily accessible and ethically sound alternative source of stem cells. Liver cells derived from umbilical cord blood (UBC) can potentially be exploited for gene therapy, cellular transplant, bio-artificial liver-assisted devices, drug toxicology testing and use as an *in vitro* model to understand the developmental biology of the liver. Here we review the latest scientific developments relevant for future liver cell therapy.

5.2 The Liver—A Unique Organ

The liver is an intricate organ composed of cells such as hepatocytes, hepatocyte precursor cells,[1] stellate cells, Kupffer cells, epithelial cells, sinusoidal epithelial cells, biliary epithelial cells and fibroblast.[2] The liver is involved in over 150 different vital functions including metabolism, detoxification and maintenance of homeostasis in the body. The liver also possesses endocrine functions, production of many serum proteins into the blood, blood clotting factors and an exocrine function, secreting large amounts of bile into the digestive tract.[2,3] Correct liver functions are fundamental to human health and loss of these can be severely compromising. The liver is an exceptional organ; in an event of parenchymal cell loss, the mammalian liver can cite at least three apparently distinct cell lineages to contribute to regeneration and repair after damage (Figure 5.1).

The first line of defense is provided by hepatoblasts/hepatocytes, which are themselves believed to be functional stem cells of the liver.[4,5] More severe liver damage calls upon activation of a potential stem cell compartment located within the intrahepatic biliary tree, giving rise to cords of bipotential cells that ultimately differentiate to hepatic or biliary epithelial cells. These hepatic stem/progenitor cells are referred to by different names including 'oval cells' and 'small hepatocyte-like progenitors' (SHPCs) in rodents[6,7] and 'intermediate cells' in humans.[7,8] This cell population has been shown to have 'bipotential characteristics' expressing morphological and immunophenotypical features typical of both hepatic and biliary epithelial cells. In 1958 Wilson and Leduc were the first to describe activation of this 'reserve cell compartment', which in humans is described as 'ductular activation' or 'progenitor cell activation'.[9] A third population of stem cells with hepatic potential resides outside the liver in the bone marrow. These hematopoietic stem cells may contribute to the low renewal rate of hepatocytes, but in the face of severe and/or extensive damage can contribute significantly to regeneration under very strong positive selective

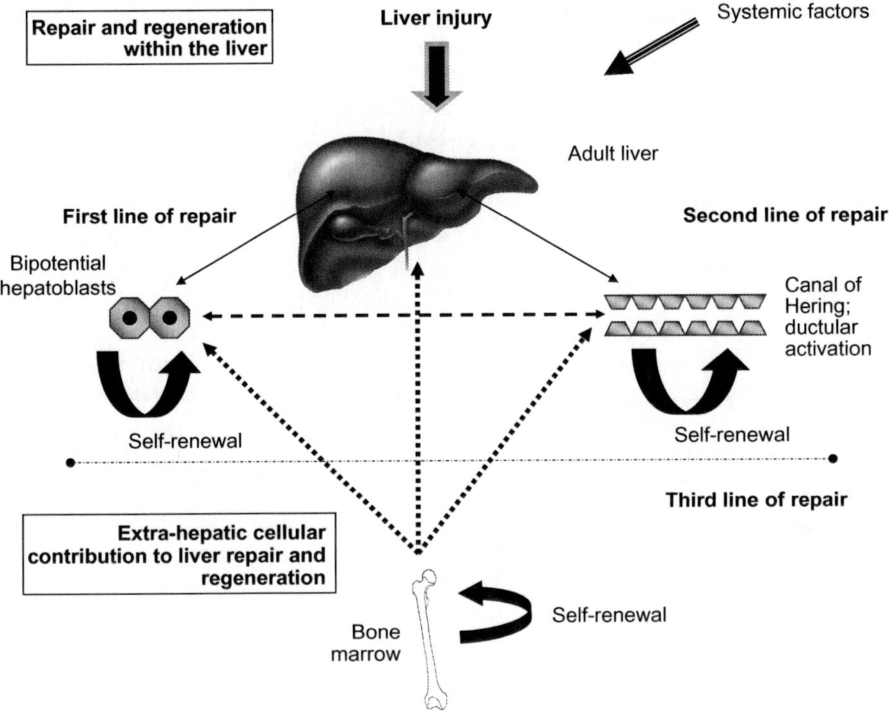

Figure 5.1 Three different cell lineages that contribute to liver regeneration and repair after injury. The first line of defense in the face of injury is provided by bipotential hepatoblast which are capable of differentiating into hepatocytes and cholangiocytes. These are believed to be functional stem cells of the liver and are capable of self-renewal after loss. When hepatocyte renewal is compromised or the liver damage is more severe, this calls upon activation of bipotential stem/progenitor cells in the canal of Hering, located in the intra-hepatic biliary tree, to take over the burden of regeneration. Under conditions where the insult to the liver is extensive and/or chronic or the intrinsic proliferative capacity of the liver is compromised, extrinsic stem cells from the bone marrow harbor to the site of injury and provide a third line of repair and regeneration. It should be noted, however, that systemic factors also play a pivotal role in modulating signaling pathways critical to tissue regeneration.

pressure.[10–14] Under conditions where the insult to the liver is extensive and/or chronic, the liver is crippled by hepatocyte senescence due to telomere shortening that is associated with ongoing proliferation during the prolonged chronic history of disease. During incidents where the intrinsic proliferative and clonogenic capacity of inherent hepatocytes/hepatoblasts is affected, extrinsic stem cells with hepatogenic potential are chemo-attracted towards to the site of liver injury, finding a niche and attempt to salvage the liver. It should be noted, however, that in the diseased liver, there may be lack of the major growth stimulus and/or absence of substantial growth advantage for hematopoietic stem cells to correct and repair the damaged tissue. Another important factor is the intrinsic genetic regulation of stem cells.[15] In addition, there are many other regulatory systems involved which dictate the replacement of damaged, aged or diseased tissue with new cells. Stem cells secrete factors which act through a paracrine system and play a role in regulating tissue regeneration. Systemic factors are also important in modulating signaling pathways critical to activate tissue specific stem cells.[16]

Liver cells (hepatocytes, cholangiocytes and progenitor cells), mesenchymal cells (Kupffer cells, endothelial cells, hepatic stellate cells) and the liver stroma collectively form the liver 'stem cell niche' that regulates stem cell proliferation, maintenance and cell fate decisions. The importance of such cellular and non-cellular interactions was demonstrated by Kon *et al.*, who elegantly illustrated *in vitro* differentiation of small rat liver progenitor cells into hepatocytes upon coculture with non-parenchymal supporting cells.[17] Additionally, the importance of such niches is reinforced by the fact that hepatocytes, which have a great growth potential *in vivo*, when isolated and put in culture pose a huge challenge to maintain alive and differentiated.[18] Even if stem cells did tend to retain much of their intrinsic proliferative potential when old, age-related changes in the systemic environment and niche in which stem cells reside preclude full activation of these cells for productive tissue regeneration.

5.3   Current Management of Liver Disorders

Liver disease is highly prevalent worldwide. The majority of the approximately 20 million people worldwide suffering from cirrhosis of the liver and/or mitotic lesions of the liver arise from those 500 million people (approximately 10% of the world population) who are victims of hepatitis B (HBV) or hepatitis C (HCV) viral infections.[19] Hepatic failure alone accounts for 1–2 million deaths per year and is nominated the fifth leading cause of death around the globe. Decompensated liver function and ultimate failure has many culprits including:[19]

- excessive alcohol consumption;
- aggressive forms of fatty liver disease;
- fibrosis;
- inflammatory liver conditions; and
- unregulated ingestion of common over-the-counter medication such as acetaminophen (Tylenol).

Within the UK, acetaminophen over dosage contributes to 48% of poisoning admissions to hospitals and is involved in an estimated 100–200 death per year.[20]

Medical advancement so far allows successful treatment of compromised liver function through resection and transplantation surgeries.[21,22] This is an invasive procedure and is limited by the availability of donor organs. Non-treatable and/or curable liver conditions impending transplantation are subject to supportive and palliative modes of management. These comprise non-biological and biological liver support systems. Non-biological systems primarily provide detoxification/purification by removing toxins of hepatic failure, but their utility is limited by their inability to provide missing liver functions. Bio-artificial or hepatocyte-based devices represent advancement in the management of hepatic disorders. These cellular-dependent (animal or human cells) systems are able to provide whole liver functions including detoxification, biosynthesis and biotransformation. However, it is worth noting that, in addition to the risk of xenosis in the animal cell based systems (mainly of porcine origin), additional well-conducted studies are warranted to better demonstrate safety and proof-of-concept of these devices.[19] Thus far, development of an effective liver assist technology has proven challenging due to the complexity of liver functions that must be replaced, as well as the heterogeneity of the patient population.

Alternatively, cellular therapies including hepatocyte transplantation can be used either to replace or increase the number of functional hepatocytes. The current source of hepatocytes is from discarded livers not suitable for whole organ transplantation, yet again limiting its accessibility. These cells are also used to establish primary cultures. Such primary cultures of hepatocytes have been hindered by their short lifespan and the rapid loss of hepatic function under *in vitro* conditions. There is still a great need for new sources of stem/progenitor cells with an ability to differentiate into functional liver cells. Isolation of an expandable population of adult human pluripotent stem cells will be an attractive alternative for current therapies. Before that, however, we need to understand the cellular and molecular (genetic and epigenetic) mechanisms responsible for liver cell differentiation. Therefore, adequate *in vitro* platforms and animal models of liver disease are of great importance.

5.4   Stem Cell-based Therapy—An Alternative Approach in Healthcare

Based on the Clinical Trials Database (www.clinicaltrials.gov), the present day is witness to more than 2400 stem cell-based therapies in early human trials around the world. The beginning of stem cell research dates back to 1866 when Ernst Neumann first postulated the bone marrow as a blood-forming organ with a common stem cell for all hematopoietic cells. Further studies followed to increase the understanding of early mammalian development and more recently substantial work has been done in the area of stem cell biomedical research to help improve our understanding of how stem cells influence the organism to maintain its homeostasis.

Research looking at the role of stem cell application in liver disease has, unfortunately, been subject to various confines (Table 5.1). Liver progenitor cells have been identified in embryonic and fetal livers, and have been successfully differentiated into hepatocyte or bile epithelial cells.[23–25] Such studies have been restricted by the low yield of cells and the inadequate *in vitro*

Table 5.1 Limitations of various stem/progenitor cells with potential for management of liver disease.

Embryonic/fetal liver progenitor cells	Immortalized hepatocytes Viruses/enzyme (telomerase)	Embryonic stem cells	Induced pluripotent stem cells	Adult stem cells Bone marrow/peripheral blood
a) Low yield of cells b) Inadequate *in vitro* expansion Ref. 23–26	a) Compromised by phenotypic changes and karyotype abnormality over prolonged culture durations. Ref. 27–30	a) Problems of genetic instability b) High risk if teratoma c) Ethical issues d) Difficulty in producing large quantities of homogenous cells/tissue e) Complications associated with feeder layers Ref. 31–35	a) Genetic instability b) Low efficiency of cell production Ref. 36	a) Invasive collection procedures b) Differentiation potential widely studied but proponents question the very existence of the process, claiming that cell fusion is responsible for the phenomena. Ref. 86,87

expansion after isolation.[26] To overcome this problem, attempts were made by several groups to immortalize hepatocytes using viruses[27,28] or the enzyme telomerase.[29] These approaches allowed expansion of hepatocytes, but were compromised by phenotypic changes and karyotypic abnormalities over prolonged culture durations.[30]

The introduction of embryonic stem cells (ESC) provided another potential source of human hepatocytes. Mouse embryonic stem cells (mESC) have been shown to be able to differentiate down the hepatocyte lineage, which can integrate into liver tissue and produce albumin.[31–33] Similar work has been demonstrated in human embryonic stem cells (hESC) as well.[34,35]

Although embryonic stem cells have good proliferation and differentiation properties they cannot be used in the clinic because of their genetic instability and a high risk of tumor formation. Embryonic stem cells are also limited by the ethical issues that surround the use of human embryos associated with the fact that their retrieval requires destruction of an embryo. The scientific realism surrounding embryonic stem cells is that the production of large quantities of homogenous cells/tissue for clinical application are difficult and there could be potential complications associated with the animal feeder layers on which hESC tend to rely in addition to the risk of teratomas. Hence it is unlikely that ESC-derived treatments will be available for clinical use anytime soon.

Recently developed techniques by Takahashi and Yamanaka for pluripotency induced in adult fibroblast offer an interesting alternative to embryonic stem cells.[36] These induced pluripotent stem cells can be generated without ethical concerns, but their genome instability and low efficiency of cell production raises the same concerns as for embryonic stem cells when considering their clinical use. More acceptable sources of stem/progenitor cells that exhibit pluripotency and can be used for differentiation down the hepatocyte lineage are adult stem cells (ADS) and umbilical cord and cord blood stem cells.

5.5 Pluripotent Stem Cells Derived from Cord and Cord Blood

Adult stem cells can be derived from bone marrow, adipose tissue, mobilized peripheral blood and some other sources. Their potential for differentiation towards liver has already been reviewed.[37] We focus here on umbilical cord and cord blood (UCB) derived stem cells which enjoy an intermediate niche between ESC and ADS with the added advantages of being ethically sound and associated with completely non-invasive methods for collection of the cord and cord blood units. Until the advent of cord blood banking in the mid 1990s, placenta and umbilical cord were considered clinical by-products but ever since cord blood has created hope as a new alternative in the management of over 85 different disease conditions for which there is no cure or the current treatment options are inefficient.

Term and preterm umbilical cord blood contains an equal or significantly higher number of early and committed progenitor cells compared with bone marrow and greatly surpasses that of adult peripheral blood.[38] With a global population of six billion people and a global birth rate in access of 130 million per year, umbilical cord and cord blood possibly provide the most readily accessible and ethically sound alternative source of stem cells.[39] UCB cells have longer telomeres, a higher proliferation potential than bone marrow stem cells, and increased capacity for self-renewal.[40] UCB units also have a reduced risk of viral contaminations (cytomegalovirus or Epstein Barr virus).[41–43] Hematological transplantation using UCB has further demonstrated the lower incidence of graft-*versus*-host disease (GVHD) for allogeneic graft.[44,45] UCB induces better tolerance for human leukocyte antigen (HLA) mismatches compared with bone marrow, probably due to the immaturity of the immune cells or control by UCB dendritic cells and/or natural killer cells.[46,47] They may be the only source of allogeneic hematopoietic stem cells (HSC) available to patients with rare HLA types and hence to ethnic minorities, to siblings suffering from diagnosed hematological disorders, and for urgent unrelated donor transplants.[48] Cord blood provides an easily available and rich source of hematopoietic and non-hematopoietic stem cells. Much work has been done in the area of exploring the therapeutic potential of stem cells in cord blood.

One of such exciting studies is the work done by McGuckin's group who investigated the multipotential capability of cord blood-derived embryonic-like (CBE) stem cells for differentiation into tissues originating from all three germ layers including liver tissue.[49,50] CBE cells, which are very small highly immature stem cells, also reside in other organs.[48] Freshly isolated CBE cells express the transcription markers characteristic for pluripotent ESC (Oct4, Sox2, c-Myc, Nanog) and, during differentiation, the activity of these genes decreases allowing expression of lineage-specific genes for hepatic differentiation.[39]

Another population of non-hematopoietic multipotent stem cells termed mesenchymal stem cells (MSC) has also been derived from umbilical cord[51–53] and UCB.[54–58] These cells grow adherent to plastic and express specific pattern of cell surface determinants (CD105, CD90, CD73 and CD44). This cell fraction is distinct from the hematopoietic and pluripotent stem cells present in cord blood and does not express blood cell determinants (CD34, CD45). Mesenchymal cells can be expanded very efficiently *in vitro*. These multipotent cells[59] have unique immunoregulatory features that suppress lymphocyte proliferation *in vitro*[60] and exhibit high self-renewal potential.[61] MSC have the potential to differentiate into tissues of all three germ layers including bone, cartilage, fat, muscle, endothelial cells, neuronal, glial cells and liver.[54,62–81]

5.6 Hepatic Differentiation Properties of Adult Stem Cells

Work on adult stem cells has been in the context of ageing, cell repair strategies and hematopoiesis. Previous studies have demonstrated the possibility of turning bone marrow (BM) and umbilical cord blood (UCB) stem cells into hepatic-like cells *in vivo* and *in vitro*.[34,35,39,59,80–86] The ability of bone marrow to contribute to hepatocytes was first demonstrated by Petersen *et al.*, who showed that bone marrow cells transplanted into lethally irradiated mice engrafted in the recipient's liver and differentiated into liver stem cells (called oval cells in mice) or mature hepatocytes.[12] Gordon *et al.* reported that human CD34 + cells mobilized in peripheral blood by administration of G-CSF (granulocyte-colony stimulating factor), followed by leukapheresis and re-infusion of these cells in patients with liver insufficiency, resulted in improved liver function in patients.[87] Although the exact mechanism for the effect on liver function is not clear and little is known as to which biomolecular and bio-chemical pathways regulate such differentiation, such data may reflect activation of genes corresponding to a hepatocyte differentiation program upon exposure to the injured liver environment.

Although proponents see ADS as an attractive alternative to the use of ESC in regenerative medicine, opponents have questioned the very existence of the process claiming that cell fusion is probably responsible for the phenomenon. Several critics have challenged the concept of stem cell plasticity. Issues have included the inability to reproduce data and the suggestion that some apparently reprogrammed ADS could be engrafted cells fusing with cells in their new location. This opinion is based on experiments exploring the outcome of coculturing BM with highly volatile ESC but not ADS, and noting occasional tetraploid cells from the fusion of the two cell types.[88,89]

A better understanding of the mechanisms of lineage specific differentiation and plasticity of pluripotent stem cells would provide critical clues for the use of stem cells in regenerative medicine. For this purpose the scientific community needs to develop reliable *in vitro* and animal models that will allow for better understanding of how to efficiently differentiate ADS and UCB stem cells into functional hepatic tissue. The new approach of liver tissue-engineered constructs in a three-dimensional (3D) environment *in vitro* gives a unique tool for preclinical and toxicological study of ADS and UCB for their clinical applications in a future. Hence, it may be safely concluded that cord and cord blood-derived stem cells offer multiple advantages over truly adult stem cells and over embryonic stem cells for liver differentiation.

5.7 Can Cord and Cord Blood Stem Cells Regenerate Liver Tissue?

Amongst the first evidence of stem cell differentiation to hepatocytes were reports by Peterson *et al.*, Alison *et al.* and Theise *et al.*[10,12,14] They were among the first to illustrate, in rodents and humans, that hepatobiliary cells can be derived from bone marrow. With Y-chromosome staining and liver specific markers, they detected bone marrow-derived hepatocytes in livers of irradiated mice and humans after gender-mismatched bone marrow transplantation. These interesting results, however, were not reproducible by several groups including Wagers *et al.*, Kanazawa *et al.* and Cantz *et al.*[90–92] Even when results were replicable by groups (*e.g.* Lagasse *et al.*[93]), they were noted not to be the result of direct differentiation but rather occurring due to fusion of hematopoietic stem cells with recipient hepatocytes in the animal models.[94,95]

Taking into explanation the superiority of umbilical cord blood stem cells over bone marrow derived cells;[12,59,61,71,77] several groups explored the potential of human umbilical cord blood to generate hepatocyte and biliary epithelial cells. These experiments were performed using mononucleated cells from cord blood for *in vivo* transplantation in animal models. *In vivo* studies was first performed in sheep by Almeida-Poroda *et al.*, and then in rodents by Newsome *et al.*, Wang *et al.* and Ishikawa *et al.*, by transplanting cells into sublethally irradiated NOD-SCID mice.[96–99]

Although these recent publications highlight the differentiation potential of human UCB cells and were able to produce hepatocytes, further characterization of these differentiated cells was demanded. In 2005, Sharma and colleagues for the first time characterized these human UCB-derived hepatocyte-like cells after *in vivo* experiments in mouse animal models and demonstrated the ability of these cells to express human albumin and human hepatocyte-specific antigen, Hep Par1.[100] They compared BM cells with human UCB mononucleated cells in parallel controlled studies and showed that cord blood was superior to BM in its differentiation potential. It could have been argued that, because neonatal UCB stem cells are different from BM stem cells and may have a reservoir of preformed hepatic progenitors which may not be present in the BM preparation, they demonstrated better results.[101,102] To avoid such questions, Sharma *et al.* performed the experiments using both adult and neonatal BM cells for transplantation but were unable to illustrate an increase in frequency of BM-derived hepatocytes in the model livers of neonatal BM transplantation.[100]

Until 2005 all the research done on human cord blood was performed using cord blood-derived mononucleated cells and the studies were executed in animal models. Mononucleated cells constitute the entire white blood cell/leukocyte compartment of blood, of which stem cells are a component. Although UCB demonstrates a better tolerance for HLA mismatch compared to ADS (perhaps due to the immaturity of its immune cells), the production of a purified population of stem cells from cord blood for transplantation would further reduce any risk of immune-related graft rejection. In 2005, our group reported the world's first reproducible production of cells, expressing embryonic stem cell markers, from cord blood. These cells were termed cord blood-derived embryonic-like (CBE) stem cells and were produced by exposing UCB to an immunomagnetic separation technology that allows sequential removal of nucleated granulocytes, hematopoietic myeloid/lymphoid progenitors and erythrocytes, leaving behind a purified population of very immune naive stem/progenitor cells.[39] Our work is supported by other reports concerning the existence of circulating embryonic stem cell-like cells during fetal development.[103,104] We went on to demonstrate the differentiation potential of these cells and showed that they cells were capable of generating endodermal tissues, including hepatocyte-like cells in two-dimensional and three-dimensional culture systems, expressing hepatocyte-specific markers[39] as well as pancreatic-like cells testing positive for insulin and C-peptide.[105]

To our knowledge, we were the first to demonstrate production of hepato-biliary cells in culture. This was achieved by a thorough study of classic liver biology and development and creating an artificial culture system that closely resembles the natural microenvironment necessary for normal development of the liver. Stimulation with a range of differentaitive hepatic signals resulted in successful commitment of UCB-derived undifferentiated cells along hepatic and biliary lineages. A three-step protocol encompassing: short-term expansion with mesoendoderm commitment, hepatobiliary differentiation and short-term maturation was applied to effect hepatic differentiation from UCB-derived undifferentiated primitive cells and cord-derived mesenchymal stem cells (unpublished work). We were able to successfully generate niche-like cellular colonies after culture. Similar staged protocols have been reported in human embryonic stem cell-derived hepatic cells but not in cord blood studies.[34] A few other research groups have also been able to successfully differentiate cord MSC into hepatobiliary cells.[80,106] Campard and colleagues are among the few research groups that have demonstrated differentiation of cord matrix (Wharton's jelly) MSC into hepatocyte-like cells.[107]

It is worth noting, that since the isolation of MSC from cord blood is inefficient, extraction of hematopoietic stem/progenitor cells may prove more resourceful when using cord blood, as irrespective of the size of the cord blood units, these cells can be isolated from nearly every cord blood specimen processed. The umbilical cord, on the other hand, is a brilliant resource for MSC. We were able to illustrate a 100% efficiency in isolation of MSC from umbilical cords ($n = 10$) (data not published), thus emphasizing the importance of every component of human term placenta which, not so long ago, used to be discarded postpartum. However, one of the limitations of MSC is their strict dependence on selected lots of fetal bovine serum which means limited clinical applicability for *ex vivo* expanded MSC. This drawback, however, may be overcome by resorting to low serum or serum-free culture systems as shown by Reinisch *et al.*, who for the first time demonstrated propagation of both BM and UCB MSCs in bovine serum-free systems using human platelet lysate-conditioned medium.[108] Both these aspects (generation of artificial liver tissue *in vitro* and use of defined culture media for stem cells differentiation) are crucial for establishing reliable *in vitro* platform for studying cellular and molecular mechanisms of stem cell differentiation towards liver.

5.8 Molecular Mechanisms and Stem Cell Culture Protocols for Liver Differentiation

5.8.1 Growth Factors Involved in the Induction of Liver Differentiation

During embryonic development, early liver development from embryonic endoderm requires a series of inductive signals from at least three different mesodermal cell types. Induction of hepatic genes requires fibroblast growth factor (FGF) signaling from the adjacent cardiogenic mesoderm, bone morphogenic protein (BMP) signaling from nearby septum transversum mesenchyme, and interaction with endothelial cells.[3]

When hepatic endoderm is specified and the liver bud begins to grow, the cells are referred to as hepatoblasts. These hepatoblasts are bipotent cells and eventually give rise to definitive hepatocytes and bile duct epithelial cells (cholangiocytes). Hepatoblasts have a phenotype intermediate between hepatocytes and cholangiocytes. Our differentiation protocols reflect an analogy to the mechanisms involved in normal mammalian liver development. We were able to generate stem cell colonies from UCB similar to the liver 'stem cell niche' described by Roskam[8] using exogenous differentiative cues that mimic the physiological conditions needed for liver development. It has been shown that exogenous fibroblast growth factor (FGF) and hepatocyte growth factor (HGF) could mimic the hepatic inductive effects of cardiogenic mesoderm.[34,35,59,81,106,109–111] Both acidic and basic FGF have been shown to constitute autocrine and paracrine regulation of liver development.[112] It is known that FGF is essential for the induction of endoderm in the liver.[113] Mouse embryonic stem cells grown in medium supplemented with FGF could differentiate into cells expressing hepatocyte-specific genes and antigens.[112,114,115] HGF was first identified as a blood-derived mitogen for hepatocytes. HGF and its receptor c-Met are the key factor for liver growth and function. In addition to FGF and HGF, epidermal growth factor (EGF) was applied in this study for hepatic induction in culture. EGF has been reported to increase the expression of all liver-specific genes and stimulate differentiation of fetal liver epithelial cells.[113]

We have been able to successfully demonstrate initial endoderm commitment of UCB-derived stem/progenitor cell populations using acidic-FGF, HGF and stem cell factor (SCF) in low-serum culture conditions for a week followed by differentiation using by a cocktail of acidic-FGF, HGF and EGF in low serum culture conditions which was eventually tapered to achieve a serum-free system. Using post-endoderm induction and cell differentiation we were able to generate bipotent liver progenitor cells from cord blood (unpublished data). Our findings validate results of previous work this area[39,81,106,111] and additionally illustrate novelty by exploring low serum and eventual serum-free conditions (abiding by the clinical-grade standards of tissue culture).

5.8.2 Morphogens Responsible for Liver Differentiation and Maturation

Maturation of UCB-derived hepatocyte progenitors was achieved using a combination of dexamethasone, Oncostatin M and ITS + 3. Oncostatin M is a member of the interleukin 6 family of cytokines and was originally identified by its ability to inhibit growth of A375 melanoma cells.[116] Later studies showed that primary cultures stimulated with oncostatin M showed a progression of hepatocytic development towards maturation.[117–119]

5.8.3 Stage-specific Markers of Stem Cell Differentiation Towards Liver Tissue

In vitro characterization assays were performed on the cells at the various stages of differentiation. The differentiated cells expressed various liver markers—both hepatocyte specific and biliary epithelial specific (Table 5.2), as detected by immunocytochemistry: albumin, AFP, hepatocyte growth factor receptor (HGFR), glucose-6-phosphate dehydrogenase (G-6PD), Ck8, Ck18, Ck19 and Ck7.

Differences in the expression of cytokeratins subtypes between human hepatocytes and biliary cells have been well studied.[120–122] Hepatocytes express Ck-8 and Ck-18, and intrahepatic bile duct epithelial cells also express these in addition to Ck-7 and Ck-19. Ck-7, Ck-8, Ck-18 and Ck-19 were used to characterize differentiated cells in this study. Ck-7 is expressed specifically in bile ductular epithelial cells and is often co-expressed with Ck-19.[123] Ck-19 is expressed by bipotential hepatic progenitor cells but is lost as these cells differentiate along hepatic lineage. However, expression of Ck-19 should be persistent when progenitor cells develop along biliary epithelium.[124] Ck-19 is expressed in immature as well as remodeled bile duct cells.[123] Ck-19 is also expressed on hepatic progenitor cells but disappears as development progresses. Sasaki *et al.* reported that Ck-20 is another keratin detected only in developing bile duct cells in human fetal liver and is absent in late fetal stages.[123]

Our results indicate that the differentiated cells from UCB are liver progenitor cells with bipotential capabilities, expressing both hepatocyte and cholangiocyte/biliary epithelial cell markers (Figure 5.2). These cells are similar in morphology and immunophenotype to the 'intermediate cells' described by Roskam[8] and the 'small hepatocytes' characterized by Kon *et al.*[17]

Table 5.2 Hepatic and biliary markers that aid in the identification of differentiated stem/progenitor cells.

Markers of human hepatic stem cells	Markers of human hepatoblasts	Mature hepatocytes	Cholangiocytes (biliary epithelial cells)
EpCAM	AFP (strongly	Albumin	Ck19
NCAM	positive	HGFR	Ck18
Ck19	EpCAM	GGT	Ck8
CD133/1	ICAM-1 (strongly	G-6PD	Ck7
CD44	positive	Ck8	Ck20
HCD44	CK19	Ck18	
HCLDN-3	Albumin	Cytochrome P450	
Indian and sonic hedgehog telomerase			
Albumin (weakly positive)			
Ck8			
Ck18			

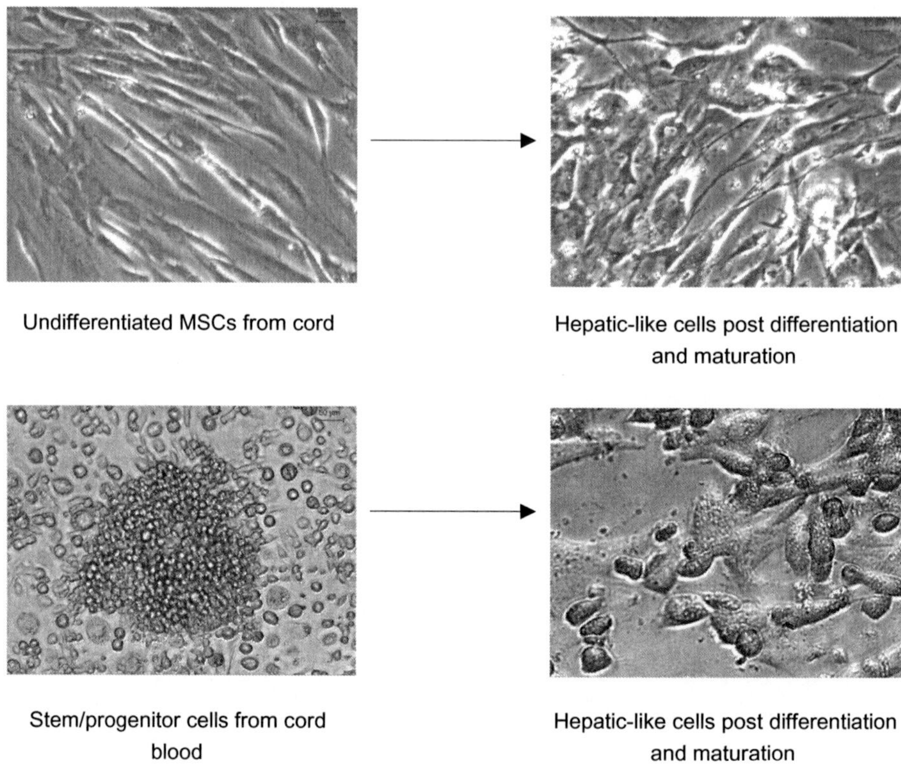

Figure 5.2 Differential interference contrast images of UCB-derived stem/progenitors cells and cord-derived MSCs before and after exposure to hepatocyte differentiative signals. Images are taken at magnifications of ×5, ×10 and ×20. Cells exhibit morphological changes consistent with their commitment along hepatic lineage.

The importance of the society of liver cells was demonstrated by the culture systems applied in our research by the fact that, once the surrounding supporting cells of the niche were lost, some of the centrally located differentiated stem cells escaped from the colony and became suspended in the medium or underwent apoptosis (Figure 5.3).

In addition to illustrating morphological and immunophenotypical characteristics of differentiated hepatic cells, a few groups including ours have been able to exhibit *in vitro* functional activity. Various assays have been used including those that determine:[109,125,126]

- glycogen storage (periodic acidic Schiff staining);
- functional multidrug resistant protein (MDR) receptors (Indocyanine green uptake and release test);
- urea production; and
- metabolic activity (Cytochrome P450 activity).

A

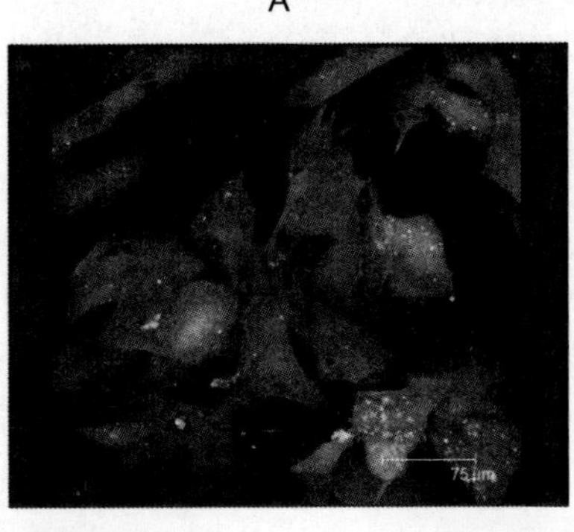

B Albumin and alpha-fetoprotein

C Ck18/Ck8/coexpressing cells

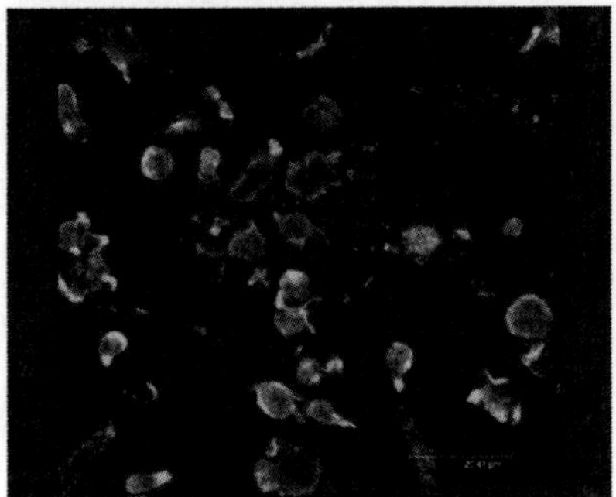

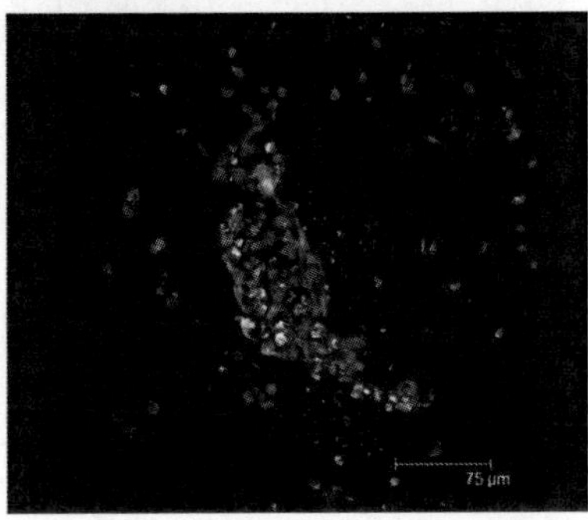

Figure 5.3   Immunocytochemistry illustrating the expression of some hepatic and biliary markers after *ex vivo* differentiation of UCB-derived stem/progenitor cells (B and C) and cord MSCs (A). Confocal images are obtained at magnification ×20.

5.9   Liver Tissue Engineering

It is worth noting that most *in vitro* hepatic tissue models available for research and development use two-dimensional (2D) culture systems. These systems fail to represent the physical cell–cell interactions of a three-dimensional human liver and do not always produce scientific data that can be fully translated to physiological interpretation. The quest for an efficient cellular technology that can be exploited for the management of end-stage liver diseases has paved the way for 3D tissue engineering. Such technology employs synthetic biodegradable porous scaffolds and rotational cell culture systems, also referred to as bioreactors, which allow cells to grow and differentiate in a 3D environment facilitating cell–cell interaction. Groups such as Baharvand *et al.* have utilized scaffolds to effect 3D hepatic differentiation of human embryonic stem cells[34] and we have demonstrated similar results from umbilical cord blood by employing microgravity bioreactors.[39]

5.10 Clinical Trials

After several reports in animal models, Theise *et al.* was the first to confirm the ability of bone marrow to generate hepatocytes and cholangiocytes in humans by evaluating archival autopsy and biopsy liver specimens obtained from gender mismatched therapeutic BM transplantations and from orthotropic liver transplantations.[14] Approximately 11 BM stem cell transplantation clinical trials for treatment of liver diseases in humans have been published since.[37] These include work by Mohamadnejad and colleagues in 2007, which carried out autologous bone marrow enriched CD34+ hematopoietic stem cell transplantation into hepatic arteries of four decompensated liver cirrhosis patients. Although a mild degree of improvement of liver function was noted in two out of four patients, one died due to radiocontrast nephropathy and hepatorenal failure. Thus they concluded that the intra-hepatic artery route is not a safe mode of stem cell transplantation, but could not preclude the use of CD34+ cells *via* other routes of administration.[127] Conversely, clinical trials performed on similar patients by Lyra *et al.*[128,129] and Gordon *et al.*[87] using mononucleated cell- enriched BM cells reported the intra-hepatic artery route of infusion as safe and feasible.

Mohamadnejad *et al.* further reported on the safety and feasibility of autologous bone marrow mesenchymal stem cell transplantation in patients with liver decompensation due to cirrhosis. This time, however, the stem cells were administered *via* a peripheral vein in four patients. No post-operative side effects were noted and all patients showed evidence of improvements of liver function and a better quality of life.[130] The efficiency of intravenous infusion was confirmed by studies performed in mice by Kou *et al.*[131] Terai *et al.* and Gordon *et al.* reported transplantation of autologous BM stem cells and human CD34+ stem/progenitor cell populations mobilized into the blood by granulocyte colony-stimulating factor, respectively, and both noted improvement of liver functions in patients post transplantation.[87,132]

It seems clear that trials of BM stem cells treatment in patients with liver disease are still at a preliminary stage and a better understanding of the physiology and mechanism of action of BM stem cell in animal models of liver disease is needed. The results of these clinical trials are, however, very exciting and open the road to explore the potential of umbilical cord and cord blood stem cells for cellular therapy of liver disease. In addition to intra-hepatic delivery of stem cells for treatment of liver disease, tissue engineering offers novel opportunities for the generation of extracorporeal liver devices. Such technology would allow temporary replacement of liver function, buying time till a suitable organ is available for transplantation.

Thus far the research into exploring the potential of the umbilical cord and cord blood for regenerative management of end-stage disease is commendable. Although the outcome of the various studies highlights extra-embryonic tissues as an indispensable reserve with immense potential for liver therapy, limitations such as lack of consensus in the immunophenotype of liver progenitor cells, uncertainty about the physiological role of reported candidate stem/progenitor cells, long-term efficacy and safety challenge the use of these cells in humans. Current molecular techniques of stem cell identification are confounded by cell fusion, horizontal gene transfer, incomplete differentiation and chimera formation. It is exciting to note that stem cell transplantation and phase 1 trials of bone marrow transplantation in humans for liver diseases[127,130] are underway but require more robust verification. More research is demanded to help identity the best source of stem/progenitor cells that can transferred from the bench to the bedside for the management of patients with severe or life-threatening liver disorders.

List of Abbreviations

ADS	adult stem cell
AFP	alpha-fetoprotein
BM	bone marrow
BMP	bone morphogenic protein
CBE	cord blood-derived embryonic like stem cells
Ck	cytokeratin
EpCAM	epithelial cell adhesion molecule
ESC	embryonic stem cells
FGF	fibroblast growth factor
GCSF	granulocyte colony stimulating factor
HBV	hepatitis B virus
HCV	hepatitis C virus
hESC	human embryonic stem cells
HGF	hepatocyte growth factor
HLA	human leukocyte antigen
HSC	hematopoietic stem cells
ICAM	inter-cellular adhesion molecule
MDR	multidrug resistant protein
MPB	mobilized peripheral blood
MSC	mesenchymal stem cells
NCAM	neural cell adhesion molecule
SCF	stem cell factor
SHPC	small hepatocyte-like progenitor cell
UCB	umbilical cord blood

References

1. D. T. Scadden, The stem-cell niche as an entity of action, *Nature*, 2006, **441**, 1075–1079.
2. V. Sandig and U. Marx, *Drug Testing In vitro: Breakthroughs and Trends in Cell Culture Technology*, Wiley-VCH, Weinheim, 2007.

3. F. Lemaigre and K. S. Zaret, Liver development update, new embryo models, cell lineage control, and morphogenesis, *Curr. Opin. Genet. Dev.*, 2004, **14**, 582–590.

4. H. C. Fiegel, J. J. Park, M. V. Lioznov, A. Martin, S. Jaeschke-Melli, P. M. Kaufmann, B. Fehse, A. R. Zander and D. Kluth, Characterization of cell types during rat liver development, *Hepatology*, 2003, **37**, 148–154.

5. D. A. Shafritz and M. D. Dabeva, Liver stem cells and model systems for liver repopulation, *J. Hepatol.*, 2002, **36**, 552–564.

6. G. J. Gordon, W. B. Coleman, D. C. Hixson and J. W. Grisham, Liver regeneration in rats with retrorsine-induced hepatocellular injury proceeds through a novel cellular response, *Am. J. Pathol.*, 2000, **156**, 607–619.

7. T. A. Roskams, N. D. Theise, C. Balabaud, G. Bhagat, P. S. Bhathal, P. Bioulac-Sage, E. M. Brunt, J. M. Crawford, H. A. Crosby, V. Desmet, M. J. Finegold, S. A. Geller, A. S. Gouw, P. Hytiroglou, A. S. Knisely, M. Kojiro, J. H. Lefkowitch, Y. Nakanuma, J. K. Olynyk, Y. N. Park, B. Portmann, R. Saxena, P. J. Scheuer, A. J. Strain, S. N. Thung, I. R. Wanless and A. B. West, Nomenclature of the finer branches of the biliary tree, canals, ductules, and ductular reactions in human livers, *Hepatology*, 2004, **39**, 1739–1745.

8. T. Roskams, Different types of liver progenitor cells and their niches, *J. Hepatol.*, 2006, **45**, 1–4.

9. J. W. Wilson and E. H. Leduc, Role of cholangioles in restoration of the liver of mouse after dietary injury, *J. Pathol. Bacteriol.*, 1958, **76**, 441–449.

10. M. R. Alison, R. Poulsom, R. Jeffery, A. P. Dhillon, A. Quaglia, J. Jacob, M. Novelli, G. Prentice, J. Williamson and N. A. Wright, Hepatocytes from non-hepatic adult stem cells, *Nature*, 2000, **406**, 257.

11. W. Kleeberger, T. Rothamel, S. Glockner, P. Flemming, U. Lehmann and H. Kreipe, High frequency of epithelial chimerism in liver transplants demonstrated by microdissection and STR-analysis, *Hepatology*, 2002, **35**, 110–116.

12. B. E. Petersen, W. C. Bowen, K. D. Patrene, W. M. Mars, A. K. Sullivan, N. Murase, S. S. Boggs, J. S. Greenberger and J. P. Goff, Bone marrow as a potential source of hepatic oval cells, *Science*, 1999, **284**, 1168–1170.

13. N. D. Theise, S. Badve, R. Saxena, O. Henegariu, S. Sell, J. M. Crawford and K. S. Krause, Derivation of hepatocytes from bone marrow cells in mice after radiation-induced myeloablation, *Hepatology*, 2000, **31**, 235–240.

14. N. D. Theise, M. Nimmakayalu, R. Gardner, P. B. Illei, G. Morgan, L. Teperman, O. Henegariu and D. S. Krause, Liver from bone marrow in humans, *Hepatology*, 2000, **32**, 11–16.

15. H. Geiger, G. Rennebeck and G. van Zant, Regulation of hematopoietic stem cell aging *in vivo* by a distinct genetic element, *Proc. Natl. Acad. Sci. U. S. A.*, 2005, **102**, 5102–5107.

16. I. M. Conboy, M. J. Conboy, A. J. Wagers, E. R. Girma, I. L. Weissman and T. A. Rando, Rejuvenation of aged progenitor cells by exposure to a young systemic environment, *Nature*, 2005, **433**, 760–764.

17. J. Kon, H. Ooe, H. Oshima, Y. Kikkawa and T. Mitaka, Expression of CD44 in rat hepatic progenitor cells, *J. Hepatol.*, 2006, **45**, 90–98.
18. B. F. Scharschmidt, J. G. Waggoner and P. D. Berk, Hepatic organic anion uptake in the rat, *J. Clin. Invest.*, 1975, **56**, 1280–1292.
19. J. Rozga, Liver support technology—an update, *Xenotransplantation*, 2006, **13**, 380–389.
20. L. C. Hawkins, J. N. Edwards and P. I. Dargan, Impact of restricting paracetamol pack sizes on paracetamol poisoning in the United Kingdom: a review of the literature, *Drug Saf.*, 2007, **30**, 465–479.
21. S. H. Belle, M. K. Porayko, J. H. Hoofnagle, J. R. Lake and R. K. Zetterman, Changes in quality of life after liver transplantation among adults. National Institute of Diabetes and Digestive and Kidney Diseases (NIDDK) Liver Transplantation Database (LTD), *Liver Transpl. Surg.*, 1997, **3**, 93–104.
22. Y. Fong, R. L. Sun, W. Jarnagin and L. H. Blumgart, An analysis of 412 cases of hepatocellular carcinoma at a Western center, *Ann. Surg.*, 1999, **229**, 790–799, discussion, 799–800.
23. H. Kubota and L. M. Reid, Clonogenic hepatoblasts, common precursors for hepatocytic and biliary lineages, are lacking classical major histocompatibility complex class I antigen, *Proc. Natl. Acad. Sci., U. S. A.*, 2000, **97**, 12132–12137.
24. C. A. Lazaro, J. A. Rhim, Y. Yamada and N. Fausto, Generation of hepatocytes from oval cell precursors in culture, *Cancer Res.*, 1998, **58**, 5514–5522.
25. L. E. Rogler, Selective bipotential differentiation of mouse embryonic hepatoblasts *in vitro*, *Am. J. Pathol.*, 1997, **150**, 591–602.
26. J. Czyz, C. Wiese, A. Rolletschek, P. Blyszczuk, M. Cross and A. M. Wobus, Potential of embryonic and adult stem cells *in vitro*, *Biol. Chem.*, 2003, **384**, 1391–1409.
27. J. E. Allain, I. Dagher, D. Mahieu-Caputo, N. Loux, M. Andreoletti, K. Westerman, P. Briand, D. Franco, P. Leboulch and A. Weber, Immortalization of a primate bipotent epithelial liver stem cell, *Proc. Natl. Acad. Sci. U. S. A.*, 2002, **99**, 3639–3644.
28. J. Cai, M. Ito, K. A. Westerman, N. Kobayashi, P. Leboulch and I. J. Fox, Construction of a non-tumorigenic rat hepatocyte cell line for transplantation, reversal of hepatocyte immortalization by site-specific excision of the SV40 T antigen, *J. Hepatol.*, 2000, **33**, 701–708.
29. H. Wege, H. T. Le, M. S. Chui, L. Liu, J. Wu, R. Giri, H. Malhi, B. S. Sappal, V. Kumaran, S. Gupta and M. A. Zern, Telomerase reconstitution immortalizes human fetal hepatocytes without disrupting their differentiation potential, *Gastroenterology*, 2003, **124**, 432–444.
30. J. P. Delgado, A. Parouchev, J. E. Allain, G. Pennarun, L. R. Gauthier, A. M. Dutrillaux, B. Dutrillaux, J. Di Santo, F. Capron, F. D. Boussin and A. Weber, Long-term controlled immortalization of a primate hepatic progenitor cell line after Simian virus 40 T-Antigen gene transfer, *Oncogene*, 2005, **24**, 541–551.

31. R. Chinzei, Y. Tanaka, K. Shimizu-Saito, Y. Hara, S. Kakinuma, M. Watanabe, K. Teramoto, S. Arii, K. Takase, C. Sato, N. Terada and H. Teraoka, Embryoid-body cells derived from a mouse embryonic stem cell line show differentiation into functional hepatocytes, *Hepatology*, 2002, **36**, 22–29.
32. H. Yamamoto, G. Quinn, A. Asari, H. Yamanokuchi, T. Teratani, M. Terada and T. Ochiya, Differentiation of embryonic stem cells into hepatocytes, biological functions and therapeutic application, *Hepatology*, 2003, **37**, 983–993.
33. T. Teratani, G. Quinn, Y. Yamamoto, T. Sato, H. Yamanokuchi, A. Asari and T. Ochiya, Long-term maintenance of liver-specific functions in cultured ES cell-derived hepatocytes with hyaluronan sponge, *Cell Transplant.*, 2005, **14**, 629–635.
34. H. Baharvand, S. M. Hashemi, S. Kazemi Ashtiani and A. Farrokhi, Differentiation of human embryonic stem cells into hepatocytes in 2D and 3D culture systems *in vitro*, *Int. J. Dev. Biol.*, 2006, **50**, 645–652.
35. D. C. Hay, D. Zhao, A. Ross, R. Mandalam, J. Lebkowski and W. Cui, Direct differentiation of human embryonic stem cells to hepatocyte-like cells exhibiting functional activities, *Cloning Stem Cells*, 2007, **9**, 51–62.
36. K. Takahashi and S. Yamanaka, Induction of pluripotent stem cells from mouse embryonic and adult fibroblast cultures by defined factors, *Cell*, 2006, **126**, 663–676.
37. D. D. Houlihan and P. N. Newsome, Critical review of clinical trials of bone marrow stem cells in liver disease, *Gastroenterology*, 2008, **135**, 438–450.
38. H. Mayani, M. Gutierrez-Rodriguez, L. Espinoza, E. Lopez-Chalini, A. Huerta-Zepeda, E. Flores, E. Sanchez-Valle, F. Luna-Bautista, I. Valencia and O. T. Ramirez, Kinetics of hematopoiesis in Dexter-type long-term cultures established from human umbilical cord blood cells, *Stem Cells*, 1998, **16**, 127–135.
39. C. P. McGuckin, N. Forraz, M. O. Baradez, S. Navran, J. Zhao, R. Urban, R. Tilton and L. Denner, Production of stem cells with embryonic characteristics from human umbilical cord blood, *Cell Prolif.*, 2005, **38**, 245–255.
40. S. G. Emerson, *Ex vivo* expansion of hematopoietic precursors, progenitors, and stem cells, the next generation of cellular therapeutics, *Blood*, 1996, **87**, 3082–3088.
41. F. Bertolini, L. Lazzari, E. Lauri, C. Corsini, C. Castelli, F. Gorini and G. Sirchia, Comparative study of different procedures for the collection and banking of umbilical cord blood, *J. Hematother.*, 1995, **4**, 29–36.
42. R. S. Chang and D. S. Seto, Perinatal infection by Epstein-Barr virus, *Lancet*, 1979, **2**, 201.
43. S. Stagno, R. F. Pass, G. Cloud, W. J. Britt, R. E. Henderson, P. D. Walton, D. A. Veren, F. Page and C. A. Alford, Primary cytomegalovirus infection in pregnancy. Incidence, transmission to fetus, and clinical outcome, *JAMA*, 1986, **256**, 1904–1908.

44. E. Gluckman and V. Rocha, Donor selection for unrelated cord blood transplants, *Curr. Opin. Immunol.*, 2006, **18**, 565–570.
45. P. Rubinstein, Why cord blood?, *Hum. Immunol.*, 2006, **67**, 398–404.
46. E. Liu, H. K. Law and Y. L. Lau, Tolerance associated with cord blood transplantation may depend on the state of host dendritic cells, *Br. J. Haematol.*, 2004, **126**, 517–526.
47. Y. Cohen and A. Nagler, Cord blood biology and transplantation, *Isr. Med. Assoc. J.*, 2004, **6**, 39–46.
48. M. Z. Ratajczak, E. K. Zuba-Surma, B. Machalinski, J. Ratajczak and M. Kucia, Very small embryonic-like (VSEL) stem cells, purification from adult organs, characterization, and biological significance, *Stem Cell Rev.*, 2008, **4**, 89–99.
49. C. P. McGuckin, N. Forraz, Q. Allouard and R. Pettengell, Umbilical cord blood stem cells can expand hematopoietic and neuroglial progenitors *in vitro*, *Exp. Cell Res.*, 2004, **295**, 350–359.
50. N. Forraz, R. Pettengell and C. P. McGuckin, Haemopoietic and neuroglial progenitors are promoted during cord blood *ex vivo* expansion, *Br. J. Haematol.*, 2002, **119**, 888.
51. Y. A. Romanov, V. A. Svintsitskaya and V. N. Smirnov, Searching for alternative sources of postnatal human mesenchymal stem cells, candidate MSC-like cells from umbilical cord, *Stem Cells*, 2003, **21**, 105–110.
52. R. Sarugaser, D. Lickorish, D. Baksh, M. M. Hosseini and J. E. Davies, Human umbilical cord perivascular (HUCPV) cells: a source of mesenchymal progenitors, *Stem Cells*, 2005, **23**, 220–229.
53. H. S. Wang, S. C. Hung, S. T. Peng, C. C. Huang, H. M. Wei, Y. J. Guo, Y. S. Fu, M. C. Lai and C. C. Chen, Mesenchymal stem cells in the Wharton's jelly of the human umbilical cord, *Stem Cells*, 2004, **22**, 1330–1337.
54. A. Erices, P. Conget and J. J. Minguell, Mesenchymal progenitor cells in human umbilical cord blood, *Br. J. Haematol.*, 2000, **109**, 235–242.
55. K. Mareschi, E. Biasin, W. Piacibello, M. Aglietta, E. Madon and F. Fagioli, Isolation of human mesenchymal stem cells, bone marrow *versus* umbilical cord blood, *Haematologica*, 2001, **86**, 1099–1100.
56. G. Kogler, S. Sensken, J. A. Airey, T. Thorten, M. Markus, N. Feldhahm, S. Liedtke, V. S. Sorg, J. Fischer, C. Rosenbaum, S. Greschat, A. Knipper, J. Bender, O. Degistirici, J. Gao, I. A. Caplan, J. E. Colletti, G. Almeida-Porada, W. H. Muller, E. Zanjani and P. Wernet, A new human somatic stem cell from placental cord blood with intrinsic pluripotent differentiation potential, *J. Exp. Med.*, 2004, **200**, 123–135.
57. K. Bieback, S. Kern, H. Kluter and H. Eichler, Critical parameters for the isolation of mesenchymal stem cells from umbilical cord blood, *Stem Cells*, 2004, **22**, 625–634.
58. V. Markov, K. Kusumi, M. G. Tadesse, D. A. William, D. M. Hall, V. Lounev, A. Carlton, J. Leonard, R. I. Cohen, E. F. Rappaport and B. Saitta, Identification of cord blood-derived mesenchymal stem/stromal

cell populations with distinct growth kinetics, differentiation potentials, and gene expression profiles, *Stem Cells Dev.*, 2007, **16**, 53–73.

59. X. Q. Kang, W. J. Zang, L. J. Bao, D. L. Li, X. L. Xu and X. J. Yu, Differentiating characterization of human umbilical cord blood-derived mesenchymal stem cells *in vitro*, *Cell Biol. Int.*, 2006, **30**, 569–575.

60. S. L. Gerson, Mesenchymal stem cells: no longer second class marrow citizens, *Nat. Med.*, 1999, **5**, 262–264.

61. A. Bartholomew, C. Sturgeon, M. Siatskas, K. Ferrer, K. McIntosh, S. Patil, W. Hardy, S. Devine, D. Ucker, R. Deans, A. Moseley and R. Hoffman, Mesenchymal stem cells suppress lymphocyte proliferation *in vitro* and prolong skin graft survival *in vivo*, *Exp. Hematol.*, 2002, **30**, 42–48.

62. C. Campagnoli, I. A. Roberts, S. Kumar, P. R. Bennett, I. Bellantuono and N. M. Fisk, Identification of mesenchymal stem/progenitor cells in human first-trimester fetal blood, liver, and bone marrow, *Blood*, 2001, **98**, 2396–2402.

63. M. F. Pittenger, A. M. Mackay, S. C. Beck, R. K. Jaiswal, R. Douglas, J. D. Mosca, M. A. Moorman, D. W. Simonetti, S. Craig and D. R. Marshak, Multilineage potential of adult human mesenchymal stem cells, *Science*, 1999, **284**, 143–147.

64. K. O'Donoghue, M. Choolani, J. Chan, J. de la Fuente, S. Kumar, C. Campagnoli, P. R. Bennett, I. A. Roberts and N. M. Fisk, Identification of fetal mesenchymal stem cells in maternal blood: implications for non-invasive prenatal diagnosis, *Mol. Hum. Reprod.*, 2003, **9**, 497–502.

65. S. P. Bruder, N. Jaiswal and S. E. Haynesworth, Growth kinetics, self-renewal, and the osteogenic potential of purified human mesenchymal stem cells during extensive subcultivation and following cryopreservation, *J. Cell. Biochem.*, 1997, **64**, 278–294.

66. D. Woodbury, E. J. Schwarz, D. J. Prockop and I. B. Black, Adult rat and human bone marrow stromal cells differentiate into neurons, *J. Neurosci. Res.*, 2000, **61**, 364–370.

67. G. C. Kopen, D. J. Prockop and D. G. Phinney, Marrow stromal cells migrate throughout forebrain and cerebellum, and they differentiate into astrocytes after injection into neonatal mouse brains, *Proc. Natl. Acad. Sci. U. S. A.*, 1999, **96**, 10711–10716.

68. E. M. Horwitz, P. L. Gordon, W. K. Koo, J. C. Marx, M. D. Neel, R. Y. McNall, L. Muul and T. Hofmann, Isolated allogeneic bone marrow-derived mesenchymal cells engraft and stimulate growth in children with osteogenesis imperfecta: implications for cell therapy of bone, *Proc. Natl. Acad. Sci. U. S. A.*, 2002, **99**, 8932–8937.

69. M. Krampera, S. Glennie, J. Dyson, D. Scott, R. Laylor, E. Simpson and F. Dazzi, Bone marrow mesenchymal stem cells inhibit the response of naive and memory antigen-specific T cells to their cognate peptide, *Blood*, 2003, **101**, 3722–3729.

70. C. Lange, H. Bruns, D. Kluth, A. R. Zander and H. C. Fiegel, Hepatocytic differentiation of mesenchymal stem cells in cocultures with fetal liver cells, *World J. Gastroenterol.*, 2006, **12**, 2394–2397.

71. H. H. Lu, G. J. Teng, S. H. Ju, J. H. Sun, A. M. Li and A. F. Zhang, Committed differentiation of transplanted bone derived mesenchymal stem cells and their potential to amend damaged liver functions: *in vivo* experiment with mice, *Zhonghua Yi Xue Za Zhi*, 2007, **87**, 223–227.

72. X. H. Lu, Y. L. Dong and P. Liu, Differentiation of bone marrow mesenchymal stem cells of rats suffering from acute liver damage, *Zhonghua Gan Zang Bing Za Zhi*, 2008, **16**, 65–66.

73. T. Tamagawa, S. Oi, I. Ishiwata, H. Ishikawa and Y. Nakamura, Differentiation of mesenchymal cells derived from human amniotic membranes into hepatocyte-like cells *in vitro*, *Hum. Cell*, 2007, **20**, 77–84.

74. V. Paunescu, E. Deak, D. Herman, I. R. Siska, G. Tanasie, C. Bunu, S. Anghel, C. A. Tatu, I. I. Oprea, R. Henschler, B. Ruster, R. Bistrian and E. Seifried, *In vitro* differentiation of human mesenchymal stem cells to epithelial lineage, *J. Cell. Mol. Med.*, 2007, **11**, 502–508.

75. S. Y. Ong, H. Dai and K. W. Leong, Hepatic differentiation potential of commercially available human mesenchymal stem cells, *Tissue Eng.*, 2006, **12**, 3477–3485.

76. S. Snykers, T. Vanhaecke, P. Papeleu, T. Henkens, M. Vinken, G. Elaut, I. van Riet and V. Rogiers, *In vitro* multipotency of human bone marrow (mesenchymal) stem cells, *ALTEX*, 2006, **23**(Suppl), 400–405.

77. B. L. Yen, C. C. Chien, Y. C. Chen, J. T. Chen, J. S. Huang, F. K. Lee and H. I. Huang, Placenta-derived multipotent cells differentiate into neuronal and glial cells *in vitro*, *Tissue Eng.*, 2008, **14**, 9–17.

78. C. C. Chien, B. L. Yen, F. K. Lee, T. H. Lai, Y. C. Chen, S. H. Chan and H. I. Huang, *In vitro* differentiation of human placenta-derived multipotent cells into hepatocyte-like cells, *Stem Cells*, 2006, **24**, 1759–1768.

79. B. L. Li, Q. Qu, Y. P. Zhao, X. D. He, L. Wang, C. Z. Chen and Z. Y. Liu, Expression of albumin during hepatocyte differentiation by human bone marrow stem cells, *Zhonghua Wai Ke Za Zhi*, 2005, **43**, 713–715.

80. S. H. Hong, E. J. Gang, J. A. Jeong, C. Ahn, S. H. Hwang, I. H. Yang, H. K. Park, H. Han and H. Kim, *In vitro* differentiation of human umbilical cord blood-derived mesenchymal stem cells into hepatocyte-like cells, *Biochem. Biophys. Res. Commun.*, 2005, **330**, 1153–1161.

81. K. D. Lee, T. K. Kuo, J. Whang-Peng, Y. F. Chung, C. T. Lin, S. H. Chou, J. R. Chen, Y. P Chen and O. K. Lee, *In vitro* hepatic differentiation of human mesenchymal stem cells, *Hepatology*, 2004, **40**, 1275–1284.

82. H. Aurich, S. Koenig, C. Schneider, J. Walldorf, P. Krause, W. E. Fleig and B. Christ, Functional characterization of serum-free cultured rat hepatocytes for downstream transplantation applications, *Cell Transplant.*, 2005, **14**, 497–506.

83. C. Di Campli, A. C. Piscaglia, L. Pierelli, S. Rutella, G. Bonanno, M. R. Alison, A. Mariotti, F. M. Vecchio, M. Nestola, G. Monego, F. Michetti, S. Mancuso, P. Pola, G. Leone, G. Gasbarrini and A. Gasbarrini, A human umbilical cord stem cell rescue therapy in a murine model of toxic liver injury, *Dig. Liver Dis.*, 2004, **36**, 603–613.

84. K. Nonome, X. K. Li, T. Takahara, Y. Kitazawa, N. Funeshima, Y. Yata, F. Xue, M. Kanayama, E. Shinno, C. Kuwae, S. Saito, A. Watanabe and T. Sugiyama, Human umbilical cord blood-derived cells differentiate into hepatocyte-like cells in the Fas-mediated liver injury model, *Am. J. Physiol. Gastrointest. Liver Physiol.*, 2005, **289**, G1091–1099.

85. X. P. Tang, M. Zhang, X. Yang, L. M. Chen and Y. Zeng, Differentiation of human umbilical cord blood stem cells into hepatocytes *in vivo* and *in vitro*, *World J. Gastroenterol.*, 2006, **12**, 4014–4019.

86. Y. Zhan, Y. Wang, L. Wei, H. Chen, X. Cong, R. Fei, Y. Gao and F. Liu, Differentiation of hematopoietic stem cells into hepatocytes in liver fibrosis in rats, *Transplant. Proc.*, 2006, **38**, 3082–3085.

87. M. Y. Gordon, N. Levicar, M. Pai, P. Bachellier, I. Dimarakis, F. Al-Allaf, H. M'Hamdi, T. Thalji, J. P. Welsh, S. B. Marley, J. Davies, F. Dazzi, F. Marelli-Berg, P. Tait, R. Playford, L. Jiao, S. Jensen, J. P. Nicholls, A. Ayav, M. Nohandani, F. Farzaneh, J. Gaken, R. Dodge, M. Alison, J. F. Apperley, R. Lechler and N. A. Habib, Characterization and clinical application of human CD34 + stem/progenitor cell populations mobilized into the blood by granulocyte colony-stimulating factor, *Stem Cells*, 2006, **24**, 1822–1830.

88. M. R. Alison, R. Poulsom, W. R. Otto, P. Vig, M. Brittan, N. C. Direkze, S. L. Preston and N. A. Wright, Plastic adult stem cells:π will they graduate from the school of hard knocks?, *J. Cell. Sci.*, 2003, **116**, 599–603.

89. S. L. Preston, M. R. Alison, S. J. Forbes, N. C. Direkze, R. Poulsom and N. A. Wright, The new stem cell biology: something for everyone, *Mol. Pathol.*, 2003, **56**, 86–96.

90. A. J. Wagers, R. I. Sherwood, J. L. Christensen and I. L. Weissman, Little evidence for developmental plasticity of adult hematopoietic stem cells, *Science*, 2002, **297**, 2256–2259.

91. Y. Kanazawa and I. M. Verma, Little evidence of bone marrow-derived hepatocytes in the replacement of injured liver, *Proc. Natl. Acad. Sci. U. S. A.*, 2003, **100**(Suppl 1), 11850–11853.

92. T. Cantz, A. D. Sharma, A. Jochheim-Richter, L. Arseniev, C. Klein, M. P. Manns and M. Ott, Reevaluation of bone marrow-derived cells as a source for hepatocyte regeneration, *Cell Transplant.*, 2004, **13**, 659–666.

93. E. Lagasse, H. Connors, M. Al-Dhalimy, M. Reitsma, M. Dohse, L. Osborne, X. Wang, M. Finegold, I. L. Weissman and M. Grompe, Purified hematopoietic stem cells can differentiate into hepatocytes *in vivo*, *Nat. Med*, 2000, **6**, 1229–1234.

94. X. Wang, H. Willenbring, Y. Akkari, Y. Torimaru, M. Foster, M. Al-Dhalimy, E. Lagasse, M. Finegold, S. Olson and M. Grompe, Cell

fusion is the principal source of bone-marrow-derived hepatocytes, *Nature*, 2003, **422**, 897–901.

95. G. Vassilopoulos, P. R. Wang and D. W. Russell, Transplanted bone marrow regenerates liver by cell fusion, *Nature*, 2003, **422**, 901–904.

96. G. Almeida-Porada, C. D. Porada, J. Chamberlain, A. Torabi and E. D. Zanjani, Formation of human hepatocytes by human hematopoietic stem cells in sheep, *Blood*, 2004, **104**, 2582–2590.

97. P. N. Newsome, I. Johannessen, S. Boyle, E. Dalakas, K. A. McAulay, K. Samuel, F. Rae, L. Forrester, M. L. Turner, P. C. Hayes, D. J. Harrison, W. A. Bickmore and J. N. Plevris, Human cord blood-derived cells can differentiate into hepatocytes in the mouse liver with no evidence of cellular fusion, *Gastroenterology*, 2003, **124**, 1891–1900.

98. X. Wang, S. Ge, G. McNamara, Q. L. Hao, G. M. Crooks and J. A. Nolta, Albumin-expressing hepatocyte-like cells develop in the livers of immune-deficient mice that received transplants of highly purified human hematopoietic stem cells, *Blood*, 2003, **101**, 4201–4208.

99. F. Ishikawa, C. J. Drake, S. Yang, P. Fleming, H. Minamiguchi, R. P. Visconti, C. V. Crosby, W. S. Argraves, M. Harada, L. L. Key Jr., A. G. Livingston, J. R. Wingard and M. Ogawa, Transplanted human cord blood cells give rise to hepatocytes in engrafted mice, *Ann. N. Y. Acad. Sci.*, 2003, **996**, 174–185.

100. A. D. Sharma, T. Cantz, R. Richter, K. Eckert, R. Henschler, L. Wilkens, A. Jochheim-Richter, L. Arseniev and M. Ott, Human cord blood stem cells generate human cytokeratin 18-negative hepatocyte-like cells in injured mouse liver, *Am. J. Pathol.*, 2005, **167**, 555–564.

101. H. Kubota, R. W. Storms and L. M. Reid, Variant forms of alpha-fetoprotein transcripts expressed in human hematopoietic progenitors. Implications for their developmental potential towards endoderm, *J. Biol. Chem.*, 2002, **277**, 27629–27635.

102. H. Mayani and P. M. Lansdorp, Biology of human umbilical cord blood-derived hematopoietic stem/progenitor cells, *Stem Cells*, 1998, **16**, 153–165.

103. M. Korbling and Z. Estrov, Adult stem cells for tissue repair—a new therapeutic concept?, *N. Engl. J. Med.*, 2003, **349**, 570–582.

104. Y. Jiang, B. Vaessen, T. Lenvik, M. Blackstad, M. Reyes and C. M. Verfaillie, Multipotent progenitor cells can be isolated from postnatal murine bone marrow, muscle, and brain, *Exp. Hematol.*, 2002, **30**, 896–904.

105. L. Denner, Y. Bodenburg, J. G. Zhao, M. Howe, J. Cappo, R. G. Tilton, J. A. Copland, N. Forraz, C. McGuckin and R. Urban, Directed engineering of umbilical cord blood stem cells to produce C-peptide and insulin, *Cell Prolif.*, 2007, **40**, 367–380.

106. X. Q. Kang, W. J. Zang, L. J. Bao, D. L. Li, T. S. Song, X. L. Xu and X. J. Yu, Fibroblast growth factor-4 and hepatocyte growth factor induce differentiation of human umbilical cord blood-derived mesenchymal stem cells into hepatocytes, *World J. Gastroenterol.*, 2005, **11**, 7461–7465.

107. D. Campard, P. A. Lysy, M. Najimi and E. M. Sokal, Native umbilical cord matrix stem cells express hepatic markers and differentiate into hepatocyte-like cells, *Gastroenterology*, 2008, **134**, 833–848.
108. A. Reinisch, C. Bartmann, E. Rohde, K. Schallmoser, V. Bjelic-Radisic, G. Lanzer, W. Linkesch and D. Strunk, Humanized system to propagate cord blood-derived multipotent mesenchymal stromal cells for clinical application, *Regen. Med.*, 2007, **2**, 371–382.
109. S. Kakinuma, Y. Tanaka, R. Chinzei, M. Watanabe, K. Shimizu-Saito, Y. Hara, K. Teramoto, S. Arii, C. Sato, K. Takase, T. Yasumizu and H. Teraoka, Human umbilical cord blood as a source of transplantable hepatic progenitor cells, *Stem Cells*, 2003, **21**, 217–227.
110. K. S. Kang, S. W. Kim, Y. H. Oh, J. W. Yu, K. Y. Kim, H. K. Park, C. H. Song and H. Han, A 37-year-old spinal cord-injured female patient, transplanted of multipotent stem cells from human UC blood, with improved sensory perception and mobility, both functionally and morphologically: a case study, *Cytotherapy*, 2005, **7**, 368–373.
111. K. Q. Kang, W. J. Zang, T. S. Song, X. L. Xu, X. J. Yu, D. L. Li, K. W. Meng, S. L. Wu and Z. Y. Zhao, Rat bone marrow mesenchymal stem cells differentiate into hepatocytes *in vitro*, *World J. Gastroenterol.*, 2005, **11**, 3479–3484.
112. M. Kan, J. S. Huang, P. E. Mansson, H. Yasumitsu, B. Carr and W. L. McKeehan, Heparin-binding growth factor type 1 (acidic fibroblast growth factor): a potential biphasic autocrine and paracrine regulator of hepatocyte regeneration, *Proc. Natl. Acad. Sci. U. S. A.*, 1989, **86**, 7432–7436.
113. T. Kawasaki, S. Tamura, S. Kiso, Y. Doi, Y. Yoshida, Y. Kamada, A. Saeki, Y. Saji and Y. Matsuzawa, Effects of growth factors on the growth and differentiation of mouse fetal liver epithelial cells in primary cultures, *J. Gastroenterol. Hepatol.*, 2005, **20**, 857–864.
114. J. Jung, M. Zheng, M. Goldfarb and K. S. Zaret, Initiation of mammalian liver development from endoderm by fibroblast growth factors, *Science*, 1999, **284**, 1998–2003.
115. M. Ruhnke, H. Ungefroren, G. Zehle, M. Bader, B. Kremer and F. Fandrich, Long-term culture and differentiation of rat embryonic stem cell-like cells into neuronal, glial, endothelial, and hepatic lineages, *Stem Cells*, 2003, **21**, 428–436.
116. J. M. Zarling, M. Shoyab, H. Marquardt, M. B. Hanson, M. N. Lioubin and G. J. Todaro, Oncostatin M: a growth regulator produced by differentiated histiocytic lymphoma cells, *Proc. Natl. Acad. Sci. U. S. A.*, 1986, **83**, 9739–9743.
117. T. Kinoshita, T. Sekiguchi, M. J. Xu, Y. Ito, A. Kamiya, K. Tsuji, T. Nakahata and A. Miyajima, Hepatic differentiation induced by oncostatin M attenuates fetal liver hematopoiesis, *Proc. Natl. Acad. Sci. U. S. A.*, 1999, **96**, 7265–72670.
118. A. Kamiya, T. Kinoshita, Y. Ito, T. Matsui, Y. Morikawa, E. Senba, K. Nakashima, T. Taga, K. Yoshida, T. Kishimoto and A. Miyajima,

Fetal liver development requires a paracrine action of oncostatin M through the gp130 signal transducer, *EMBO J.*, 1999, **18**, 2127–2136.

119. A. Miyajima, T. Kinoshita, M. Tanaka, A. Kamiya, Y. Mukouyama and T. Hara, Role of Oncostatin M in hematopoiesis and liver development, *Cytokine Growth Factor Rev.*, 2000, **11**, 177–183.

120. M. Osborn, G. van Lessen, K. Weber, G. Kloppel and M. Altmannsberger, Differential diagnosis of gastrointestinal carcinomas by using monoclonal antibodies specific for individual keratin polypeptides, *Lab. Invest.*, 1986, **55**, 497–504.

121. F. Ramaekers, A. Huysmans, G. Schaart, O. Moesker and P. Vooijs, Tissue distribution of keratin 7 as monitored by a monoclonal antibody, *Exp. Cell Res.*, 1987, **170**, 235–249.

122. P. S. R. van Eyken, B. van Damme, C. Wolf-Peeters and V. J. Desmet, Keratin-immunochemistry in normal human liver. Cytokeratin pattern of hepatocyte, bile ducts and acinar gradient, *Virchows Arch. A: Pathol. Anat. Histopathol.*, 1987, **412**, 63–72.

123. H. Sasaki, M. Nio, D. Iwami, N. Funaki, R. Ohi and H. Sasano, Cytokeratin subtypes in biliary atresia: immunohistochemical study, *Pathol. Int.*, 2001, **51**, 511–518.

124. S. Haque, Y. Haruna, K. Saito, M. A. Nalesnik, E. Atillasoy, S. N. Thung and M. A. Gerber, Identification of bipotential progenitor cells in human liver regeneration, *Lab. Invest.*, 1996, **75**, 699–705.

125. Y. J. Moon, M. W. Lee, H. H. Yoon, M. S. Yang, I. K. Jang, J. E. Lee, H. E. Kim, Y. W. Eom, J. S. Park, H. C. Kim, Y. J. Kim and K. H. Lee, Hepatic differentiation of cord blood-derived multipotent progenitor cells (MPCs) *in vitro*, *Cell Biol. Int.*, 2008, **32**, 1293–1301.

126. S. Sensken, S. Waclawczyk, A. S. Knaupp, T. Trapp, J. Enczmann, P. Wernet and G. Kogler, *In vitro* differentiation of human cord blood-derived unrestricted somatic stem cells towards an endodermal pathway, *Cytotherapy*, 2007, **9**, 362–378.

127. M. Mohamadnejad, M. Namiri, M. Bagheri, S. M. Hashemi, H. Ghanaati, N. Zare Mehrjardi, S. Kazemi Ashtiani, R. Malekzadeh and H. Baharvand, Phase 1 human trial of autologous bone marrow-hematopoietic stem cell transplantation in patients with decompensated cirrhosis, *World J. Gastroenterol.*, 2007, **13**, 3359–3363.

128. A. C. Lyra, M. B. Soares, R. R. dos Santos and L. G. Lyra, Bone marrow stem cells and liver disease, *Gut*, 2007, **56**, 1640, author reply, 1640–1641.

129. A. C. Lyra, M. B. Soares, L. F. da Silva, M. F. Fortes, A. G. Silva, A. C. Mota, S. A. Oliveira, E. L. Braga, W. A. de Carvalho, B. Genser, R. R. dos Santos and L. G. Lyra, Feasibility and safety of autologous bone marrow mononuclear cell transplantation in patients with advanced chronic liver disease, *World J. Gastroenterol.*, 2007, **13**, 1067–1073.

130. M. Mohamadnejad, K. Alimoghaddam, M. Mohyeddin-Bonab, M. Bagheri, M. Bashtar, H. Ghanaati, H. Baharvand, A. Ghavamzadeh and R. Malekzadeh, Phase 1 trial of autologous bone marrow mesenchymal

stem cell transplantation in patients with decompensated liver cirrhosis, *Arch. Iran Med.*, 2007, **10**, 459–466.
131. T. K. Kuo, S. P. Hung, C. H. Chuang, C. T. Chen, Y. R. Shih, S. C. Fang, V. W. Yang and O. K. Lee, Stem cell therapy for liver disease, parameters governing the success of using bone marrow mesenchymal stem cells, *Gastroenterology*, 2008, **134**, 2111–21121, 2121 e1-3.
132. S. Terai, T. Ishikawa, K. Omori, K. Aoyama, Y. Marumoto, Y. Urata, Y. Yokoyama, K. Uchida, T. Yamasaki, Y. Fujii, K. Okita and I. Sakaida, Improved liver function in patients with liver cirrhosis after autologous bone marrow cell infusion therapy, *Stem Cells*, 2006, **24**, 2292–2298.

Induced Pluripotent Stem Cells: Their Role in Modeling Disease and Regenerative Medicine

YONATAN STELZER AND MARJORIE PICK

Stem Cell Unit, Department of Genetics, Institute of Life Sciences, The Hebrew University, Edmund Safra Campus, Givat Ram, Jerusalem, 91904, Israel

6.1   Introduction

6.1.1   The Importance of Reprogramming and Comparing iPS to ES Cells

Just over three years ago a revolution in reprogramming of somatic cells occurred. Shinya Yamanaka showed that he could reprogram embryonic and adult mouse fibroblasts by just introducing four genes—*Oct3/4*, *Sox2*, *c-Myc* and *Klf4*—*via* retroviral infection into embryonic-like stem cells.[1] This stunned the scientific community. Since that day many laboratories have begun reprogramming human and mouse somatic cells using his original protocol and introducing improvements. Even entire research departments have been established to generate and study induced pluripotent stem (iPS) cells and, furthermore, *Science* named reprogramming cells with the 'stemness' genes as the 'breakthrough of the year' for 2008.

Stem Cell-Based Tissue Repair
Edited by Raphael Gorodetsky and Richard Schäfer
© The Royal Society of Chemistry 2011
Published by the Royal Society of Chemistry, www.rsc.org

One of the most interesting aspects of this research is that it is much easier to generate an iPS cell line from species such as monkey,[2] rat[3,4] and pig[5] compared with generating embryonic stem (ES) cells from the same species. Monkey ES cells were established a while ago but rat ES lines were generated only recently[6,7] and pig ES cells have not been established by any means. Additionally, some mouse strains such as non-obese diabetic mice, which were considered to be 'nonpermissive' to ES cell derivation, have been recently established by providing constitutive expression of Klf4 or c-Myc during the derivation of the cell lines.[8] It is still an open question how Oct3/4, Sox2, c-Myc, Klf4, Lin28 and Nanog can reprogram somatic cells to behave as stem cells and become pluripotent. Future studies combining knowledge on epigenetic processes controlling gene expression and the mechanism of reprogramming adult somatic cells should shed more light on this process.

6.1.2 Other Types of Reprogramming

In mammals, differentiation is considered to be a unidirectional pathway so that reprogramming is observed rarely. Before the advent of reprogramming using Oct3/4, Sox2, c-Myc, Klf4, Lin28 and Nanog three other types of reprogramming existed. The establishment of several methods that reset adult cells to an embryonic state demonstrated the plasticity of the nucleus of most, if not all, somatic cells.

The first and most successful, although only performed in animal cells, is somatic cell nuclear transfer (SCNT). Nuclei from somatic cells are injected into an oocyte from which the nucleus and DNA was previously removed. This process is considered to mimic a more 'natural' phenomenon by which the reprogramming is enhanced by the oocyte factors rather than through exogenous factors. Nevertheless, the generation of pluripotent stem cells by nuclear transfer is an extremely inefficient process (Figure 6.1).[9] Moreover, by generating animals using this assay most embryos die soon after implantation, and those that survive generate animals with severe abnormalities such as large birth weight, premature death and defects in imprinted genes.[10,11] However, this form of reprogramming requires only two cell divisions, which is much faster then generating iPS cells.

In order to study the plasticity of different differentiated cells, fusion between different somatic cell types and ES cells was preformed to generate a tetraploid cell that acquired a pluripotent state (Figure 6.1).[12] It has been shown that, in different hybrids, the less differentiated cell is dominant over the more committed cell, thus determining the characters of the established pluripotent cell.[13,14] At the molecular level, reactivations of silenced X chromosome, demethylation and emergence of known pluripotent markers[14] demonstrated the intrinsic character of ES cells to induce reprogramming in mature cells by fusion.[15]

Success in reprogramming cells *via* fusion raised the question as to whether the nucleus or the cytoplasm of the ES cell is required for the enhancement of

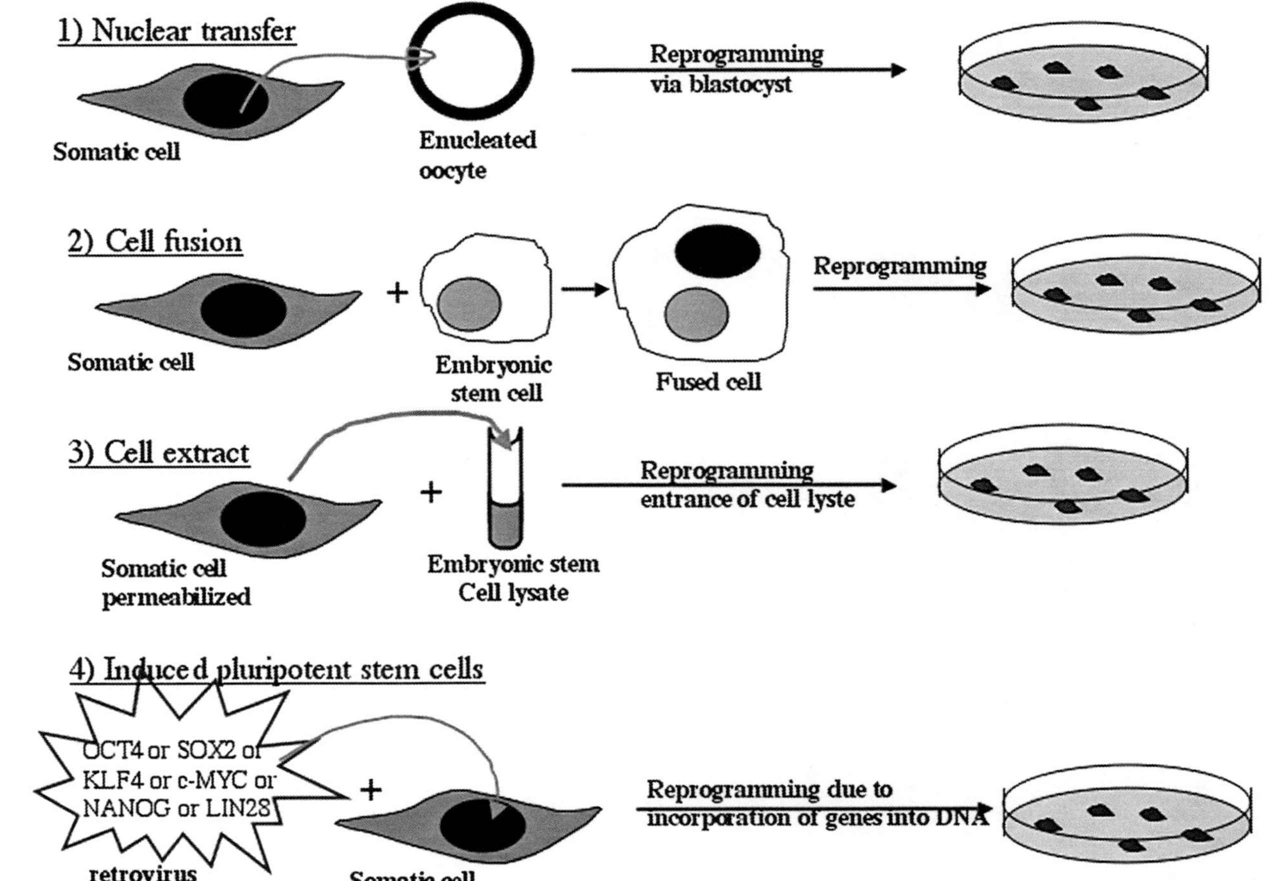

Figure 6.1 Diagramic representation of available methods for reprogramming somatic cells.

reprogramming. This question has been addressed as nuclear compartments (karyoblast) delivered to neuronal cells enhanced the reactivation of Oct4, suggesting that nuclear factors are necessary for reprogramming. In contrast, cytoplasmic compartments (cytoblasts) seemed not to affect this process.[16] Reprogramming by cell extracts offers an attractive method for studying the mechanism at the basis of reprogramming by enabling study of the role of purified proteins involved in this process. A major shortcoming of cell fusion is presented by the fact that the pluripotent cell generated is tetraploid (4N), thus limiting the possibilities for use in cell therapy and a model for diseases.

Short-term incubation of permeabilized fibroblasts exposed to an extract of mouse embryonic stem cells resulted in a biphasic activation of Oct4, with the first transient rise of Oct4 upregulation being necessary for the second, long-term activation of Oct4 and a genome-wide reprogramming. Retinoic acid triggers downregulation of Oct4, *de novo* activation of A-type lamins and of nestin. Furthermore, the cells can be induced to differentiate toward neurogenic, adipogenic, osteogenic and endothelial lineages but convincing evidence for reprogramming of the somatic cell genome (*e.g.* teratoma formation or germline transmission) is still lacking.[17]

6.1.3   First Steps in Generating Induced Pluripotent Cells

Twenty-four genes were initially selected as candidate genes to enable reprogramming of mouse fibroblasts. This list was compiled due to their high expression levels in ES cells and their involvement in their maintenance.[1] In the initial studies, four genes were found to induce the reprogramming of mouse fibroblasts: *Oct3/4*, *Sox2*, *c-Myc* and *Klf4*. Induction of reprogramming of the mouse fibroblast cells was performed using insertion of a *βgeo* cassette (a fusion of the β – galactosidase and neomycin resistance gene)[11] into the mouse *Fbx15* gene by homologous recombination; the four genes were then introduced into these cells using the retroviral vectors pMXs.[1] After transplantation in mice, the iPS cell lines produced were able to generate teratomas comprising all three embryonic germ layers. When injected into mouse blastocysts, these teratomas could contribute to embryonic development.[1] In addition, these iPS cells expressed stem cell markers similar to ES cells and the genomic DNA showed demethylation at the promoter sites of the *Oct3/4* and *Nanog* genes.[1] However, these iPS cells failed to produce chimeric mice and varied significantly in their gene expression patterns to mouse ES cells.[18] Changing the selection criteria by using a more stringent marker, the upregulation of expression of *Nanog* instead of *Fbx15*, *via* reporter gene green fluorescent protein (GFP) into the internal ribosome entry site of the *Nanog* gene improved the reprogramming of the fibroblasts and established the findings that the generated iPS cell lines were germline-competent.[18]

Back-to-back with this work was a paper by Rudolf Jaenisch's group where iPS cells were generated from different types of mouse fibroblasts.[19] They used the same four factor combination to reprogram cells, but the selection criterion for the iPS cells was the activation of endogenous Oct3/4 as well as Nanog with the use of GFP as a reporter.[19] GFP-positive clones were selected and both morphology and epigenetic status were shown to be similar to ES cells and, when injected into mouse blastocysts of mice, generated viable full-term chimeric mice.[19] A few months later, genetically unmodified mouse fibroblasts were induced into a pluripotent state using ES-like morphology to select the clones.[20] Additionally, the reprogrammed mouse iPS cells showed reactivation of a somatically silenced X chromosome; they underwent random X inactivation upon differentiation and analysis of two key histone modifications genome-wide indicated that these iPS were highly similar to ES cells.[21]

A year later human somatic cells were shown to reprogram in a similar way to mouse cells using the same four factors[22] or the combination Oct4, Sox2, Lin28 and Nanog.[23] Fetal fibroblasts, newborn foreskin fibroblasts and adult human dermal fibroblasts were readily reprogrammed and generated ES-like cells which expressed stem cell markers, had similar gene expression array patterns to ES cells, showed demethylation at the promoter site of stem cell genes and correct histone modifications. In addition, the human iPS cell lines showed a high level of telomerase activity as has been seen in ES cells. Furthermore, the human iPS cells could differentiate either *via* embryoid body formation or *via* teratomas formation into all three embryonic germ layers. The cell lines also showed a normal karyotype and downregulated transgene expression, which is critical for the iPS cells to differentiate.[22–24]

6.2 Applications for the Reprogramming of Mammalian Cells

6.2.1 Modeling of Diseases—Generating iPS Cells from Fibroblasts Obtained from Patients

Creating disease models *in vitro* using iPS cell technology should enable a better understanding of the diseases that is modeled. This could assist in the discovery of new drugs to treat the diseases and, if correction of the diseased cells could occur, might be useful for cell replacement therapy. However, choosing the correct disease to model is important and thus it might be much easier to study early onset diseases rather than those that develop later in life, since it might be difficult to simulate the pathogenesis of late onset diseases. The other hurdle that needs to be overcome if iPS cells are to be used to model diseases is knowing which cell types to generate so that the disease can be manifested *in vitro* for study.

However, even with these previously mentioned hurdles, there have been articles published on the generation of iPS from patients with various diseases. These include (Table 6.1):

- amyotrophic lateral sclerosis;[24]
- adenosine deaminase deficiency-related severe combined immunodeficiency;[25]

Table 6.1 Derivation of induced pluripotent stem cells from patients with the genetic diseases.

Disease	Molecular defect	Reprogramming	Age	Ref.
Amyotrophic lateral sclerosis	Heterozygous for L144F allele of super-oxide dismutase	OCT4, SOX2, KLF4, cMyc	82 years	16
Amyotrophic lateral sclerosis	Heterozygous for L144F allele of super-oxide dismutase	OCT4, SOX2, KLF4, cMyc	89 years	17
ADA-SCID	GGG→AGG, exon 7 and Del(GAAGA) exon 10, ADA gene	OCT4, SOX2, KLF4, cMyc	3 months	17
Gaucher disease type III	AAC→AGC, exon 9, and G-insertion, nucleotide 84 of cDNA	OCT4, SOX2, KLF4, $(+/-$ cMyc)	20 years	17
Duchenne muscular dystrophy	Deletion of exon 45–52, dystrophin gene	OCT4, SOX2, KLF4, cMyc	6 years	17
Becker muscular dystrophy	Unidentified mutation in dystrophin	OCT4, SOX2, KLF4, $(+/-$ cMyc)	38 years	17
Down syndrome	Trisomy 21	OCT4, SOX2, KLF4, $(+/-$ cMyc)	1 years	17
Down syndrome	Trisomy 21	OCT4, SOX2, KLF4, $(+/-$ cMyc)	1 month	17
Parkinson disease	Multifactorial	OCT4, SOX2, KLF4, $(+/-$ cMyc)	57 years	17
AG20443 Parkinson's disease	Idiopathic	(OCT4, SOX2, KLF4)	71 years	18
AG20442 Parkinson's disease	Idiopathic	(OCT4, SOX2, KLF4)	53 years	18

AG20442 Parkinson's disease	Idiopathic	(OCT4, SOX2, KLF4, c-MYC)	53 years	18
AG20446 Parkinson's disease	Idiopathic	(OCT4, SOX2, KLF4)	57 yrs	18
AG20445 Parkinson's disease	Idiopathic	(OCT4, SOX2, KLF4)	60 years	18
AG20445 Parkinson's disease	Idiopathic	(OCT4, SOX2, KLF4, c-MYC)	60 years	18
AG08395 Parkinson's disease	Idiopathic	(OCT4, SOX2, KLF4)	65 years	18
Juvenile diabetes mellitus	Multifactorial	OCT4, SOX2, KLF4, $(+/-$ cMyc?)	42 years	17
Shwachman–Bodian–Diamond syndrome	IV2 + 2T → C and IVS3-1G > A, SBDS gene	OCT4, SOX2, KLF4, $(+/-$ cMyc?)	4 months	17
Huntington disease	72 CAG repeats, Huntington gene	OCT4, SOX2, KLF4, $(+/-$ cMyc?)	20 years	17
Lesch–Nyhan syndrome (carrier)	Heterozygosity of HPRT1	OCT4, SOX2, KLF4, cMyc, NANOG	34 years	17
Spinal muscular dystrophy	Mutation in the survival motor neuron gene 1	OCT4, SOX2, LIN28, NANOG	3 years	19
Mother of above	Wild type	OCT4, SOX2, LIN28, NANOG	?	19

- Schwachman–Bodian–Diamond syndrome;[25]
- Gaucher disease type III;[25]
- Duchenne muscular dystrophy;[25]
- Becker muscular dystrophy;[25]
- Parkinson's disease;[25,26]
- Huntington disease;[25]
- juvenile diabetes mellitus;[25]
- Down syndrome/Trisomy 21;[25]
- Lesch–Nyhan syndrome;[25] and
- spinal muscular atrophy[27]

In the first study, fibroblasts from two elderly patients (82 and 89 years old) suffering from amyotrophic lateral sclerosis were reprogrammed. Cells from the elderly patients were used to determine if the age of the patient could affect the efficiency of cells' reprogramming.[24] The rates of efficiency were 0.04% and 0.01%, respectively, for the two patients and even lower percentages were seen once stable cell lines had been established.[24] These rates are slightly lower than those seen in recent articles with cells from younger individuals and thus it is important to prepare a high number of somatic cells for each reprogramming experiment, so that enough iPS cell lines could be generated. DNA finger-printing was performed to establish whether the iPS cells were derived from patient origin. The iPS cell lines were tested and believed to fulfill all criteria of pluripotency including the expression of stem cell markers both at the mRNA and protein level, and the ability to differentiate into various cell types *via* embryoid body formation.[24] However, although silencing of SOX2 and KLF4 transgene was observed, it did not occur for the OCT4 and c-MYC transgene.[24]

Amyotrophic lateral sclerosis is a progressive neurodegenerative disorder in which the motor neuron are lost in the spinal cord and degeneration of the motor cortex in the patient leads to progressive paralysis, which may eventually lead to death.[24] The cells that are thought to be involved in this degenerative disease are motor neurons and glia. Thus the iPS cell lines generated from the two patients were induced to differentiate towards the neuronal lineage. The iPS cell lines were able to generate similar neuronal cell types to that of differentiated human ES cells. However, no other anomalies were noticed which made it impossible to study the mechanism of disease presented in these patients using the differentiation protocol established.[24] The problem is that it is a late onset disease and thus it is difficult to mimic the disease *in vitro*. Although the disease in these patients involved a mutation in their DNA, unfortunately over 90% of amyotrophic lateral sclerosis cases are presented sporadically. Although this published article states that iPS cell lines allow for patient specific cells, unraveling the specific process of the disease will be difficult if the reasons for the onset of the disease are both variable and unknown.

The derivation of iPS cell lines from disease-specific patients could potentially complement the information previously generated from studies of mouse models and human ES cell lines generated from pre-implantation genetic disorders.[28] There is still a lack of suitable models for various diseases. This derives from the lack of adequate mouse models due to differences in the physiology of mouse and man.

It is also hard to maintain the disease cell types in culture for long time intervals to study the disease due to the failure to generate relevant human ES cell lines. For example, a direct study could be performed to compare iPS cell lines generated from existing individuals with Down syndrome and human ES cell lines generated from embryos with Trisomy 21. This is important since it is estimated that over 80% embryos with Trisomy 21 do not reach full term. If the cell lines are compared, the genetic and developmental mechanisms could be studied between two types of cell lines to determine why most embryos do not reach full term.

To date it seems quite easy to generate iPS cells from fibroblasts of patients with either Mendelian or complex inheritance diseases.[25] In a cohesive study of ten different genetic disorders, fibroblasts or bone marrow- derived mesenchymal cells were obtained from patients with a prior diagnosis of a specific disease. With the use of stem reprogramming factors Oct4, Sox2, Klf4, c-Myc and Nanog iPS cells lines were generated from these patients (Table 6.1).[25] The iPS cell lines were evaluated to confirm, where possible, disease-specific genotype (*e.g.* Trisomy 21 from patients with Down syndrome) and for the typical characteristics of reprogramming including stem cells markers, expression of pluripotency genes upon differentiation and silencing of transgene.[25] No further studies were completed on the patient-specific iPS cell lines generated in this research.

Spinal muscular atrophy is an autosomal recessive genetic disorder and one of the most common inherited forms of neuronal diseases leading to infant mortality.[27] Due to a mutation in the survival of the motor neuron 1 gene, a significant reduction in the level of this protein occurs resulting in a selective degeneration of lower motor neurons that eventually leads to muscle weakness, paralysis and death.[27] The mouse model for spinal muscular atrophy does not complete the entire spectrum of the disease since the mice lack the second form of this protein–survival motor neuron 2, which is central for the onset of the disease in man.[27] Generating iPS cell lines from patients suffering from this disease would enable a better disease model and, hopefully, the ability to screen new drugs and thus improve the treatment of the disease. For the latter to occur, the correct cell type needs to be produced that can mimic the deficient phenotype seen in the patients.

Motor neuron differentiation has been achieved using human ES cells and, due to iPS cells having similar pluripotential properties to ES cells, the differentiation protocols could easily be transferred. The iPS cell lines generated from the patients' fibroblasts retained the low transcript levels of survival motor neuron 1 observed in the patients' fibroblasts.[27] Spinal muscular atrophy iPS cells were able to generate motor neurons similar in numbers and size to wild type iPS cells at four weeks of culture.[27] However, a further two weeks of differentiation showed a significant decrease in the number choline acetyltransferase and neuron-specific class III beta-tubulin positive (ChAt[+-] Tuji[+]) and the size of motor neurons generated by the spinal muscular atrophy iPS cells.[27] Consequently, iPS cells generated from spinal muscular atrophy patient fibroblasts can generate neurons but have difficulty with multiplying.[27]

To further demonstrate the usefulness of these iPS cell lines generated from spinal muscular atrophy patient fibroblast, two drugs (valproic acid and tobramycin) were supplemented to the cultures and levels of survival motor neuron production was assessed.[27] These two compounds have been shown to increase the level of survival motor neuron protein and the addition to the culture resulted in the increase production of survival motor neuron by Western blot analysis.[27] It is crucial to note, however, the practicality of using valproic acid and tobramycin in a clinical setting. Valproic acid is a chemical compound that has found clinical use as an anticonvulsant and mood-stabilizing drug; it is also known to change the efficiency of iPS cells by inhibiting histone deacetylase by inhibiting the acetylation of lysine. Tobramycin has been used as an antibiotic to treat various types of bacterial infections, particularly Gram-negative infections by binding to a site on the bacterial 30S and 50S ribosome, preventing formation of the 70S complex. However, it is hard to estimate what effect these drugs will have on the overall differentiation of the iPS cells.

The ability to reprogram somatic cells without the need for virus encoding the reprogramming factors to integrate—thus potentially altering the differentiation patterning of the reprogrammed cells or causing the cells to transform—would be very useful. It has been observed that, when transplanted into mouse, iPS cells showed a high incidence of tumor development.[26] Using doxycycline-inducible lentiviral vectors that can be excised using Cre-recombinase after reprogramming can generate iPS cell lines free of viral vectors.[26]

All iPS cell lines generated from seven different Parkinson's patients uniformly expressed pluripotency markers tested both by immunohistochemistry and reverse transcriptase polymerase chain reaction (RT-PCR) analysis, demethylation of Oct4 promoter and produced teratomas containing all three embryonic germ layers.[26] Using Southern blot analysis, each iPS cell line contained between three and ten virus integration sites and, after Cre-recombinase treatment, most integrated vectors were excised.[26] iPS cell lines that were treated retained their ES cell characteristics and genome-wide transcription analysis showed that the iPS cell lines, which were vector-free, more closely resembled human ES cells compared with the same iPS cells lines previous to the vector's excision.[26] Dopaminergic neurons were generated from patient-specific iPS and human ES cells with no difference in the percentage production of either TUJI or tyrosine hydroxlase labeled neurons.[26] Although no significant differences were found in the dopaminergic neurons generated, these cells can still be used to study the molecular and cellular mechanisms of sporadic Parkinson's disease even though recapitulation of disease was not established. This could be due to the fact that Parkinson's disease is a late onset disease rather then a developmental one.

6.2.2 Regenerative Medicine Potential—Phenotype Correction of Genetic Diseased Cells

Treatment of patients with genetic diseases is a major goal of regenerative medicine. The recent advent of reprogramming somatic cells to an embryonic-stem cell state using various stemness factors has allowed this to become a possibility to the degree that patient-specific major histocompatible (MHC) iPS cell lines can be generated. In addition, the issue of tissue rejection is overcome due to the transplanted cells being of host origin—an issue still present in the human ES cells scenario.

There are two major strategies that would therapeutically be helpful. One is generate iPS cells from the somatic cells from the patients, repair the genetic disorder, differentiate the cell type that was effected by the genetic disorder and transplant. The second approach is to generate iPS cells from MHC matched donor, then differentiate them towards the cell type effected in the recipient and transplant them to replace the cells with the defect. Both strategies are close to being accomplished (Table 6.2).[29–33]

6.2.2.1 Sickle Cell Anemia

Sickle cell anemia and thalassemia are genetic diseases that affect red blood cells and are caused by a point mutation in the hemoglobins. In the case of sickle cell anemia, a mutation in the β-globin chain of hemoglobin causes distortion in the shape of red blood cells, turning them sickle shaped. Sickle cell anemia is a qualitative problem of synthesis of incorrectly functioning globin and only the homozygous mutation of the globin gene causes the disease. Mouse iPS cells were generated from humanized knock-in mouse model of sickle cell anemia in which the mouse α-globin genes were replaced with a human version and the mouse β-globin genes with human $A\gamma \beta^s$ (sickle) globin genes.[29] Homozygous mice for the human β^s allele remain viable for up to 18 months, but develop typical disease such as severe anemia due to erythrocyte sickling and have overall poor health.[29] Retroviral c-Myc that was integrated into various sites of the mouse genome was excised using Cre-recombinase in the hope that, if iPS cells are to be transplanted, fewer tumors will be formed.[29]

To correct the sickle cell anemia mutation in the mouse iPS cells, homologous recombination was performed using DNA encoded with the non-mutated human β^A wild type globin gene and iPS cell lines were screened for the presence of the corrected gene. Hematopoietic progenitors ($CD41^+c\text{-}Kit^+$) were generated from the mouse iPS cell lines containing the corrected β – globin gene and transplanted into $h\beta^S/h\beta^S$ sickle cell mice until engraftment of the donor cells was established.[29] Morphological examination of the red blood cells from blood smears of sickle cell mice transplanted with genetically corrected hematopoietic progenitors differentiated from iPS cells showed significant improvement with less sickle-shaped cells and abnormal decreased levels of reticulocytes.[29] Furthermore, the blood cells counts showed increased numbers of red blood cells and normalized values of other parameters relating

Table 6.2   Disease treated using induced pluripotent stem cells.

Strategy	Disease	Source of iPS	Process to repair genetic disease	Cell type	Species	Ref.
1	Sickle cell anemia	Autologous	Homologous recombination	Hematopoietic	Mouse	21
1	Fanconi anemia	Autologous	Lenti virus encoding correct gene	Hematopoietic	Human	22
1	Thalassemia	Autologous	Lenti virus encoding correct gene	Hematopoietic	Human	25
2	Hemophilia A	Allogeneic	Normal iPS cells transplanted	Endothelial	Mouse	24
2	Parkinson disease	Allogeneic	Normal iPS cells transplanted	Neuron	Mouse	23

to red blood cells.[29] These parameters were measured up to 12 weeks post-transplant; however, it would be interesting to know what happened to these recipient mice long term and if the hematopoietic progenitors (CD41$^+$c-Kit$^+$) contained the long-term initiating cells critical for sustained hematopoietic engraftment. This issue is a critical one since transplanting the correct cell type will determine how successful the long-term treatment will be. This issue remains a problem for future treatments using both human ES and iPS cell lines.

6.2.2.2 Fanconi Anemia

There are various inherited diseases that affect the hematopoietic system of which Fanconi anemia is the most common. This is a recessive autosomal or X-linked genetic disorder caused by mutations in any of the genes identified in the Fanconi anemia pathway.[31] The mutations cause chromosomal instability, resulting in cellular sensitivity to DNA damaging agents and some patients show an increased predisposition to developing malignancies.[31] The mutations cause short stature, skeletal anomalies and dysplasia in the bone marrow of patients, and the treatment of choice is allogeneic bone marrow transplantation to correct the abnormal bone marrow.

Samples of fibroblasts from six Fanconi anemia patients were obtained, but iPS cell lines were generated only from three patients since in this disorder iPS cell lines were much more difficult and/or impossible to generate probably due too many chromosomal abnormalities in these patients' cells.[31] Thus genetic modification is important even in genetic disorders that present a mosaic of chromosomal changes. The iPS cell lines that could be generated fulfilled the criteria of pluripotency by expression of stem cells markers such as Oct4 (both at RNA and protein levels), silencing of the transgene, demethylation of Oct4 and Nanog promoters, and teratoma formation containing the three embryonic germ layers.[31]

The iPS cell lines generated from corrected Fanconi anemia patient fibroblasts were tested for the expression of the *FANCA* or *D2* gene (the Fanconi anemia complementation group genes A and D2) *via* Western blot analysis. It has been observed in the past that lentiviral transgenes are silenced, but these corrected iPS cell lines still expressed the protein and the Fanconi anemia pathway function remained intact.[31] Hematopoietic cells (CD34$^+$CD45$^+$) were generated in equal numbers from human ES, cord blood and bone marrow cells of healthy individuals compared with the iPS cell lines generated from Fanconi anemia patient cells.[31] Furthermore, when mitomycin C (a compound to which Fanconi anemia cells are sensitive) was added to the culture during differentiation, no effect on the output of hematopoietic cells generated by the Fanconi anemia iPS cells was observed.[31] Unfortunately, engraftment of the hematopoietic cells from these iPS cell lines in non-obese diabetic/SCID mice failed and thus it was impossible to assess the effectiveness of the cells generated. This could be due to the general technical difficulty that was also recorded in the attempts to transplant hematopoietic cells generated from human ES cells successfully.

6.2.2.3 *Thalassemia*

Thalassemia is a complex genetic autosomal recessive disorder in which mutations occur in both the alpha and beta chains of the globin gene, causing too few globins to be synthesized and anemia in patients; the patients require life-long blood transfusions.[32] A cure for this disease may be bone marrow transplantation, but finding matched donors for every patient remains impossible. Numerous iPS cell lines were generated from the fibroblast of an affected β-thalassemic patient and these lines expressed stem cells markers (*e.g.* Tra-1-60), expressed protein levels of Nanog similar to human ES cells and formed teratomas containing differentiated cells from the mesoderm, endoderm and ectoderm lineages.[32] Genomic DNA from the iPS cell lines was sequenced and each line was correctly derived from the fibroblasts of the β-thalassemia patient.[32] The cells were then differentiated to generate hematopoietic cells including hemoglobinized red blood cells carrying the fetal version of globin.[32] Potentially these fetal hemoglobin carrying red blood cells could be transfused into the patient and function just like adult hemoglobinized red blood cells. As yet generating red blood cells with adult hemoglobin has not been possible.

6.2.2.4 *Hemophilia A*

Hemophilia A is inherited as an X-linked recessive trait caused by a mutation within the factor VIII gene which leads to depleted production of this protein, resulting in problems with blood clotting and bleeding.[30] Various attempts at treating patients using gene therapy have failed, mainly due to immune rejection of the virus encoding the correct gene.[30] Thus the main form of treatment is transfusions of recombinant factor VIII protein, which is both expensive and inefficient. Since only a small amount of protein production is needed to alleviate the effects of the lack of circulating factor VIII, only a small number of factor VIII producing cells may need to be transplanted to obtain the anticipated effect. IPS cell lines were generated from mice and then differentiated to endothelial cells as these are the cells thought to secrete factor VIII.[30] Both the iPS cells generated from tail tip fibroblasts and the differentiated endothelial cells were tested to fulfill the criteria of pluripotency and characterization of endothelial cells, respectively.[30] Factor VIII producing endothelial cells were then injected into the livers of sublethally irradiated MHC compatible mice (model of hemophilia A with a mutation in the factor VIII gene) and, one week after transplantation, recipient hemophilia A mice were tested for the presence of transplanted cells. Blood clotting was seen, plasma levels of factor VIII were 8–12% of levels seen in controls and the donor cells were detected in these mice up to three months post-transplant.[30] This strategy did not need to use iPS cells but would have worked just as well with the use of MHC compatible mouse ES cells or even endothelial cells generated from the bone marrow of MHC compatible mice.

Generating neural cells from adult tissue is virtually impossible. Thus the advent of ES cell lines that could potentially differentiate into neurons due to their pluripotency was a scientific breakthrough and led to a flood of potential protocols to produce neurons. The success of this research as led to the US Food and Drug Administration (FDA) recently approving the first clinical trial using these cells. The only pitfall is that neurons generated from ES cells need to be immunogenically compatible to the recipient and thus the source of ES cell lines available may not encompass every individual. The ability to reprogram somatic cells using the 'stemness' markers has enabled a further leap by enabling patient/recipient specific iPS to be generated for potentially any individual and thus a source of immunogenically compatible neurons.

6.2.2.5 Parkinson's Disease

Parkinson's disease is a degenerative disease of the brain that impairs motor skills and speech. It occurs due to a decreased stimulation of the motor cortex by the basal ganglia as a result of insufficient formation and activity of dopamine, which is produced in the dopaminergic neurons of the brain. Since this disease involves a specific neural type cell replacement, therapy with implanted dopaminergic neurons would be an ideal treatment regime.

Nanog-selected (GFP) mouse iPS cell lines[19] were differentiated to produce clinically relevant dopamine-producing neurons of the midbrain and accessed according to markers characterizing this cell type.[33] These neuronal precursors were then transplanted into the striate of the brain of a rat model of Parkinson's disease.[33] Four weeks after surgery, engraftments of the transplanted cells were assessed and found to be numerous, showing complex morphology. They were positive for various of dopamine neurons and rats showed substantial improvement in their behavior.[33] Histological investigation of the grafted area showed many non-neural cell types, suggesting teratoma formation from the contaminating undifferentiated iPS cells.[33] Cell sorting was performed to decrease the presence of undifferentiated iPS cells; however, this is an impractical option for human studies and thus the difficulty with this treatment, as in the ES model, is the presence of undifferentiated stem cells.

6.3 Reducing the Risks Associated with the Use of iPS Cells

6.3.1 Increasing Safety in the Generation of iPS Cells

Induced pluripotent stem cells hold a promising potential for future use in regenerative medicine and other therapeutic applications, as it will be possible to generate patient-specific autologous cell types for transplantation.[34]

They have also been demonstrated to be a powerful tool in the study of diseases and the development of pathological condition.[29–33]

Nonetheless, there are still some issues that need to be addressed if iPS cells are to be used in future clinical applications.

- Generation of iPS cells without the integration of viral DNA into the host genome, reprogramming the cells in the absence of the oncogenes such as *c-Myc* and *Klf4*.
- Evaluation of the similarities and differences of iPS cell lines compared with their ES cell counterparts.
- Further study is required to better understand the epigenetic modifications and general mechanism taking place during nuclear reprogramming.
- Even a few undifferentiated cells transplanted can generate teratoma formation. Thus evaluation of iPS cells tumorigenicity must be completed and compared with current knowledge on ES cells.

6.3.2 Decreasing the Number of Factors Involved in Reprogramming

The promise of somatic reprogramming was first introduced by Yamanaka's group, which performed a retroviral transduction of four transcription factors—Oct4, Sox2, Klf4 and c-Myc—into the genome of a mouse fibroblast, transforming it to a pluripotent stem cell-like state.[1]

However, integration of the DNA into the genome of the host by retroviral infection causes several obstacles that need to be accounted for. The majority of iPS cell lines reported have been generated using retroviruses since it was thought that expression of the stemness genes and reprogramming could only occur with integration of the DNA. Since this process involves integration into the host cell genome, it raises the possibility of multiple and random integration sites some of which might be in coding regions of the DNA.

In addition and despite the fact that transgenes are largely silenced during reprogramming, there is a chance that spontaneous reactivation of the viral transgene could occur—especially mediated by c-Myc oncogene—and form tumors.[18] c-Myc is a pleiotropic factor involved in many cellular pathways, including control over cell cycle.[35] Although the role of c-Myc during reprogramming is mostly unknown, it seems to increase the overall efficiency, apparently by increasing the pace of stochastic events, which leads to rapid proliferation.[20,35] Nevertheless, as it was recently shown in mouse fibroblast, direct reprogramming is possible even in the absence of this factor[36,37] and thus it seems that excluding c-Myc from direct reprogramming is an efficient way of increasing the safety of the procedure.

Still even 'leaky' transcription of one of the transgenes within iPS cells and any of their progeny may affect the cells' molecular signature, their differentiation and developmental potential. Consequently, several published works address this safety concern by altering the number of factors involved in the reprogramming procedure. These especially include oncogenes such Klf4 and c-Myc. New protocols involving the induction of pluripotency of human primary fibroblasts using only Oct4 and Sox2 with addition of small molecules,[38] together with reports of successive reprogramming in mouse neural progenitor either by excluding Sox2[39] or by exogenous expression of only Oct4 have been published. The most notable protocol solely using Oct4 to reprogram somatic cells was shown to be possible in neural progenitor cells due to an endogenous expression of Sox2, c-Myc and Klf4.[40] Other systems that provided a powerful tool for studying pluripotent stem cells markers using Dox-inducible transgenes were introduced by the groups of Hochedlinger and Jaenisch.[21,41] These studies provide the possibility of:

- achieving a better understanding of the role of each of the factors involved in reprogramming; and
- developing a powerful tool with which to study the overall mechanism of reprogramming.

6.3.3 Integration-free iPS Cell Lines

The overall inefficiency of the reprogramming process is due to the small fraction of cells expressing the correct dosage of each of the pro-viral insertions. Recently, it was demonstrated that reprogramming efficiency is considerably improved by adding small molecule compounds such as valproic acid (histone deacetylase inhibitor) or 5′-azacytidine (which plays a role in global DNA demethylation).[38] A novel system allowing simultaneous insertion of four factors using a polycistronic vector showed a lower reprogramming efficiency, arguing that the stoichiometry of the factors might be suboptimal to induce reprogramming.[42] However, even though polycistronic vector system is clearly reducing the number of integrating sites, it still does not address the safety matter.

Adenoviruses can be considered a good candidate for improving safety of iPS cell lines since the virus can be efficiently transfected into the host without the need of viral integration. It has been reported by several groups that it is feasible to induce reprogramming and pluripotency of somatic cells without the integration of the transgenes into the host genome. One group introduced the genes into mouse iPS cells using adenoviral vectors that allow transient, high level expression of the genes without integrating the DNA into the host genome.[43] It was shown that transgene expression could be maintained for up to 12 days post-infection in order to commence reprogramming in the somatic cell, and thereafter allow the cells to self-sustain pluripotency.[43,44] Notably, no tumor formation was observed in chimera mice and their progeny generated from these iPS cell lines, suggesting a safer method. In order to overcome the transient expression of adenovirus vectors, the cells were continuously re-infected with the adenoviruses carrying the factors up to the point at which self-sustaining pluripotency and self-renewal was achieved (reprogramming).

Although this demonstrates the feasibility of generating iPS cells without integration of the viral proteins with the 'stemness' genes, the drawback is the extremely low efficiency of the procedure (0.0001–0.001%) comparing with the frequency obtained with integrating viruses (0.01–0.1%).

A further attempt was made to reprogram both human and mouse somatic cells with a non-integrating adenoviruses plasmid containing the cDNA of all four reprogramming factor—Oct4, Sox2, Klf4 and c-Myc.[45] due to the transient expression of adenoviruses, a continuous delivery of the plasmid was essential for the cells to fully reprogram and become pluripotent.[45] Following the selection of colonies similar in morphology and pluripotent markers to ES cells, chimeric mice and teratomas were established.[45] No integration of the plasmids into the genome was observed but again the yield/efficiency was extremely low.[45]

Recently two different groups have demonstrated elegant methods to induce pluripotency of somatic cells—both in mouse and human—again without viral integration[26,46] using the 'PiggyBac' transposon vector to introduce the four reprogramming factors Oct3, Sox2, Klf4 and c-Myc.[46] The PiggyBac transposon system requires inverted terminal repeats flanking a transgene and transient expression of another plasmid containing the transposase enzyme, which transfected with the PiggyBac transposon containing reprogramming factors, catalyze insertion or excision events.[46] By taking advantage of the natural propensity of the PiggyBac system, the inserted DNA can then be removed from established iPS cell lines. Moreover, the transposon plasmid containing the four factors under the control of Dox-inducible promoter allows the expression of the genes only once doxorubicin is supplemented to the culture,[46] which allows for an extra degree of safety to the procedure. This method holds further advantages as it eliminates the use of retroviruses, thus increasing the range of somatic cell types to those that are not susceptible to viral infection. Reprogramming of somatic cells can thus be performed using non-integrating plasmids and has shown to be as effective as viral integrating plasmids. Consequently this new plasmid holds a great potential for the future of regenerative medicine.

The group from Jaenisch's laboratory have demonstrated another method for generating integration-free human iPS cells, this time using Cre-recombinase excisable viruses.[26] In this method, a doxycycline-inducible promoter followed the four factors, replaced the 3''LTR containing the loxP site and was delivered by a lentiviral vector to the cells. Following the generation of human iPS cells, the iPS cells were transiently transfected with an expression vector containing Cre-recombinase which excised the fragment from the host cell.[26]

Interestingly, this study also compared the similarities between three populations:

- human ES cells;
- human iPS cell before introduction to Cre-recombinase; and
- the human iPS cell derivatives which were integrated-free.[26]

Using genome-wide gene expression analysis, it was surprisingly shown that integrated-free human iPS cells were more similar in their gene expression to human ES cells, as compared with their parental human iPS cell containing viral integration.[26] In addition to the safety concern, iPS cell lines that contain integrated vectors raise the possibility that there are somewhat intrinsic genomic dissimilarities between these iPS cells and ES cells. These differences might result from somatic memory, global demethylation or stochastic events during reprogramming that could also affect their differentiation potential.[26]

Several groups have studied the use of reprogramming somatic cells using small molecules and chemicals in an alternative method to avoid viral integration.[47,48] It was thought that, although exogenous production of Oct4 has been shown to be a crucial factor essential for reprogramming, that reprogramming might be possible by activating the endogenous *Oct4* gene present in the somatic cells by chemically enhancing its transcription.[47,48] An similar concept was used to deliver the reprogramming factors as proteins in an episomal vector[49,50] to enhance reprogramming without the need of viral integration to the host cell genome. These recent studies illustrate the fact that generated iPS cell lines that were viral free were closely related in genomic imprinting to ES cells and a possibility for use in regenerative medicine in the near future. The ability to generate iPS cell lines without viral integration is a major step in achieving cell lines applicable for future use in regenerative medicine and cell therapy.

6.4 Conclusions

One of the crucial issues facing iPS cell technology is to determine the criteria needed to affirm that the iPS cell line is truly reprogrammed. Each clone must be evaluated for various standard criteria, but the question is what should these criteria be? It may be necessary to sequence the entire genome of the iPS cell line to make sure the genome is correct.[34] This issue has been discussed recently by the forerunners of iPS cell research.[51–53] The criteria that need to be assessed include:

1. The morphology of the colonies picked must have ES-like properties.
2. The cell lines should be expandable for at least 12 passages.
3. The cells must display a gene expression profile indistinguishable from ES cell lines.
4. The retrovirus genes should be silenced.
5. The cells should be epigenetically similar to ES cell lines (*e.g.* DNA methylation of stem cell genes).
6. The cells must be able to differentiate into three embryonic germ line lineages either by teratoma formation or differentiation *via* embryoid body formation.
7. The cells should express key pluripotent makers such as Oct4 and Nanog at the protein level.
8. Mouse iPS cell lines should specifically demonstrate germline transmission.

But as there are many variables to consider that could influence the resulting iPS cell line, each somatic cell type might need a different combination of factors.[34] The method of delivery with non-integrating vectors has resulted in cell lines more similar in genome expression to human ES cell lines than iPS cell lines generated for vectors that integrate into the host genome.[26] The main issue that needs to be overcome for iPS cells to be used in a clinical setting is removal of the retroviral vector integration into the host genome.

Two critical issues that have been observed with human ES cells can potentially be overcome with the advent of iPS cells associated with immune rejection since donor-specific iPS cells could potentially be generated and the ethical issue concerning the use of human embryos would be removed. However, there are also some similar issues to overcome both with iPS cells and human ES cells which include:

1. How to differentiate these cells to a specific lineage to make them useful for regenerative medicine.
2. As ES cells, undifferentiated iPS cells might still have the ability to form teratomas if transplanted.
3. More importantly and exclusive to iPS cells, the abnormal expression of genes in the iPS cell generated lines.

Although it has been observed that the genes (introduced to the cells *via* retroviral vectors) are silenced early on during the reprogramming event, they could be reactivated in the cells over time.

Furthermore, only recently has the first clinical trial using differentiated human ES cells to treat patients with spinal cord injury been approved by the US FDA. To use iPS cells for clinical purpose could be far off, especially with the methods available to date to generate these cells, *i.e.* using retroviral and lentiviral vectors to introduce the genes into the cells. Much work is being done to remove the problem either by introducing the genes using plasmids, or replacing or removing one or more of the four genes.

References

1. K. Takahashi and S. Yamanaka, Induction of pluripotent stem cells from mouse embryonic and adult fibroblast cultures by defined factors, *Cell*, 2006, **126**, 663–76.
2. H. Liu, F. Zhu, J. Yong, P. Zhang, P. Hou, H. Li, W. Jiang, J. Cai, M. Liu, K. Cui, X. Qu, T. Xiang, D. Lu, X. Chi, G. Gao, W. Ji, M. Ding and H. Deng, Generation of induced pluripotent stem cells from adult rhesus monkey fibroblasts., *Cell Stem Cell*, 2008, **3**, 587–90.
3. W. Li, W. Wei, S. Zhu, J. Zhu, Y. Shi, T. Lin, E. Hao, A. Hayek, H. Deng and S. Ding, Generation of rat and human induced pluripotent stem cells by combining genetic reprogramming and chemical inhibitors., *Cell Stem Cell*, 2009, **4**, 16–9.

4. J. Liao, C. Cui, S. Chen, J. Ren, J. Chen, Y. Gao, H. Li, N. Jia, L. Cheng, H. Xiao and L. Xiao, Generation of induced pluripotent stem cell lines from adult rat cells., *Cell Stem Cell*, 2009, **4**, 11–5.
5. M. A. Esteban, J. Xu, J. Yang, M. Peng, D. Qin, W. Li, Z. Jiang, J. Chen, K. Deng, M. Zhong, J. Cai, L. Lai and D. Pei, Generation of induced pluripotent stem cell lines from tibetan miniature pig., *J Biol Chem*, 2009.
6. S. Ueda, M. Kawamata, T. Teratani, T. Shimizu, Y. Tamai, H. Ogawa, K. Hayashi, H. Tsuda and T. Ochiya, Establishment of rat embryonic stem cells and making of chimera rats., *PLoS ONE*, 2008, **3**, e2800.
7. H. Suemori, T. Tada, R. Torii, Y. Hosoi, K. Kobayashi, H. Imahie, Y. Kondo, A. Iritani and N. Nakatsuji, Establishment of embryonic stem cell lines from cynomolgus monkey blastocysts produced by IVF or ICSI., *Dev Dyn*, 2001, **222**, 273–9.
8. J. Hanna, S. Markoulaki, M. Mitalipova, A. W. Cheng, J. P. Cassady, J. Staerk, B. W. Carey, C. J. Lengner, R. Foreman, J. Love, Q. Gao, J. Kim and R. Jaenisch, Metastable Pluripotent States in NOD-Mouse-Derived ESCs., *Cell Stem Cell*, 2009.
9. R. Jaenisch and R. Young, Stem cells, the molecular circuitry of pluripotency and nuclear reprogramming., *Cell*, 2008, **132**, 567–82.
10. K. L. Tamashiro, T. Wakayama, H. Akutsu, Y. Yamazaki, J. L. Lachey, M. D. Wortman, R. J. Seeley, D. A. D'Alessio, S. C. Woods, R. Yanagimachi and R. R. Sakai, Cloned mice have an obese phenotype not transmitted to their offspring., *Nat Med*, 2002, **8**, 262–7.
11. N. Ogonuki, K. Inoue, Y. Yamamoto, Y. Noguchi, K. Tanemura, O. Suzuki, H. Nakayama, K. Doi, Y. Ohtomo, M. Satoh, A. Nishida and A. Ogura, Early death of mice cloned from somatic cells., *Nat Genet*, 2002, **30**, 253–4.
12. H. M. Blau and B. T. Blakely, Plasticity of cell fate: insights from heterokaryons., *Semin Cell Dev Biol*, 1999, **10**, 267–72.
13. T. Tada and M. Tada, Toti-/pluripotential stem cells and epigenetic modifications., *Cell Struct Funct*, 2001, **26**, 149–60.
14. J. T. Do, D. W. Han, L. Gentile, I. Sobek-Klocke, M. Stehling, H. T. Lee and H. R. Scholer, Erasure of cellular memory by fusion with pluripotent cells., *Stem Cells*, 2007, **25**, 1013–20.
15. K. Hochedlinger and R. Jaenisch, Nuclear reprogramming and pluripotency., *Nature*, 2006, **441**, 1061–7.
16. J. T. Do and H. R. Scholer, Nuclei of embryonic stem cells reprogram somatic cells., *Stem Cells*, 2004, **22**, 941–9.
17. J. M. Lemaitre, E. Danis, P. Pasero, Y. Vassetzky and M. Mechali, Mitotic remodeling of the replicon and chromosome structure., *Cell*, 2005, **123**, 787–801.
18. K. Okita, T. Ichisaka and S. Yamanaka, Generation of germline-competent induced pluripotent stem cells., *Nature*, 2007, **448**, 313–7.
19. M. Wernig, A. Meissner, R. Foreman, T. Brambrink, M. Ku, K. Hochedlinger, B. E. Bernstein and R. Jaenisch, In vitro reprogramming of fibroblasts into a pluripotent ES-cell-like state., *Nature*, 2007, **448**, 318–24.

20. A. Meissner, M. Wernig and R. Jaenisch, Direct reprogramming of genetically unmodified fibroblasts into pluripotent stem cells., *Nat Biotechnol*, 2007, **25**, 1177–81.
21. N. Maherali, R. Sridharan, W. Xie, J. Utikal, S. Eminli, K. Arnold, M. Stadtfeld, R. Yachechko, J. Tchieu, R. Jaenisch, K. Plath and K. Hochedlinger, Directly reprogrammed fibroblasts show global epigenetic remodeling and widespread tissue contribution., *Cell Stem Cell*, 2007, **1**, 55–70.
22. K. Takahashi, K. Tanabe, M. Ohnuki, M. Narita, T. Ichisaka, K. Tomoda and S. Yamanaka, Induction of pluripotent stem cells from adult human fibroblasts by defined factors., *Cell*, 2007, **131**, 861–72.
23. J. Yu, M. A. Vodyanik, K. Smuga-Otto, J. Antosiewicz-Bourget, J. L. Frane, S. Tian, J. Nie, G. A. Jonsdottir, V. Ruotti, R. Stewart, Slukvin, II and J. A. Thomson, Induced pluripotent stem cell lines derived from human somatic cells., *Science*, 2007, **318**, 1917–20.
24. J. T. Dimos, K. T. Rodolfa, K. K. Niakan, L. M. Weisenthal, H. Mitsumoto, W. Chung, G. F. Croft, G. Saphier, R. Leibel, R. Goland, H. Wichterle, C. E. Henderson and K. Eggan, Induced pluripotent stem cells generated from patients with ALS can be differentiated into motor neurons., *Science*, 2008, **321**, 1218–21.
25. I. H. Park, N. Arora, H. Huo, N. Maherali, T. Ahfeldt, A. Shimamura, M. W. Lensch, C. Cowan, K. Hochedlinger and G. Q. Daley, Disease-specific induced pluripotent stem cells., *Cell*, 2008, **134**, 877–86.
26. F. Soldner, D. Hockemeyer, C. Beard, Q. Gao, G. W. Bell, E. G. Cook, G. Hargus, A. Blak, O. Cooper, M. Mitalipova, O. Isacson and R. Jaenisch, Parkinson's disease patient-derived induced pluripotent stem cells free of viral reprogramming factors., *Cell*, 2009, **136**, 964–77.
27. A. D. Ebert, J. Yu, F. F. Rose, Jr., V. B. Mattis, C. L. Lorson, J. A. Thomson and C. N. Svendsen, Induced pluripotent stem cells from a spinal muscular atrophy patient., *Nature*, 2009, **457**, 277–80.
28. N. Lavon, K. Narwani, T. Golan-Lev, N. Buehler, D. Hill and N. Benvenisty, Derivation of euploid human embryonic stem cells from aneuploid embryos., *Stem Cells*, 2008, **26**, 1874–82.
29. J. Hanna, M. Wernig, S. Markoulaki, C. W. Sun, A. Meissner, J. P. Cassady, C. Beard, T. Brambrink, L. C. Wu, T. M. Townes and R. Jaenisch, Treatment of sickle cell anemia mouse model with iPS cells generated from autologous skin., *Science*, 2007, **318**, 1920–3.
30. D. Xu, Z. Alipio, L. M. Fink, D. M. Adcock, J. Yang, D. C. Ward and Y. Ma, Phenotypic correction of murine hemophilia A using an iPS cell-based therapy., *Proc Natl Acad Sci U S A*, 2009, **106**, 808–13.
31. A. Raya, I. Rodriguez-Piza, G. Guenechea, R. Vassena, S. Navarro, M. J. Barrero, A. Consiglio, M. Castella, P. Rio, E. Sleep, F. Gonzalez, G. Tiscornia, E. Garreta, T. Aasen, A. Veiga, I. M. Verma, J. Surralles, J. Bueren and J. C. Belmonte, Disease-corrected haematopoietic progenitors from Fanconi anaemia induced pluripotent stem cells., *Nature*, 2009.

32. L. Ye, J. C. Chang, C. Lin, X. Sun, J. Yu and Y. W. Kan, Induced pluripotent stem cells offer new approach to therapy in thalassemia and sickle cell anemia and option in prenatal diagnosis in genetic diseases., *Proc Natl Acad Sci U S A*, 2009, **106**, 9826–30.
33. M. Wernig, J. P. Zhao, J. Pruszak, E. Hedlund, D. Fu, F. Soldner, V. Broccoli, M. Constantine-Paton, O. Isacson and R. Jaenisch, Neurons derived from reprogrammed fibroblasts functionally integrate into the fetal brain and improve symptoms of rats with Parkinson's disease., *Proc Natl Acad Sci U S A*, 2008, **105**, 5856–61.
34. S. Yamanaka, A fresh look at iPS cells., *Cell*, 2009, **137**, 13–7.
35. P. S. Knoepfler, Why myc? An unexpected ingredient in the stem cell cocktail., *Cell Stem Cell*, 2008, **2**, 18–21.
36. M. Wernig, A. Meissner, J. P. Cassady and R. Jaenisch, c-Myc is dispensable for direct reprogramming of mouse fibroblasts., *Cell Stem Cell*, 2008, **2**, 10–2.
37. M. Nakagawa, M. Koyanagi, K. Tanabe, K. Takahashi, T. Ichisaka, T. Aoi, K. Okita, Y. Mochiduki, N. Takizawa and S. Yamanaka, Generation of induced pluripotent stem cells without Myc from mouse and human fibroblasts., *Nat Biotechnol*, 2008, **26**, 101–6.
38. D. Huangfu, K. Osafune, R. Maehr, W. Guo, A. Eijkelenboom, S. Chen, W. Muhlestein and D. A. Melton, Induction of pluripotent stem cells from primary human fibroblasts with only Oct4 and Sox2., *Nat Biotechnol*, 2008, **26**, 1269–75.
39. S. Eminli, J. Utikal, K. Arnold, R. Jaenisch and K. Hochedlinger, Reprogramming of neural progenitor cells into induced pluripotent stem cells in the absence of exogenous Sox2 expression., *Stem Cells*, 2008, **26**, 2467–74.
40. J. B. Kim, V. Sebastiano, G. Wu, M. J. Arauzo-Bravo, P. Sasse, L. Gentile, K. Ko, D. Ruau, M. Ehrich, D. van den Boom, J. Meyer, K. Hubner, C. Bernemann, C. Ortmeier, M. Zenke, B. K. Fleischmann, H. Zaehres and H. R. Scholer, Oct4-induced pluripotency in adult neural stem cells., *Cell*, 2009, **136**, 411–9.
41. D. Hockemeyer, F. Soldner, E. G. Cook, Q. Gao, M. Mitalipova and R. Jaenisch, A drug-inducible system for direct reprogramming of human somatic cells to pluripotency., *Cell Stem Cell*, 2008, **3**, 346–53.
42. B. W. Carey, S. Markoulaki, J. Hanna, K. Saha, Q. Gao, M. Mitalipova and R. Jaenisch, Reprogramming of murine and human somatic cells using a single polycistronic vector., *Proc Natl Acad Sci U S A*, 2009, **106**, 157–62.
43. M. Stadtfeld, M. Nagaya, J. Utikal, G. Weir and K. Hochedlinger, Induced pluripotent stem cells generated without viral integration., *Science*, 2008, **322**, 945–9.
44. T. Brambrink, R. Foreman, G. G. Welstead, C. J. Lengner, M. Wernig, H. Suh and R. Jaenisch, Sequential expression of pluripotency markers during direct reprogramming of mouse somatic cells., *Cell Stem Cell*, 2008, **2**, 151–9.

45. K. Okita, M. Nakagawa, H. Hyenjong, T. Ichisaka and S. Yamanaka, Generation of mouse induced pluripotent stem cells without viral vectors., *Science*, 2008, **322**, 949–53.
46. K. Woltjen, I. P. Michael, P. Mohseni, R. Desai, M. Mileikovsky, R. Hamalainen, R. Cowling, W. Wang, P. Liu, M. Gertsenstein, K. Kaji, H. K. Sung and A. Nagy, piggyBac transposition reprograms fibroblasts to induced pluripotent stem cells., *Nature*, 2009, **458**, 766–70.
47. D. Huangfu, R. Maehr, W. Guo, A. Eijkelenboom, M. Snitow, A. E. Chen and D. A. Melton, Induction of pluripotent stem cells by defined factors is greatly improved by small-molecule compounds., *Nat Biotechnol*, 2008, **26**, 795–7.
48. Y. Shi, C. Desponts, J. T. Do, H. S. Hahm, H. R. Scholer and S. Ding, Induction of pluripotent stem cells from mouse embryonic fibroblasts by Oct4 and Klf4 with small-molecule compounds., *Cell Stem Cell*, 2008, **3**, 568–74.
49. D. Kim, C. H. Kim, J. I. Moon, Y. G. Chung, M. Y. Chang, B. S. Han, S. Ko, E. Yang, K. Y. Cha, R. Lanza and K. S. Kim, Generation of human induced pluripotent stem cells by direct delivery of reprogramming proteins., *Cell Stem Cell*, 2009, **4**, 472–6.
50. J. Yu, K. Hu, K. Smuga-Otto, S. Tian, R. Stewart, Slukvin, II and J. A. Thomson, Human induced pluripotent stem cells free of vector and transgene sequences., *Science*, 2009, **324**, 797–801.
51. G.Q. Daley, M.W. Lensch, R. Jaenisch, A. Meissner, K. Plath and S. Yamanaka, Broader implications of defining standards for the pluripotency of iPSCs. *Cell Stem Cell*, 2009, **4**, 200-1; author reply 202.
52. J. Ellis, B.G. Bruneau, G. Keller, I.R. Lemischka, A. Nagy, J. Rossant, D. Srivastava, P.W. Zandstra and W.L. Stanford, Alternative induced pluripotent stem cell characterization criteria for in vitro applications. *Cell Stem Cell*, 2009, **4**, 198-9; author reply 202.
53. N. Maherali and K. Hochedlinger, Guidelines and techniques for the generation of induced pluripotent stem cells., *Cell Stem Cell*, 2008, **3**, 595–605.

CHAPTER 7
Mesenchymal Stromal/Stem Cells from Tissue Repair to Destruction of Tumor Cells

RITA BUSSOLARI,[a] GIULIA GRISENDI,[a] LUIGI CAFARELLI,[a] PIETRO LOSCHI,[b] LAURA SCARABELLI,[a] ANTONIO FRASSOLDATI,[a] MICHELA MAUR,[a] GIORGIO DE SANTIS,[b] PAOLO PAOLUCCI,[c] PIERFRANCO CONTE[a] AND MASSIMO DOMINICI[a]

[a] Department of Oncology, Hematology and Respiratory Diseases; [b] Plastic Surgery Unit; and [c] Department of Mother and Child, University-Hospital of Modena and Reggio Emilia, Modena, Italy

7.1 Stromal Cells and Tumor

A tumor is a complex framework composed of tumor cells (TC) and stroma where the extracellular matrix (ECM) and cellular components such as immune cells, vessels cells and fibroblasts interact closely together (Figure 7.1).[1,2]

This microenvironment plays a fundamental role in tumor initiation and development as it influences tumor growth and its ability to progress and metastasize.[3,4] Moreover, the microenvironment substantially contributes to tumor escape by:[5]

- limiting the action of the immune system;
- counteracting the performance of chemotherapeutic agents;

Stem Cell-Based Tissue Repair
Edited by Raphael Gorodetsky and Richard Schäfer
© The Royal Society of Chemistry 2011
Published by the Royal Society of Chemistry, www.rsc.org

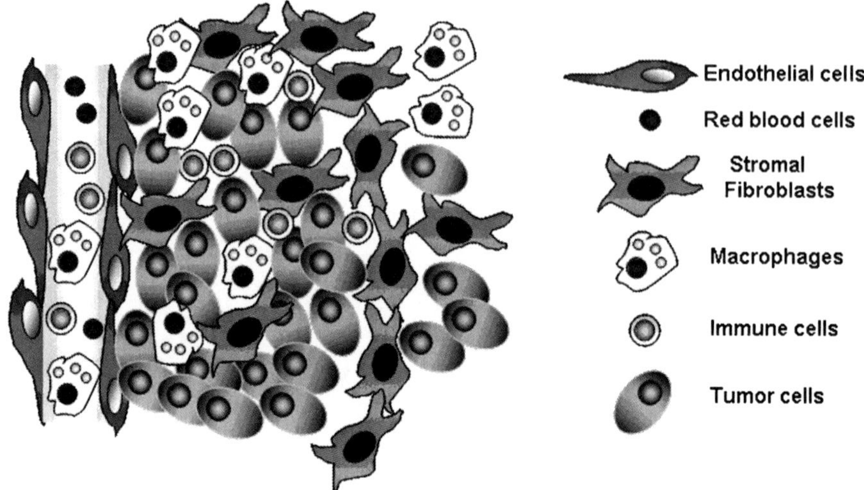

Figure 7.1   The tumor microenvironment. Representative cartoon of solid tumor bulk composed by tumor cells and non-tumorigenic normal cells such as immune cells, vessels and stroma fibroblasts.

- modifying drug metabolism; and
- supporting the development of drug resistance.

Based on this evidence, the tumor microenvironment represents an attractive therapeutic target in cancer therapy. Despite the importance of these tumor–stroma interactions, for decades scientists have focused their research almost exclusively on understanding the transformation of tumor cells themselves; only recently have stromal composition and the relationships between tumor cells and their surroundings also been investigated.[6]

Tumor cells can affect their adjacent stroma to form a permissive and supportive environment for tumor progression. This event is driven by the release of various proteases and growth factors, which act in autocrine and paracrine manners reciprocally influencing tumor cells and stromal growth.[1,2] The main stroma modulating growth factors produced by tumor cells are represented by:[2]

- basic fibroblast growth factor (bFGF);
- vascular endothelial growth factor (VEGF);
- platelet-derived growth factor (PDGF);
- epidermal growth factor receptor (EGFR)-ligand; and
- transforming growth factor β (TGFβ).

All these cytokines, but in particular TGFβ and PDGF, are able to alter the normal tissue homeostasis stimulating the formation of new blood vessels as well as the recruitment and activation of inflammatory cells and fibroblasts.[7–9]

Since fibroblasts constitute the majority of stromal cells within the primary tumor burden in various types of human cancer, these cells are also referred as tumor or cancer-associated fibroblast (TAF or CAF, respectively).[10–13] Inside the tumor, TAFs acquire a modified so called 'activated phenotype'—similar to the fibroblasts associated with wound healing—and characterized by high production of α-smooth muscle actin (α-SMA).[14–16]

Activated TAFs may promote tumor progression in several different ways. They release pro-migratory ECM components and upregulate the expression of ECM-degrading proteases, promoting ECM turnover and remodelling.[17,18] Both these events are fundamental for angiogenesis and tumor cell invasion. In addition, TAFs express a wide variety of cytokines and growth factors which promote TC proliferation, survival, migration and invasion.[17,19–21] Finally, TAFs stimulate the generation of new blood vessels secreting themselves VEGF, and driving the recruitment of inflammatory cells by the release of monocyte chemo-attractant protein 1 (MCP1) and interleukin-1 (IL-1).[22–23] This latter aspect is not irrelevant if one considers the role played by cancer-infiltrating leucocytes in promoting tumor growth. These immune cells, attracted inside tumor burden, release different proteases and pro-angiogenic growth factors including VEGF, angiopoietin 1, bFGF, TGFβ, PDGF and tumor necrosis factor α (TNFα)—all fundamental in supporting and maintaining tumor progression.[24–25]

7.2 Role of Mesenchymal Stromal/Stem Cells in a Tumor Microenvironment

The ontogeny of TAFs remains controversial. Even though some studies have confirmed that stroma fibroblasts are recruited from local healthy tissue fibroblasts,[26–28] recent findings have suggested that, during the tumor stroma development, TAFs could be equally recruited from a putative circulating fibroblast pool such as the one derived from bone marrow (BM) and defined as mesenchymal stromal/stem cells (MSCs).[29–30]

There are two main items of evidence in support of this assumption. First, human BM-MSCs exposed *in vitro* to tumor-conditioned medium assume a TAF-like myofibroblastic phenotype, including sustained expression of stromal-derived factor-1 (SDF-1) and α-SMA.[31–32] Secondly, BM-MSCs are able *in vivo* to migrate into tumor burden after systemic or local infusion.[27,30,33,34] More recently, Hall and colleagues[35] hypothesized that, once inside the tumor microenvironment, MSCs could proliferate and acquire the main biological properties of TAFs. Despite this early evidence, the topic requires further clarification and insights can be derived from the understanding of the biology of MSCs.

MSCs are fibroblastic spindle-shaped cells which exhibit a robust self-renewal capacity and retain the ability to differentiate into adipocytes, chondrocytes and osteoblasts.[36,37] They were found initially in bone marrow, but recently they have also been described in adipose tissue, umbilical cord and in other tissues.[38–40] Phenotypically, MSCs express a combination of different antigens such as CD73, CD90, CD105 and CD146, and are characterized by a lack of hematopoietic and endothelial-specific markers.[36,37] Unfortunately, no exclusive MSC-specific markers has been yet discovered and thus their *in vivo* identification is still to be clarified.

Comparing MSCs with TAFs, we and others have observed that TAFs (isolated from different primary tumor samples) show some MSCs-like features including the ability to adhere to plastic, spindle shape morphology (Figure 7.2), antigen expression and *in vitro* differentiation potentials.[41–43]

It has been hypothesized that, under normal conditions, BM-MSCs play an essential role in maintaining adult tissue homeostasis.[44] Accumulating evidence has also demonstrated that, when systematically injected into healthy animals, MSCs migrate preferentially in lung, liver and bone while they have been found to a lesser extent in other tissue.[45–47] Simultaneously, an increasing number of studies have underlined the participation of MSCs in the course of wound repair. As observed in several pathological animal models, MSCs preferentially migrate into injured sites, independent of the type of damaged organ. The reasons of this event are still unclear, but MSCs may be attracted by chemotactic gradients of cytokines and chemokines released after damage.[48–53] Once there, MSCs start to proliferate and take part in tissue remodeling, providing both structural support and secreting stimulatory factors for tissue repair.[54–55] Starting from these assumptions, we and others assumed that a similar

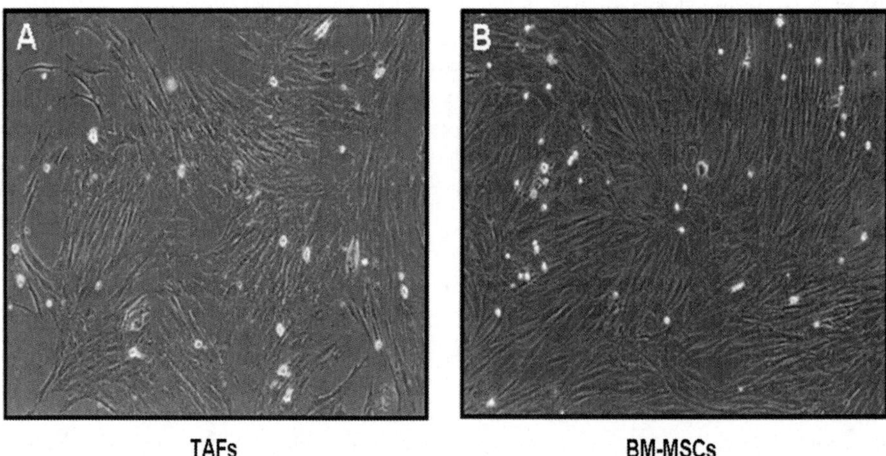

Figure 7.2 *In vitro* culture of primary tumor associated fibroblasts (TAFs) and mesenchymal stromal/stem cells (MSCs). (A) TAFs obtained from primary tumor lung biopsy, released after enzymatic digestion showing a spindle-shaped and fibroblastic morphology after ten days of culture. (B) *In vitro* typical features of bone marrow MSCs 13 days after isolation. In both cases cells adhering to plastic showed an overlapping morphology with elongated spindle-shaped cells.

behavior may be taking place in presence of a tumor which represents 'a wound that never heals'.[56]

In cancer, the cells themselves proliferate rapidly, the ECM is invaded, connective tissue is remodeled, epithelial and stromal cells migrate, and new blood vessels are formed. Within this context, many of the same factors involved in wound healing are released, probably providing MSC homing.

Several studies have demonstrated the recruitment and engraftment of intravenously derived MSC into tumor sites of melanoma, breast and brain tumor xenograft models.[29,30,33,35] The mechanisms and the chemotactic factors that underline MSCs homing to tumors are not yet fully understood and researchers have so far focused on known effectors of leukocyte trafficking such as the stromal-derived factor 1 (SDF1), a chemokine implicated in the regulation of MSC migration in response to tumor-conditioned medium.[57] The insulin-like growth factor 1 (IGF1), which enhances MSCs migration through upregulation of chemokine receptors, has been also considered.[58,59]

In addition, toll-like receptor ligands such as the pro-inflammatory peptide LL-37 (leucine, leucine 37) have been recently linked to the tumor homing of MSCs. Several studies have uncovered how different tumor types (including ovarian, breast and lung cancer) exhibit a high expression of LL-37.[60–62] It has been demonstrated that LL-37 acts as a proliferative signal, pro-angiogenic factor and chemo-attractant for various immune cells through activation of formyl peptide receptor like-1 (FPRL-1) as a member of the toll-like receptor family.[63–65]

Coffelt and colleagues proposed that LL-37, expressed on ovarian cancer cells, could be also involved in the recruitment of MSCs by tumors.[66] They demonstrated the expression of FPRL-1 on an MSC population, confirming that LL-37 activates the migration of MSCs in a dose-dependent manner. Their *in vivo* data have demonstrated a homing of MSCs into tumors demonstrating that an inhibition of MSCs engraftment into tumor cells results in disorganization of the fibroblast–vascular network as well as a reduction in tumor growth.

Based on these early findings, we could hypothesize that the migration of MSCs is due to a complex network of factors which, released by tumor cells, may mobilize MSCs from bone marrow or from other mesenchymal organs such as the adipose tissue.[67]

Once inside the tumor microenvironment, MSCs could also act in two different ways, secreting pro-angiogenic factors as well as becoming themselves a structural component of tumor. Accordingly to this hypothesis, MSCs appear to support tumor angiogenesis by the production of VEGF, FGF, PDGF and secreting growth factors acting on tumor and endothelial cells.[68,69] In addition, the MSCs themselves may also assume an endothelial-like phenotype as shown both *in vitro*[70] and *in vivo* after implantation in Matrigel plugs.[71] In this latter study, Al-Khaldi and his team demonstrated the spontaneous angiogenic and vasculogeneic activities of MSCs using an *in vivo* murine Matrigel model. Matrigel-embedded MSCs elicit a robust angiogenic response just two weeks after implantation, leading to a significant increase in vessel density. These findings were strengthened by immunohistochemical analyses of

Matrigel implants which revealed that implanted MSCs can acquire an endo-thelial-like phenotype resulting positive for CD31 and VEGF.[71]

To further validate this concept, Ishii and colleagues demonstrated that BM-MSCs contribute to both tumor-associated vasculature and TAFs in a syngeneic mouse model.[72] The authors transplanted BM-derived cells obtained from β-galactosidase transgenic mice into sublethally irradiated NOD-SCID mice. A few weeks after transplantation, a human pancreatic cancer cell line was injected subcutaneously and stromal components were analyzed. Histology revealed the presence of X-gal positive cells inside the tumor and the staining demonstrated that these cells were also double positive for CD31 and α-SMA, suggesting that marrow progenitors might have been recruited to constitute endothelial cells and myofibroblasts.

7.3 MSCs and Tumors: Between Growth and Inhibition

All this evidence suggests that MSCs could take part in the tumor micro-environment, hypothetically contributing to cancer progression. However, studies that aimed to investigate the *in vitro* and *in vivo* influence of MSCs on tumor growth provide conflicting results.

One of the first studies underlining a double role played by MSCs in carci-nogenesis was performed by Zipori and collegues.[73] This pioneering work tested 11 different tumor cell lines seeded onto either murine or human BM-MSCs. They showed that *in vitro* MSCs caused a dramatic increase in the growth of human lung and colon carcinoma cell lines. In contrast, they inhibited the *in vitro* cloning of both human and murine sarcoma cell lines. Thus, these opposite effects seem to depend on the tumor type.

More recently, based on evidence indicating an immunomodulatory property of MSCs, researchers have described an immunosuppressive effect which may favor tumor growth. Djouad and colleagues *in vitro* confirmed an immuno-suppressive effect of both mouse and human MSCs on activated T lympho-cytes, allogeneic splenocytes and professional antigen presenting cells.[74] Subsequently, they verified an *in vivo* immunosuppressive effect of MSCs, revealing the growth and progression of an allogeneic melanoma cell line in immunocompetent mice after both subcutaneously and systemic co-injections of mouse MSCs. According to the hypothesis that MSCs may also favor tumor progression, Karnoub and colleagues suggested a critical role of MSCs in the development of breast cancer metastasis.[75] All four tested human breast cancer cell lines showed an increased metastatic potential to the lungs when sub-cutaneously co-injected with MSCs. However, only one cell line showed accelerated primary tumor growth when co-injected with MSCs. Interestingly, the co-injection of TC with other types of mesenchymal cells (*e.g.* BJ fibro-blasts) did not enhance tumor growth or metastasis. In a search for cytokines, chemokines and growth factors differentially expressed in cocultures of human breast cancer cells and MSCs, these authors identified the chemokine CCL5

(RANTES) to be upregulated by 60-fold in cocultures compared to mono-cultures. It seems that CCL5, released from MSCs stimulated by tumor cells, induces the recruitment of tumor-associated macrophages and endothelial cells in spite of primary tumor growth enhancing cancer metastasis.[76,77]

To further validate this concept, other authors have described how MSCs are able to differentiate in cells resembling TAFs and favoring adenocarcinoma growth *in vivo*; the co-injection of ovarian cancer cell line and MSCs favored tumor growth in relationship with a release of IL-6 by MSCs.[78] Beside BM-derived MSCs, adipose tissue (AT) MSCs have also been described to enhance tumor progression. Subcutaneously injected AD-MSCs can migrate into stromal tumor compartment, promoting cancer cells proliferation.[67]

Taken together, these studies revealed that different tumors can recruit *ex vivo* expanded MSCs and that this recruitment facilitates tumor growth and a metastatic spread through the release of soluble factors. Although the properties exhibited by MSCs in the mentioned studies may lead us to think these cells to be involved in a tumor-supporting function, other reports have demonstrated the opposite describing a tumor suppression.

Maestroni and colleagues reported experimental models of Lewis lung carcinoma and B16 melanoma, where co-injection of mouse MSCs and tumor cells inhibited tumor growth and metastasis.[79] An anti-proliferative action by MSCs was also reported by Ohlsson and colleagues[80] who investigated the inhibitory effect of MSCs on colon carcinoma using gel loaded with tumor cells alone or tumor cells mixed with MSCs. During their studies they discovered that MSCs inhibited the outgrowth of colon carcinoma both *in vitro* and *in vivo*, reaching a complete inhibition when the number of MSCs was at least equal to the number of tumor cells.

Similarly rat primary MSCs, after inoculation into the controlateral hemisphere, showed a migratory capacity and an inhibitory effect against rat glioma cells.[81] Co-injection of glioma cells with wild type MSCs prolonged survival of tumor-bearing rats by inhibiting tumor progression. In another elegant experimental model of Kaposi Sarcoma (KS), it was observed that intravenously injection of MSCs potently inhibited tumor growth.[82] The inhibition was observed not only when MSCs and KS cells were co-injected, but also when MSCs were administered in mice bearing already established tumors. The *in vivo* tumor-suppressive effect of MSCs against KS was found to depend on E-cadherin mediated cell-to-cell contact and by the inhibition of Akt activation within tumor cells. Again this anti-tumorigenic effect seems to be strictly tumor related and it was not reproduced for other tumor cells types such as breast cancer (MCF-7) and prostate (PC-3) tested in the same study.

Taken together, these studies reveal that the effect of MSCs on tumor cells remains controversial and further studies are demanded. These conflicting results could be explained on the basis of different experimental conditions applied by investigators including:

- the tumor histology;
- the source of MSCs;

- the protocol adopted for MSCs *in vitro* expansion;
- the ratio of MSCs to tumor cells; and
- the *in vivo* delivery approach.

All these factors introduce biases which could affect the possible translation of MSCs into more advanced therapeutic approaches. Thus, it would be advisable to specifically standardize experimental models in order to minimize confounding factors. With these fundamental studies we now know that MSCs can participate in the tumor microenvironment and, although still far away from knowing their exact role, we could take advantage of these properties in the context of preclinical models of cell and gene therapy.

7.4   Mesenchymal Stromal/Stem Cells as Vehicles for Cancer Therapy

The introduction of innovative approaches for cancer treatment characterized by better pharmacokinetic, biodistribution and availability is one of the most important topics in the battle against cancer. In order to improve these aspects, researchers have developed several approaches such as drug delivery by liposomes, conjugated antibodies, nanoparticles or polymers and protein-DNA complexes. Unfortunately, the use of these strategies may be limited by toxicity, poor absorption and release, short half-life and lack of tissue specificity.

Therefore, new promising therapeutic tools have recently been tested. These include biological agents (*e.g.* bacteria or viruses) and mammalian cells capable of inducing tumor cell death by delivering cytotoxic compounds. In this respect, MSCs have gained particular interest thanks to their biological and immunological features. As described above, it is now established that MSCs may play a pivotal role in tumor stroma *via* an inherent ability to migrate to clinically relevant target tissues and to preferentially locate into tumors. These properties make MSCs carriers able to delivery therapeutic agents—possibly bypassing any presumed tumor-promoting ability.

Gene therapy is a promising treatment for a large number of congenital and acquired severe diseases. In this setting one can replace, inactivate, upregulate or introduce a missing gene into a cellular target.[83] Cell-based gene therapies can rely on two different cell types as vehicles:

- terminally differentiated cells such as macrophage or natural killer cells; and
- progenitor cells such as hematopoietic stem cells, endothelial progenitors and MSCs.

As described previously, all these cells are part of the tumor microenvironment and therefore can all be potentially used to delivery the desired gene to cancer cells.

Terminally differentiated cells are generally difficult to handle *ex vivo* having also a suboptimal gene modification efficiency. In addition, a short survival after transplantation for these committed cells may negatively affect treatment outcome. On the contrary, progenitor cells can be more easily transduced by retroviral and lentiviral vectors, maintaining both proliferative and differentiative attitudes. Molecular-engineered progenitors present a better survival profile in comparison with terminally differentiated gene-modified cells and their ability to migrate into tumor makes them very attractive to deliver proteins with a poor pharmacokinetics profile.[92,100] In particular, gene-modified MSCs can be incorporated into the tumor microenvironment and express the desired therapeutic function, thus optimizing the therapeutic approach.

The application of MSCs in cellular therapies is also based on a number of favorable properties (summarised in Table 7.1). MSCs are readily accessible from different sources and particularly from bone marrow and subcutaneous adipose tissue.[84] In addition they show a high proliferative potential without specially defined nutritional requirements.[85,86] MSCs are also able to differentiate *in vitro* into one of many cell phenotypes and to migrate and engraft into injured sites.[87,88] More interestingly, MSCs may be genetically modified by several gene transfer vectors to obtain a robust level of transduction coupled with an efficient metabolic activity for the generation of wanted therapeutic agents.[89] MSCs are additionally characterized by a poor immunogenicity due to an intermediate expression of the class I major histocompatibility complex, low levels of MHC class II molecules and the absence of co-stimulatory molecules such as CD40, CD80 and CD86.[90]

Based on these properties, MSCs can become vehicles to express a therapeutic gene to be then transplanted *in vivo* (Figure 7.3). Generally speaking there are two main approaches to gene modify progenitor cells including MSCs:

- a transient modification; and
- a stable gene introduction.

In the first case, the transgene expression has a short half-life. This could be based on modifying agents able to allow gene penetration through cell membrane by physical (*i.e.* electroporation), chemical (*i.e.* calcium phosphate and polycations) or biological (adenoviral vectors) agents.[91] On the contrary, a stable modification of progenitor cells has been constantly obtained by retro- or HIV-lentiviral based plasmid vectors.[92] In these latter approaches, the

Table 7.1   Mesenchymal stromal/stem cells advantages.

Simple isolation and expansion methods
Easy manipulation
Poor immunogenicity
Easy gene modification
Systemic delivery and homing capability

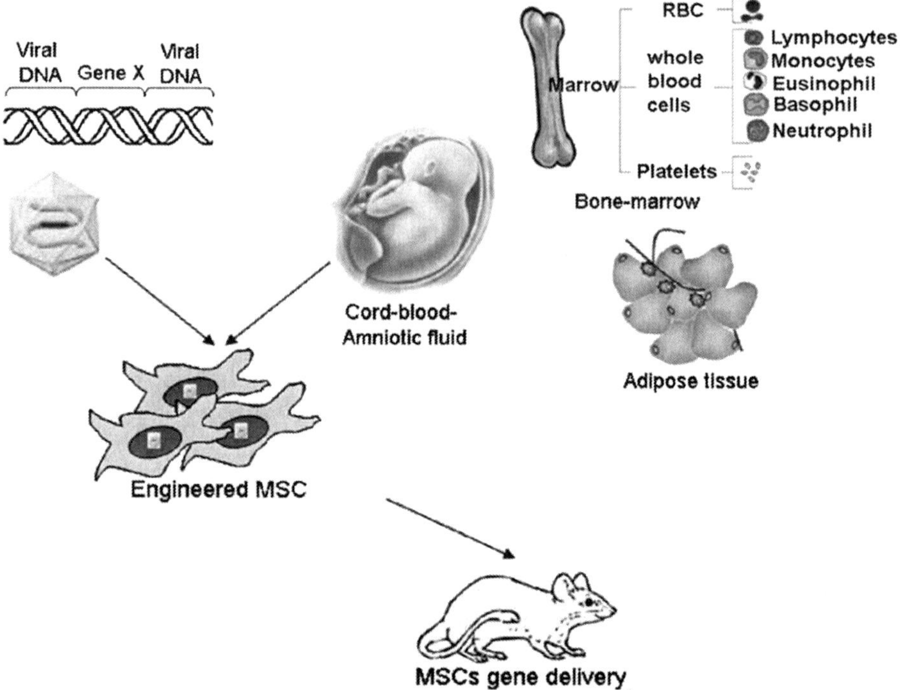

Figure 7.3   Representative strategy for delivering a therapeutic transgene X by vector engineered mesenchymal stromal/stem cells (MSCs). In these models, MSCs can be derived from prenatal tissues such as cord blood amniotic fluid, adipose tissue and bone marrow.

transgene integrates into the progenitor cell genome allowing the propagation of the infection into the cell progeny after proliferation.[93]

Both approaches has been used to gene modify MSCs. Adenoviral vectors have been used to modify BM-MSC to deliver interferon-β (INF-β).[94] This molecule is known to be effective for the treatment of several tumors,[95] though INF-β *in vivo* delivery has been limited by its excessive toxicity—in particular at high doses. We have engineered MSCs to induce INF-β expression; in this manner MSCs could more specifically deliver the drug, reducing undesired side effects. Dealing with retroviral-stable modification of MSCs for cancer therapy, Elzaouk *et al.* stably transduced MSCs with a retroviral vector expressing murine IL-12, as a potent anti-tumor cytokine.[96] Gene-modified MSCs also deliver IL-12 efficiently in a mouse model of melanoma.[97] Besides these two models, others have used MSCs to delivery cytotoxic compounds targeting different tumor types. Table 7.2 presents a summary of published preclinical anti-cancer therapies based on gene-modified MSCs obtained from different sources.

Although protocols for gene and cell manipulation have been well standardized in cancer gene therapy projects based on MSCs, the manipulation of

Table 7.2 Mesenchymal stromal/stem cells based therapies.

Mesenchymal stem cells source	Tumor source	Tumor host	Vector	Therapeutic transgene	Reference
BM-MSC (human)	human melanoma	mouse	Adenoviral vector	INF-β	29
BM-MSC (rat)	rat brain glioma	rat	Adenoviral vector	IL-2	81
BM-MSC (human)	human breast cancer, human melanoma	mouse	Adenoviral vector	INF-β	30
BM-MSC (human)	murine brain glioma	mouse	Adenoviral vector	INF-β	33
BM-MSC (human)	murine melanoma	rat	Retroviral vector	IL-12	97
BM-MSC (mouse)	murine Colon-26 lung metastasis	mouse	Adenoviral vector	NK4	107
AT-MSC (human)	human colon adenocarcinoma	mouse	Retroviral vector	Cytosine deaminase (CD)	99
BM-MSC (mouse)	murine lung	mouse	Adenoviral vector	CX3CL1	108
UCB-MSC (human)	human glioma	mouse	Adenoviral vector	stTRAIL	104
BM-MSC (mouse)	murine lung	mouse	Adenoviral vector	INF-β	94
BM-MSC (human)	human lung	mouse	Adenoviral vector	TRAIL	102
AT-MSC (human)	human prostate	mouse	Retroviral vector	Cytosine deaminase (CD)	109
BM-MSC (human)	murine brain glioma	mouse	Adenoviral vector	S-TRAIL	110
BM-MSC (human)	human glioma	mouse	Adenoviral vector	Delta24-RGD	111
BM-MSC (human)	human glioma	mouse	Lentiviral vector	S-TRAIL	103
AT-MSC (human)	human cervical carcinoma	mouse	Retroviral vector	TRAIL	37

MSCs by viral vectors for gene therapy application requires safety controls related to possible side-effects such as a genomic insertion followed by cell transformation. A possible strategy to overcome these unwanted events can be the use of suicide genes encoding vectors, such as the herpes simplex virus thymidine kinase gene (HSV-TK). This represents a system capable of killing host cells by suicide. In this strategy, the suicide gene uses host cellular machinery to convert a non-toxic pro-drug into a DNA inhibiting agent causing tumor cell apoptosis.[98]

Next to the use of HSV-TK, as a suicide gene able to control both unwanted side-effects and to kill bystander tumor cells (a double benefit), other authors have introduced approaches based on gene-modified MSC capable of activating pro-drugs into active cytotoxic compounds. This is the case with Kucerova and colleagues who showed that AT-MSC expressing the fusion cytosine deaminase : uracil phosphoribosyltransferase (CD : UPRT) gene in combination with prodrug 5-FC increased the potent cytotoxic effect over colon adenocarcinoma cell lines. The administration of the anticancer drug 5-FU can be associated with relevant side effects and high doses are required for therapeutic response. The gene manipulation of AT-MSCs by retroviral vector encoding for CD: UPRT augmented the bystander effect and the selective cytotoxicity against tumor cells both *in vitro* and *in vivo*.[99]

7.5 Mesenchymal Stromal/Stem Cells can Efficiently Delivery Death Ligands

Recently, members of the TNF superfamily including FAS ligand, TNF and TNF-related apoptosis-inducing ligand (TRAIL) have been identified as important bullets for cancer therapy.[100] These molecules—also referred as death ligands (DL)—are able to transmit apoptosis signals by binding their specific death receptors on the surface of tumor cells.

Among these death ligands, TRAIL has been considered the most attractive candidate for cancer therapy because of its ability to induce apoptosis in several tumor cell lines sparing healthy tissue and cells, including bone marrow and adipose tissue derived MSCs. Although several studies have shown an anti-tumor activity of recombinant TRAIL, its use *in vivo* is limited due to a short half-life in plasma and to unwanted side-effects due to high doses. As an alternative, the use of trimerized or non-tagged form of TRAIL as well as an agonistic anti-TRAIL receptors antibody has been introduced. However, even in these cases, undesired toxicity against hepatocytes and keratinocytes has been reported.[101]

To overcome these limitations of TRAIL pharmacokinetic, toxicity profile and drug resistance, over the last few years several gene therapy approaches based on gene-modified MSCs expressing TRAIL have been developed. In these studies, MSCs obtained from different sources were as cellular vehicles and many tumor types were tested.

Mohr *et al.* first described the use of MSCs expressing TRAIL against a lung cancer cell line. They genetically modified BM-MSCs with an adenoviral vector expressing the full length human TRAIL gene, obtaining a transient anti-tumor effect both *in vitro* and *in vivo*.[102] Sasportas *et al.* described a therapeutic approach in a glioma mouse model where MSCs, expressing a secretable TRAIL (S-TRAIL), induce CASPASE-3 mediated apoptosis in tumor cells.[103]

Human umbilical cord blood MSCs (UCB-MSCs) are also reported as suitable source of cellular vector for TRAIL delivery systems, as reported by Kim *et al.* In this study, UCB-MSCs were resistant to TRAIL-mediated apoptosis, revealing a marked migratory ability and a potent anti-tumoral activity toward glioma.[104]

Very recently, a TRAIL delivery approach has been also performed by our group and, for the first time, stably transduced AT-MSCs expressing-TRAIL showed a strong cytotoxic effect against human cervical carcinoma *in vivo*.[105] Collectively, these data demonstrate the efficiency of TRAIL and its secretable forms in the inhibition of tumor cell growth[106] suggesting that, despite the still unclear role of MSCs in influencing tumor cell growth, these cells can be translated from the regenerative medicine to specifically target cancer.

Acknowledgements

This work has been supported in part by Ministero Italiano Istruzione Università e Ricerca-PRIN 2008 (MD, PP, GD), Regione Emilia Romagna (PP, PC, MD), the Associazione ASEOP (PP) and Fondazione Cassa di Risparmio di Modena (MD).

References

1. M. J. Bissel and D. Radiski, *Nat. Rev. Cancer*, 2001, **1**, 46.
2. M. M. Mueller and N. F. Fusenig, *Nat. Rev. Cancer*, 2004, **4**, 839.
3. R. Kalluri, *Nat. Rev. Cancer*, 2003, **3**, 422.
4. D. Hanahan and R. A. Weinberg, *Cell*, 2000, **100**, 57.
5. O. Trédan, C. M. Galmarini, K. Patel and I. F. Tannock, *J. Natl. Cancer Inst.*, 2007, **99**, 1441.
6. N. A. Bhowmick and H. L. Moses, *Curr. Opin. Genet. Dev.*, 2005, **15**, 97.
7. G. Bergers and L. E. Benjamin, *Nat. Rev. Cancer*, 2003, **3**, 401.
8. L. M. Coussens and Z. Werb, *Nature*, 2002, **420**, 860.
9. O. De Wever and M. Mareel, *J. Path.*, 2003, **200**, 429.
10. A. P. Sappino, O. Skalli, B. Jackson, W. Schürch and G. Gabbiani, *It. J. Cancer*, 1988, **41**, 707.
11. R. Kalluri and M. Zeisberg, *Nat. Rev. Cancer*, 2006, **6**, 392.
12. L. A. Kunz-Schugahart and R. Knuechel, *Histol. Histopathol.*, 2002, **17**, 599.
13. L. A. Kunz-Schugahart and R. Knuechel, *Histol. Histopathol.*, 2002, **17**, 623.

14. A. F. Olumi, G. D. Grossfeld, S. W. Hayward, P. R. Carroll, T. D. Tlsty and G. R. Cunha, *Cancer. Res.*, 1999, **59**, 5002.
15. T. Tsukada, M. A. McNutt, R. Ross and A. M. Gown, *Am. J. Pathol.*, 1987, **127**, 384.
16. G. Gabbiani, *Int. J. Cancer*, 1988, **41**, 707.
17. O. De Wever, Q. D. Nguyen, L. Van Hoorde, M. Bracke, E. Bruyneel, C. Gespach and M. Mareel, *FASEB. J.*, 2004, **18**, 1016.
18. H. P. Rodemann and G. A. Muller, *Am. J. Kidney Dis.*, 1991, **17**, 684.
19. R. K. Akhurst, *J. Clin. Invest.*, 2002, **109**, 1533.
20. G. Li, K. Satyamoorthy, F. Meier, C. Berking, T. Bogenrieder and M. Herlyn, *Oncogene*, 2033, **22**, 3162.
21. M. P. Lewis, K. A. Lygoe, M. L. Nystrom, W. P. Anderson, P. M. Speight, J. F. Marshall and G. J. Thomas, *Br. J. Cancer*, 2004, **90**, 822.
22. A. Orimo, Y. Tomioka, Y. Shimizu, M. Sato, S. Oigawa, K. Kamata, Y. Nogi, S. Inoue, M. Takahashi, T. Hata and M. Muramatsu, *Clin. Cancer Res.*, 2001, **7**, 3097.
23. R. M. Strieter, R. Wiggins, S. H. Phan, B. L. Wharram, H. J. Showell, D. G. Remick, S. W. Chensue and S. L. Kunkel, *Biochem. Biophys. Res. Commun.*, 1998, **162**, 695.
24. J. W. Pollard, *Nat. Rev. Cancer*, 2004, **4**, 71.
25. P. Carmeliet and R. K. Jain, *Nature*, 2000, **407**, 295.
26. L. Ronnov-Jessen, O. W. Petersen, V. E. Koteliansky and M. J. Bissell, *J. Clin. Invest.*, 1995, **95**, 859.
27. H. Kiaris, I. Chatzistamou and G. Trimis, *Cancer Res.*, 2005, **65**, 1627.
28. E. Sivridis, A. Giantromanolaki and M. I. Koukourakis, *J. Clin. Pathol.*, 2005, **58**, 1033.
29. M. Studeny, F. C. Marini, R. E. Champlin, C. Zompetta, I. J. Filder and M. Andreeff, *Cancer. Res.*, 2002, **62**, 3603.
30. M. Studeny, F. C. Marini, J. L. Dembiniski, C. Zompetta, M. Cabreira-Hansen, B. N. Bekele, R. E. Champlin and M. Andreeff, *J. Natl. Cancer Inst.*, 2004, **96**, 1593.
31. P. J. Mishra, P. J. Mishra and R. Humeniuk, *Cancer Res.*, 2008, **68**, 4331.
32. M. Emura, A. Ochiai, M. Horino, W. Arndt, K. Kamino and S. Hirohashi, *In vitro Cell. Dev. Biol. Anim.*, 2000, **36**, 77.
33. A. Nakamizo, F. Marini, T. Amano, A. Khan, M. Studeny, J. Gumin, J. Chen, S. Hentschel, G. Vecil, J. Dembinski, M. Andreeff and F. F. Lang, *Cancer Res.*, 2005, **65**, 3307.
34. M. De Palma, M. A. Venneri, C. Roca and L. Naldini, *Nat. Med.*, 2003, **9**, 789.
35. B. Hall, J. Dembinski, A. K. Sasser, M. Studeny, M. Andreeff and F. Marini, *Int. J. Hematol.*, 2007, **86**, 8.
36. M. Dominici K. Le Blanc, I. Mueller, I. Slaper-Cortenbach, F. Marini, D. Krause, R. Deans, A. Keating, D. J. Prockop and E. Horwitz, *Cytotherapy*, 2006, **8**, 315.

37. G. Grisendi, C. Annerén, L. Cafarelli, R. Sternieri, E. Veronesi, G. Cervo, S. Luminari, M. Maur, A. Frassoldati, G. Palazzi, F. Bambi, S. Otsuru, P. Paolucci, P. Conte and M. Dominici, *Cytotherapy*, 2010, **12**, 466.

38. A. J. Friedenstein, U. F. Deriglasova, N. N. Kulagina, A. F. Panasuk, S. F. Rudakowa, E. A. Luriá and I. A. Ruadkow, *Exp. Hematol.*, 1974, **2**, 83.

39. P. A. Zuk, M. Zhu, P. Ashjian, P. A. Zuk, M. Zhu, H. Mizuno, J. Huang, J. W. Futrell, A. J. Katz, P. Benhaim, H. P. Lorenz and M. H. Hedrick, *Mol. Biol. Cell.*, 2002, **13**, 4279.

40. P. Moretti, T. Hatlapatka, D. Marten, A. Lavrentieva, I. Majore, R. Hass and C. Kasper, *Adv. Biochem. Eng. Biotechnol.* Epub ahead of print.

41. H. Cao, W. Xu, H. Qian, W. Zhu, Y. Yan, H. Zhou, X. Zhang, X. Xu, J. Li, Z. Chen and X. Xu, *Cancer Lett.*, 2009, **274**, 61.

42. T. M. Lin, H. W. Chang, A. P. Wang, A. P. Kao, C. C. Chang, C. H. Wen, C. S. Laiv and S. D. Lin, *Biochem. Biophys. Res. Commun.*, 2007, **361**, 883.

43. C. P. Gibbs, V. G. Kukekov, J. D. Reith, O. Tchigrinova, O. N. Suslov, E. W. Scott, S. C. Ghivizzani, T. N. Ignatova and D. A. Steindler, *Neoplasia*, 2005, **7**, 967.

44. D. G. Phinney and D. J. Prockop, *Stem Cells*, 2007, **25**, 2896.

45. C. Allers, W. D. Sierralta, S. Neubauer, F. Rivera, J. J. Minguell and P. A. Conget, *Transplantation*, 2004, **78**, 503.

46. A. A. Erices, C. I. Allers, P. A. Conget, C. V. Rojas and J. J. Minguell, *Cell. Transplant.*, 2003, **12**, 555.

47. R. J. Deans and A. B. Moseley, *Exper Hematol*, 2000, **28**, 875.

48. K. Natsu, M. Ochi, Y. Mochizuki H. Hachisuka, S. Yanada and Y. Yasunaga, *Tissue Eng.*, 2004, **10**, 1093.

49. G. V. Silva, S. Litovsky, J. A. Assad, A. L. Sousa, B. J. Martin, D. Vela, S. C. Coulter, J. Lin, J. Ober, W. K. Vaughn, R. V. Branco, E. M. Oliveira, R. He, Y. J. Geng, J. T. Willerson and E. C. Perin, *Circulation*, 2005, **111**, 150.

50. C. Lange, F. Togel, H. Ittrich, F. Clayton, C. Nolte-Ernsting, A. R. Zander and C. Westenfelder, *Kidney Int.*, 2005, **68**, 1613.

51. M. Rojas, J. Xu, C. R. Woods, A. L. Mora, W. Spears, J. Roman and K. L. Brigham, *Am. J. Resp. Cell. Mol. Biol.*, 2005, **33**, 145.

52. D. G. Phinney and I. Isakova, *Curr. Pharm. Des.*, 2005, **11**, 1255.

53. Y. Sato, H. Araki, J. Kato, K. Nakamura, Y. Kawano, M. Kobune, T. Sato, K. Miyanishi, T. Takayama, M. Takahashi, R. Takimoto, S. Iyama, T. Matsunaga, S. Ohtani, A. Matsuura, H. Hamada and Y. Niitsu, *Blood*, 2005, **106**, 756.

54. E. M. Horwitz and M. Dominici, *Cytotherapy*, 2008, **10**, 771.

55. Y. Wu, L. Chen, P. G. Scott and E. E. Tredget, *Stem Cells*, 2007, **25**, 2648.

56. H. F. Dvorak, *N. Engl. J. Med.*, 1986, **315**, 1650.

57. L. G. Menon, S. Picinich, R. Koneru, H. Gao, S. Y. Lin, M. Koneru, P. Mayer-Kuckuk, J. Glod and D. Banerjee, *Stem Cells*, 2007, **25**, 520.

58. Y. Li, X. Yu, S. Lin, X. Li, S. Zhang and Y. H. Song, *Biochem. Biophys. Res. Commun.*, 2007, **356**, 780.
59. E. Mira, R. A. Lacalle, M. A. Gonzalez, C. Gómez-Moutón, J. L. Abad, A. Bernad, C. Martínez-A and S. Mañes, *EMBO Rep.*, 2001, **2**, 151.
60. S. B. Coffelt, R. S. Waterman, L. Florez, K. Höner zu Bentrup, K. J. Zwezdaryk, S. L. Tomchuck, H. L. LaMarca, E. S. Danka, C. A. Morris and A. B. Scandurro, *Int. J. Cancer*, 2008, **122**, 1030.
61. J. D. Heilborn, M. F. Nilsson, C. I. Jimenez, B. Sandstedt, N. Borregaard, E. Tham, O. E. Sørensen, G. Weber and M. Ståhle, *Int. J. Cancer*, 2005, **114**, 713.
62. J. von Haussen, R. Koczulla, R. Shaykhiev, C. Herr, O. Pinkenburg, D. Reimer, R. Wiewrodt, S. Biesterfeld, A. Aigner, F. Czubayko and R. Bals, *Lung Cancer*, 2008, **59**, 12.
63. R. Koczulla, G. von Degenfeld, C. Kupatt, F. Krötz, S. Zahler, T. Gloe, K. Issbrücker, P. Unterberger, M. Zaiou, C. Lebherz, A. Karl, P. Raake, A. Pfosser, P. Boekstegers, U. Welsch, P. S. Hiemstra, C. Vogelmeier, R. L. Gallo, M. Clauss and R. Bals, *J. Clin. Invest.*, 2003, **111**, 1665.
64. R. Shaykhiev, C. Beisswenger, K. Kändler, J. Senske, A. Püchner, T. Damm, J. Behr and R. Bals, *Am. J. Physiol.*, 2005, **289**, L842.
65. D. Yang, Q. Chen, A. P. Schmidt, G. M. Anderson, J. M. Wang, J. Wooters, J. J. Oppenheim and O. Chertov, *J. Exp. Med.*, 2000, **192**, 1069.
66. S. Coffelt, F. Marini, K. Watson, K. J. Zwezdaryk, J. L. Dembinski, H. L. LaMarca, S. L. Tomchuck, K. Honer zu Bentrup, E. S. Danka, S. L. Henkle and A. B. Scandurro, *PNAS*, 2009, **10**, 3806.
67. Y. Zhang, A. Daquinag, D. O. Traktuev, F. Amaya-Manzanares, P. J. Simmons, K. L. March, R. Pasqualini, W. Arap and M. G. Kolonin, *Cancer Res.*, 2009, **69**, 5259.
68. T. Kinnaird, E. Stabile, M. S. Burnett, C. W. Lee, S. Barr, S. Fuchs and S. E. Epstein, *Circ. Res.*, 2004, **94**, 678.
69. I. A. Potapova, G. R. Gaudette, P. R. Brink, R. B. Robinson, M. R. Rosen, I. S. Cohen and S. V. Doronin, *Stem Cells*, 2007, **25**, 1761.
70. K. Reddy, Z. Zhou, K. Schadler, S. F. Jia and E. S. Kleinerman, *Mol. Cancer Res.*, 2008, **6**, 929–36.
71. A. Al-Khaldi, N. Eliopoulos, D. Martineau, L. Lejeune, K. Lachapelle and J. Galipeau, *Gene Ther.*, 2003, **10**, 621.
72. G. Ishii, T. Sangai, T. Oda, Y. Aoyagi, T. Hasebe, N. Kanomata, Y. Endoh, C. Okumura, Y. Okuhara, J. Magae, M. Emura, T. Ochiya and A. Ochiai, *Biochem. Biophys. Res. Commun.*, 2003, **309**, 232.
73. D. Zipori, M. Krupsky and P. Resnitzky, *Cancer*, 1987, **60**, 1757.
74. F. Djouad, P. Plence, C. Bony, P. Tropel, F. Apparailly, J. Sany, D. Noël and C. Jorgensen, *Blood*, 2003, **102**, 3837.
75. A. E. Karnoub, A. B. Dash, A. P. Vo, A. Sullivan, M. W. Brooks, G. W. Bell, A. L. Richardson, K. Polyak, R. Tubo and R. A. Weinberg, *Nature*, 2007, **449**, 557.

76. S. C. Robinson, K. A. Scott, J. L. Wilson, R. G. Thompson, A. E. Proudfoot and F. R. Balkwill, *Cancer Res.*, 2003, **63**, 8360.

77. E. Azenshtein, G. Luboshits, S. Shina, E. Neumark, D. Shahbazian, M. Weil, N. Wigler, I. Keydar and A. Ben-Baruch, *Cancer Res.*, 2002, **62**, 1093.

78. E. L. Spaeth, J. L. Dembinski, A. K. Sasser, K. Watson, A. Klopp, B. Hall, M. Andreeff and F. Marini, *PloS ONE*, 2009, **4**, 1.

79. G. J. Maestroni, E. Hertens and P. Galli, *Cell. Mol. Life Sci.*, 1999, **55**, 663.

80. L. Ohlsson, L. Varas, C. Kjellman, K. Edvardsen and M. Lindvall, *Exper. Mol. Pathol.*, 2003, **75**, 248.

81. K. Nakamura, Y. Ito, Y. Kawano, K. Kurozumi, M. Kobune, H. Tsuda, A. Bizen, O. Honmou, Y. Niitsu and H. Hamada, *Gene Ther.*, 2004, **11**, 1155.

82. A. Y. Khakoo, S. Pati, S. A. Anderson, W. Reid, M. F. Elshal, I. I. Rovira, A. T. Nguyen, D. Malide, C. A. Combs, G. Hall, J. Zhang, M. Raffeld, T. B. Rogers, W. Stetler-Stevenson, J. A. Frank, M. Reitz and T. Finkel, *J. Exp. Med.*, 2006, **203**, 1235.

83. W. Qasim, H. B. Gaspar and A. J. Thrasher, *Gene Ther.*, 2009, **16**, 1285.

84. W. Wagner, F. Wein, A. Seckinger, M. Frankhauser, U. Wirkner, U. Krause, J. Blake, C. Schwager, V. Eckstein, W. Ansorge and A. D. Ho, *Exp Hematol.*, 2005, **33**, 1402.

85. J. J. Minguell, P. Conget and A. Erices, *Braz. J. Med. Biol. Res.*, 2000, **33**, 881.

86. M. E. Bernardo, N. Zaffaroni, F. Novara, A. M. Cometa, M. A. Avanzini, A. Moretta, D. Montagna, R. Maccario, R. Villa, M. G. Daidone, O. Zuffardi and F. Locatelli, *Cancer Res.*, 2007, **1**, 9142.

87. R. F. Pereira, K. W. Halford, M. D. O'Hara, D. B. Leeper, B. P. Sokolov, M. D. Pollard, O. Bagasra and D. J. Prockop, *Proc. Natl. Acad. Sci. U.S.A.*, 1995, **92**, 4857.

88. Y. Sato, H. Araki, J. Kato, K. Nakamura, Y. Kawano, M. Kobune, T. Sato, K. Miyanishi, T. Takayama, M. Takahashi, R. Takimoto, S. Iyama, T. Matsunaga, S. Ohtani, A. Matsuura, H. Hamada and Y. Niitsu, *Blood*, 2005, **106**, 756.

89. J. C. Marx, J. A. Allay, D. A. Persons, S. A. Nooner, P. W. Hargrove, P. F. Kelly, E. F. Vanin and E. M. Horwitz, *Hum. Gene Ther.*, 1999, **10**, 1163.

90. M. F. Pittenger, A. M. Mackay, S. C. Beck, R. K. Jaiswal, R. Douglas, J. D. Mosca, M. A. Moorman, D. W. Simonetti, S. Craig and D. R. Marshak, *Science*, 1999, **284**, 143.

91. T. Helledie, V. Nurcombe and S. M. Cool, *Stem Cells Dev.*, 2008, **17**, 837.

92. K. S. Aboody, J. Najbauer and M. K. Danks, *Gene Ther.*, 2008, **15**, 739.

93. L. Naldini, U. Blömer, P. Gallay, D. Ory, R. Mulligan, F. H. Gage, I. M. Verma and D. Trono, *Science*, 1996, **272**, 263.

94. C. Ren, S. Kumar, D. Chanda, L. Kallman, J. Chen, J. D. Mountz and S. Ponnazhagan, *Gene Ther.*, 2008, **1**.

95. A. Lokshin, J. E. Mayotte and M. L. Levitt, *J. Natl. Cancer Inst.*, 1995, **87**, 206.
96. K. Nakamura, Y. Ito, Y. Kawano, K. Kurozumi, M. Kobune, H. Tsuda, A. Bizen, O. Honmou, Y. Niitsu and H. Hamada, *Gene Ther.*, 2004, **11**, 1155.
97. L. Elzaouk, K. Moelling and J. Pavlovic, *Exp. Dermatol.*, 2006, **15**, 865.
98. C. Fillat, M. Carrió, A. Cascante and B. Sangro, *Curr. Gene Ther.*, 2003, **3**, 13.
99. L. Kucerova, V. Altanerova, M. Matuskova, S. Tyciakova and C. Altaner, *Cancer Res.*, 2007, **67**, 6304.
100. C. Carlo-Stella, C. Lavazza, M. Di Nicola, L. Cleris, P. Longoni, M. Milanesi, M. Magni, D. Morelli, A. Gloghini, A. Carbone and A.M. Gianni, *Hum. Gene Ther.*, 2006, **17**, 1225.
101. H. Walczak, R. E. Miller, K. Ariail, B. Gliniak, T. S. Griffith, M. Kubin, W. Chin, J. Jones, A. Woodward, T. Le, C. Smith, P. Smolak, R. G. Goodwin, C. T. Rauch, J. C. Schuh and D. H. Lynch, *Nat. Med.*, 1999, **5**, 157.
102. A. Mohr, M. Lyons, L. Deedigan, T. Harte, G. Shaw, L. Howard, F. Barry, T. O'Brien and R. Zwacka, *J. Cell Mol. Med.*, 2008, **12**, 2628.
103. L. S. Sasportas, R. Kasmieh, H. Wakimoto, S. Hingtgen, J. A. van de Water, G. Mohapatra, J. L. Figueiredo, R. L. Martuza, R. Weissleder and K. Shah, *Proc. Natl. Acad. Sci. U. S. A.*, 2009, **106**, 4822.
104. S. M. Kim, J. Y. Lim, S. I. Park, C. H. Jeong, J. H. Oh, M. Jeong, W. Oh, S. H. Park, Y. C. Sung and S. S. Jeun, *Cancer Res.*, 2008, **68**, 9614.
105. G. Grisendi, R. Bussolari, L. Cafarelli, I. Petak, V. Rasini, E. Veronesi, G. De Santis, C. Spano, M. Tagliazzucchi, H. Barti-Juhasz, L. Scarabelli, F. Bambi, A. Frassoldati, G. Rossi, C. Casali, U. Morandi, E. M. Horwitz, P. Paolucci, P.F. Conte and M. Dominici, *Cancer Res.*, 2010, in press.
106. A. K. Sasser, B. L. Mundy, K. M. Smith, A. W. Studebaker, A. E. Axel, A. M. Haidet, S. A. Fernandez and B. M. Hall, *Cancer Lett.*, 2007, **254**, 255.
107. M. Kanehira, H. Xin, K. Hoshino, M. Maemondo, H. Mizuguchi, T. Hayakawa, K. Matsumoto, T. Nakamura, T. Nukiwa and Y. Saijoo, *Cancer Gene Ther.*, 2007, **14**, 894.
108. H. Xin, M. Kanehira, H. Mizuguchi, T. Hayakawa, T. Kikuchi, T. Nukiwa and Y. Saijo, *Stem Cells*, 2007, **25**, 1618.
109. I. T. Cavarretta, V. Altanerova, M. Matuskova, L. Kucerova, Z. Culig and C. Altaner *Mol Ther.*, 2009, Oct 20.
110. L. G. Menon, K. Kelly, H. W. Yang, S. K. Kim, P. M. Black and R. S. Carroll, *Stem Cells*, 2009, **27**, 2320.
111. R. L. Yong, N. Shinojima, J. Fueyo, J. Gumin, G. G. Vecil, F. C. Marini, O. Bogler, M. Andreeff and F. F Lang, *Cancer Res.*, 2009, Dec 1.

Fibrin-based Matrices to Support Stem Cell-Based Tissue Regeneration

RAPHAEL GORODETSKY,[a] IRIS MIRONI-HARPAZ[b] AND DROR SELIKTAR[b]

[a] Hadassah Hebrew University Medical Center, Jerusalem, Israel; [b] Faculty of Biomedical Engineering, Technion-Israel Institute of Technology, Haifa, Israel

8.1 Background: The Pitfalls of Matrix-based Three-dimensional Tissue Engineering

Due to the complex structure of different tissues and their inability to spontaneously regenerate after birth, the practice of tissue engineering is not straightforward.[1–10] Proposed techniques in this field are based on building cellular *in vitro* scaffolds from synthetic or biological polymeric materials, either with the differentiated cells of the damaged tissue or with adequate multipotent stem cells and the addition of growth factors. Both the cells and matrix of the implanted scaffold may be rejected. But even with an optimal choice of cells and scaffold, in most tissues, the integration of a construct with the wound bed requires the immediate formation of a vascular network so that the implanted living cells within the scaffold will be exposed to the circulation immediately, and will be at least 100–150 µm in proximity to a blood capillary. This cannot be achieved in three-dimensional (3D) implants because induction

Stem Cell-Based Tissue Repair
Edited by Raphael Gorodetsky and Richard Schäfer
© The Royal Society of Chemistry 2011
Published by the Royal Society of Chemistry, www.rsc.org

of angiogenesis takes time, and the cells within the matrix may be subjected to severe hypoxia in certain cases of construct integration.

8.2 Fibrin Structure, its Formation and Mechanism of Action in Damaged Tissue

As a matrix that is naturally formed in wounds, fibrin seems to serve as a raw material of choice for building scaffolds for tissue engineering. Normal fibrinogen (fib_{340}) is synthesized by the liver and secreted into the blood and lymph systems.[11,12] It consists of two sets of three chains (α, β and γ). In addition, a variant of fibrinogen with an extended α chain, αE,[13] whose function is yet not clear, appears in a small percentage of the total adult human fibrinogen. Fibrinogen circulates in the blood vessels in concentrations ranging between 2–$5\,mg\,mL^{-1}$ and it forms a fibrin clot upon its activation. Its major role is to block blood flow after an injury and reduce hemorrhage upon injury.

Blood coagulation occurs in two stages. Initially, a soft fibrin gel, based on a linear fibrillar structure, is self-assembled upon thrombin cleavage of fibrinopeptides A and B. Then the gel is reinforced by laterally associated fibers that cross-link by intrinsic thrombin activated factor XIII and tissue transglutaminases.[14,15] Circulating platelets, trapped in the fibrin mesh, become activated and secrete fibrinogen, as well as factor XIII from their alpha granules. Factor XIII is also a transglutaminase that cross-links the fibrin, and hence further contributes to clot stabilization.[16] Highly potent growth factors such as platelet derived growth factor (PDGF) are secreted upon platelet activation. Red blood cells are trapped in the clot and leukocytes are recruited to it as well. The first aid role of the fibrin clot in physiological conditions is to control bleeding immediately after injury. Eventually the clot is rapidly degraded in the tissues by the efficient fibrinolytic enzyme plasmin. The clot is cleaved into soluble fibrin-degradation products, which are cleared away by the blood. The longest circulating domain after fibrinolysis is the D–D dimer which is constituted by the 3D structure made of the combined C-terminal structures of all the fibrin chains.[17–19]

Besides being a hemostatic agent, fibrin can serve as an interim matrix to recruit cells participating in the formation of granulation tissue and initiate the process of tissue regeneration.[20] The pro-migratory (chemotactic) and cell binding (haptotactic) properties of fibrin can modulate the migration of progenitors within the normal tissues into the clot.[19,21–25] Cells bound to fibrin can induce clot retraction and secrete new collagen, as well as other extracellular matrix (ECM) molecules.[26] Platelet derived growth factor (PDGF) and vascular endothelial growth factor (VEGF) are the two major factors that attract endothelial cells into the granulation tissue and induce accelerated repopulation while inducing neo-vascularization in the area of the damage. Fibrin and its degradation products as well as fibrin-based composite materials and matrices also have been described as pro-angiogenic and may participate in tissue repair.[19,27–30]

Though fibrin-based materials were proposed for cell delivery in epidermal repair,[31–34] it appears that in a normal situation of wound healing, keratinocytes are not attracted to fibrin,[35,36] although they may help in fibrinolysis.[37] The keratinocytes seem to regenerate the damaged skin by crawling underneath the clot on the surface of the dermis and eject it when the recovery of the epidermis is complete.[36]

8.3 Proposed Mechanisms for Cell Interactions with Fibrin(ogen)

Fibrin was shown to attach matrix dependent cell types, mostly from mesenchymal origin, including fibroblasts (HF), endothelial cells (EC) and smooth muscle cells (SMC).[20,21,25,26] As the degradation of the fibrin clot occurs, these cells replace the fibrin-based provisional matrix with a new extracellular matrix of the healed tissue and/or of a new scar tissue.

The most established cellular interaction of fibrin is through platelet integrin $\alpha IIb\beta3$ through the γ-chain C-terminal epitope containing Ala–Gly–Asp–Val (AGDV).[21,38,39] Epitopes on fibrin were also described as binding sites for other hematopoietic cells such macrophages and other leukocytes.[40,41] A number of cell attachment and adhesion epitopes have been proposed for fibrin(ogen). Each α-chain in human fibrinogen contains two arginine–glycine–aspartic acid (RGD) sequences that were regarded as candidates for integrin-mediated cell binding to fibrinogen.[21,26,38] The RGD binding sites on fibrinogen that interact with $\alpha v\beta3$ on endothelial cells were claimed to have a major contribution to the adhesion of these and other cells, and to be indirectly responsible for the regeneration of blood vessels in the fibrin clot.[42,43] However, these RGD-containing domains seem to be of low conservation between different species, even within mammals (as evident from the sequences in the gene bank), which bring to question their importance in cell binding. Moreover, the ability of RGD-depleted recombinant fibrinogen to interact with endothelial cells[26] suggests that other cell-binding domains on fibrin(ogen) are involved. It has been suggested that the minute quantities of residual fibronectin, which may co-purify with fibrinogen, might be responsible for the major cell binding activity of the fibrin clot. However, other reports on cell binding to highly purified fibrin(ogen) seem to rule out this possibility.[20]

Another possible cell-interacting mechanism to fibrin that can be considered is based on the family of membranal proteins of the cell marker CD44. These multifunctional set of cell surface molecules are primarily hyaluronate-binding protein that bind to chondroitin sulfate proteoglycans. CD44 has been reported to be involved in cell proliferation, differentiation and migration, as well as signaling for cell survival.[44–47] Binding of CD44 with fibrin may also reduce apoptotic processes.[48–50] The expression of CD44 specifically in mesenchymal cells may contribute to the binding of these cells to fibrin substrates.[51,52] It is still open for further investigation whether binding of cells to CD44 also triggers amplification of CD44 expression on their membrane.

Recently, alternative cell-binding sites to fibrinogen have been proposed.[53] Highly conserved ~20mer sequences in the C-terminal ends of the fibrinogen β-chains and their homologous sequence preceding the C-terminal end of γ chains were found to strongly bind different types of anchorage-dependent cells, as does the whole molecule. The C-terminus of the extended α chain (αE) also has such a homologous cell binding sequence. Peptides of these sequences form a family of new cell-binding epitopes in fibrinogen that were termed 'haptides'. The haptides on the chains β, γ and αE were termed Cβ, Cγ and CαE, respectively.[53] The preferential affinity of the haptides to cells of mesenchymal origin, or other cells secreting extracellular matrix, was evident.[53] Haptides could attach to matrices and augment their performance for tissue engineering through reducing the immune response towards the coated matrices.[54] Cell binding by sequences homologous to fibrin-derived haptides in the C-termini of other non-fibrin molecules with fibrin-related domains (FRED) has been noted, which suggests that the haptide peptidic family appear not only in fibrin, and that homologous sequences of other proteins can interact with cells in a similar manner.

8.4 Utilization of Products Derived from Fibrinogen and Thrombin Purified from Blood Plasma

Some breakthrough technologies that can purify, sterilize and viral-inactivate pooled plasma derived proteins provided safe use of plasma proteins for different clinical applications. Formation of a fibrin clot as a two-component biological glue led to the development of a variety of fibrin glues and sealants for clinical use. These products are based on much higher concentrations of fibrinogen (>40–$115\,\mathrm{mg\,L^{-1}}$) and their activation is performed by thrombin. Such products are currently available for different applications in surgery and wound healing.[43,55–58] The inclusion of naturally occurring growth factors that co-purify with fibrinogen and findings on the potent interaction of fibrin(ogen) with cells further contributed to the benefit of such approaches.

8.5 Tissue Engineering with Modified Fibrin Matrices

The biodegradable, non-toxic and non-immunogenic properties of fibrin glue itself with no stabilizing agents makes it a good basic material for producing bio-engineered matrices.[59–61] *In vitro* responses of cultured cells to the fibrin and its components have been evaluated.[20] For example, a fibrin clot recruited cultured human fibroblasts and endothelial cells from the surrounding area. Fibrinogen only slightly increased fibroblast proliferation while thrombin enhanced their proliferation by a factor of 1.5–1.8.

The disadvantage of fibrin sealants as the sole component of a cell-binding matrix for tissue regeneration is the fast rate of fibrinolysis once introduced into an *in vivo* environment. A few approaches that try to design fibrin matrices have been adopted to overcome this problem. Such solutions include the formation of soft beads from alginate and fibrinogen composites as cell carriers. The alginate is used to stabilize the fibrin matrix and to form the beads onto which cells could be loaded.[62,63] Others designed composite scaffolds made of whole or separated fibrinogen chains, bound covalently to synthetic polymer cross-linkers, and formed into hydrogels. These can form a more durable and controlled pore size meshwork for culturing cells in 3D conditions and for tissues regeneration matrices for implantation.[64–66]

8.5.1 Fibrinogen Conjugated to Synthetic Polymers as a Scaffold Material

Sophisticated fibrin-based tissue engineering scaffolds can be designed as instructive biomaterials of a biomimetic nature.[67] Such engineered scaffolds can be designed to degrade slower than fibrin clots and still preserve the desired natural cell-binding properties of fibrinogen. A biomimetic fibrin scaffold design necessitates not only insight into how the biological building blocks in the natural ECM are arranged to function properly, it also requires technical know-how in manipulating these building blocks by using man-made technologies, *i.e.* polymer chemistry.

Proteins such as fibrinogen are ideal candidates for biological building blocks in scaffolds that include polymer chemistry methodologies for introducing control features.[68] They has been widely used to make hydrogel scaffolds for tissue engineering due to their unique biomimetic 3D architecture coupled with intrinsic cell signaling that promote extensive cellular remodeling.[67,68] Reconstituted proteins such as fibrinogen and collagen typically assemble into fibrillar hydrogel networks that are supported by covalent and non-covalent protein interactions[69,70] and can be physically cross-linked with additional factors.[70] Even though the 3D microenvironment created by a network of fibrin mesh has a highly variable micrometer pore size, which is not easily controlled, cells are still able to remodel this environment with ease.[71] Matrix architecture may be altered somewhat by varying the composition or changing the cross-linking conditions,[72,73] yet the pore size, permeability, proteolytic susceptibility and mechanical properties that proactively and favorably affect tissue regeneration may be far from reach in such systems.[74,75]

There are a number of techniques to improve or modify the physicochemical properties of reconstituted protein hydrogels,[76–80] but most processing of pre-cast biological materials involves harsh or toxic conditions that do away with the benefits of *in situ* cell seeding during gelation.[70,81] An alternative approach is to use a biosynthetic hybrid material for tissue engineering. Hybrid biomaterials incorporate biological macromolecules with structurally versatile synthetic polymers to create a cross-linked hydrogel network.[67,82–92] These hybrid biomaterials can be used to create a biomimetic cellular environment by balancing the structural and biofunctional elements. Control over structural properties including porosity, compliance, bulk density, mechanical properties and degradability are directed through the synthetic polymer network,[93–96] while the biological cell signaling is controlled through the incorporation of biological macromolecules, which may include protein fragments,[97] growth factors[98–100] or biologically active peptide sequences.[101–103] In this regard, both the biochemical and biomechanical features of the scaffold may be used to initiate important cellular remodeling events including cell migration, proliferation and guided differentiation. Such materials could readily be customized with micro-architecture, matrix stiffness and proteolytic resistance specifically designed for guiding the remodeling and cell phenotype towards specific tissue engineering end-points.[104]

One such hybrid biomaterial can be created using the natural fibrinogen monomers and synthetic polymers as the building blocks of the hydrogel matrix (protein–polymer adducts). Using this protein–polymer approach, the natural fibrinogen molecules contain and provide the necessary cell signaling domains within their amino acid sequence, including cell adhesion sites and a protease degradation substrate. The fibrinogen also serves as the structural backbone of the polymeric network, thereby rendering the hydrogel naturally biodegradable *via* the inherent degradation sites on the protein sequence. Most of the structural properties of the protein–polymer hydrogel network are controlled through the synthetic polymer constituent. One of the unique properties of this biomaterial is that it can pseudo-independently alter its biochemistry and physical properties.[64,105] This material has been rigorously validated with *in vitro* techniques, to demonstrate the utility of the protein–polymer adduct approach.[64–66,105–111] The polymer–protein hydrogel system was also validated as a cell culture matrix in studies aimed at understanding how a precisely controlled 3D cellular microenvironment—including specific changes in matrix stiffness, proteolytic resistance and ligand density—impacts cell phenotype and remodeling.

8.5.2 Heat Stabilized Fibrin Matrices and Fibrin Microbeads for Tissue Engineering

A simple dehydrothermal approach for stabilizing fibrin and transforming it into a slowly degrading matrix without compromising its cell binding properties was proposed. This technique is based on dry heating the different forms of fibrin in controlled temperatures that do not exceed 80 °C, which further condensates the factor XIII cross-linked fibrin by heat-induced massive crosslinking.[54,112] The cell-binding capabilities to the surface of such fibrin dense matrices are preserved and may even be enhanced.[61,112] The hardened, slowly biodegradable fibrin matrices are therefore good materials for tissue engineering.

Fibrin microbeads (FMB), which were developed based on moderate heat fibrin condensation, may also be used for tissue engineering. They are hard and stable beads in a size range of 50–200 μm in diameter. They address some of the major problems associated with the high rate of degradation of fibrin-based material. These small particles can be loaded with cells in suspension culture.

Such cells could eventually be delivered with cells for tissue regeneration without the need for large 3D scaffolds.[43,112–116] FMB allow the isolation of cells with higher yields because growth and differentiation of the matrix-dependent mesenchymal stem cells (MSCs) from different sources can be implanted directly with the beads for tissue regeneration.[43,112,113,115–120]

FMB appear to possess unique properties that can be exploited for a wide variety of applications which require manipulation of cells. In particular, as cells-on-FMB are cultured in suspension, they can be transferred without trypsinization and can act as a substrate for the growth of tissues *in vitro*. In their preparation, FMB undergo excessive exposure to oils, organic solvents and ethanol without losing their fibrin-based cell-binding activity. FMB can also be sterilized by ethanol and high dose γ-irradiation without losing significantly their cell binding properties.[112] FMB bind a large variety of cell types, mostly from mesenchymal origin, including fibroblastic cells and MSCs from different sources, osteoblasts, chondrocytes, endothelial and smooth muscle cells, and transformed cells of similar origin (Table 8.1). Interestingly, human

Table 8.1 Cell binding properties to FMB.[a]

Cell types tested	Binding efficiency (grade 0–4)	Suspension culture (grade 0–4)	Possible applications
Mesenchymal tissues: fibroblasts, chondrocytes, osteoblasts, adipocytes	4	3–4	Wound healing Cell based cosmetic treatments
Embryonic stem cells	0–2	1	Exclusion of mesenchymal cells from pure embryonic stem cell cultures
Mesenchymal stem cells: from bone marrow, GCSF mobilized blood, endothelial progenitors	4	4	Tissue regeneration, mainly bone, cartilage and skin Research models of 3D cultures Research models for culturing cells on beads in suspension.
Mature endothelial and smooth-muscle cells	4	4	Cell implant for vascular regeneration
Leukocytic cells	1	0	Negative selection: exclusion of mesenchymal cells to purify leukocytic cultures.
Normal keratinocytes	1–2	0	–
Transformed keratinocytes (HaKat)	3–4	4	Research models of 3D cultures
Tumor cell lines matrix-dependent and transformed	3–4	3–4	Research models of 3D cultures

[a]Based on ref. 20, 43, 53, 112 and 123, and unpublished data.

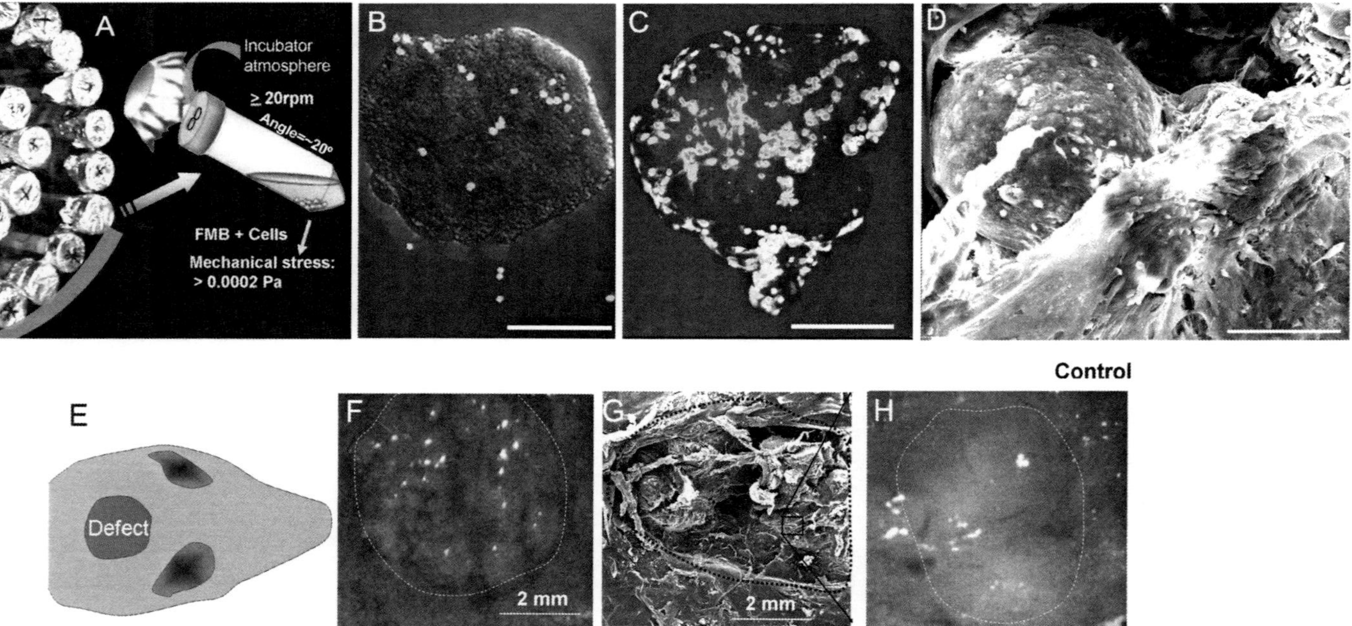

Figure 8.1 FMB-based stem cells growth and differentiation in suspension. (A) A slow rotating device within the CO_2 incubator for the growth of multiple samples in one run. The small cells separated from bone marrow initially (B) grow on them so that within about 1–2 weeks they cover them (C) (nuclei stained in fluorescence microscopy). Eventually when induced to differentiate to bone tissue, a new cellular hard tissue is formed *in vitro* between the cell-coated residues of the initial FMB scaffolds (D). (E)–(H) show an example of bone repair model in mice. A defect of ∼5 mm was induced as shown schematically (E). The defect was filled with FMB+ cells or left untreated. After a month, a hard bone-like tissue was formed where the FMB were installed with bone differentiated cells as seen on the exposed skulls (F), and also as shown by scanning electron microscopy (G), while the control skulls were left with only minimal repaired soft tissue without any bone filling (B). (C) shows low magnification scanning electron microscopy (SEM) of the area of the defect where new bone tissue is organizing within the defect. (D) shows a SEM magnification of a small fraction of the repaired tissue of (C) where cells are seen to weave a new calcified tissue. Based on ref. 43, 116, 126.

keratinocytes bind very poorly to FMB (as shown in Table 8.1), which explains their interaction with fibrin *in vivo*.[43] Cells adhere to FMB and proliferate thereon to reach a very high density of up to many millions of cells per mL of packed beads. FMB can also serve as an experimental tool for cell culture in 3D conditions, in which cell behavior changes dramatically compared with two-dimensional (2D) cultures. This can serve as a simple model for studies in tumor biology and cell behavior in suspension cultures.[112,118,119]

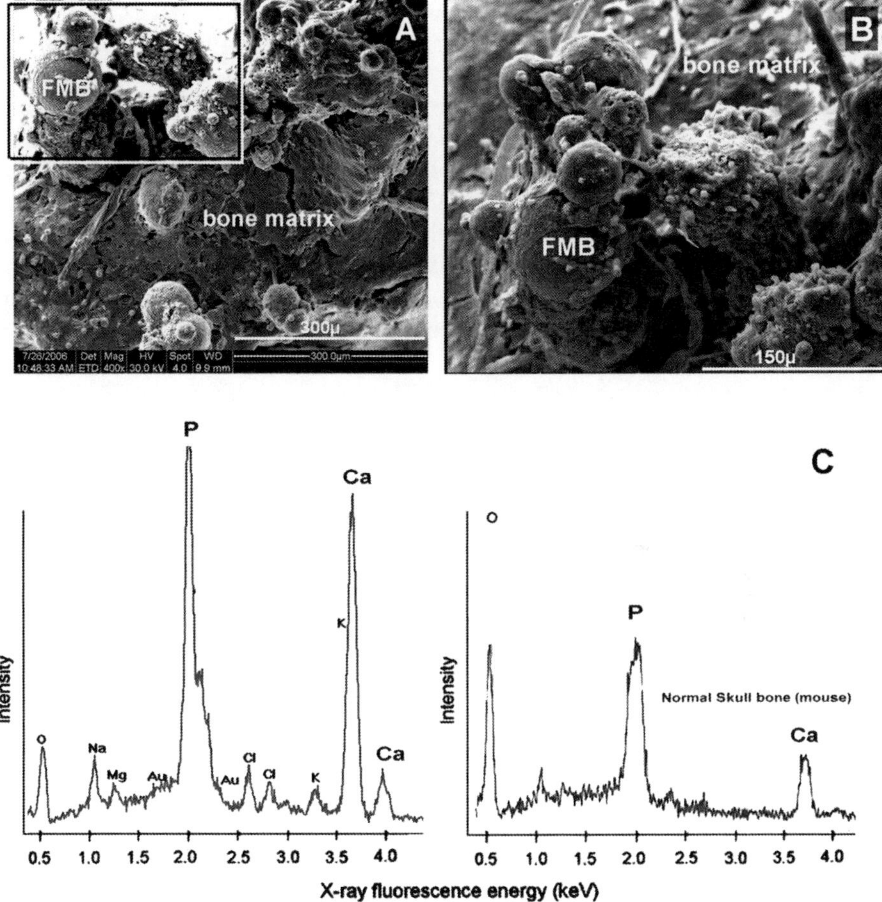

Figure 8.2 Formation of bone-like matrix from cells on FMB. Scanning electron microscopy (SEM) of bone-like tissue formed *in vitro* by MSCs isolated by FMB that are cultured and driven to differentiate to bone in slow rotation for prolonged time (30 days). (A) Low magnification (×400). (B) The frame in (A) is further magnified for the details (×600). The FMB loaded with cells still show proliferating cells. The cells migrate from the FMB onto the matrix that they secrete to form hard calcified bone-like structures as can be verified by the deposited Ca ratio of Ca/P by microprobe X-ray spectrometry attached to the SEM (C).

Fast vascularization is a main requirement for normal tissue survival, and for successful cell based tissue regeneration on one hand and for tumor viability on the other hand. Mesenchymal cells bind to FMB even better than to fibrinogen and fibrin gels.[43] Endothelial cells were shown to respond to fibrinogen in terms of attachment, migration and formation of capillary structures.[25,120–122] When the cells were loaded in high density on FMB and were introduced onto a wound bed, they seemed to secrete factors such as VEGF that could further contribute to the induction of growth of capillaries into the implanted area to support the formation of the new tissue. The basis for such a mechanism is not clear, although it has been previously described for their activity within fibrin gels.[123–125]

Figure 8.1 describes some features of FMB technology. Initially, the MSCs are separated and grown in suspension culture in rotation. Once the cells load on the FMB, they begin to proliferate on them and with an adequate pro-differentiation matrix, these expanded cells form *in vitro* bone-like structures (Figure 8.2). In bone regeneration models such as a critical defect model in the mouse skull, the cells were induced to differentiate to osteoblasts on FMB and were implanted to form a new bone filling towards full recovery of the bone defect that did not heal by itself in control animals. The fact that the implant was made of small beads with cells allowed nutrition of the cells until neo-vascularization occurred and allowed the expansion of the tissue built by the implanted cells.[116,126,127]

For cartilage formation the use of FMB alone is not sufficient, since upon implantation, FMB promote angiogenesis. Therefore the combination of FMB for the isolation and collagen matrices was used to produce a cartilage-like tissue.[127]

8.6 Conclusions

Attempts to use fibrin-based materials as a platform for tissue regeneration together with stem cells are well-established in the scientific literature. Nevertheless, unmodified fibrin as a sole constituent is not ideally suited for many applications requiring long-term stability and biodegradability from the implantable matrix. To address this difficulty, a number of techniques were recently introduced to produce fibrin-based materials for stem cell tissue regeneration with similar biocompatibility and the added advantage of enhanced *in vivo* residence. These so-called second generation fibrin matrices may still be employed for *in vivo* implantation using minimally invasive techniques, similar to conventional fibrin materials. A couple of these second generation fibrin materials have already been demonstrated efficacious in advancing the overall efforts of using stem cells for tissue regeneration.

References

1. R. Zdrahala and I. Zdrahala, *In vivo* tissue engineering: Part I. Concept genesis and guidelines for its realization, *J. Biomater. Appl.*, 1999, **14**, 192–209.

2. L. Bonassar and C. Vacanti, Tissue engineering: the first decade and beyond, *J. Cell. Biochem. Suppl.*, 1998.
3. J. Young, J. Teumer, P. Kemp and N. Parenteau, Approaches to transplanting engineered cells and tissues, in *Principles of Tissue Engineering*, R. Lanza, R. Langer and W. Chick, (ed.), R.G. Landes Company, Austin, TX, 1997, pp. 297–307.
4. D. Michaeli and M. McPherson, Immunologic study of artificial skin used in the treatment of thermal injuries, *J. Burn Care. Rehabil.*, 1990, **11**, 21–26.
5. A. T. Truong, A. Kowal-Vern, B. A. Latenser, D. E. Wiley and R. J. Walter, Comparison of dermal substitutes in wound healing utilizing a nude mouse model, *J. Burns Wounds*, 2005, **4**, e4.
6. C. Pham, J. Greenwood, H. Cleland, P. Woodruff and G. Maddern, Bioengineered skin substitutes for the management of burns: a systematic review, *Burns*, 2007, **33**, 946–957.
7. M. Ehrenreich and Z. Ruszczak, Update on tissue-engineered biological dressings, *Tissue Eng.*, 2006, **12**, 2407–2424.
8. G. D. Gentzkow, *et al.*, Use of dermagraft, a cultured human dermis, to treat diabetic foot ulcers, *Diabetes Care*, 1996, **19**, 350–354.
9. L. Griffith, Emerging design principles in biomaterials and scaffolds for tissue engineering, *Ann. N. Y. Acad. Sci.*, 2002, **961**, 83–95.
10. Z. Lokmic, F. Stillaert, W. A. Morrison, E. W. Thompson and G. M. Mitchell, An arteriovenous loop in a protected space generates a permanent, highly vascular, tissue-engineered construct, *FASEB J.*, 2007, **21**, 511–522.
11. R. F. Doolittle, Z. Yang and I. Mochalkin, Crystal structure studies on fibrinogen and fibrin, *Ann. N. Y. Acad. Sci.*, 2001, **936**, 31–43.
12. R. F. Doolittle, G. Spraggon and S. J. Everse, Crystal structures of fragments D and double-D from fibrinogen and fibrin, *Thromb. Haemost.*, 1999, **82**, 271–276.
13. V. K. Lishko, V. P. Yakubenko, K. M. Hertzberg, G. Grieninger and T. P. Ugarova, The alternatively spliced alpha(E)C domain of human fibrinogen-420 is a novel ligand for leukocyte integrins alpha(M)beta(2) and alpha(X)beta(2), *Blood*, 2001, **98**, 2448–2455.
14. P. A. Owren and H. Stormorken, The mechanism of blood coagulation, *Ergeb. Physiol.*, 1973, **68**, 1–53.
15. G. C. Troy, An overview of hemostasis, *Vet. Clin. North Am. Small Anim. Pract.*, 1988, **18**, 5–20.
16. G. Marx, G. Korner, X. Mou and R. Gorodetsky, Packaging zinc, fibrinogen, and factor XIII in platelet alpha-granules, *J. Cell. Physiol.*, 1993, **156**, 437–442.
17. G. P. B. McNicol, The fibrinolytic mechanism. The mechanism of fibrinolysis, *Proc. R. Soc. Lond. B Biol. Sci.*, 1969, **173**, 285–291.
18. C. W. Francis and V. J. Marder, Concepts of clot lysis, *Annu. Rev. Med.*, 1986, **37**, 187–204.
19. W. D. Thompson, C. M. Stirk, W. T. Melvin and E. B. Smith, Plasmin, fibrin degradation and angiogenesis, *Nat. Med.*, 1996, **2**, 493.

20. R. Gorodetsky, A. Vexler, J. An, Z. Mou and G. Marx, Haptotactic and growth stimulatory effects of fibrin(ogen) and thrombin on cultured fibroblasts, *J. Lab. Clin. Med.*, 1998, **131**, 269–280.
21. K. Suehiro, *et al.*, Fibrinogen binds to integrin alpha(5)beta(1) *via* the carboxyl-terminal RGD site of the Aalpha-chain, *J. Biochem.*, 2000, **128**, 705–710.
22. T. Browder, J. Folkman and S. Pirie-Shepherd, The hemostatic system as a regulator of angiogenesis, *J. Biol. Chem.*, 2000, **275**, 1521–1524.
23. E. Dejana, A. Zanetti and G. Conforti, Biochemical and functional characteristics of fibrinogen interaction with endothelial cells, *Haemostasis*, 1988, **18**, 262–270.
24. P. A. Janmey, J. P. Winer and J. W. Weisel, Fibrin gels and their clinical and bioengineering applications, *J. R. Soc. Interface*, 2009, **6**, 1–10.
25. M. Ge, G. Tang, T. J. Ryan and A. B. Malik, Fibrinogen degradation product fragment D induces endothelial cell detachment by activation of cell-mediated fibrinolysis, *J. Clin. Invest.*, 1992, **90**, 2508–2516.
26. R. A. Smith, *et al.*, The role of putative fibrinogen Aalpha-, Bbeta-, and GammaA-chain integrin binding sites in endothelial cell-mediated clot retraction, *J. Biol. Chem.*, 1997, **272**, 22080–22085.
27. G. Zhang, Q. Hu, E. A. Braunlin, L. J. Suggs and J. Zhang, Enhancing efficacy of stem cell transplantation to the heart with a PEGylated fibrin biomatrix, *Tissue Eng. Part A*, 2008, **14**, 1025–1036.
28. H. Hall, Modified fibrin hydrogel matrices: both, 3D-scaffolds and local and controlled release systems to stimulate angiogenesis, *Curr. Pharm. Des.*, 2007, **13**, 3597–3607.
29. N. Laurens, P. Koolwijk and M. P. de Maat, Fibrin structure and wound healing, *J. Thromb. Haemost.*, 2006, **4**, 932–939.
30. W. D. Thompson and M. A. Kazmi, Angiogenic stimulation compared with angiogenic reaction to injury: distinction by focal and general application of trypsin to the chick chorioallantoic membrane, *Br. J. Exp. Pathol.*, 1989, **70**, 627–635.
31. R. E. Horch, A. M. Munster and B. M. Achauer, *Cultured Human Keratinocytes and Tissue Engineered Skin Substitutes*, Thieme, Stuttgart, 2001.
32. C. A. Acevedo, *et al.*, A mathematical model for the design of fibrin microcapsules with skin cells, *Bioprocess. Biosyst. Eng.*, 2009, **32**, 341–351.
33. J. Hunyadi, B. Farkas, C. Bertenyi, J. Olah and A. Dobozy, Keratinocyte grafting: covering of skin defects by separated autologous keratinocytes in a fibrin net, *J. Invest. Dermatol.*, 1987, **89**, 119–120.
34. L. J. Currie, R. Martin, J. R. Sharpe and S. E. James, A comparison of keratinocyte cell sprays with and without fibrin glue, *Burns*, 2003, **29**, 677–685.
35. E. Weiss, *et al.*, Un-cross-linked fibrin substrates inhibit keratinocyte spreading and replication: correction with fibronectin and factor XIII cross-linking, *J. Cell. Physiol.*, 1998, **174**, 58–65.

36. M. Kubo, *et al.*, Fibrinogen and fibrin are anti-adhesive for keratinocytes: a mechanism for fibrin eschar slough during wound repair, *J. Invest. Dermatol.*, 2001, **117**, 1369–1381.

37. D. J. Donaldson, J. T. Mahan, D. L. Amrani, D. H. Farrell and J. H. Sobel, Further studies on the interaction of migrating keratinocytes with fibrinogen, *Cell Adhes. Commun.*, 1994, **2**, 299–308.

38. J. Gailit, *et al.*, Human fibroblasts bind directly to fibrinogen at RGD sites through integrin alpha(v)beta3, *Exp. Cell Res.*, 1997, **232**, 118–126.

39. R. A. Clark, J. Q. An, D. Greiling, A. Khan and J. E. Schwarzbauer, Fibroblast migration on fibronectin requires three distinct functional domains, *J. Invest. Dermatol.*, 2003, **121**, 695–705.

40. D. I. Simon, *et al.*, Mac-1 (CD11b/CD18) and the urokinase receptor (CD87) form a functional unit on monocytic cells, *Blood*, 1996, **88**, 3185–3194.

41. E. Wojtecka-Lukasik and S. Maslinski, Fibronectin and fibrinogen degradation products stimulate PMN-leukocyte and mast cell degranulation, *J. Physiol. Pharmacol.*, 1992, **43**, 173–181.

42. X. Guo, *et al.*, Repair of large articular cartilage defects with implants of autologous mesenchymal stem cells seeded into b-tricalcium phosphate in a sheep model, *Tissue Eng.*, 2004, **10**, 1818–1829.

43. R. Gorodetsky, *et al.*, Fibrin microbeads (FMB) as biodegradable carriers for culturing cells and for accelerating wound healing, *J. Invest. Dermatol.*, 1999, **112**, 866–872.

44. E. A. Turley, P. W. Noble and L. U. Bourguignon, Signaling properties of hyaluronan receptors, *J. Biol. Chem.*, 2002, **277**, 4589–4592.

45. M. Hofmann, *et al.*, Identification of IHABP, a 95 kDa intracellular hyaluronate binding protein, *J. Cell. Sci.*, 1998, **111**, 1673–1684.

46. J. Bajorath, Molecular organization, structural features, and ligand binding characteristics of CD44, a highly variable cell surface glycoprotein with multiple functions, *Protein. Struct. Funct. Genet.*, 2000, **39**, 103–111.

47. L. Sherman, D. Wainwright, H. Ponta and P. Herrlich, A splice variant of CD44 expressed in the apical ectodermal ridge presents fibroblast growth factors to limb mesenchyme and is required for limb outgrowth, *Genes Dev.*, 1998, **12**, 1058–1071.

48. L. Eshkar Sebban, *et al.*, The involvement of CD44 and its novel ligand galectin-8 in apoptotic regulation of autoimmune inflammation, *J. Immunol.*, 2007, **179**, 1225–1235.

49. C. A. Henke, U. Roongta, D. J. Mickelson, J. R. Knutson and J. B. McCarthy, CD44-related chondroitin sulfate proteoglycan, a cell surface receptor implicated with tumor cell invasion, mediates endothelial cell migration on fibrinogen and invasion into a fibrin matrix, *J. Clin. Invest.*, 1996, **97**, 2541–2552.

50. C. Henke, P. Bitterman, U. Roongta, D. Ingbar and V. Polunovsky, Induction of fibroblast apoptosis by anti-CD44 antibody: implications for the treatment of fibroproliferative lung disease, *Am. J. Pathol.*, 1996, **149**, 1639–1650.

51. V. Nehls and W. Hayen, Are hyaluronan receptors involved in three-dimensional cell migration?, *Histol. Histopathol.*, 2000, **15**, 629–636.

52. R. A. Clark, F. Lin, D. Greiling, J. An and J. R. Couchman, Fibroblast invasive migration into fibronectin/fibrin gels requires a previously uncharacterized dermatan sulfate-CD44 proteoglycan, *J. Invest. Dermatol.*, 2004, **122**, 266–277.

53. R. Gorodetsky, A. Vexler, M. Shamir, J. An, L. Levdansky, I. Shimeliovich and G. Marx, New cell attachment peptide sequences from conserved epitopes in the carboxy termini of fibrinogen, *Exp. Cell Res.*, 2003, **287**, 116–129.

54. G. Marx, A. Hotovely-Salomon, L. Levdansky, E. Gaberman, G. Snir, Z. Sievner, Y. Klauzner, M. Silberklang, D. Thomas, N. Hoffman, S. Luke, D. Lesnoy and R. Gorodetsky, Haptide-coated collagen sponge as a bioactive matrix for tissue regeneration, *J. Biomed. Mater. Res. B Appl. Biomater.*, 2008, **84**, 571–583.

55. M. Radosevich, H. I. Goubran and T. Burnouf, Fibrin sealant: scientific rationale, production methods, properties, and current clinical use, *Vox Sang.*, 1997, **72**, 133–143.

56. B. M. Alving, M. J. Weinstein, J. S. Finlayson, J. E. Menitove and J. C. Fratantoni, Fibrin sealant: summary of a conference on characteristics and clinical uses, *Transfusion*, 1995, **35**, 783–790.

57. G. Marx, X. Mou, R. Freed, E. Ben-Hur, C. Yang and B. Horowitz, Protecting fibrinogen with rutin during UVC irradiation for viral inactivation, *Photochem. Photobiol.*, 1996, **63**, 541–546.

58. S. Cox, M. Cole and B. Tawil, Behavior of human dermal fibroblasts in three-dimensional fibrin clots: dependence on fibrinogen and thrombin concentration, *Tissue Eng.*, 2004, **10**, 942–954.

59. I. Catelas, J. D. Bobyn, J. B. Medley, J. J. Krygier, D. J. Zukor and O. L. Huk, Size, shape, and composition of wear particles from metal-metal hip simulator testing: effects of alloy and number of loading cycles, *J. Biomed. Mater. Res. A*, 2003, **67**, 312–327.

60. D. Ho, S. Chang and C. D. Montemagno, Fabrication of biofunctional nanomaterials *via* Escherichia coli OmpF protein air/water interface insertion/integration with copolymeric amphiphiles, *Nanomedicine*, 2006, **2**, 103–112.

61. G. Marx, X. Mou, A. Hotovely-Salomon, L. Levdansky, E. Gaberman, D. Belenky and R. Gorodetsky, Heat denaturation of fibrinogen to develop a biomedical matrix, *J. Biomed. Mater. Res. B Appl. Biomater.*, 2008, **84**, 49–57.

62. C. Perka, U. Arnold, R. S. Spitzer and K. Lindenhayn, The use of fibrin beads for tissue engineering and subsequential transplantation, *Tissue Eng.*, 2001, **7**, 359–361.

63. K. F. Almqvist, L. Wang, J. Wang, D. Baeten, M. Cornelissen, R. Verdonk, E. M. Veys and G. Verbruggen, Culture of chondrocytes in alginate surrounded by fibrin gel: characteristics of the cells over a period of eight weeks, *Ann. Rheum. Dis.*, 2001, **60**, 781–790.

64. L. Almany and D. Seliktar, Biosynthetic hydrogel scaffolds made from fibrinogen and polyethylene glycol for 3D cell cultures, *Biomaterials*, 2005, **26**, 2467–2477.
65. O. Schmidt, J. Mizrahi, J. Elisseeff and D. Seliktar, Immobilized fibrinogen in PEG hydrogels does not improve chondrocyte-mediated matrix deposition in response to mechanical stimulation, *Biotechnol. Bioeng.*, 2006, **95**, 1061–1069.
66. E. Peled, J. Boss, J. Bejar, C. Zinman and D. Seliktar, A novel poly (ethylene glycol)-fibrinogen hydrogel for tibial segmental defect repair in a rat model, *J. Biomed. Mater. Res. A*, 2007, **80**, 874–884.
67. M. P. Lutolf and J. A. Hubbell, Synthetic biomaterials as instructive extracellular microenvironments for morphogenesis in tissue engineering, *Nat. Biotechnol.*, 2005, **23**, 47–55.
68. J. A. Pedersen and M. A. Swartz, Mechanobiology in the third dimension, *Ann. Biomed. Eng.*, 2005, **33**, 1469–1490.
69. S. Herrick, O. Blanc-Brude, A. Gray and G. Laurent, Fibrinogen, *Int. J. Biochem. Cell. Biol.*, 1999, **31**, 741–746.
70. W. Friess, Collagen--biomaterial for drug delivery, *Eur. J. Pharm. Biopharm.*, 1998, **45**, 113–136.
71. G. P. Raeber, M. P. Lutolf and J. A. Hubbell, Molecularly engineered PEG hydrogels: a novel model system for proteolytically mediated cell migration, *Biophys. J.*, 2005, **89**, 1374–1388.
72. W. Hayen, M. Goebeler, S. Kumar, R. Riessen and V. Nehls, Hyaluronan stimulates tumor cell migration by modulating the fibrin fiber architecture, *J. Cell. Sci.*, 1999, **112**, 2241–2251.
73. R. M. Kuntz and W. M. Saltzman, Neutrophil motility in extracellular matrix gels: mesh size and adhesion affect speed of migration, *Biophys. J.*, 1997, **72**, 1472–1480.
74. E. Alsberg, E. Feinstein, M. P. Joy, M. Prentiss and D. E. Ingber, Magnetically-guided self-assembly of fibrin matrices with ordered nanoscale structure for tissue engineering, *Tissue Eng.*, 2006, **12**, 3247–3256.
75. A. G. Mikos, G. Sarakinos, M. D. Lyman, D. E. Ingber, J. P. Vacanti and R. Langer, Prevascularization of porous biodegradable polymer, *Biotechnol. Bioeng.*, 1993, **42**, 716–723.
76. S. N. Park, J. C. Park, H. O. Kim, M. J. Song and H. Suh, Characterization of porous collagen/hyaluronic acid scaffold modified by 1-ethyl-3-(3-dimethylaminopropyl)carbodiimide cross-linking, *Biomaterials*, 2002, **23**, 1205–1212.
77. A. Alhadlaq and J. J. Mao, Mesenchymal stem cells: isolation and therapeutics, *Stem Cells Dev.*, 2004, **13**, 436–448.
78. H. Schoof, J. Apel, I. Heschel and G. Rau, Control of pore structure and size in freeze-dried collagen sponges, *J. Biomed. Mater. Res.*, 2001, **58**, 352–357.
79. L. Buttafoco, P. Engbers-Buijtenhuijs, A. A. Poot, P. J. Dijkstra, W. F. Daamen, T. H. van Kuppevelt, I. Vermes and J. Feijen, First steps towards tissue engineering of small-diameter blood vessels: preparation of

flat scaffolds of collagen and elastin by means of freeze drying, *J. Biomed. Mater. Res. B Appl. Biomater.*, 2006, **77**, 357–368.

80. M. Pieters, J. C. Jerling and J. W. Weisel, Effect of freeze-drying, freezing and frozen storage of blood plasma on fibrin network characteristics, *Thromb. Res.*, 2002, **107**, 263–269.

81. M. E. Nimni, D. Cheung, B. Strates, M. Kodama and K. Sheikh, Chemically modified collagen: a natural biomaterial for tissue replacement, *J. Biomed. Mater. Res.*, 1987, **21**, 741–771.

82. C. Wang, R. J. Stewart and J. Kopecek, Hybrid hydrogels assembled from synthetic polymers and coiled-coil protein domains, *Nature*, 1999, **397**, 417–420.

83. C. Wang, J. Kopecek and R. J. Stewart, Hybrid hydrogels cross-linked by genetically engineered coiled-coil block proteins, *Biomacromolecules*, 2001, **2**, 912–920.

84. J. B. Leach, K. A. Bivens, C. N. Collins and C. E. Schmidt, Development of photocrosslinkable hyaluronic acid-polyethylene glycol-peptide composite hydrogels for soft tissue engineering, *J. Biomed. Mater. Res.*, 2004, **70A**, 74–82.

85. S. Kim and K. E. Healy, Synthesis and characterization of injectable poly(N-isopropylacrylamide-co-acrylic acid) hydrogels with proteolytically degradable cross-links, *Biomacromolecules*, 2003, **4**, 1214–1223.

86. S. Kim, E. H. Chung, M. Gilbert and K. E. Healy, Synthetic MMP-13 degradable ECMs based on poly(N-isopropylacrylamide-co-acrylic acid) semi-interpenetrating polymer networks. I. Degradation and cell migration, *J. Biomed. Mater. Res. A*, 2005, **75**, 73–88.

87. B. K. Mann, A. S. Gobin, A. T. Tsai, R. H. Schmedlen and J. L. West, Smooth muscle cell growth in photopolymerized hydrogels with cell adhesive and proteolytically degradable domains: synthetic ECM analogs for tissue engineering, *Biomaterials*, 2001, **22**, 3045–3051.

88. A. S. Gobin and J. L. West, Cell migration through defined, synthetic ECM analogs, *FASEB J.*, 2002, **16**, 751–753.

89. S. Halstenberg, A. Panitch, S. Rizzi, H. Hall and J. A. Hubbell, Biologically engineered protein-graft-poly(ethylene glycol) hydrogels: a cell adhesive and plasmin-degradable biosynthetic material for tissue repair, *Biomacromolecules*, 2002, **3**, 710–723.

90. S. J. Bryant and K. S. Anseth, Controlling the spatial distribution of ECM components in degradable PEG hydrogels for tissue engineering cartilage, *J. Biomed. Mater. Res.*, 2003, **64A**, 70–79.

91. C. R. Nuttelman, M. C. Tripodi and K. S. Anseth, Synthetic hydrogel niches that promote hMSC viability, *Matrix Biol.*, 2005, **24**, 208–218.

92. Y. Luo and M. S. Shoichet, A photolabile hydrogel for guided three-dimensional cell growth and migration, *Nat. Mater.*, 2004, **3**, 249–253.

93. N. A. Peppas, Y. Huang, M. Torres-Lugo, J. H. Ward and J. Zhang, Physicochemical foundations and structural design of hydrogels in medicine and biology, *Annu. Rev. Biomed. Eng.*, 2000, **2**, 9–29.

94. V. L. Tsang and S. N. Bhatia, Three-dimensional tissue fabrication, *Adv. Drug Deliv. Rev.*, 2004, **56**, 1635–1647.
95. J. Baier Leach, K. A. Bivens, C. W. Patrick Jr. and C. E. Schmidt, Photocrosslinked hyaluronic acid hydrogels: natural, biodegradable tissue engineering scaffolds, *Biotechnol. Bioeng.*, 2003, **82**, 578–589.
96. R. A. Stile, C. Chung, W. R. Burghardt and K. E. Healy, Poly(*N*-isopropylacrylamide)-based semi-interpenetrating polymer networks for tissue engineering applications. Effects of linear poly(acrylic acid) chains on rheology, *J. Biomater. Sci. Polym. Ed.*, 2004, **15**, 865–878.
97. S. M. Cutler and A. J. Garcia, Engineering cell adhesive surfaces that direct integrin alpha5beta1 binding using a recombinant fragment of fibronectin, *Biomaterials*, 2003, **24**, 1759–1770.
98. D. Seliktar, A. H. Zisch, M. P. Lutolf, J. L. Wrana and J. A. Hubbell, MMP-2 sensitive, VEGF-bearing bioactive hydrogels for promotion of vascular healing, *J. Biomed. Mater. Res.*, 2004, **68A**, 704–716.
99. A. H. Zisch, M. P. Lutolf, M. Ehrbar, G. P. Raeber, S. C. Rizzi, N. Davies, H. Schmokel, D. Bezuidenhout, V. Djonov, P. Zilla and J. A. Hubbell, Cell-demanded release of VEGF from synthetic, biointeractive cell ingrowth matrices for vascularized tissue growth, *FASEB J.*, 2003, **17**, 2260–2262.
100. S. A. Delong, J. J. Moon and J. L. West, Covalently immobilized gradients of bFGF on hydrogel scaffolds for directed cell migration, *Biomaterials*, 2005, **26**, 3227–34.
101. B. K. Mann, R. H. Schmedlen and J. L. West, Tethered-TGF-beta increases extracellular matrix production of vascular smooth muscle cells, *Biomaterials*, 2001, **22**, 439–444.
102. M. P. Lutolf, J. L. Lauer-Fields, H. G. Schmoekel, A. T. Metters, F. E. Weber, G. B. Fields and J. A. Hubbell, Synthetic matrix metalloproteinase-sensitive hydrogels for the conduction of tissue regeneration: engineering cell-invasion characteristics, *Proc. Natl. Acad. Sci. U. S. A.*, 2003, **100**, 5413–5418.
103. R. A. Stile and K. E. Healy, Thermo-responsive peptide-modified hydrogels for tissue regeneration, *Biomacromolecules*, 2001, **2**, 185–194.
104. K. E. Healy, Control of cell function with tunable hydrogel networks, *Conf. Proc. IEEE Eng. Med. Biol. Soc.*, 2004, **7**, 5035.
105. M. Gonen-Wadmany, I. Oss-Ronen and D. Seliktar, Protein-polymer conjugates for forming photopolymerizable biomimetic hydrogels for tissue engineering, *Biomaterials*, 2007, **28**, 3876–3886.
106. N. Livnat, O. Sarig-Nadir, R. Zajdman, D. Seliktar and S. Shoham, in Proceedings of the 3rd International IEEE EMBS Conference on Neural Engineering, Kohala Coast, Hawaii, USA, 2007.
107. K. Shapira-Schweitzer and D. Seliktar, Matrix stiffness affects spontaneous contraction of cardiomyocytes cultured within a PEGylated fibrinogen biomaterial, *Acta Biomater.*, 2007, **3**, 33–41.
108. D. Dikovsky, H. Bianco-Peled and D. Seliktar, Defining the role of matrix compliance and proteolysis in three-dimensional cell spreading and remodeling, *Biophys. J.*, 2008, **94**, 2914–2925.

109. O. Sarig-Nadir and D. Seliktar, Compositional alterations of fibrin-based materials for regulating *in vitro* neural outgrowth, *Tissue Eng. Part A*, 2008, **14**, 401–411.

110. D. Dikovsky, H. Bianco-Peled and D. Seliktar, The effect of structural alterations of PEG-fibrinogen hydrogel scaffolds on 3-D cellular morphology and cellular migration, *Biomaterials*, 2006, **27**, 1496–1506.

111. D. Seliktar, Extracellular stimulation in tissue engineering, *Ann. N.Y. Acad. Sci.*, 2005, **1047**, 386–394.

112. R. Gorodetsky, A. Vexler, L. Levdansky and G. Marx, Fibrin microbeads (FMB) as biodegradable carriers for culturing cells and for accelerating wound healing, *Methods Mol. Biol.*, 2004, **238**, 11–24.

113. R. Rivkin, A. Ben-Ari, I. Kassis, L. Zangi, E. Gaberman, L. Levdansky, G. Marx and R. Gorodetsky, High-yield isolation, expansion, and differentiation of murine bone marrow-derived mesenchymal stem cells using fibrin microbeads (FMB), *Cloning Stem Cells*, 2007, **9**, 157–175.

114. O. Gurevich, A. Vexler, G. Marx, T. Prigozhina, L. Levdansky, S. Slavin, I. Shimeliovich and R. Gorodetsky, Fibrin microbeads for isolating and growing bone marrow-derived progenitor cells capable of forming bone tissue, *Tissue Eng.*, 2002, **8**, 661–672.

115. L. Zangi, R. Rivkin, I. Kassis, L. Levdansky, G. Marx and R. Gorodetsky, High-yield isolation, expansion, and differentiation of rat bone marrow-derived mesenchymal stem cells with fibrin microbeads, *Tissue Eng.*, 2006, **12**, 2343–2354.

116. R. Gorodetsky, The use of fibrin based matrices and fibrin microbeads (FMB) for cell based tissue regeneration, *Expert Opin. Biol. Ther.*, 2008, **8**, 1831–1846.

117. I. Kassis, L. Zangi, R. Rivkin, L. Levdansky, S. Samuel, G. Marx and R. Gorodetsky, Isolation of mesenchymal stem cells from G-CSF-mobilized human peripheral blood using fibrin microbeads, *Bone Marrow Transplant.*, 2006, **37**, 967–976.

118. N. Shimony, R. Gorodetsky, G. Marx, D. Gal, R. Rivkin, A. Ben-Ari, A. Landsman and Y. S. Haviv, A 3D rotary renal and mesenchymal stem cell culture model unveils cell death mechanisms induced by matrix deficiency and low shear stress, *Nephrol. Dial. Transplant.*, 2008, **23**, 2071–2080.

119. N. Shimony, R. Gorodetsky, G. Marx, D. Gal, R. Rivkin, A. Ben-Ari, A. Landsman and Y. S. Haviv, Fibrin microbeads (FMB) as a 3D platform for kidney gene and cell therapy, *Kidney Int.*, 2006, **69**, 625–633.

120. D. G. Chalupowicz, Z. A. Chowdhury, T. L. Bach, C. Barsigian and J. Martinez, Fibrin II induces endothelial cell capillary tube formation, *J. Cell Biol.*, 1995, **130**, 207–215.

121. V. Trochon, C. Mabilat, P. Bertrand, Y. Legrand, F. Smadja-Joffe, C. Soria, B. Delpech and H. Lu, Evidence of involvement of CD44 in endothelial cell proliferation, migration and angiogenesis *in vitro*, *Int. J. Cancer*, 1996, **66**, 664–668.

122. V. Nehls and D. Drenckhahn, A novel, microcarrier-based *in vitro* assay for rapid and reliable quantification of three-dimensional cell migration and angiogenesis, *Microvasc. Res.*, 1995, **50**, 311–322.
123. L. Martineau and C. J. Doillon, Angiogenic response of endothelial cells seeded dispersed *versus* on beads in fibrin gels, *Angiogenesis*, 2007, **10**, 269–277.
124. K. L. Christman, A. J. Vardanian, Q. Fang, R. E. Sievers, H. H. Fok and R. J. Lee, Injectable fibrin scaffold improves cell transplant survival, reduces infarct expansion, and induces neovasculature formation in ischemic myocardium, *J. Am. College Cardiol.*, 2004, **44**, 654–660.
125. V. W. van Hinsbergh, A. Collen and P. Koolwijk, Role of fibrin matrix in angiogenesis, *Ann. N. Y. Acad. Sci.*, 2001, **936**, 426–437.
126. A. Ben-Ari, R. Rivkin, M. Frishman, E. Gaberman, L. Levdansky and R. Gorodetsky, Isolation and implantation of bone narrow-derived mesenchymal stem cells with fibrin micro beads to repair a critical-size bone defect in mice, *Tissue Eng. Part A.*, 2009, **15**, 2537–2546.
127. R. Shainer, E. Gaberman, L. Levdansky and R. Gorodetsky, Efficient isolation and chondrogenic differentiation of adult mesenchymal stem cells with fibrin microbeads and micronized collagen sponges, *Regen. Med.*, **5**, 255–265.

CHAPTER 9

Culturing Non-hematopoietic Mesenchymal Stromal Cells and Requirements of GMP in Stem Cell-based Therapies

KAREN BIEBACK,[a] MARIANNA KARAGIANNI,[a] GERLINDE SCHMIDTKE-SCHREZENMEIER,[b] NATALIE FEKETE[b] AND HUBERT SCHREZENMEIER[b]

[a] Institute of Transfusion Medicine and Immunology, Medical Faculty Mannheim, Heidelberg University, DRK-Blutspendedienst, Ludolf-Krehl-Strasse 13-17, D-68167 Mannheim, Germany; [b] Institute of Clinical Transfusion Medicine and Immunogenetics Ulm, German Red Cross Blood Service Baden-Württemberg – Hessia and Institute of Transfusion Medicine, University of Ulm, Ulm, Germany

9.1   Introduction

Regenerative medicine is of growing interest in biomedical research.[1] The role of stem cells in this context is under intense scrutiny, on the one hand side to define principles of organ regeneration and on the other to develop innovative novel methods to treat organ failure.[2] In general, organ injuries or defects induce a mobilization of immature progenitor cells either locally or systemically. Triggered by the milieu, regulated by factors of the extracellular matrix, cellular components or soluble mediators, the progenitor cells differentiate

Stem Cell-Based Tissue Repair
Edited by Raphael Gorodetsky and Richard Schäfer
© The Royal Society of Chemistry 2011
Published by the Royal Society of Chemistry, www.rsc.org

along a hierarchy of committed to mature cells to functionally regenerate the cellular compartment of the organ.[3]

Mesenchymal stromal cells (MSC) emerge as key candidates for cellular therapies. Manufacturing of these cellular products adherent to 'current good manufacturing practice' (cGMP) guidelines poses a critical challenge. The definition of cell type, criteria for autologous/allogeneic donor selection, source of adventitious material during the manufacturing process with a special focus on material of human or animal origin, manufacturing processes, cellular integrity and purity, cell banking, formulation and lot release testing of the final product, characterization of the final product, validation of all procedures and quality assurance are all of utmost necessity. These topics are addressed in the rest of this chapter.

9.1.1 Skeletal Stem Cells, Mesenchymal Stem Cells, Mesenchymal Stromal Cells: Alternatives or Evolution of a Term?

Early studies by Friedenstein and coworkers formed the foundation for the current concept of mesenchymal stromal cells (MSC).[4] They were able to define a minor subpopulation in bone marrow capable of osteogenic differentiation. These were easily distinguishable from the hematopoietic proportion of cells by their fibroblastoid phenotype and ability to quickly adhere to plastic surfaces. These features resulted in the establishment of the colony-forming unit fibroblastic (CFU-F) assay to determine the frequency of MSC precursors. *In vivo* transplantation revealed that a variety of skeletal tissues including bone, cartilage, adipose and fibrous tissue can be generated from a single CFU-F. Consequently, these cells were named skeletal or stromal stem cells. Subsequently, highlighting the difference with hematopoietic stem cells, these cells were called non-hematopoietic stem cells. In analogy to the hematopoietic system, a mesengenic process was suggested by Caplan, who introduced the term 'mesenchymal stem cells'.[5] However, this term implies that a variety of mesodermal lineages can emerge from a single MSC. The nonskeletal differentiation potential remains a controversial issue.[6] In addition, classical stem cell features established to assess the 'stemness' of hematopoietic stem cells do not strictly attribute to MSC. Accordingly, the term 'mesenchymal stromal cell' has been proposed as a potential alternative.[7]

9.2 What Are MSC?

9.2.1 *Ex Vivo*

MSC are highly interesting candidates for various cellular therapeutic applications. They can be easily isolated from a variety of tissues by their adherence to plastic surfaces. Seeded under clonal conditions, they form colonies of fibroblastoid cells with varying potency. Seeding the cells at higher densities results in a heterogeneous population, including multipotent and more committed cells.[6] Even clonal cultures demonstrate heterogeneity as stem cells may either self-renew or more probably mature upon proliferation, undergo spontaneous differentiation and senescence.

MSC have a great appeal for cell therapy and tissue engineering for numerous reasons:

1. They are relatively easy to procure.
2. They expand rapidly in cell culture.
3. They show only minor spontaneous differentiation during *ex vivo* expansion.
4. They are multipotential.
5. They form supportive stroma for hematopoiesis.
6. They seem to be largely immunologically inert, paving the way for allogeneic transplantations.
7. They are immunosuppressive.
8. They secrete numerous trophic factors which modulate inflammation, remodeling and apoptosis.

MSC can be isolated from a variety of adult tissues as specified in more detail below. Their immune phenotype, as identified to date, is identical with only minor variations. The current consensus states that MSC do not express hematopoietic lineage markers. Based on the minimal criteria defined by the International Society for Cellular Therapies, 'MSC must express CD105, CD73 and CD90, and lack expression of CD45, CD34, CD14 or CD11b, CD79alpha or CD19 and HLA-DR surface molecules.'[7]

The lack of knowledge of an exclusive positive marker for MSC hampers the discrimination from other cell types, especially endothelial cells and fibroblasts. Most groups characterizing MSC use a large panel of surface markers. Unfortunately, most groups use different subsets of markers and reagents.[8] Thus the interpretations are difficult and better standardizations are mandatory. Nevertheless, recent publications have reported a set of markers which can be used for prospective isolation for mouse MSC (PDGFRalpha +, Sca1 +, CD45-, Ter119-)[9] or human MSC (human mesenchymal stem cell antigen-1 (MSCA-1), CD56).[10]

Furthermore MSC are plastic-adherent and should *in vitro* exert at least trilineage differentiation potential, *i.e.* towards the adipogenic, osteogenic and chondrogenic lineage. In addition, MSC may differentiate into other lineages as well, but this remains a controversially discussed issue.[11]

Controversies may arise because different laboratories employ not only different tissue sources, but also extraction methods, culture protocols and characterization tools. Any variation may result in the isolation and expansion of different subpopulations of cells, or may change the characteristics of the cells. Given that even MSC used in clinical trials are produced using a variety of different protocols, the results may not be interpretable or reproducible. Accordingly standardized protocols have to be developed which assure that the manufactured cells behave solely in the clinical intended purpose and do not exert adverse effects by, for example, uncontrolled differentiation or transformation. Typically MSC can be cultured to 40–50 population doublings until the growth rate declines significantly and cells undergo replicative senescence. Under certain conditions MSC, however, seem to undergo spontaneous transformation.[12]

An essential requirement therefore is that all steps of MSC manufacturing from starting material up to potency testing in the intended indication have to be highly standardized to assure a required and reproducible cellular quality and potency (see Section 9.3). The challenge for scientists aiming to produce MSC for clinical trials is to define optimal cell culture conditions to expand homogenous MSC efficiently *ex vivo* whilst maintaining the cellular qualities required for the intended clinical application and minimizing the risk of adverse events.

9.2.2 *In vivo*

Although the presence of these cells has been documented in a variety of tissues, MSC are mainly characterized as culture-expanded cells. Thus in contrast to the hematopoietic system, where hematopoietic stem cells can be prospectively isolated using, for example, CD34 antibodies, in most of the studies MSC have only been characterized retrospectively.

The origin, native distribution and anatomical localization of MSC have only recently been identified.[13] A perivascular niche has been proposed for MSC but awaits confirmation. Shi and Gronthos have described how Stro-1-positive cells from both the bone marrow and dental pulp express alpha smooth muscle actin, CD146, and to a varying degree the pericyte marker 3G5.[13] Pericytes are known as structural components of blood vessels regulating multiple stages of vascular development and differentiation.[14,15]

Perivascular localization has been further supported by subsequent studies. These revealed that MSC can be isolated from a large variety of postnatal organs and tissues, and importantly large and small vessels, whereas they were not detected in the peripheral blood.[16] Crisan *et al.* elegantly demonstrated that use of markers of pericytes enables the prospective isolation of cells expressing MSC markers and characteristics.[17] Even noncultivated cells from various organs express MSC markers and feature MSC characteristics. These findings are supported by Covas *et al.*, who defined CD146 + cells from diverse human tissues as MSC.[18]

As pericytes seem to represent the *in vivo* correlate of MSC, the question arises whether these serve as a natural, local reservoir of mesodermal progenitor cells in development and regeneration. Further studies will elucidate, for example, whether MSC can be mobilized and recruited in case of tissue injury.

9.2.3 Tissue Sources

As mentioned above, MSC can be isolated from virtually all postnatal organs and tissues. In addition, various fetal sources have been identified. MSC originating from the bone marrow (BM) stroma were the first identified as osteoprogenitor cells by Friedenstein. Based on this they are the ones best described and most advanced in clinical settings. In most comparative studies, bone marrow MSC therefore serve as the gold standard.

With a therapeutic goal in mind, easily accessible and highly abundant sources would be advantageous. Adipose tissue (AT), most often obtained as lipoaspirate, has emerged as an alternative tissue where cells occur at high frequency and procurement is less invasive than that of bone marrow aspiration.[19] Blood is obviously the most accessible source for MSC in adult tissues. However, peripheral blood does not contain MSC in a non-pathological setting.[20] Umbilical cord blood yields MSC, albeit at very low frequencies, inversely correlated to the gestational age.[21] The umbilical cord matrix or Wharton's jelly also contains MSC.[22] Fetal tissues are believed to contain comparatively immature MSC, as they express pluripotency markers such as SSEA- 3 and 4, Oct-4, Sox-2 and Nanog.[23]

9.3   Clinical Potential of MSC

MSC are increasingly used in many pre-clinical as well as in some clinical settings for immunomodulation or tissue repair. Table 9.1 summarizes the clinical trials with MSC announced on the US National Institutes of Health (NIH) website, www.clinicaltrials.gov (by August 2009). Several of these disorders are characterized by both inflammation and tissue defects. Often it can not be dissected whether efficacy of MSC is due to their production of trophic factors which stimulate endogenous repair mechanism, their direct differentiation into various cell types or their immunomodulatory effect.[24] Whether efficacy of MSC requires long-term persistence of MSC remains to be elucidated. In some of the models, factors released by the MSC are sufficient to mount a substantial part of the effect.[25]

9.3.1   Stromal Support for Hematopoiesis

Since MSC secrete many growth factors stimulating hematopoiesis, provide a scaffold for hematopoiesis and support primitive progenitor cells *in vivo*, it was hypothesized that they might enhance engraftment after stem cell transplantation. It has been successfully demonstrated in animal models that MSC can support reconstitution of lymphoid, myeloid and megakaryocytic lineages.[135]

After co-infusion of autologous peripheral blood stem cells and culture-expanded BM-derived MSC in 28 advanced breast cancer patients receiving high-dose chemotherapy, rapid hematopoietic engraftment was noted. The median time to reach more than 500 neutrophils per µl was eight days. In a study of MSC given to three patients as treatment of graft failure and four patients in a pilot study, the co-transplantation of MSC resulted in fast engraftment and 100% donor chimerism.[26] Donor-derived *ex vivo* expanded MSC might restore medullary function in some patients with poor

Table 9.1   Clinical trials with MSC registered at www.clinicaltrials.gov (as of 28 August 2009).

Indication	Number of registered trials
GvHD treatment or prophylaxis	15
Co-transplantation with hematopoietic stem cells	4
Renal allograft (treatment of rejection, allograft nephropathy, induction of tolerance)	4
Myocardial infarction	7
Ischemic heart failure/chronic myocardial ischemia	4
Congestive heart failure	2
Stroke	1
Critical limb ischemia	1
Cartilage defects	5
Degenerative disc disease	1
Osteogenesis imperfacta	2
Osteonecrosis	2
Bone fracture/pseudoarthrosis	2
Periodontitis	1
Liver cirrhosis/liver failure	3
Inflammatory bowel diseases	4
Multiple sclerosis	3
System Sclerosis	1
System lupus erythematodes	2
Sjögren's syndrome	1
Diabetes mellitus	4
Wound healing	2
Chronic obstructive pulmonary emphysema	1

hematopoietic recovery after allogeneic transplantion.[27] Co-transplantation of third-party MSC increased engraftment from double cord transplantation in NOD/SCID mice.[28] In a phase I/II study it could be demonstrated that co-transplantation of parental haploidentical MSC did promote engraftment in pediatric recipients of unrelated donor umbilical cord blood.[29] In another trial, allogeneic MSC together with human leukocyte antigen (HLA) disparate CD34+ cells were well-tolerated and all patients showed durable engraftment.[30] In a study which added third-party donor MSC to double cord blood and third-party donor mobilized hematopoietic cells, a positive effect of MSC on engraftment could not be confirmed.[31] Thus, in clinical experience MSC so far also hold some promise to enhance hematopoietic engraftment.

9.3.2   MSC for Treatment and Prevention of Graft-versus-host disease

Severe acute graft-versus-host disease (aGvHD) after allogeneic stem cell transplantation (allo-SCT) is associated with high morbidity and mortality.[32]

This is particularly true for patients who do not respond to corticosteroids, which are still the gold standard for initial treatment of aGvHD.

Based on the profound immunomodulatory effects of MSC *in vivo* and *in vitro*, it was obvious to use *ex vivo* expanded MSC for the treatment or prophylaxis of steroid-resistant GvHD. In the initial case report, Le Blanc *et al.* could demonstrate rapid improvement of grade IV steroid-resistant aGvHD of the gut and liver after infusion of BM-derived MSC.[33] Following this first report, further non-randomized open-label studies suggested that infusion of *ex vivo* expanded MSC can improve aGvHD; however, the response rates reported so far differ substantially ranging from 15–83% (Table 9.2).[34–38] Preliminary results from phase III double-blind, placebo-controlled trials for use of MSC for either first-line treatment of aGvHD or treatment of steroid-resistant aGvHD were announced in September 2009. Neither trial reached its primary end-point (durable complete response)[39]. However, MSC showed significant improvements in response rates in steroid-refractory liver and gastrointestinal GvHD;[39] however, in children, responses seems to be higher.[i,36] Most studies used BM-derived MSC;[36] however AT-derived MSC were also effective.[34] Successful treatment has been reported with expansion protocols using either fetal calf serum (FCS),[34,36] or platelet lysate as supplement.[37,38] Thus, infusion of MSC expanded *in vitro*, irrespective of the donor, might be an effective therapy at least for some subgroups of steroid-resistant, acute GVHD. Further trials on the impact of cell source and the different *ex vivo* expansion protocols (see Section 9.3.1.2) on the clinical efficacy will be necessary.

Some studies also reported that co-transplantation of MSC with hemato-poietic stem cells can reduce the incidence of aGvHD.[31,40,41] Further studies addressing this in a randomized setting in large patient cohorts are underway.[42]

There were no side effects during or immediately after infusion of MSC.[36,40] However, there is concern that MSC treatment of aGvHD might be a double-edged sword.[43] The potential late adverse events include transformation of MSC, allosensitization by mismatched MSC, increase incidence of infections and increased risk of relapse. The impact of T cell depletion and other immu-nomodulatory interventions on the risk of relapse, in particular in myeloid neoplasias, has been well established. Also as consequence of their 'niche function', the MSC might exert deleterious effects. In one randomized study, the incidence of aGvHD after HLA-identical sibling stem cell transplantation was lower in a group of ten patients who had received a co-transplantation of MSC compared with 15 patients who did not receive MSC prior to the allograft.[41] However, the risk of risk of relapse was increased and the survival was decreased in the MSC group.[41] The conclusions from this trial have been con-troversially discussed because of an unequal distribution of bone marrow and peripheral blood stem cells as the stem cell source over the two study groups.[44]

Nevertheless, this observation as well as the failure to reach the primary end-point in the randomized, placebo-controlled clinical trials announced by Osiris

---

[i]See press release from Osiris Theraputics Inc., www.osiristx.com/pdf/PR%20123%2008Sep09%20Phase%20III%20GvHD%20Topline%20Results.pdf [accessed June 2010].

Table 9.2 Published trials on *ex vivo* expanded MSC for treatment or prevention of graft-versus-host disease (GvHD).[a]

GvHD classification	Number of patients	MSC (source donor supplement)	Dose	Outcome	Reference
aGvHD grade III–IV	6	AT-derived 4 mismatch UD 2 haploidentical FD FCS	$1 \times 10^6\,\mathrm{kg}^{-1}$	5/6 response 4/6 alive	34
Randomized Prospective Clinical trial	10 MSC 15 no MSC	BM-derived HLA-identical sibling	$0.3\text{–}15 \times 10^5\,\mathrm{kg}^{-1}$	1/10 aGvHD II–IV 6/10 relapse 8/10 aGvHD II–V 3/10 relapse	41
aGvHD 5 grade II 25 grade III 25 grade IV	55	BM-derived 5 HLA-identical siblings 18 haploidentical donors 69 third-party UD FCS	$1.4 \times 10^6\,\mathrm{kg}^{-1}$	30/55 CR 9/55 improvement 53% survival in responders 16% in PR/CR	36
aGvHD cGvHD	2 3	BM-derived 5 SCT donors 2 haploidentical parental donors Platelet lysate	$0.4\text{–}3.0 \times 10^6\,\mathrm{kg}^{-1}$	1 improvement 1 no progression of aGvHD to cGvHD	37
aGvHD 2 grade III 11 grade IV	13	BM-derived UD Platelet lysate	$0.9 \times 10^6\,\mathrm{kg}^{-1}$	2/13 respnse 5/13 alive	38
aGvHD 21 grade II 8 grade III 3 grade IV	32	BM-derived FCS (Prochymal®)	$2 \text{ or } 8 \times 10^6\,\mathrm{kg}^{-1}$	77% CR 16% PR	35

[a]Abbreviations: aGvHD = acute graft-verus-host disease; AT = adipose tissue; BM = bone marrow; CR = complete remission; FCS = fetal calf serum; FD = family donor; PR = partial remission; STC = stem cell transplantation; UD = unrelated donor.

Therapeutics clearly demonstrated that further well-defined studies—including large patient numbers and appropriate control groups—are necessary to define the role of MSC among other agents for GvHD treatment such as monoclonal antibodies or new immunosuppressive drugs.[32] In particular it is necessary to study the impact of cell source and the different *ex vivo* expansion protocols (see Sections 9.1.1.3 and 9.3.1.2), dose and application mode on the clinical efficacy of MSC.

9.3.3 MSC for Other Immune Interventions

Since MSC migrate in response to inflammatory cytokines, they can specifically home to the inflamed tissue. This might also help to restrict the immune-regulatory effects on both innate and adaptive immunity to the site of inflammation and tissue damage. Effects of MSC treatment have been demonstrated in animal models of autoimmune disorders such as Type I diabetes, multiple sclerosis and rheumatoid arthritis, glomerulonephritis and Crohn's disease (see below).[45–48] A number of clinical trials in these indications are ongoing (Table 9.1).

9.3.4 Hard Tissue Repair (Bone and Cartilage)

The ability of MSC to differentiate to chondrocytes and osteoblast is one of the hallmarks of the functional characterization of MSC. Many studies demonstrated chondrogenesis or osteogenesis of MSC when incubated with appropriate inducers to occur in various scaffolding materials.[49]

Osteoblastic potential of MSC has been demonstrated in children with osteogenesis imperfecta who showed osteopoietic engraftment of MSC and improved growth.[50] MSC might also be used for treatment of non-union bone fractures[51] and repair of bone defects.[52]

Differentiation to chondrocytes as assessed by phenotypic markers does not necessarily correlate with mechanical function.[49] There are functional differences between chondrogenically differentiated MSC and fully differentiated chondrocytes. Recent work demonstrated that mechanical stimulation can improve MSC chondrogenesis.[53] MSC remain a promising candidate for generation of replacement cartilage for the repair of damaged joints,[54,55] or stimulation of endogenous repair mechanisms.[56]

9.3.5 Soft Tissue Repair

9.3.5.1 *Vascular Potential*

Numerous clinical and basic studies have demonstrated beneficial effects of MSC during tissue ischemia and tumour vascularization in either myocardial infarction or cerebral ischemia or tumor angiogeneis.[57–59]

While multiple studies have demonstrated improvements in parameters of local tissue damage and function following MSC delivery, the underlying mechanisms seem to be complex. Some studies indicate angiogenesis as one mechanism. *In vitro*, direct endothelial differentiation has been demonstrated mediated by vascular endothelial growth factor (VEGF).[22,60,61] *In vivo*, MSC transplantation significantly increased capillary density in the myocardium.[62] After intramyocardial injection of MSCs, histopathological and immunohistochemical analyses revealed differentiation into cardiomyocytes, endothelial cells and smooth muscle cells.[61] *In vitro*, MSC secrete a variety of angiogenic, anti-apoptotic and mitogenic factors including VEGF, basic fibroblast growth factor (bFGF), hepatocyte growth factor (HGF) and insulin-like growth factor 1 (IGF-1).[25] The sustained expression of these growth factors can prevent vessel regression and promote stabilization of the vessels.

9.3.5.2 Epithelial Regeneration

9.3.5.2.1 Skin. MSC have shown potential to improve healing of skin defects in animal models[63–67] and in humans.[68–70] Effects have been demonstrated for acute incisional and excisional wounds, diabetic skin ulcers, thermal injury and radiation of the skin. Recent data suggest that MSC loaded on amniotic membranes (AM) are more effective than injected cells.[71] Furthermore, bFGF can enhance wound healing by MSC.[65] Autologous and allogeneic MSC seem to be equally effective for wound repair.[71] Animal studies suggested that radiation-induced tissue injury—including radiation-induced skin lesions—can be mitigated by MSC.[72–74]

Enhancement of wound healing might be due to factors released by MSC which recruit macrophages and endothelial lineage cells,[64,67] or due to the recruitment of MSC into wounds and trans-differentiation in multiple skin cell types,[67,75] or improved neovascularization.[63,67]

A shortcoming of most of the published studies on the use of MSC for wound repair is the heterogeneity of methods used for MSC isolation and culture as well as lack of extensive immunophenotype and functional characterization of MSC used. Thus, there is some encouraging experience on MSC for wound healing. However, still many open questions remain on optimal culture conditions, dose and route of application, combination with scaffolds and type of MSC.

9.3.5.2.2 Gut. Both adipose tissue and bone marrow-derived MSC have been used for the treatment of inflammatory bowel disease. Healing of fistula and perianal wounds as well as improvement of Crohn's disease activity index has been reported.[31,76–80] In a phase II trial, about one third of patients achieved clinical remission.[81] According to preliminary results of a randomized, placebo-controlled phase III trial, no significant improvement difference was observed between MSC and the control group.[39]

9.3.5.2.3 Kidney. In models of acute ischemia and reperfusion and acute tubular injury, MSC have been shown to improve tissue repair of both glomerular and tubular compartments.[82,83] This is accompanied by an improved functional recovery of renal injury. VEGF is one factor critical for mediating renal protection.[84]

9.3.5.3 Neurological Repair

MSC can adopt phenotypic and morphological characteristics of astrocytes and neurons after injection into the central nervous system of mice.[85] Many *in vitro* studies have demonstrated that neural-specific markers [*e.g.* nestin, glial fibrillary acidic protein (GFAP), neurofilament heavy chain and many others] and some morphological and functional features of neural cells can be induced during *in vitro* culture of MSC. However, convincing evidence is still missing that MSC can acquire the full repertoire of neural proteins and all the functional properties of neurons. MSC expressing some neuronal markers might just comprise a subtype of MSC not necessarily implying that they can adopt the full repertoire of bona fide neutrons or glial cells. In animal models, MSC showed some efficacy in hypoxic–ischemia neural damage, retinal injury and Parkinson's disease.[86,87]

9.4 Good Manufacturing Practice (GMP) and Standardization in Manufacturing

As indicated above, MSC research has made remarkable progress leading to early and late stage clinical trials. However, the worldwide regulatory framework seems to lag behind. Most regulations are country or even region specific. A variety of organizations have responded to this threat by proposing a minimal set of standards or consensus guidelines.[7,88] These are especially critical when developing quality control assays. As well as checking viral, bacterial, fungal and mycoplasma contamination, it is also necessary to consider safety, potency, efficacy and lot-to-lot variation or donor-specific variations.

As indicated above, MSC can be isolated from various tissues. Each tissue source encompasses individual characteristics which have to be taken into account for GMP production. The most relevant criterion for the selection of manufacturing protocols should be to isolate and maintain cells with properties close to primary cells (predicting that these cells play a regenerative role in physiological tissue turnover).

9.4.1 European Union and US Legal Framework

9.4.1.1 European Union

In Europe, MSC are classified as advanced therapy medicinal products (ATMP). They include gene therapy medicinal products, somatic cell therapy products (as defined in Directive 2001/83/EC) and tissue-engineered products. This means they contain or consist of engineered cells or tissues, and are presented as having properties for, or are used in or administered to human beings with a view to regenerating, repairing or replacing a human tissue. A tissue-engineered product may contain cells or tissues of human or animal origin, or both. The cells or tissues may be viable or non-viable. The product may also contain additional substances such as cellular products, biomolecules, biomaterials, chemical substances, scaffolds and/or matrices.

ATMP are defined by Regulation (EC) No. 1394/2007.[ii] This Regulation, which has applied since 30 December 2008, is binding in its entirety and directly applicable in all Member States. According to the Regulation, cells or tissues shall be considered 'engineered' if they fulfill at least one of the following conditions:

- The cells or tissues have been subjected to substantial manipulation, resulting in a change of biological characteristics, physiological functions or structural properties relevant for the intended regeneration, repair or replacement.
- The cells or tissues are not intended to be used for the same essential function or functions in the recipient as in the donor.

This means that MSC can be considered as somatic cell therapy products or tissue-engineered products depending on the indication and the manipulation during the manufacturing process.

For clinical trials with MSC, the rules set out in Article 6(7) and Article 9(4) and (6) of Directive 2001/20/EC in respect of gene therapy and somatic cell therapy medicinal products apply to tissue-engineered products.

This EU directive complies with Directive 2004/23/EC on the donation, procurement and testing of human cells and tissues and with Directive 2002/98/EC on human blood and blood components. The Committee for Advanced Therapies, which was established in accordance with the Regulation, has produced special procedures such as the certification procedure for small- and medium-sized enterprises (SMEs) and revised the technical requirements for Marketing Authorization Applications (quality, non-clinical and clinical). For more details see ref. 89.

9.4.1.2 USA

In the USA, MSC are classified as HCT/Ps, *i.e.* human cells, tissues, or cellular and tissue-based products. MSC production must comply with Current Good Tissue Practice (cGTP) under the Code of Federal Regulation (CFR). In May 2009 the Food and Drug Administration (FDA) released *Guidance for Industry on Current Good Tissue Practice (CGTP) and Additional Requirements for*

[ii] http://eur-lex.europa.eu/LexUriServ/LexUriServ.do?uri = OJ:L:2007:324:0121:0137:en:PDF
[Accessed June 2010]

Manufacturers of Human Cells, Tissues and Cellular and Tissue-based Products.[iii] This provides recommendations to manufacturers of HCT/Ps.

9.4.2   Culture Conditions

Several critical parameters for *ex vivo* expansion of MSC are important to ensure a good expansion rate as well as maintenance of the multipotency of MSC. Among these parameters are:

- the starting material;
- the method used for enrichment or separation;
- plating density;
- devices used for MSC culture;
- media;
- supplements and growth factors; and
- passage number or population doublings.

9.4.2.1   Enrichment and Separation

In the absence of specific markers capable of prospectively isolating MSC, numerous attempts to purify MSC after isolation have been performed. Common protocols utilize the selective adherence of MSC to plastic surfaces in FCS supplemented minimal medium.[7] Further enrichment can be achieved by:

- density gradient centrifugationl[90] and
- depletion of contaminating cells such as erythrocytes[91] or lineage-positive cells.[92]

Immune selection has used markers such as STRO-1, CD49a, CD105, CD133, CD146, CD271, SSEA-4, antifibrin microbeads, aptamers and aldehyde dehydrogenase activity.[7,10,92–95]

9.4.2.2   Cell Seeding

Plating densities have emerged as critical issues for MSC isolation and expansion. Low seeding densities in primary culture seem to be associated with the emergence of more immature progenitor subsets.[96] Moreover, seeding at low densities allows higher expansion rates. Accordingly two different protocols are proposed for scale-up. In one protocol, cells are seeded at nearly clonal levels, which allows expanding the cells to high cumulative population doublings within one passage, but which necessitates a high use of cell culture plastic.[97] The other protocol seeds cells at higher concentrations to harvest high cell numbers, but due to reduced possible population doublings, this procedure most probably necessitates a second passage.[98]

9.4.2.3   Media + Supplements

Culture conditions should retain or even accelerate the regenerative and trophic properties of MSC. This is considered to coincide with self-renewal. The classical media composition consists of a basal medium and 10–20% supplement. Here most often FCS is used. The variety in protocols is immense and a standard has not yet been defined. The most often used basal media are alpha-MEM or DMEM.

The pivotal compound seems to be the supplement as initial data indicate that only selected FCS batches retain MSC stem cell properties.[99] The ongoing debate regarding xenogenic proteins (especially ruminant proteins in pharmaceuticals) also applies to MSC. FCS bears the risk of transferring xenogenic, potentially infectious or immunogenic proteins. Immunogenicity against FCS proteins has been demonstrated to compromise the therapeutic benefit.[50,100] Thus although GMP-compliant FCS batches are available and used in clinical-grade manufacturing, the regulatory authorities ask manufacturers to replace FCS with a non-xenogenic alternative if this is an option.

Up to now, no completely serum-free media formulation in clinical grade is available which allows for both the isolation and expansion of MSC. Serum proteins provide not only nutrients but also essential attachment factors. Several laboratories have proposed the use of human components to supplement a MSC growth medium.

In the main, attempts can be grouped into an approach using autologous serum or platelet factors, and an allogeneic approach aiming especially at large batch production of pooled human serum or platelet-derived factors. Pooled human platelet lysate has emerged as an interesting supplement. Beside its own regenerative properties it significantly accelerates MSC proliferation whilst maintaining differentiation and immune-suppressive potential.[101–106,92,97]

Serum and platelet lysate are very crude protein cocktails. Essential growth factors for optimal MSC culture have not yet been defined. Platelet-derived growth factor (PDGF), epidermal growth factor (EGF), transforming growth factor beta (TGF-ß) and IGF have been subject to investigation. Basic FGF has demonstrated most promising effects in expanding MSCs whilst maintaining stem cell properties and reducing replicative senescence.[107,108]

Any significant change in the production process may affect cellular functions. It is accordingly necessary to analyze the qualities of MSC in comparability studies to ensure that cellular qualities are not compromised.

9.4.2.4   Devices for Expansion

MSC grow as adherent cells until reaching confluency. Therefore the number of cells which can be harvested in an *ex vivo* expansion culture is determined by the surface area. Typically they are cultivated in conventional monolayer cultures. In order to achieve a large surface area, multi-layered cell factories are used.[97–98,106] Closed systems for large-scale MSC expansion in conventional two-dimensional (2D) cell culture are not yet available. This approach is labor-intensive and expensive.

Therefore other approaches which allow automation of the cell culture process are in development. It has been demonstrated that MSC can also be expanded in a three-dimensional (3D) system using a rotary bioreactor[109] or microcarriers in a spinner flask bioreactor.[110] It has been further shown that adult human bone marrow contains a population of MSC that can be expanded in non-adherent, cytokine-dependent, suspension culture conditions in spinner flasks for at least 42 days.[111] The cells generated during suspension culture lacked detectable levels of gene expression associated with differentiated mesenchymal cell types (including bone, muscle and fat), suggesting that suspension culture maintains MSC in an uncommitted state.[111]

The matrix is also crucial for the functional properties of the expanded cells. It has been demonstrated that the elasticity of the matrix directs the MSC lineage specification.[112]

The optimal route of administration of MSC for clinical therapy is still unclear. Whether systemic intravenous infusion (*e.g.* as in the trials so far on the treatment of GvHD) is the best route needs to be analyzed. In particular when MSC are used for local repair of damaged tissue (*e.g.* skin wounds, tendon repair and bone defects), it might be advantageous to expand the MSC on carriers which can be used for direct transfer of expanded cells to the site of treatment. Various such carrier systems are currently under investigation such as hydrogels, collagen gels, biphasic calcium phosphate nanoparticles, hydroxyapatite, polymers and amniotic membrane.[113–116,71]

Further developments will aim at improved attachment and expansion in 3D-culture systems which, in an optimal setting, would implement a closed and largely automated system.

9.4.3 Quality Assessment

It is a prerequisite for pharmaceutical grade products to ensure that the product fulfils all predefined quality criteria. For cellular products, a sufficient number of viable, high quality cells is demanded. Furthermore the product must not transmit infections or transfer malignancy. Basic guidelines[iv] on how to define the minimal quality are published by the respective regulatory authorities.

[iv] Examples include: *Points to Consider on the Manufacture and Quality Control of Human SOmatic Cell Therapy Medicinal Products*, www.emea.europa.eu/pdfs/human/bwp/4145098EN.pdf; and *Scientific Guideline on the Minimum Quality and Non-clinical Data for the Certification of Advanced Therapy Medicinal Products*, www.emea.europa.eu/pdfs/human/cat/ 48683108en.pdf [Accessed June 2010].

9.4.3.1 Impurities

Product-related impurities have to be determined and specified. In MSC cultures, cellular impurities can be quantified by performing flow cytometric analyses to characterize the proportion of hematopoietic, endothelial, epithelial and potentially other contaminating cell types. Product-related impurities relating to, for example, degradation products from structural or matrix components must be specified, as well as process-related impurities derived from added bioactive components.

9.4.3.2 Safety

Using human material bears the intrinsic risk of contamination with potential infectious agents. Independent of the use in autologous or allogeneic settings, sterility of cellular products is a perquisite and thus bacterial and fungal sterility has to be controlled for each product. The analysis of potential contaminations during the manufacturing procedure (*i.e.* control of mycoplasma contamination and endotoxin levels) is essential.

MSC as adherent cells grow and have to be reseeded to further expand cells upon reaching a subconfluent stage. Like most adherent cells, MSC show growth inhibition upon confluence. This implies that MSC have to be passaged successively until the targeted cell amount can be obtained. Admittedly, so far no systematic dose finding studies have been performed. Target doses now used in most studies refer to the initial experience in the first case reports and clinical studies on MSC.[33,117]

MSC show replicative senescence and reach the Hayflick limit like many other cell types after approximately 40–50 population doublings. Late cells demonstrate loss of proliferative activity and subsequent loss of multipotentiality. Thus the current consensus proposes to use cells up to approximately 15 population doublings to ensure potency and safety. Safety may be affected as prolonged *ex vivo* culture can accumulate aberrations. Up to now, only anecdotal studies indicate that MSC may undergo spontaneous transformation, associated with chromosomal aberrations, induction of oncogenes and tumorigenicity after transplantation.[118–120] Recent data revealed contaminations as 'source' of 'spontaneous' transformation.[136,137] Moreover Ewing's tumors or sarcoma have been shown to arise from MSC cultures.[121] The process of transformation, like in other cells, is a multistep process. Besides chromosomal abnormalities, loss of tumor suppressor genes followed by telomerase expression was observed in long-term culture spontaneously immortalized MSC. While this process has not so far been demonstrated in short-term cultivated MSC, this aspect is still discussed controversially.[12]

Current testing systems assaying transformation events imply karyotype analyses, FISH, or comparative genomic hybridization to analyze copy number changes or polymerase chain reaction (PCR) to check for, for example, c-Myc expression. Currently, all these assays are not sensitive enough to detect the expected low proportion of affected cells. A single transformed cell may be sufficient to transfer unwanted reactions to the recipient. To provide a better basis for quality control, detection systems must be optimized and methodologies adapted for clinical applications.

Clinically safety concerns relate to observations that the possibility of ectopic tissue formation cannot be neglected despite, in general, no records of adverse reactions. Mice treated locally with MSC for myocardial infarction developed calcifications.[122] In addition, the intravascular transplantation may cause pulmonary sequestration and embolism due to the relatively big cell size.[123]

Although therapeutically wanted in some clinical settings (*e.g.* preventing or treating GvHD), the immunomodulatory capacities may favor tumor growth or formation of metastasis as indicated in a few studies.[124,125]

9.4.3.3 Potency

The minimal criteria of the International Society for Cellular Therapy (ISCT) demand control of MSC for their capacity to adhere to normal plastic culture surfaces, generating cells with a fibroblastoid phenotype which express/fail to express a typical set of surface markers and exhibit multilineage differentiation potential at least in the osteo-, adipo- and chondrogenic lineage.[7]

Although every laboratory employs such assays, they are far from being standardized so comparison between laboratories and also the read-out of clinical data is hampered. As mentioned above, development of pre-clinical efficacy tests in the investigated indication are highly desirable as MSC appear to employ different modes of action in specific indications.

9.4.3.3.1 Clonogenicity. The CFU-F assay is a suitable but not standardized tool for the quantification of precursor frequencies. Analysis demands appropriate dilution to clonal levels as CFU-F frequencies do not follow a linear regression correlated with the input cell number.[97,126]

9.4.3.3.2 Differentiation Potential. The multilineage differentiation potential is a hallmark of MSC, but discussed in detail elsewhere.[6,11,127] *In vitro* assays can be performed using self-made or commercially available induction media. If connective tissue repair is intended, these differentiation assays should be performed irrespective of the open question as to how *in vitro* data correlate to the *in vivo* differentiation potential.[6,128]

9.4.3.3.3 Immunomodulatory Capacities. The perspective of modulating immune responses against allo- and possible also autoantigens has rendered MSC an attractive population of cells for immune therapies. *In vitro*, assays have been established to quantify the expression of surface molecules such as like HLA class I and II and co-stimulatory molecules. In cocultures with peripheral blood mononuclear cells, MSC do not elicit an alloreactive response. Furthermore, when added as third party in mixed lymphocyte reactions or mitogen driven cultures, MSC dose-dependently inhibit immune cell responses. Very low concentrations of MSC, however, can stimulate immune responses.[129] While this has not yet been observed *in vivo*, too low numbers of MSC transplanted may accelerate the immune response rather than mitigate it in GvHD or autoimmune settings.

9.4.3.3.4 Hematopoiesis/Stromal Support. The beneficial effects of co-transplanting MSC in hematological settings have been demonstrated.[117,130] This effect can be assayed *in vitro* in coculture experiments using hematopoietic stem cells and MSC, and thus may be an adequate quality control system for this indication.[131,132]

9.4.3.3.5 Trophic Support. In a variety of settings, MSC have shown promising therapeutic effects even though the transfused cells were—if at all—only barely detectable in the injured organs. Recent data have further demonstrated that especially secreted factors actively modulate debilitating local inflammatory reactions. Reduction of apoptosis and fibrotic tissue remodeling, as well as recruitment of local resident regenerative cells, contributed to the beneficial effects.[25,133] Accordingly some studies have already demonstrated therapeutic effects when infusing MSC conditioned medium instead of cells.[134] Depending on the therapeutical setting, quantifying levels of chemo- or cytokines may subsequently emerge as an additional potency assay.[105,92]

References

1. E. Lagasse, J. A. Shizuru, N. Uchida, A. Tsukamoto and I. L. Weissman, *Immunity*, 2001, **14**, 425–436.
2. T. A. Rando, *Nature*, 2006, **441**, 1080–1086.
3. S. Ding and P. G. Schultz, *Nat. Biotechnol.*, 2004, **22**, 833–840.
4. A. J. Friedenstein, U. F. Deriglasova, N. N. Kulagina, A. F. Panasuk, S. F. Rudakowa, E. A. Luria and I. A. Ruadkow, *Exp. Hematol.*, 1974, **2**, 83–92.
5. A. I. Caplan, *Clin. Plast. Surg.*, 1994, **21**, 429–435.
6. P. Bianco, P. G. Robey and P. J. Simmons, *Cell Stem Cell*, 2008, **2**, 313–319.
7. M. Dominici, K. Le Blanc, I. Mueller, I. Slaper-Cortenbach, F. Marini, D. Krause, R. Deans, A. Keating, D. Prockop and E. Horwitz, *Cytotherapy*, 2006, **8**, 315–317.
8. M. T. Rojewski, B. M. Weber and H. Schrezenmeier, *Transfus. Med. Hemother.*, 2008, **35**, 168–184.
9. S. Morikawa, Y. Mabuchi, Y. Kubota, Y. Nagai, K. Niibe, E. Hiratsu, S. Suzuki, C. Miyauchi-Hara, N. Nagoshi, T. Sunabori, S. Shimmura, A. Miyawaki, T. Nakagawa, T. Suda, H. Okano and Y. Matsuzaki, *J. Exp. Med.*, 2009, **206**, 2483–2496.

10. V. L. Battula, S. Treml, P. M. Bareiss, F. Gieseke, H. Roelofs, P. de Zwart, I. Müller, B. Schewe, T. Skutella, W. E. Fibbe, L. Kanz and H. J. Buhring, *Haematologica*, 2009, **94**, 173–184.
11. D. G. Phinney and D. J. Prockop, *Stem Cells*, 2007, **25**, 2896–2902.
12. G. Lepperdinger, R. Brunauer, A. Jamnig, G. Laschober and M. Kassem, *Exp. Gerontol.*, 2008, **43**, 1018–1023.
13. S. Shi and S. Gronthos, *J. Bone Miner. Res.*, 2003, **18**, 696–704.
14. L. Diaz-Flores, R. Gutierrez, J. F. Madrid, H. Varela, F. Valladares, E. Acosta, P. Martin-Vasallo and L. Diaz-Flores Jr, *Histol. Histopathol.*, 2009, **24**, 909–969.
15. K. K. Hirschi and P. A. D'Amore, *Cardiovasc. Res.*, 1996, **32**, 687–698.
16. L. da Silva Meirelles, P. C. Chagastelles and N. B. Nardi, *J. Cell. Sci.*, 2006, **119**, 2204–2213.
17. M. Crisan, S. Yap, L. Casteilla, C. W. Chen, M. Corselli, T. S. Park, G. Andriolo, B. Sun, B. Zheng, L. Zhang, C. Norotte, P. N. Teng, J. Traas, R. Schugar, B. M. Deasy, S. Badylak, H. J. Buhring, J. P. Giacobino, L. Lazzari, J. Huard and B. Peault, *Cell Stem Cell*, 2008, **3**, 301–313.
18. D. T. Covas, R. A. Panepucci, A. M. Fontes, W. A. Silva Jr, M. D. Orellana, M. C. Freitas, L. Neder, A. R. Santos, L. C. Peres, M. C. Jamur and M. A. Zago, *Exp. Hematol.*, 2008, **36**, 642–654.
19. S. Kern, H. Eichler, J. Stoeve, H. Kluter and K. Bieback, *Stem Cells*, 2006, **24**, 1294–1301.
20. S. A. Wexler, C. Donaldson, P. Denning-Kendall, C. Rice, B. Bradley and J. M. Hows, *Br. J. Haematol.*, 2003, **121**, 368–374.
21. M. J. Javed, L. E. Mead, D. Prater, W. K. Bessler, D. Foster, J. Case, W. S. Goebel, M. C. Yoder, L. S. Haneline and D. A. Ingram, *Pediatr. Res.*, 2008, **64**, 68–73.
22. M. Y. Chen, P. C. Lie, Z. L. Li and X. Wei, *Exp. Hematol.*, 2009, **37**, 629–640.
23. O. Parolini, F. Alviano, G. P. Bagnara, G. Bilic, H. J. Buhring, M. Evangelista, S. Hennerbichler, B. Liu, M. Magatti, N. Mao, T. Miki, F. Marongiu, H. Nakajima, T. Nikaido, C. B. Portmann-Lanz, V. Sankar, M. Soncini, G. Stadler, D. Surbek, T. A. Takahashi, H. Redl, N. Sakuragawa, S. Wolbank, S. Zeisberger, A. Zisch and S. C. Strom, *Stem Cells*, 2008, **26**, 300–311.
24. L. Sensebe, M. Krampera, H. Schrezenmeier, P. Bourin and R. Giordano, *Vox Sang.*, 2009.
25. E. M. Horwitz and W. R. Prather, *Isr. Med. Assoc. J.*, 2009, **11**, 209–211.
26. K. Le Blanc, H. Samuelsson, B. Gustafsson, M. Remberger, B. Sundberg, J. Arvidson, P. Ljungman, H. Lonnies, S. Nava and O. Ringden, *Leukemia*, 2007, **21**, 1733–1738.
27. N. Meuleman, G. Vanhaelen, T. Tondreau, P. Lewalle, J. Kwan, J. Bennani, P. Martiat, L. Lagneaux and D. Bron, *Haematologica*, 2008, **93**, e11–13.
28. D. W. Kim, Y. J. Chung, T. G. Kim, Y. L. Kim and I. H. Oh, *Blood*, 2004, **103**, 1941–1948.

29. M. L. Macmillan, B. R. Blazar, T. E. DeFor and J. E. Wagner, *Bone Marrow Transplant.*, 2009, **43**, 447–454.

30. L. M. Ball, M. E. Bernardo, H. Roelofs, A. Lankester, A. Cometa, R. M. Egeler, F. Locatelli and W. E. Fibbe, *Blood*, 2007, **110**, 2764–2767.

31. R. Gonzalo-Daganzo, C. Regidor, T. Martin-Donaire, M. A. Rico, G. Bautista, I. Krsnik, R. Fores, E. Ojeda, I. Sanjuan, J. A. Garcia-Marco, B. Navarro, S. Gil, R. Sanchez, N. Panadero, Y. Gutierrez, M. Garcia-Berciano, N. Perez, I. Millan, R. Cabrera and M. N. Fernandez, *Cytotherapy*, 2009, **11**, 278–288.

32. H. J. Deeg, *Blood*, 2007, **109**, 4119–4126.

33. K. Le Blanc, I. Rasmusson, B. Sundberg, C. Gotherstrom, M. Hassan, M. Uzunel and O. Ringden, *Lancet*, 2004, **363**, 1439–1441.

34. B. Fang, Y. Song, L. Liao, Y. Zhang and R. C. Zhao, *Transplant. Proc.*, 2007, **39**, 3358–3362.

35. P. Kebriaei, L. Isola, E. Bahceci, K. Holland, S. Rowley, J. McGuirk, M. Devetten, J. Jansen, R. Herzig, M. Schuster, R. Monroy and J. Uberti, *Biol. Blood Marrow Transplant.*, 2009, **15**, 804–811.

36. K. Le Blanc, F. Frassoni, L. Ball, F. Locatelli, H. Roelofs, I. Lewis, E. Lanino, B. Sundberg, M. E. Bernardo, M. Remberger, G. Dini, R. M. Egeler, A. Bacigalupo, W. Fibbe and O. Ringden, *Lancet*, 2008, **371**, 1579–1586.

37. I. Müller, S. Kordowich, C. Holzwarth, G. Isensee, P. Lang, F. Neunhoeffer, M. Dominici, J. Greil and R. Handgretinger, *Blood Cells Mol. Dis.*, 2008, **40**, 25–32.

38. M. von Bonin, F. Stolzel, A. Goedecke, K. Richter, N. Wuschek, K. Holig, U. Platzbecker, T. Illmer, M. Schaich, J. Schetelig, A. Kiani, R. Ordemann, G. Ehninger, M. Schmitz and M. Bornhauser, *Bone Marrow Transplant.*, 2009, **43**, 245–251.

39. M. Allison, *Nat. Biotechnol.*, 2009, **27**, 966–967.

40. H. M. Lazarus, O. N. Koc, S. M. Devine, P. Curtin, R. T. Maziarz, H. K. Holland, E. J. Shpall, P. McCarthy, K. Atkinson, B. W. Cooper, S. L. Gerson, M. J. Laughlin, F. R. Loberiza Jr., A. B. Moseley and A. Bacigalupo, *Biol. Blood Marrow Transplant.*, 2005, **11**, 389–398.

41. H. Ning, F. Yang, M. Jiang, L. Hu, K. Feng, J. Zhang, Z. Yu, B. Li, C. Xu, Y. Li, J. Wang, J. Hu, X. Lou and H. Chen, *Leukemia*, 2008, **22**, 593–599.

42. P. Taupin, *Curr. Opin. Investig. Drugs*, 2006, **7**, 473–481.

43. F. Vianello and F. Dazzi, *Leukemia*, 2008, **22**, 463–465.

44. G. Behre, S. Theurich, T. Weber and M. Christopeit, *Leukemia*, 2009, **23**, 178author reply, 179–180.

45. F. Djouad, V. Fritz, F. Apparailly, P. Louis-Plence, C. Bony, J. Sany, C. Jorgensen and D. Noel, *Arthritis Rheum.*, 2005, **52**, 1595–1603.

46. U. Kunter, S. Rong, Z. Djuric, P. Boor, G. Muller-Newen, D. Yu and J. Floege, *J. Am. Soc. Nephrol.*, 2006, **17**, 2202–2212.

47. V. S. Urban, J. Kiss, J. Kovacs, E. Gocza, V. Vas, E. Monostori and F. Uher, *Stem Cells*, 2008, **26**, 244–253.

48. E. Zappia, S. Casazza, E. Pedemonte, F. Benvenuto, I. Bonanni, E. Gerdoni, D. Giunti, A. Ceravolo, F. Cazzanti, F. Frassoni, G. Mancardi and A. Uccelli, *Blood*, 2005, **106**, 1755–1761.
49. A. H. Huang, M. J. Farrell and R. L. Mauck, *J. Biomech.*, 2009.
50. E. M. Horwitz, P. L. Gordon, W. K. Koo, J. C. Marx, M. D. Neel, R. Y. McNall, L. Muul and T. Hofmann, *Proc. Natl. Acad. Sci. U. S. A.*, 2002, **99**, 8932–8937.
51. R. Quarto, M. Mastrogiacomo, R. Cancedda, S. M. Kutepov, V. Mukhachev, A. Lavroukov, E. Kon and M. Marcacci, *N. Engl. J. Med.*, 2001, **344**, 385–386.
52. P. Hernigou, A. Poignard, F. Beaujean and H. Rouard, *J. Bone Joint Surg. Am.*, 2005, **87**, 1430–1437.
53. D. Schumann, R. Kujat, M. Nerlich and P. Angele, *Biomed. Mater. Eng.*, 2006, **16**, S37–52.
54. F. Djouad, C. Bouffi, S. Ghannam, D. Noel and C. Jorgensen, *Nat. Rev. Rheumatol.*, 2009, **5**, 392–399.
55. K. Pelttari, E. Steck and W. Richter, *Injury*, 2008, **39**(Suppl 1), S58–65.
56. C. De Bari and F. Dell'accio, *Clin. Sci. (Lond.)*, 2007, **113**, 339–348.
57. M. Fatar, M. Stroick, M. Griebe, I. Marwedel, S. Kern, K. Bieback, F. L. Giesel, C. Zechmann, S. Kreisel, F. Vollmar, A. Alonso, W. Back, S. Meairs and M. G. Hennerici, *Neurosci. Lett.*, 2008, **443**, 174–178.
58. C. Nesselmann, N. Ma, K. Bieback, W. Wagner, A. Ho, Y. T. Konttinen, H. Zhang, M. E. Hinescu and G. Steinhoff, *J. Cell. Mol. Med.*, 2008, **12**, 1795–1810.
59. B. Sun, S. W. Zhang, C. S. Ni, D. F. Zhang, Y. X. Liu, W. Z. Zhang, X. L. Zhao, C. H. Zhao and M. X. Shi, *Stem Cells Dev.*, 2005, **14**, 292–298.
60. J. Oswald, S. Boxberger, B. Jorgensen, S. Feldmann, G. Ehninger, M. Bornhauser and C. Werner, *Stem Cells*, 2004, **22**, 377–384.
61. D. Zisa, A. Shabbir, G. Suzuki and T. Lee, *Biochem. Biophys. Res. Commun.*, 2009.
62. N. Nagaya, T. Fujii, T. Iwase, H. Ohgushi, T. Itoh, M. Uematsu, M. Yamagishi, H. Mori, K. Kangawa and S. Kitamura, *Am. J. Physiol. Heart Circ. Physiol.*, 2004, **287**, H2670–H2676.
63. A. T. Badillo, R. A. Redden, L. Zhang, E. J. Doolin and K. W. Liechty, *Cell Tissue Res.*, 2007, **329**, 301–311.
64. L. Chen, E. E. Tredget, P. Y. Wu and Y. Wu, *PLoS One*, 2008, **3**, e1886.
65. H. Nakagawa, S. Akita, M. Fukui, T. Fujii and K. Akino, *Br. J. Dermatol.*, 2005, **153**, 29–36.
66. H. Satoh, K. Kishi, T. Tanaka, Y. Kubota, T. Nakajima, Y. Akasaka and T. Ishii, *Cell Transplant.*, 2004, **13**, 405–412.
67. Y. Wu, L. Chen, P. G. Scott and E. E. Tredget, *Stem Cells*, 2007, **25**, 2648–2659.
68. V. Falanga, S. Iwamoto, M. Chartier, T. Yufit, J. Butmarc, N. Kouttab, D. Shrayer and P. Carson, *Tissue Eng.*, 2007, **13**, 1299–1312.

69. J. J. Lataillade, C. Doucet, E. Bey, H. Carsin, C. Huet, I. Clairand, J. F. Bottollier-Depois, A. Chapel, I. Ernou, M. Gourven, L. Boutin, A. Hayden, C. Carcamo, E. Buglova, M. Joussemet, T. de Revel and P. Gourmelon, *Regen. Med.*, 2007, **2**, 785–794.

70. J. Vojtassak, L. Danisovic, M. Kubes, D. Bakos, L. Jarabek, M. Ulicna and M. Blasko, *Neuro. Endocrinol. Lett.*, 2006, **27**(Suppl 2), 134–137.

71. S. S. Kim, C. K. Song, S. K. Shon, K. Y. Lee, C. H. Kim, M. J. Lee and L. Wang, *Cell Tissue Res.*, 2009, **336**, 59–66.

72. A. Chapel, J. M. Bertho, M. Bensidhoum, L. Fouillard, R. G. Young, J. Frick, C. Demarquay, F. Cuvelier, E. Mathieu, F. Trompier, N. Dudoignon, C. Germain, C. Mazurier, J. Aigueperse, J. Borneman, N. C. Gorin, P. Gourmelon and D. Thierry, *J. Gene Med.*, 2003, **5**, 1028–1038.

73. S. Francois, M. Mouiseddine, N. Mathieu, A. Semont, P. Monti, N. Dudoignon, A. Sache, A. Boutarfa, D. Thierry, P. Gourmelon and A. Chapel, *Ann. Hematol.*, 2007, **86**, 1–8.

74. M. Mouiseddine, S. Francois, A. Semont, A. Sache, B. Allenet, N. Mathieu, J. Frick, D. Thierry and A. Chapel, *Br. J. Radiol.*, 2007, **80**(Spec No 1), S49–55.

75. M. Sasaki, R. Abe, Y. Fujita, S. Ando, D. Inokuma and H. Shimizu, *J. Immunol.*, 2008, **180**, 2581–2587.

76. D. Garcia-Olmo, M. Garcia-Arranz, L. G. Garcia, E. S. Cuellar, I. F. Blanco, L. A. Prianes, J. A. Montes, F. L. Pinto, D. H. Marcos and L. Garcia-Sancho, *Int. J. Colorectal Dis.*, 2003, **18**, 451–454.

77. D. Garcia-Olmo, M. Garcia-Arranz and D. Herreros, *Expert Opin. Biol. Ther.*, 2008, **8**, 1417–1423.

78. D. Garcia-Olmo, M. Garcia-Arranz, D. Herreros, I. Pascual, C. Peiro and J. A. Rodriguez-Montes, *Dis. Colon Rectum*, 2005, **48**, 1416–1423.

79. D. Garcia-Olmo, D. Herreros, I. Pascual, J. A. Pascual, E. Del-Valle, J. Zorrilla, P. De-La-Quintana, M. Garcia-Arranz and M. Pascual, *Dis. Colon Rectum*, 2009, **52**, 79–86.

80. D. Garcia-Olmo, D. Herreros, M. Pascual, I. Pascual, P. De-La-Quintana, J. Trebol and M. Garcia-Arranz, *Int. J. Colorectal Dis.*, 2009, **24**, 27–30.

81. G. Lanzoni, G. Roda, A. Belluzzi, E. Roda and G. P. Bagnara, *World J. Gastroenterol.*, 2008, **14**, 4616–4626.

82. M. Morigi, M. Introna, B. Imberti, D. Corna, M. Abbate, C. Rota, D. Rottoli, A. Benigni, N. Perico, C. Zoja, A. Rambaldi, A. Remuzzi and G. Remuzzi, *Stem Cells*, 2008, **26**, 2075–2082.

83. H. Qian, H. Yang, W. Xu, Y. Yan, Q. Chen, W. Zhu, H. Cao, Q. Yin, H. Zhou, F. Mao and Y. Chen, *Int. J. Mol. Med.*, 2008, **22**, 325–332.

84. F. Togel, A. Cohen, P. Zhang, Y. Yang, Z. Hu and C. Westenfelder, *Stem Cells Dev.*, 2009, **18**, 475–485.

85. G. C. Kopen, D. J. Prockop and D. G. Phinney, *Proc. Natl. Acad. Sci. U. S. A.*, 1999, **96**, 10711–10716.

86. J. Chen, Y. Li, R. Zhang, M. Katakowski, S. C. Gautam, Y. Xu, M. Lu, Z. Zhang and M. Chopp, *Brain Res.*, 2004, **1005**, 21–28.
87. G. Munoz-Elias, D. Woodbury and I. B. Black, *Stem Cells*, 2003, **21**, 437–448.
88. L. Ahrlund-Richter, M. De Luca, D. R. Marshak, M. Munsie, A. Veiga and M. Rao, *Cell Stem Cell*, 2009, **4**, 20–26.
89. V. Jekerle, C. Schroder and E. Pedone, *Bundesgesundheitsblatt Gesundheitsforschung Gesundheitsschutz*, 2010, **53**, 4–8.
90. S. Carrancio, N. Lopez-Holgado, F. M. Sanchez-Guijo, E. Villaron, V. Barbado, S. Tabera, M. Diez-Campelo, J. Blanco, J. F. San Miguel and M. C. Del Canizo, *Exp. Hematol.*, 2008, **36**, 1014–1021.
91. P. Horn, S. Bork, A. Diehlmann, T. Walenda, V. Eckstein, A. D. Ho and W. Wagner, *Cytotherapy*, 2008, **10**, 676–685.
92. K. Bieback, A. Hecker, A. Kocaomer, H. Lannert, K. Schallmoser, D. Strunk and H. Kluter, *Stem Cells*, 2009, **27**, 2331–2341.
93. R. Schafer, J. Wiskirchen, K. Guo, B. Neumann, R. Kehlbach, J. Pintaske, V. Voth, T. Walker, A. M. Scheule, T. O. Greiner, U. Hermanutz–Klein, C. D. Claussen, H. Northoff, G. Ziemer and H. P. Wendel, *Rofo*, 2007, **179**, 1009–1015.
94. F. Deschaseaux, F. Gindraux, R. Saadi, L. Obert, D. Chalmers and P. Herve, *Br. J. Haematol.*, 2003, **122**, 506–517.
95. R. Gorodetsky, *Expert Opin. Biol. Ther.*, 2008, **8**, 1831–1846.
96. I. Sekiya, B. L. Larson, J. R. Smith, R. Pochampally, J. G. Cui and D. J. Prockop, *Stem Cells*, 2002, **20**, 530–541.
97. K. Schallmoser, E. Rohde, A. Reinisch, C. Bartmann, D. Thaler, C. Drexler, A. C. Obenauf, G. Lanzer, W. Linkesch and D. Strunk, *Tissue Eng. Part C Methods*, 2008, **14**, 185–196.
98. L. Sensebe, *Biomed. Mater. Eng.*, 2008, **18**, S3–10.
99. E. J. Caterson, L. J. Nesti, K. G. Danielson and R. S. Tuan, *Mol. Biotechnol.*, 2002, **20**, 245–256.
100. M. Sundin, O. Ringden, B. Sundberg, S. Nava, C. Gotherstrom and K. Le Blanc, *Haematologica*, 2007, **92**, 1208–1215.
101. M. A. Avanzini, M. E. Bernardo, A. M. Cometa, C. Perotti, N. Zaffaroni, F. Novara, L. Visai, A. Moretta, C. Del Fante, R. Villa, L. M. Ball, W. E. Fibbe, R. Maccario and F. Locatelli, *Haematologica*, 2009.
102. C. Bartmann, E. Rohde, K. Schallmoser, P. Purstner, G. Lanzer, W. Linkesch and D. Strunk, *Transfusion*, 2007, **47**, 1426–1435.
103. N. Chevallier, F. Anagnostou, S. Zilber, G. Bodivit, S. Maurin, A. Barrault, P. Bierling, P. Hernigou, P. Layrolle and H. Rouard, *Biomaterials*, **31**, 270–278.
104. C. Doucet, I. Ernou, Y. Zhang, J. R. Llense, L. Begot, X. Holy and J. J. Lataillade, *J. Cell. Physiol.*, 2005, **205**, 228–236.
105. A. Kocaoemer, S. Kern, H. Kluter and K. Bieback, *Stem Cells*, 2007, **25**, 1270–1278.

106. K. Schallmoser, C. Bartmann, E. Rohde, A. Reinisch, K. Kashofer, E. Stadelmeyer, C. Drexler, G. Lanzer, W. Linkesch and D. Strunk, *Transfusion*, 2007, **47**, 1436–1446.
107. S. K. Lee, Y. Kim, S. S. Kim, J. H. Lee, K. Cho, S. S. Lee, Z. W. Lee, K. H. Kwon, Y. H. Kim, H. Suh–Kim, J. S. Yoo and Y. M. Park, *Proteomics*, 2009, **9**, 4389–4405.
108. S. Tsutsumi, A. Shimazu, K. Miyazaki, H. Pan, C. Koike, E. Yoshida, K. Takagishi and Y. Kato, *Biochem. Biophys. Res. Commun.*, 2001, **288**, 413–419.
109. X. Chen, H. Xu, C. Wan, M. McCaigue and G. Li, *Stem Cells*, 2006, **24**, 2052–2059.
110. S. Frauenschuh, E. Reichmann, Y. Ibold, P. M. Goetz, M. Sittinger and J. Ringe, *Biotechnol. Prog.*, 2007, **23**, 187–193.
111. D. Baksh, P. W. Zandstra and J. E. Davies, *Biotechnol. Bioeng.*, 2007, **98**, 1195–1208.
112. A. J. Engler, S. Sen, H. L. Sweeney and D. E. Discher, *Cell*, 2006, **126**, 677–689.
113. T. Cordonnier, P. Layrolle, J. Gaillard, A. Langonne, L. Sensebe, P. Rosset and J. Sohier, *J. Mater. Sci. Mater. Med.*, 2009.
114. N. Kimelman–Bleich, G. Pelled, D. Sheyn, I. Kallai, Y. Zilberman, O. Mizrahi, Y. Tal, W. Tawackoli, Z. Gazit and D. Gazit, *Biomaterials*, 2009, **30**, 4639–4648.
115. D. Sakai, J. Mochida, Y. Yamamoto, T. Nomura, M. Okuma, K. Nishimura, T. Nakai, K. Ando and T. Hotta, *Biomaterials*, 2003, **24**, 3531–3541.
116. B. G. Santoni, G. E. Pluhar, T. Motta and D. L. Wheeler, *Biomed. Mater. Eng.*, 2007, **17**, 277–289.
117. O. N. Koc, S. L. Gerson, B. W. Cooper, S. M. Dyhouse, S. E. Haynesworth, A. I. Caplan and H. M. Lazarus, *J. Clin. Oncol.*, 2000, **18**, 307–316.
118. D. Rubio, J. Garcia-Castro, M. C. Martin, R. de la Fuente, J. C. Cigudosa, A. C. Lloyd and A. Bernad, *Cancer Res.*, 2005, **65**, 3035–3039.
119. G. V. Rosland, A. Svendsen, A. Torsvik, E. Sobala, E. McCormack, H. Immervoll, J. Mysliwietz, J. C. Tonn, R. Goldbrunner, P. E. Lonning, R. Bjerkvig and C. Schichor, *Cancer Res.*, 2009, **69**, 5331–5339.
120. D. Rubio, S. Garcia, M. F. Paz, T. De la Cueva, L. A. Lopez-Fernandez, A. C. Lloyd, J. Garcia-Castro and A. Bernad, *PLoS One*, 2008, **3**, e1398.
121. F. Tirode, K. Laud-Duval, A. Prieur, B. Delorme, P. Charbord and O. Delattre, *Cancer Cell*, 2007, **11**, 421–429.
122. M. Breitbach, T. Bostani, W. Roell, Y. Xia, O. Dewald, J. M. Nygren, J. W. Fries, K. Tiemann, H. Bohlen, J. Hescheler, A. Welz, W. Bloch, S. E. Jacobsen and B. K. Fleischmann, *Blood*, 2007, **110**, 1362–1369.
123. D. Furlani, M. Ugurlucan, L. Ong, K. Bieback, E. Pittermann, I. Westien, W. Wang, C. Yerebakan, W. Li, R. Gaebel, R. K. Li, B. Vollmar, G. Steinhoff and N. Ma, *Microvasc. Res.*, 2009, **77**, 370–376.

124. A. E. Karnoub, A. B. Dash, A. P. Vo, A. Sullivan, M. W. Brooks, G. W. Bell, A. L. Richardson, K. Polyak, R. Tubo and R. A. Weinberg, *Nature*, 2007, **449**, 557–563.

125. P. J. Mishra, J. W. Glod and D. Banerjee, *Cancer Res*, 2009, **69**, 1255–1258.

126. K. Bieback, K. Schallmoser, H. Klutera and D. Strunk, *Transfus. Med. Hemother.*, 2008, **35**, 286–294.

127. M. F. Pittenger, A. M. Mackay, S. C. Beck, R. K. Jaiswal, R. Douglas, J. D. Mosca, M. A. Moorman, D. W. Simonetti, S. Craig and D. R. Marshak, *Science*, 1999, **284**, 143–147.

128. C. Yang, H. Frei, F. M. Rossi and H. M. Burt, *J. Tissue Eng. Regen. Med.*, 2009, **3**, 601–614.

129. K. Le Blanc, L. Tammik, B. Sundberg, S. E. Haynesworth and O. Ringden, *Scand. J. Immunol.*, 2003, **57**, 11–20.

130. O. N. Koc, J. Day, M. Nieder, S. L. Gerson, H. M. Lazarus and W. Krivit, *Bone Marrow Transplant.*, 2002, **30**, 215–222.

131. A. Reinisch, C. Bartmann, E. Rohde, K. Schallmoser, V. Bjelic-Radisic, G. Lanzer, W. Linkesch and D. Strunk, *Regen. Med.*, 2007, **2**, 371–382.

132. T. Walenda, S. Bork, P. Horn, F. Wein, R. Saffrich, A. Diehlmann, V. Eckstein, A. D. Ho and W. Wagner, *J. Cell. Mol. Med*, 2010, **14**, 337–350.

133. A. I. Caplan and J. E. Dennis, *J. Cell. Biochem.*, 2006, **98**, 1076–1084.

134. J. Y. Oh, M. K. Kim, M. S. Shin, H. J. Lee, J. H. Ko, W. R. Wee and J. H. Lee, *Stem Cells*, 2008, **26**, 1047–1055.

135. S. M. Devine and R. Hoffman, *Curr. Opin. Hematol.*, 2000, **7**, 358–363.

136. A. Torsvik *et al.*, *Cancer Research.*, 2010, [Epub].

137. S. Garcia, *et al.*, *Exp. Cell. Res.*, 2010, **316**, 1648–1650.

(Stem) Cell Based Therapy for Neurological Disorders

O. EINSTEIN AND T. BEN-HUR

Department of Neurology, The Agnes Ginges Center for Human Neurogenetics, Hadassah–Hebrew University Medical Center, Jerusalem, Israel

10.1 Introduction

Recent progress in stem cell biology has raised new hopes for the utilization of stem cells in the treatment of various neurological disorders such as Parkinson's disease (PD), Huntington's disease (HD), amyotrophic lateral sclerosis (ALS), stroke, spinal cord injuries (SCI), demyelinating diseases and, in particular, multiple sclerosis (MS).

A key issue in these neurological diseases is that injurious processes are often irreversible. The adult central nervous system (CNS) does in fact harbor a variety of cells with the potential to perform regeneration, but this process most often fails. Multiple factors contribute to this failure, arising both from the inactivity of regenerating cells, leading to delay and reduced recruitment and from lack of environmental support and even blockage of their action.[1–5] The design of therapeutic approaches should address these issues either by manipulating the CNS tissue to enhance the number and function of endogenous regenerating cells, or alternatively by introducing regenerating cells into the diseased CNS by means of transplantation. In this chapter we consider the potential application of cell-based therapeutic strategies for CNS diseases, highlighting the promises as well as the problems and future prospects of stem cell therapy in neurological disorders.

Stem Cell-Based Tissue Repair
Edited by Raphael Gorodetsky and Richard Schäfer
© The Royal Society of Chemistry 2011
Published by the Royal Society of Chemistry, www.rsc.org

10.2 Mechanisms of Action of Transplanted Stem Cell in the Host CNS

The accumulation of data on the beneficial effects of stem cell transplantation (SCT) in different animal models of neurological diseases has shed light on the mechanisms by which stem cells could promote neural recovery.[6] Until recently, research on the therapeutic potential of stem cells has focused primarily on their potential to replace damaged or missing cells in CNS diseases that result in cell loss. It has become apparent that, due to the complexity of CNS structure and function, cell-replacement based strategies have encountered a number of as yet insurmountable obstacles. However, recent studies indicate that transplanted stem cells may be beneficial (Figure 10.1):

- by additionally attenuating deleterious inflammation;
- by protecting the CNS from degeneration; and
- by enhancing endogenous recovery processes.

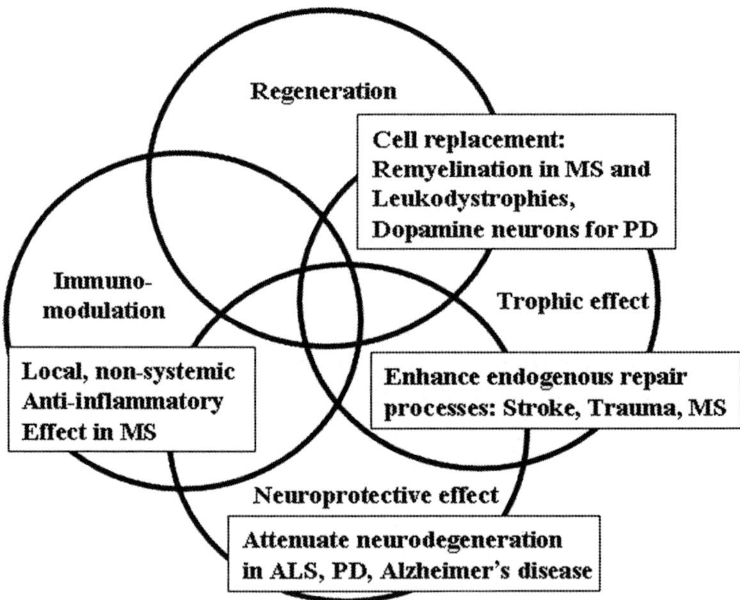

Figure 10.1 Four interacting mechanisms of action of transplanted stem cells in the host brain. The overall beneficial effect of transplanted stem cells on the diseased CNS is exerted by a combination of four interacting arms. Originally stem cells were introduced for transplantation in neurological disorders to perform CNS regeneration. Recently they were suggested to also possess neurotrophic and neuroprotective functions, as well as immunosuppressive effects on the host brain. The notion that stem cells induce beneficial effects by multiple mechanisms of action increases the likelihood of their efficacy in clinical practice.

10.2.1   Cell Replacement by Transplanted Stem Cells

The primary goal of stem cell-based therapy has been to replace missing cells and tissue. As it seems currently impossible to recapitulate the full complexity of neural circuits in the human brain, cell replacement could be more practical in diseases where a specific cell type and pathway are selectively involved. This consideration raised Parkinson's disease as a leading candidate condition for application of cell therapy. Indeed, the work in animal models of PD provided good indications that brain cells could be replaced. Experimental data from rodents and non-human primates demonstrated that transplanted dopaminergic neurons derived from fetal ventral mesencephalon formed synaptic contacts, released dopamine and ameliorated PD-like symptoms when grafted intrastriatally.[7–9]

A fundamental precondition for a cell replacement approach is the generation of specialized cells for utilization in each neurological disease. The adult CNS, and particularly in conditions of degeneration, does not direct uncommitted precursors to differentiate into the specific lineage required.[10] Therefore, the generation of sufficient amounts of specialized cells has become a major theme in the development of cell replacement therapy for PD and other neurological diseases. It should be noted that various types of uncommitted stem cells have also been successfully transplanted in several models of neurological disorders and induced remarkable functional recovery. However, this significant behavioural and clinical improvement has generally showed little correlation with the absolute numbers of terminally differentiated neural cells originated by transplanted stem cells (reviewed in ref. 11). When transplanted in rodents with experimental PD, neural precursor cells (NPCs) very scarcely differentiate into tyrosine hydroxylase (TH)-immunoreactive neurons.[12–14] Similarly, animals with HD,[15,16] SCI,[17–20] acute stroke[21] or intracerebral hemorrage[22,23] show clinical recovery regardless of transplanted NPCs terminally differentiating into neuronal cells *in vivo*.

Experimental autoimmune encephalomyelitis (EAE) is typically viewed as the most representative animal model of human multiple sclerosis, both clinically and pathologically, and has proved to be especially useful in studies on the pathogenesis and treatment of MS.[24–26] EAE is a T-cell mediated inflammatory disease of the CNS which is often accompanied by demyelination and axonal damage.[27,28] In the same context, the clinical improvement upon transplantation in EAE was accompanied with very low differentiation of NPCs into myelin-forming oligodendrocytes.[29–33] It is yet to be shown that transplanted cells can indeed remyelinate efficiently in clinically relevant models of MS. This may depend on grafting precursor cells that are already programmed to the oligodendroglial lineage[34,35] and the use of appropriate models of disease that allow remyelination to occur.

10.2.2   Immunosuppressive Effects of Transplanted Stem Cells

Although stem cells may exert their therapeutic effects by directly replacing missing cells, the scarce terminal differentiation within the host tissue has suggested that the beneficial effect of stem cells in neurological disease models may be attributable to a number of alternative, previously unrecognized mechanisms.

Recent work has highlighted important immunosuppressive properties by which transplanted NPCs exhibit therapeutic effects in models of MS. The first indication of an anti-inflammatory effect of NPCs was obtained when NPCs were transplanted intracerebroventricularly (ICV) into Lewis rats with acute EAE induced by immunization with spinal cord homogenate.[32] Transplanted NPCs significantly inhibited the inflammatory brain process and attenuated the clinical severity of disease. In this EAE model, rats exhibit acute, reversible paralytic disease that is the result of disseminated CNS inflammation without demyelination or axonal injury.[36] Therefore, it was suggested that the beneficial effect of NPC transplantation was mediated by an anti-inflammatory effect. Follow-up studies examined the effect of NPC transplantation upon either ICV or intravenous (i.v.) cell injection in the myelin oligodendrocyte glycoprotein (MOG)35-55-induced EAE in C57BL/6 mice. In this model there is an acute paralytic disease due to a T cell-mediated autoimmune process that causes severe axonal injury and demyelination. Subsequently, the mice remain with fixed neurologic sequel, the severity of which is correlated with the extent of axonal loss.[37] The use of the latter model of EAE demonstrated that NPC transplantation attenuated the inflammatory process, reduced acute and chronic axonal injury and demyelination, and improved the overall clinical performance.[31,33]

Transplanted NPC-mediated immunosuppression may take place in the CNS, at the level of the atypical perivascular niche,[29,32,38] as well as in secondary lymphoid organs such as the lymph nodes[30] or the spleen.[23] The exact mechanisms by which transplanted NPCs attenuate CNS or peripheral inflammation are not yet clear. One school of thought has suggested an immunomodulatory effect by which NPCs induce selective apoptosis of Th1 cells *via* the inflammation-driven upregulation of membrane expression of functional death receptor ligands (*e.g.* FasL, TRAIL, Apo3L) on NPCs,[38] thus shifting the inflammatory process in the CNS towards a more favorable Th2 dominant climate. Alternatively, it has been suggested that NPCs inhibit T-cell activation and proliferation by a non-specific, bystander immunosuppressive action.[30] This notion emerged when i.v. injected NPCs did not cross the blood–brain–barrier, but were transiently found in peripheral lymphoid organs where they interacted with T cells to reduce their encephalitogenicity, resulting in reduced immune cell infiltration into the CNS and consequently milder CNS damage[30] Coculture experiments that mimic the direct interactions of lymph node cells and NPCs *in vivo* showed that NPCs induced a striking inhibition of the activation and proliferation of T cells and significant suppression of pro-inflammatory cytokines.[30,32] Moreover, NPCs inhibited multiple inflammatory signals, as exemplified by attenuation of T cell receptor- IL2- and IL6- mediated immune cell activation and/or proliferation.[39] The specific molecules that mediate immunological functions of NPCs are not known. To address this issue a recent study examined the specific interactions of NPCs with antigen-presenting cells. It was found that, following subcutaneous injection to EAE mice, NPCs home to secondary lymphoid organs and modify the perivascular microenvironment by hindering the activation of myeloid dendritic cells *via* a bone morphogenetic protein (BMP)-4-dependent mechanism.[40]

Such a broad immune regulatory capacity in attenuating EAE has more recently been shown also for neural precursor cells derived from human embryonic stem cells (hESC),[29] human somatic-NPCs[41] and bone marrow mesenchymal stem cells (BMSCs)[42–44] after transplantation to EAE animals. Direct comparison has not yet been performed to study whether different populations of stem cells share common immune regulatory functions by using the same molecular mechanisms. Nevertheless, it seems that immune regulation is indeed a common characteristic of stem cells, thus widening the potential sources for cell therapy.

Together these findings provide evidence that stem cells can be manipulated to act as immune regulators in the CNS. This may be of major importance in immune-mediated neurological diseases, where cell therapy may be neuroprotective by attenuating the immune attacks against the CNS and thus preventing accumulation of tissue injury.

10.2.3 Neuroprotective Effect of Transplanted Stem Cells

A general neuroprotective effect has also been observed in other non-inflammatory experimental models of brain disease. NPCs could rescue dopaminergic neurons of the mesostriatal system in a PD model in rodents.[12] NPCs seeded on a synthetic biodegradable scaffold and grafted into the cord of hemi-sectioned rats have induced significant clinical recovery by reducing the necrosis and by preventing extensive secondary cell loss, inflammation and glial scar formation, thus inducing a permissive environment for axonal regeneration.[45] Prevention of motor neurons from dying has been observed when NPCs were transplanted in models of ALS.[46–48]

A neuroprotective effect of NPC transplantation has also been observed in other models of neurodegeneration. In mutant mice in which Purkinje neurons die in the fourth to fifth week of life, transplanted NPCs rescued host Purkinje cell function and restored motor coordination.[49] Subretinal injection of human NPCs provided almost full protection of visual function in a rat model of retinal degeneration.[50]

These findings led to the concept that NPCs are endowed with inherent mechanisms for rescuing dysfunctional neurons. NPC-induced tissue protection is generally manifested by reduction of glial scar formation and increase of survival and/or functions of endogenous glial and neuronal progenitors.

10.2.4 Neurotrophic Effects of Transplanted Stem Cells

The insufficient repair capacity of the adult CNS results from a number of failing repair programmes. This may include the apparent inability of endogenous stem and progenitor cells to respond properly to disease states, to replace damaged cells and from a lack of regenerative capacity of injured axons.[51–53] Lately, stem cell therapy is emerging as a mode of treatment that can enhance the host brain's ability to repair itself in both aspects.

It has been shown that NPC transplantation induced a permissive environment for axonal regeneration after sectioning the adult spinal cord[45] and promoted neurite growth in the optic nerve in a model of retinal degeneration.[50,54] In both cases this effect was mediated by induction of matrix metalloproteinases, which degrade the impeding extracellular matrix (ECM) and cell surface molecules, thus enabling axons to extend through the glial scar. Substantial endogenous reconstitution of the brain structural connectivity has been found following injection of NPCs in biodegradable scaffolds into regions of extensive brain degeneration caused by hypoxia[55,56] or following transplantation of NPCs after ischemia/reperfusion injury in mice.[57]

Cuprizone is a neurotoxin that induces demyelination in the adult brain. This is followed by a natural process of remyelination after withdrawal of cuprizone from the diet. In the chronic form of the model and in aged mice there is slowed remyelination,[58–62] representing some aspects of human MS. We have recently demonstrated that ICV-transplanted NPCs significantly enhance host brain progenitor cell-derived myelin regeneration.[63] The pro-regenerative effect of transplanted NPCs was mediated by inducing an increase in the proliferation of host brain oligodendrocyte progenitor cells (OPCs). *In vitro* coculture experiments also showed increased proliferation and maturation of OPCs in response to the trophic effect of NPCs; this effect was mediated by platelet-derived growth factor (PDGF)-AA and fibroblast growth factor (FGF)-2.[63]

Transplanted stem cells can enhance endogenous neurogenesis in certain physiological and pathological conditions.[64,65] Mice exposed prenatally to opioids display impaired learning associated with reduced neurogenesis. In this model, transplantation of NPCs improved learning functions, as well as host brain derived neurogenesis in the dentate gyrus of the hippocampus.[66] Similarly, while neurogenesis in the dentate gyrus declines severely by middle age, transplantation of NPCs stimulated the endogenous progenitors in the subgranular zone to produce new dentate granule cells.[64]

The underlying molecular mechanisms by which transplanted NPCs exert such broad trophic effects are still poorly understood. It has been shown that the beneficial effects of transplanted stem cells may relate in part to increased *in vivo* bioavailability of neurotrophins (*e.g.* nerve growth factor, brain-derived neurotrophic factor, ciliary neurotrophic factor, glial-derived neurotrophic factor[20,21,31,38,45,67]) and to the modulation of the host environment into more permissive for regeneration. Insulin-like growth factor (IGF) and glial growth factor-2 (GGF-2) may be released by NPCs and were shown to inhibit EAE.[68–70] Interestingly, their effect was mediated not only by enhancing oligodendrocyte survival,[71,72] but also by decreasing neuroinflammation.[69,73–75]

Thus, transplanted NPCs may enhance the capacity of the adult CNS to repair itself by restoring the ability of endogenous progenitors and stem cells to both respond properly to disease state and replace damaged CNS cells.

10.3 Candidate Stem Cells for Transplantation in Neurological Disorders

Stem cells are defined as undifferentiated cells that have the potential for continuous self-renewal and to generate a diverse range of specialized cell types. As such, stem cells are leading candidates as source for transplantation in patients with neurological diseases. Our understanding of the biology of stem and progenitor cells, and of the means by which they may be directed to differentiate into specialized cell types, has advanced tremendously over the last several years to the point where we may now reasonably consider various stem cell populations as candidates for clinical application. For each population there are specific advantages and disadvantages and possible mechanisms of action on the host brain to be utilized (summarized in Table 10.1). These can be considered for the specific requirements for therapeutic effects in the various neurological disorders (Table 10.2).

10.3.1 Neural Stem Cells (NSCs) and Neural Precursor Cells (NPCs)

Mammalian multipotential NSCs and NPCs support neurogenesis and gliogenesis within specific areas of the CNS during both development and adulthood.[76–78] NSCs and NPCs can be isolated from fetal and adult brains, can be expanded in large quantities in chemically defined culture media *in vitro*, maintain their capacity for self-renewal and generate a progeny of the three neural cell lineages.[79] NSCs retain their functional plasticity after *in vitro* passaging and after several freezing thawing cycles, and they can still be modulated *in vitro* by exposure to different growth factors. However, while adult NSCs can generate all three neural lineages, in practice they serve mainly

Table 10.1 Comparison between three candidate cell populations for transplantation in neurological disorders.

Cell type	Embryonic stem cells	Adult neural stem cells	Bone marrow stromal cells
Cell source	Allogeneic[a]	Allogeneic	Autologous
Current availability for clinical use	Available clinical grade lines	Available banks, limited source	Available immediately
Safety	Under investigation[b]	Probably safe[c]	Safe[d]
Functional properties:			
Integration in host CNS	Yes	Yes	No
Anti-inflammatory effects	Yes	Yes	Yes
Trophic effects	Yes	Yes	Yes

[a]Potentially autologous by developing induced pluripotent stem cells technology.
[b]Risk of teratomas.
[c]Possible risk of neuroepithelial tumors.
[d]Possible risk of producing ectopic mesenchymal tissue.

Table 10.2 Beneficial actions of transplanted neural stem cells in different neurological disorders.

	Cell replacement	*Anti-inflammatory effect*	*Neuro-protective effect*	*Neurotrophic effect*
Parkinson's disease	Yes	–	Yes	–
Huntington's disease	Yes	–	Yes	–
Amyotrophic lateral sclerosis	Questionable	–	Yes	–
Alzheimer's disease	–	–	Yes	–
Stroke	–	Yes	–	Questionable
Spinal cord injury	–	Yes	–	Yes
Genetic dismyelinating diseases	Yes	–	–	–
Multiple sclerosis	Yes	Yes	Yes	Yes
Age-related macular degeneration and other degenerative retinal diseases	Questionable	–	Yes	–

as source of glia as they can not be directed to specific neuronal lineages. For example, dopaminergic neurons and motor neurons are specified at early embryonic stages of CNS development[80,81] and can not be routinely enriched from adult NSCs. This is a crucial limitation in the utilization of adult NSCs for the purposes of neuronal cell replacement.

In various experimental models, transplanted NPCs integrated well in the CNS and improved clinical outcome. This was observed in experimental models of stroke,[82–84] SCI[56,85,86] and EAE.[30,32,38,87] In all these models, transplanted cells improved the clinical outcome by their immunomodulatory, neurotrophic and neuroprotective properties and not by cell replacement.

The direct regenerative effects of adult NSCs and NPCs are still highly relevant for diseases of myelin. Glial-committed NPCs have also proven to be able to generate myelinogenic oligodendrocytes following transplantation in hypomyelinated shiverer (*shi*) mice,[88–90] myelin-deficient (*md*) rats[91] and shaking (*sh*) pups,[92,93] and in retinas of young mice.[64] Expression of poly-sialylated neural cell adhesion molecule (PSA-NCAM) on the cell membrane has been associated with stem cell commitment to neuronal or glial fate, depending on time and place in development.[94] Such PSA-NCAM+ glial precursors (growing as neurospheres) remyelinated a large proportion of axons following local injection into the dorsal columns of rats[95] and migrated effi-ciently along inflamed white matter tracts of rats with EAE.[96] Similarly, adult human subventricular zone (SVZ) precursors remyelinated the demyelinated adult rat spinal cord.[97] However, to date the remyelination competence of neural stem and glial progenitor cells has been convincingly demonstrated only in focal chemical lesions and models of hereditary dysmyelination, but not in any clinically relevant model of MS.

10.3.2 Embryonic Stem Cells (ESCs)

The practical limitations on both fetal and adult cell acquisition for human allograft have driven research on deriving tissue-specific progenitor cells from human embryonic stem cells (hESCs) cells. ESCs are derived from the inner cell mass of blastocyst-stage embryos. As totipotent cells they are able to generate the entire repertoire of cell types in the body. ESC lines can be established from virtually all mammals,[98–100] and can be banked and propagated *in vitro* almost indefinitely. To date, ESCs are the best candidate for mass generation of specialized neural cells. By recapitulation of developmental conditions in culture it is possible to grow mouse and human ESCs and to generate cultures enriched with dopaminergic cells, motor neurons, oligodendrocytes and retinal cells amongst others.

Mouse ESC-derived motor neurons can establish functional synapses with muscle fibers *in vitro* and extend axons to ventral roots after transplantation into motor neuron-injured adult rats.[101] Transplantation of hESC-derived NPCs in Parkinsonian rats resulted in only partial clinical improvement due to very limited differentiation into dopaminergic neurons.[10] This highlights the need to direct the cells into the specific neuronal lineage prior to transplantation in the degenerating CNS. Although protocols for generation of dopaminergic neurons from hESC *in vitro* are available,[81,102,103] their grafting in experimental animals has been hampered by poor survival of the transplanted cells. This is currently a major obstacle in the application of cell therapy for PD.[104]

The sequential use of growth factors (including FGF2, EGF and PDGF) in a program that mimics embryonic development has been used to derive glial precursors from mouse ES cells.[105–107] The myelinogenic potential of mouse ES-derived glial progenitor cells (GPCs) that were expanded *in vitro* was demonstrated in the *md* rat brain.[106] Similarly, when transplanted into both a rodent model of chemically induced demyelination and to the spinal cords of hypomyelinated *shi* mice, mouse ES-derived GPCs were also able to differentiate into glial cells and remyelinate demyelinated axons.[107] A number of groups have now generated oligodendrocytes from hESCs. These cells appear functional and have been reported capable of myelinating demyelinated foci in spinal cord contusions.[108,109]

hESC-based transplantation approaches may prove limited by the potential for tumorogenesis, in particular by the potential for any persistent undifferentiated ESCs in the donor pool to yield teratomas,[110] or of incompletely differentiated neural cells to generate neuroepithelial tumors.[111] Indeed, while a number of groups have reported that transplantation of both mouse and human ES-derived NPCs did not result in teratoma formation, long survival times may be needed to exclude the generation of neuroepithelial tumors, as opposed to teratomas, from incompletely differentiated neuroepithelial remnants.

10.3.3 Induced Pluripotent Stem Cells (iPSCs)

Over the past three years, several groups have reported the generation of induced pluripotent stem cells (iPSCs) from mouse[112,113] and human[114] somatic cells. These iPSCs were typically generated from dermal fibroblasts, co-transduced with a number of stem cell-associated transcription factors including Oct3/4, Sox2, Myc, Klf4, in but one of several iterations.[115] iPSCs are indeed pluripotent, as defined by their ability to generate cells of all major germ layers, and teratomas *in vivo*. Importantly, initial reports of their production of dopaminergic and motor neurons have validated their ability to generate post-mitotic derivatives.[116–118]

The potential advantage of iPSCs is to obtain transplantable specialized cells from an autologous source and circumvent the bitter ethical controversy on the use of embryonic stem cells. However, the hurdles that will need to be overcome are similar to those facing hESC-derived neural and glial progenitors in terms of the need for high-yield production and purification, with the absolute abrogation of any potentially tumorigenic contaminants. The issue of safety is of special importance for iPSCs in view of their genetic modification to express potentially oncogenic genes.

10.3.4 Bone Marrow Stromal Cells (BMSCs)

A central issue in the field of stem cell biology is the suggestion that stem cells of one tissue may generate cells of other tissues.[119,120] It has been suggested that stem cells from non-neural tissue could trans-differentiate into neural progeny. A number of investigators have reported that adult mouse and human BMSCs can differentiate *in vitro* into other cell types including muscle, skin, liver, lung and neural cells.[121–123] While it is clear that BMSCs can acquire neural markers, it remains highly controversial whether they are capable of acquiring the full range of neuronal functions. Other reports have indicated that stromal cells may rather fuse with existing neurons and glia, resulting in the formation of heterokaryons.[124,125] However, BMSCs are a promising candidate for cell therapy using their powerful immunomodulatory and neurotrophic properties. These effects were evident in experimental models of stroke,[126–128] SCI[129,130] and traumatic brain injury.[127,131,132] In rats with a demyelinated lesion of the spinal cord, injection of isolated mononuclear BMSCs resulted in varying degrees of remyelination and improved axonal conduction velocity.[133,134] Transplanted BMSCs have also attenuated clinical and pathological signs of EAE.[42–44,135] BMSCs present the advantages of probable safety and of delivering autologous cells. However, it has not been studied whether they can survive and integrate well in the CNS environment and whether their overall effect is comparable to that of neural grafts.

10.4 Application of Experimental Findings to Human Neurological Diseases

Using a relevant strategy of cell-based therapy, diverse neurological disorders (*e.g.* vascular, traumatic, demyelinating and neurodegenerative diseases) may all be approachable as therapeutic targets. Each of the various neurological disorders may benefit from stem cell-based therapy by its different mechanisms of action (summarized in Table 10.2).

Cell replacement therapy in neurological diseases may be applicable in some situations such as Parkinson's disease, Huntington disease and demyelinating diseases. In patients with PD, clinical experience with human fetal mesencephalic tissue transplantation has existed for over 20 years. These early studies showed improvement in some patients, but raised many questions regarding the optimal sites of grafting, dose of cells to be delivered, graft rejection, control of non-dopaminergic pathology and adverse effects (mainly persistent dyskinesias).[8] Moreover, it became clear that fetal mesencephalic tissue, requiring up to four fetuses per patient, was not a practical source of transplantable dopamine neurons. Nevertheless, these studies provided a proof-of-principle for cell replacement in PD and boosted the search for an ethically acceptable and inexhaustible source of dopamine neurons.

HD is another candidate for cell replacement therapy since it also results from a localized degenerative process involving a specific neuronal system. The ease of generation of GABAergic compared to dopaminergic cells may suggest that clinical trials may progress faster in HD than in PD.

Cell replacement therapy is still far from practical in situations where cells must recapitulate complex neural circuitry, with multiple cellular specializations and long distance connections, within the hostile environment of the adult CNS. Degenerative diseases of the adult CNS such as Alzheimer's disease and ALS have a limited inflammatory component and a non-permissive environment for regenerative processes, rendering significant NSC migration unlikely. Thus, the prospect of therapeutic cell replacement in this scenario with appropriate cell migration and integration into existing neural circuitries is still a distant prospect.

Recent studies in ALS models have shifted the therapeutic target of NPC transplantation from neuron replacement to motor neuron salvage.[47] This came from observations that providing the proper glial environment may be sufficient to significantly delay the degeneration of motor neurons.[48,136–138]

Cell replacement therapy for stroke or SCI is an even bigger challenge as transplanted NPCs need to replace a range of neuronal types, remyelinate axons and repair complex neural circuitries. As a preliminary step towards this goal, a recent study has shown that human fetal-derived NPCs transplanted into the brains of rodents after stroke, survived, migrated and differentiated into various types of neuron and glia.[139]

With regard to diseases that currently seem amenable to cell replacement therapy, the delineation of the optimal developmental stage of transplanted cells remains a key issue. Candidate cells need to be committed to their target specialization, but also need to retain the plasticity of their precursors that is necessary for effective integration in the CNS.

A full understanding of the pros and cons of each stem cell types, of their different potential routes of delivery and of the methods available for tracking each phenotype after transplantation, will be needed in order to design clinically effective transplantation strategies. Stem cells have proven useful not only as cellular vectors for transplantation, but also as tools for understanding the signaling pathways and the control of growth and fate of progenitors *in vivo*. Thus, advances in cell biology of stem cells may aid in developing means to enhance endogenous repair processes. This may replace the need for transplantation in acquired conditions such as ischemic, traumatic and inflammatory disorders in which large accessible stores of endogenous progenitors may persist locally. In diseases involving the widespread loss of cells and disorders in which endogenous progenitor cells are themselves deficient (*e.g.* the congenital leukodystrophies), therapeutic strategies based on cell transplantation will be necessary.

10.5 Cell-based Therapy for Multiple Sclerosis (MS)

For each disease there are specific considerations regarding the therapeutic requirements of the CNS tissue, the particular characteristics of cells, the dose and the route of cell delivery that will be optimal. As an example, we discuss here the case of cell therapy for multiple sclerosis (MS). Cell therapy has been studied in particular for inducing efficient remyelination in disorders of myelin, including both the largely pediatric leukodystrophic disorders of myelin formation or its early breakdown and the acquired demyelinating diseases. The most common disorder in this group is MS.

10.5.1 Myelin Regeneration Fails in MS

MS is the most common cause of neurological disability in young adults, characterized by chronic inflammatory, demyelinating multifocal lesions within the CNS[140–143] and heterogeneous pathology.[144] A cascade of events that engage the immune system result in acute inflammatory injury of axons and glia, accompanied by frank demyelination.[142,145] Several studies have indicated that axonal pathology is the best correlate of chronic neurological impairment in EAE and MS.[37,142,146,147] The affected demyelinated regions can undergo partial remyelination, leading to structural repair and recovery of function.[148–151] However, remyelination is typically incomplete and ultimately fails in the setting of recurrent episodes, contributing to the progressive demyelination, gliosis, axonal damage and neurodegeneration, typically noted in MS. The sequential involvement of these processes underlies the clinical course, characterized by episodes of relapses, which after full remissions early in the course of disease eventually leave persistent deficits and finally deteriorate into a secondary chronic progressive phase.

It is unclear why remyelination fails over time in MS (reviewed in ref. 51). Failure of spontaneous remyelination could stem both from cell-autonomous insufficiency of endogenous remyelinating OPCs and from lack of environmental support. It has been suggested that repeated demyelinating episodes in chronic and relapsing MS cause a depletion in the endogenous pool of progenitor cells. Although progenitor cells decrease in number after experimental focal demyelination,[59,152] this was not observed in pathological specimens of chronic MS lesions.[148,153] However, in experimental focal demyelination, it has been shown that only a subpopulation of local OPCs react to injury by generating new oligodendrocytes and myelin.[152] In contrast, despite persistence of abundant OPCs in acute and chronic MS lesions, they did not expand in number in demyelinated foci relative to normal white matter.[154] This suggests that the response of the progenitor cell population to the demyelinating process in the human brain is either aborted or otherwise deficient. It has also been shown that only OPCs that reside at the margins of experimental lesions migrate into the lesion core and remyelinate it, whereas long distance migration of OPCs does not occur in the brain parenchyma.[155] Therefore, another aspect of the limited tissue support may be that mobilization of the adult OPCs is limited by insufficient supply of environmental signals in the brain. Bidirectional trophic interactions between oligodendrocytes and axons are necessary for their long-term survival. There is evidence that extensive axonal transection occurs already in acute MS lesions.[142] Remyelination can not effectively proceed in the absence of sufficient intact axonal substrate, so that it may be effectively rate-limited by the extent of underlying axonal loss.

10.5.2 Clinical Considerations in Designing Cell-based Therapy for MS

Current immunosuppressive and immunomodulatory treatments for MS have little efficacy in either preventing long-term disability or in restoring lost function.[156] Thus it is clear that new therapeutic approaches need to be developed in order to promote tissue repair. MS lesions offer the potential for remyelination either *via* enhancing the function of endogenous myelinogenic progenitor cells, or by cell transplantation.

Several issues need to be considered in the rationale design of cell therapy in MS:

1. As a multifocal disease and given the inability to graft cells in multiple distinct sites, the route of cell delivery and transplanted cell migration are crucial issues for the success of transplantation.
2. As a chronic disease, long-term survival of grafted cells is essential for their ongoing beneficial effect.
3. As an immune-mediated disease, with phases of acute inflammation followed by remissions with chronic gliosis, the bi-directional interactions between host brain and transplanted cells are highly important.
4. Considering the barriers for remyelination at the chronic gliotic phase, the apparent link between the acute inflammatory phase and myelin regeneration and the necessity to achieve remyelination before axonal damage occurs, one needs to define the time window of opportunities when remyelination is feasible. As the CNS may be considered a 'hostile environment', it is clear that successful regenerative cell

therapy in MS will also depend not only on the characteristics of transplanted cells, but also on the permissiveness of the host tissue for remyelination.

10.5.2.1 Migration of Transplanted Cells

In the lesioned CNS, spontaneous remyelination is a local event due to the limited migration of endogenous remyelinating cells.[155,157] Clearly even with an optimal cell delivery, the ability of transplanted cells to migrate to the active inflammatory, demyelinated lesions, integrate and differentiate remains a crucial requisite. While transplanted, multipotential NSCs migrate and integrate in the developing embryonic and newborn rodent CNS and adopt cellular identity according to local and temporal cues,[158,159] the normal adult brain does not permit large distance migration.[158–160] However, NPCs can migrate to the active sites of disease under the direction of inflammatory cues. In EAE animals, the inflammatory process in the CNS was found to be a powerful stimulus stimulating endogenous subventricular PSA-NCAM + cells[161,162] and attracting targeted migration of transplanted PSA-NCAM + NPCs.[96] Following ICV transplantation, NPCs migrated almost exclusively into inflamed periventricular white matter tracts, but not into gray matter. Similarly, the survival and migration of transplanted CG4 GPCs in the spinal cord were promoted by the inflammatory process.[163] Importantly, transplanted NPCs were found to possess superior migratory capabilities than endogenous NPCs.[155] Thus targeted migration of regenerating cells towards active lesions in the white matter is triggered by the inflammatory process.

Several mediators of inflammation have been implicated in stimulating NPC migration. As CNS regeneration seems to be a recapitulation of developmental processes, this prompted studies on the role of chemokines in attracting NPCs following injury. Both stromal-derived-factor 1 (SDF-1)/CXCR4 and monocyte chemoattractant protein-1 (MCP-1) and its receptor CCR2 were found as important regulators of migration of dentate granule cells,[164,165] sensory neurons[166] and cortical interneurons[167] during development, and were shown to modulate NPC migration following cerebral ischemia.[168,169] Furthermore, NPCs deficient in CCR2 failed to migrate towards focal inflammatory sites.[170] In addition, tumor necrosis factor α (TNF-α),[96] transforming growth factor β (TGF-β) and hepatocyte growth factor (HGF)[171] were implicated in NPC and OPC migration. The specific role of these cytokines and chemokines in attracting NPC migration in EAE and MS is still unknown.

The linkage between parenchymal inflammation and setting regenerative mechanisms in motion highlights the notion that the brain inflammatory process may have a dual, contrasting action in inflicting brain injury and recruiting regenerative process simultaneously. The combination of cell transplantation and immunomodulation for MS in the future will need to be developed as non-reciprocally antagonistic modes of treatments. To this end, it is important to further dissect the pro-regenerative components in the inflammatory process and target the immunomodulatory treatment without inhibiting regenerative processes.

10.5.2.2 Tracking Transplanted Cells

To develop successful clinical cell-based therapies, it is important to assess the fate and distribution of cells non-invasively. Among the various non-invasive imaging techniques that are currently available, magnetic resonance imaging (MRI) stands out in terms of resolution and whole body imaging capability. When cells are magnetically labeled *in vitro* prior to their administration to a living organism, they can be potentially traced *in vivo* by MRI to study how certain lesions target cell migration, at what speed cells migrate, and for how long they persist in the target organ. Superparamagnetic iron oxides (SPIOs) are composed of biocompatible iron that provides the targeted cell with a large magnetic moment[172] and serve as effective contrast agents. A wide variety of cells from different species can be labeled, without affecting cell viability and proliferation capacity.[173–175]

A series of recent studies have used MRI to monitor the temporal and spatial dynamics of distribution of cells after transplantation to the CNS.[172,176,177] MRI demonstrated widespread migration of transplanted magnetically labeled rodent NPCs cells along white matter tracts of EAE animals, with good correlation to histological staining. This study suggested that migration of transplanted cells was more extensive at the early, acute phase of disease. In addition, it provided the first indication that human ESC-derived NPCs respond to tissue signals in an MS model similar to rodent cells, a prerequisite for their consideration as clinical vectors.[176] More broadly, these data suggest that real-time MRI monitoring of delivered cell therapeutics may become an important tool in evaluating the efficacy of transplant-based remyelination in individual patients, as well as experimentally.

10.5.2.3 Survival of Transplanted Cells

As MS is a chronic and relapsing disease, it would be necessary to maintain long-term survival of transplanted cells through both phases of relapses and remissions. Moreover, as the time window for remyelination is considered to be narrow, it may be best to introduce remyelinating cells as early as possible in a form that will keep their survival independent of tissue support and ready for immediate mobilization upon tissue demand.

The normal adult CNS does not support the survival of transplanted cells.[160] This may be due to the especially low abundance of trophic factors in normal adult brain tissue which maintains the survival of resident cells, but is insufficient to support the survival of transplanted cells. Transplanted cells may integrate significantly better in acutely lesioned tissue. Accordingly, when OPCs were transplanted into the spinal cord of animals with EAE and an ongoing inflammatory process, they survived much better.[163] Similarly, NPCs cultured in the form of neurospheres survived for prolonged periods of time in the ventricular space of naïve animals, and responded to induction of EAE by migration into the inflamed white matter tracts and integration into the tissue.[87] Thus, culturing NPCs in the form of neurospheres may be the preferable method for clinical application.

10.5.2.4   Route of Cell Delivery

Most neurological diseases such as MS are not limited to pathology at one site or system. Therefore, another key issue for cell transplantation in MS is the appropriate route of cell delivery. It is impossible to introduce regenerating cells into all foci of disease. Moreover, it is often difficult to determine which of the multiple foci observed in the brain by MRI is most important clinically. In addition, current neuroimaging techniques do not identify the specific pathological pattern of the lesion and whether it is amenable for effective remyelination. Therefore, it is necessary to contemplate the optimal route of cell delivery that will promote efficient targeted migration of transplanted cells into multiple lesions for repair.

As most white matter tracts that are involved in MS are in close proximity to the ventricular and spinal subarachnoid spaces, transplanting the cells intraventricularly (ICV) and intrathecally is a promising approach for cell delivery. Indeed, ICV transplantation of stem cells led to widespread myelination in the genetic dysmyelinating models of the *shi* mouse[90] and the *md* rat.[106] Similarly, ICV and intrathecal transplantation resulted in widespread migration of NPCs in white matter tracts.[96] Another route of cell delivery that has been suggested was to inject NPCs into the blood stream (intravenously), from which the cells cross the blood–brain–barrier.[38,134,178] Selective homing of i.v. injected NPCs to the EAE inflamed brain was demonstrated to occur only during the acute phase of disease, and *via* membrane expression of adhesion molecules (integrins, selectins, *etc.*) such as CD44 and very late antigen-4[38] and of chemokine receptors by the transplanted cells.[38,179,180] The i.v. route of cell delivery has an apparent advantage of simplicity and of targeting the cells to the perivascular niche where their immunomodulatory properties may be most effective. However, the vast majority of cells do not enter the CNS and exhibit a profound systemic effect on the immune system.[30,39] Moreover, it is unknown whether intravenously injected cells migrate from the perivascular space further into the tissue for regenerative purposes.

10.6   Clinical Experience

In the recent years we have witnessed a spur of experimental cell-based transplantation approaches aimed at fostering biological and molecular mechanisms underlying CNS repair. Theories assuming that no (or very little) renewing potential is identified within the adult CNS have been contravened, new promising sources of stem cells for transplantation purposes have been characterized, and new cell transplantation strategies have been proposed and established. A better understanding of the dynamics of endogenous regeneration has been achieved and insights concerning the process of regeneration driven by site-specific cell transplantation have been discovered.

The first clinical trial performed in MS patients was based on autologous Schwann cell transplantation into brain areas of autoimmune demyelination. However, the first negative results of this approach have cooled down most of the expectations raised by the last 25 years of successful experimental cell-based approaches performed both in rodents and non-human primates. The negative study highlights some of the limitations of this approach:

- the limited amount of myelinating Schwann cells that could be grown *in vitro*; and
- their limited migratory capacity.

Stem cell based therapy might therefore represent a more promising strategy. Based on successful and promising results in experimental studies of animal models, there are currently two major cell-based clinical trials in the field of neurology. In January 2009, Geron Corporation received clearance from the US Food and Drug Administration (FDA) to begin the first human clinical trial of an ESC-based therapy using hESC-derived OPCs (GRNOPC1 line) for acute SCI. In August 2009, Geron's IND (Investigational New Drug application) for GRNOPC1 was placed on hold by the FDA pending its review of new animal study data submitted by the company. These new data indicated that microscopic cysts seen in an early experiment in a few animals that received GRNOPC1 were found in larger numbers of animals in a recent study.

Systemic administration of enriched autologous BMSCs has already been extensively investigated for several conditions and in particular for graft-versus-host disease (GvHD). A phase I study of intrathecal and i.v. injection of BMSC in MS and ALS patients suggested this procedure is safe.[181,182] The main side effect was of meningeal irritation, with transient headache and fever after the procedure and no long-term adverse effects. This early experience has prompted several centers to start a program of BMSC therapy for MS. These studies will hopefully pave the way for additional studies for neurological diseases, and using other cell populations as well.

References

1. Z. Cao, Y. Gao, K. Deng, G. Williams, P. Doherty and F. S. Walsh, Receptors for myelin inhibitors: structures and therapeutic opportunities, *Mol. Cell. Neurosci.*, 2010, **43**, 1–14.
2. J. Dietrich and G. Kempermann, Role of endogenous neural stem cells in neurological disease and brain repair, *Adv. Exp. Med. Biol.*, 2006, **557**, 191–220.

3. M. T. Fitch and J. Silver, CNS injury, glial scars, and inflammation: inhibitory extracellular matrices and regeneration failure, *Exp. Neurol.*, 2008, **209**, 294–301.

4. T. Kuhlmann, V. Miron, Q. Cui, C. Wegner, J. Antel and W. Bruck, Differentiation block of oligodendroglial progenitor cells as a cause for remyelination failure in chronic multiple sclerosis, *Brain*, 2008, **131**, 1749–1758.

5. G. Yiu and Z. He, Glial inhibition of CNS axon regeneration, *Nat. Rev. Neurosci.*, 2006, **7**, 617–627.

6. O. Lindvall and Z. Kokaia, Stem cells for the treatment of neurological disorders, *Nature*, 2006, **441**, 1094–1096.

7. O. Lindvall and A. Bjorklund, Cell therapy in Parkinson's disease, *NeuroRx*, 2004, **1**, 382–393.

8. C. W. Olanow, J. H. Kordower, A. E. Lang and J. A. Obeso, Dopaminergic transplantation for Parkinson's disease: current status and future prospects, *Ann. Neurol.*, 2009, **66**, 591–596.

9. L. Studer, V. Tabar and R. D. McKay, Transplantation of expanded mesencephalic precursors leads to recovery in Parkinsonian rats, *Nat. Neurosci.*, 1998, **1**, 290–295.

10. T. Ben-Hur, M. Idelson, H. Khaner, M. Pera, E. Reinhartz, A. Itzik and B. E. Reubinoff, Transplantation of human embryonic stem cell-derived neural progenitors improves behavioral deficit in Parkinsonian rats, *Stem Cells*, 2004, **22**, 1246–1255.

11. G. Martino and S. Pluchino, The therapeutic potential of neural stem cells, *Nat. Rev. Neurosci.*, 2006, **7**, 395–406.

12. J. Ourednik, V. Ourednik, W. P. Lynch, M. Schachner and E. Y. Snyder, Neural stem cells display an inherent mechanism for rescuing dysfunctional neurons, *Nat. Biotechnol.*, 2002, **20**, 1103–1110.

13. V. F. Rafuse, P. Soundararajan, C. Leopold and H. A. Robertson, Neuroprotective properties of cultured neural progenitor cells are associated with the production of sonic hedgehog, *Neuroscience*, 2005, **131**, 899–916.

14. R. M. Richardson, W. C. Broaddus, H. L. Holloway and H. L. Fillmore, Grafts of adult subependymal zone neuronal progenitor cells rescue hemiparkinsonian behavioral decline, *Brain Res.*, 2005, **1032**, 11–22.

15. J. L. McBride, S. P. Behrstock, R. J. Chen, R. J. Jakel, I. Siegel, C. N. Svendsen and J. H. Kordower, Human neural stem cell transplants improve motor function in a rat model of Huntington's disease, *J. Comp. Neurol.*, 2004, **475**, 211–219.

16. J. K. Ryu, J. Kim, S. J. Cho, K. Hatori, A. Nagai, H. B. Choi, M. C. Lee, J. G. McLarnon and S. U. Kim, Proactive transplantation of human neural stem cells prevents degeneration of striatal neurons in a rat model of Huntington disease, *Neurobiol. Dis.*, 2004, **16**, 68–77.

17. B. J. Cummings, N. Uchida, S. J. Tamaki, D. L. Salazar, M. Hooshmand, R. Summers, F. H. Gage and A. J. Anderson, Human neural stem cells differentiate and promote locomotor recovery in spinal cord-injured mice, *Proc. Natl. Acad. Sci. U. S. A.*, 2005, **102**, 14069–14074.

18. Y. Fujiwara, N. Tanaka, O. Ishida, Y. Fujimoto, T. Murakami, H. Kajihara, Y. Yasunaga and M. Ochi, Intravenously injected neural progenitor cells of transgenic rats can migrate to the injured spinal cord and differentiate into neurons, astrocytes and oligodendrocytes, *Neurosci. Lett.*, 2004, **366**, 287–291.

19. C. P. Hofstetter, N. A. Holmstrom, J. A. Lilja, P. Schweinhardt, J. Hao, C. Spenger, Z. Wiesenfeld-Hallin, S. N. Kurpad, J. Frisen and L. Olson, Allodynia limits the usefulness of intraspinal neural stem cell grafts; directed differentiation improves outcome, *Nat. Neurosci.*, 2005, **8**, 346–353.

20. P. Lu, L. L. Jones, E. Y. Snyder and M. H. Tuszynski, Neural stem cells constitutively secrete neurotrophic factors and promote extensive host axonal growth after spinal cord injury, *Exp. Neurol.*, 2003, **181**, 115–129.

21. K. Chu, M. Kim, K. I. Park, S. W. Jeong, H. K. Park, K. H. Jung, S. T. Lee, L. Kang, K. Lee, D. K. Park, S. U. Kim and J. K. Roh, Human neural stem cells improve sensorimotor deficits in the adult rat brain with experimental focal ischemia, *Brain Res.*, 2004, **1016**, 145–153.

22. S. W. Jeong, K. Chu, K. H. Jung, S. U. Kim, M. Kim and J. K. Roh, Human neural stem cell transplantation promotes functional recovery in rats with experimental intracerebral hemorrhage, *Stroke*, 2003, **34**, 2258–2263.

23. S. T. Lee, K. Chu, K. H. Jung, S. J. Kim, D. H. Kim, K. M. Kang, N. H. Hong, J. H. Kim, J. J. Ban, H. K. Park, S. U. Kim, C. G. Park, S. K. Lee, M. Kim and J. K. Roh, Anti-inflammatory mechanism of intravascular neural stem cell transplantation in haemorrhagic stroke, *Brain*, 2008, **131**, 616–629.

24. R. Gold, H. P. Hartung and K. V. Toyka, Animal models for autoimmune demyelinating disorders of the nervous system, *Mol. Med. Today*, 2000, **6**, 88–91.

25. H. Lassmann, Chronic relapsing experimental allergic encephalomyelitis, its value as an experimental model for multiple sclerosis, *J. Neurol.*, 1983, **229**, 207–220.

26. R. H. Swanborg, Experimental autoimmune encephalomyelitis in rodents as a model for human demyelinating disease, *Clin. Immunol. Immunopathol.*, 1995, **77**, 4–13.

27. L. Izikson, R. S. Klein, A. D. Luster and H. L. Weiner, Targeting monocyte recruitment in CNS autoimmune disease, *Clin. Immunol.*, 2002, **103**, 125–131.

28. V. K. Kuchroo, A. C Anderson, H. Waldner, M. Munder, E. Bettelli and L. B. Nicholson, T cell response in experimental autoimmune encephalomyelitis (EAE): role of self and cross-reactive antigens in shaping, tuning, and regulating the autopathogenic T cell repertoire, *Annu. Rev. Immunol.*, 2002, **20**, 101–123.

29. M. Aharonowiz, O. Einstein, N. Fainstein, H. Lassmann, B. Reubinoff and T. Ben-Hur, Neuroprotective effect of transplanted human embryonic stem

cell-derived neural precursors in an animal model of multiple sclerosis, *PLoS ONE*, 2008, **3**, e3145.

30. O. Einstein, N. Fainstein, I. Vaknin, R. Mizrachi-Kol, E. Reihartz, N. Grigoriadis, I. Lavon, M. Baniyash, H. Lassmann and T. Ben-Hur, Neural precursors attenuate autoimmune encephalomyelitis by peripheral immunosuppression, *Ann. Neurol.*, 2007, **61**, 209–218.

31. O. Einstein, N. Grigoriadis, R. Mizrachi-Kol, E. Reinhartz, E. Poly-zoidou, I. Lavon, I. Milonas, D. Karussis, O. Abramsky and T. Ben-Hur, Transplanted neural precursor cells reduce brain inflammation to attenuate chronic experimental autoimmune encephalomyelitis, *Exp. Neurol.*, 2006, **198**, 275–284.

32. O. Einstein, D. Karussis, N. Grigoriadis, R. Mizrachi-Kol, E. Reinhartz, O. Abramsky and T. Ben-Hur, Intraventricular transplantation of neural precursor cell spheres attenuates acute experimental allergic encephalo-myelitis, *Mol. Cell. Neurosci.*, 2003, **24**, 1074–1082.

33. S. Pluchino, A. Quattrini, E. Brambilla, A. Gritti, G. Salani, G. Dina, R. Galli, U. Del Carro, S. Amadio, A. Bergami, R. Furlan, G. Comi, A. L. Vescovi and G. Martino, Injection of adult neurospheres induces recovery in a chronic model of multiple sclerosis, *Nature*, 2003, **422**, 688–694.

34. B. Y. Hu, Z. W. Du, X. J. Li, M. Ayala and S. C. Zhang, Human oligodendrocytes from embryonic stem cells: conserved SHH signaling networks and divergent FGF effects, *Development*, 2009, **136**, 1443–1452.

35. H. S. Keirstead and W. F. Blakemore, The role of oligodendrocytes and oligodendrocyte progenitors in CNS remyelination, *Adv. Exp. Med. Biol.*, 1999, **468**, 183–197.

36. R. H. Swanborg, Experimental autoimmune encephalomyelitis in the rat: lessons in T-cell immunology and autoreactivity, *Immunol. Rev.*, 2001, **184**, 129–135.

37. J. R. Wujek, C. Bjartmar, E. Richer, R. M. Ransohoff, M. Yu, V. K. Tuohy and B. D. Trapp, Axon loss in the spinal cord determines per-manent neurological disability in an animal model of multiple sclerosis, *J. Neuropathol. Exp. Neurol.*, 2002, **61**, 23–32.

38. S. Pluchino, L. Zanotti, B. Rossi, E. Brambilla, L. Ottoboni, G. Salani, M. Martinello, A. Cattalini, A. Bergami and R. Furlan, *et al.*, Neuro-sphere-derived multipotent precursors promote neuroprotection by an immunomodulatory mechanism, *Nature*, 2005, **436**, 266–271.

39. N. Fainstein, I. Vaknin, O. Einstein, P. Zisman, S. Z. Ben Sasson, M. Baniyash and T. Ben-Hur, Neural precursor cells inhibit multiple inflammatory signals, *Mol. Cell Neurosci.*, 2008, **39**, 335–341.

40. S. Pluchino, L. Zanotti, E. Brambilla, P. Rovere-Querini, A. Capobianco, C. Alfaro-Cervello, G. Salani, C. Cossetti, G. Borsellino, L. Battistini, M. Ponzoni, C. Doglioni, J. M. Garcia-Verdugo, G. Comi, A. A. Manfredi and G. Martino, Immune regulatory neural stem/precursor cells protect from central nervous system autoimmunity by restraining dendritic cell function, *PLoS One*, 2009, **4**, e5959.

41. S. Pluchino, A. Gritti, E. Blezer, S. Amadio, E. Brambilla, G. Borsellino, C. Cossetti, U. Del Carro, G. T. Comi, B. Hart, A. Vescovi and G. Martino, Human neural stem cells ameliorate autoimmune encephalomyelitis in non-human primates, *Ann. Neurol.*, 2009, **66**, 343–354.

42. E. Gerdoni, B. Gallo, S. Casazza, S. Musio, I. Bonanni, E. Pedemonte, R. Mantegazza, F. Frassoni, G. Mancardi, R. Pedotti and A. Uccelli, Mesenchymal stem cells effectively modulate pathogenic immune response in experimental autoimmune encephalomyelitis, *Ann. Neurol.*, 2007, **61**, 219–227.

43. A. Uccelli, L. Moretta and V. Pistoia, Mesenchymal stem cells in health and disease, *Nat. Rev. Immunol.*, 2008, **8**, 726–736.

44. E. Zappia, S. Casazza, E. Pedemonte, F. Benvenuto, I. Bonanni, E. Gerdoni, D. Giunti, A. Ceravolo, F. Cazzanti, F. Frassoni, G. Mancardi and A. Uccelli, Mesenchymal stem cells ameliorate experimental autoimmune encephalomyelitis inducing T-cell anergy, *Blood*, 2005, **106**, 1755–1761.

45. Y. D. Teng, E. B. Lavik, X. Qu, K. I. Park, J. Ourednik, D. Zurakowski, R. Langer and E. Y. Snyder, Functional recovery following traumatic spinal cord injury mediated by a unique polymer scaffold seeded with neural stem cells, *Proc. Natl. Acad. Sci. U. S. A.*, 2002, **99**, 3024–3029.

46. M. Ferrer-Alcon, C. Winkler-Hirt, F. E. Perrin and A. C. Kato, Grafted neural stem cells increase the life span and protect motoneurons in pmn mice, *Neuroreport*, 2007, **18**, 1463–1468.

47. D. A. Kerr, J. Llado, M. J. Shamblott, N. J. Maragakis, D. N. Irani, T. O. Crawford, C. Krishnan, S. Dike, J. D. Gearhart and J. D. Rothstein, Human embryonic germ cell derivatives facilitate motor recovery of rats with diffuse motor neuron injury, *J. Neurosci.*, 2003, **23**, 5131–5140.

48. M. Suzuki, J. McHugh, C. Tork, B. Shelley, S. M. Klein, P. Aebischer and C. N. Svendsen, GDNF secreting human neural progenitor cells protect dying motor neurons, but not their projection to muscle, in a rat model of familial ALS, *PLoS ONE*, 2007, **2**, e689.

49. J. Li, J. Imitola, E. Y. Snyder and R. L. Sidman, Neural stem cells rescue nervous purkinje neurons by restoring molecular homeostasis of tissue plasminogen activator and downstream targets, *J. Neurosci.*, 2006, **26**, 7839–7848.

50. D. M. Gamm, S. Wang, B. Lu, S. Girman, T. Holmes, N. Bischoff, R. L. Shearer, Y. Sauve, E. Capowski, C. N. Svendsen and R. D. Lund, Protection of visual functions by human neural progenitors in a rat model of retinal disease, *PLoS ONE*, 2007, **2**, e338.

51. R. J. Franklin, Why does remyelination fail in multiple sclerosis?, *Nat. Rev. Neurosci.*, 2002, **3**, 705–714.

52. R. J. Franklin and C. Ffrench-Constant, Remyelination in the CNS: from biology to therapy, *Nat. Rev. Neurosci.*, 2008, **9**, 839–855.

53. G. Martino, How the brain repairs itself, new therapeutic strategies in inflammatory and degenerative CNS disorders, *Lancet Neurol.*, 2004, **3**, 372–378.

54. Y. Zhang, H. J. Klassen, B. A. Tucker, M. T. Perez and M. J. Young, CNS progenitor cells promote a permissive environment for neurite outgrowth *via* a matrix metalloproteinase-2-dependent mechanism, *J. Neurosci.*, 2007, **27**, 4499–4506.
55. K. I. Park, J. Ourednik, V. Ourednik, R. M. Taylor, K. S. Aboody, K. I. Auguste, M. B. Lachyankar, D. E. Redmond and E. Y. Snyder, Global gene and cell replacement strategies *via* stem cell, *Gene Ther.*, 2002, **9**, 613–624.
56. K. I. Park, Y. D. Teng and E. Y. Snyder, The injured brain interacts reciprocally with neural stem cells supported by scaffolds to reconstitute lost tissue, *Nat. Biotechnol.*, 2002, **20**, 1111–1117.
57. C. Capone, S. Frigerio, S. Fumagalli, M. Gelati, M. C. Principato, C. Storini, M. Montinaro, R. Kraftsik, M. De Curtis, E. Parati and M. G. De Simoni, Neurosphere-derived cells exert a neuroprotective action by changing the ischemic microenvironment, *PLoS ONE*, 2007, **2**, e373.
58. R. C. Armstrong, T. Q. Le, N. C. Flint, A. C. Vana and Y. X. Zhou, Endogenous cell repair of chronic demyelination, *J. Neuropathol. Exp. Neurol.*, 2006, **65**, 245–256.
59. J. L. Mason, A. Toews, J. D. Hostettler, P. Morell, K. Suzuki, J. E. Goldman and G. K. Matsushima, Oligodendrocytes and progenitors become progressively depleted within chronically demyelinated lesions, *Am. J. Pathol.*, 2004, **164**, 1673–1682.
60. S. Shen, J. Sandoval, V. A. Swiss, J. Li, J. Dupree, R. J. Franklin and P. Casaccia-Bonnefil, Age-dependent epigenetic control of differentiation inhibitors is critical for remyelination efficiency, *Nat. Neurosci.*, 2008.
61. S. A. Shields, J. M. Gilson, W. F. Blakemore and R. J. Franklin, Remyelination occurs as extensively but more slowly in old rats compared to young rats following gliotoxin-induced CNS demyelination, *Glia*, 1999, **28**, 77–83.
62. F. J. Sim, C. Zhao, J. Penderis and R. J. Franklin, The age-related decrease in CNS remyelination efficiency is attributable to an impairment of both oligodendrocyte progenitor recruitment and differentiation, *J. Neurosci.*, 2002, **22**, 2451–2459.
63. O. Einstein, Y. Friedman-Levi, N. Grigoriadis and T. Ben-Hur, Transplanted neural precursors enhance host brain-derived myelin regeneration, *J. Neurosci.*, 2009, **29**, 15694–15702.
64. B. Hattiangady, B. Shuai, J. Cai, T. Coksaygan, M. S. Rao and A. K. Shetty, Increased dentate neurogenesis after grafting of glial restricted progenitors or neural stem cells in the aging hippocampus, *Stem Cells*, 2007, **25**, 2104–2117.
65. J. R. Munoz, B. R. Stoutenger, A. P. Robinson, J. L. Spees and D. J. Prockop, Human stem/progenitor cells from bone marrow promote neurogenesis of endogenous neural stem cells in the hippocampus of mice, *Proc. Natl. Acad. Sci. U. S. A.*, 2005, **102**, 18171–18176.
66. T. L. Ben-Shaanan, T. Ben-Hur and J. Yanai, Transplantation of neural progenitors enhances production of endogenous cells in the impaired brain, *Mol. Psychiatry*, 2008, **13**, 222–231.

67. N. Kamei, N. Tanaka, Y. Oishi, T. Hamasaki, K. Nakanishi, N. Sakai and M. Ochi, BDNF, NT-3, and NGF released from transplanted neural progenitor cells promote corticospinal axon growth in organotypic cocultures, *Spine*, 2007, **32**, 1272–1278.

68. K. Akassoglou, J. Bauer, G. Kassiotis, M. Pasparakis, H. Lassmann, G. Kollias and L. Probert, Oligodendrocyte apoptosis and primary demyelination induced by local TNF/p55TNF receptor signaling in the central nervous system of transgenic mice, models for multiple sclerosis with primary oligodendrogliopathy, *Am. J. Pathol.*, 1998, **153**, 801–813.

69. B. Cannella, C. J. Hoban, Y. L. Gao, R. Garcia-Arenas, D. Lawson, M. Marchionni, D. Gwynne and C. S. Raine, The neuregulin, glial growth factor 2, diminishes autoimmune demyelination and enhances remyelination in a chronic relapsing model for multiple sclerosis, *Proc. Natl. Acad. Sci. U. S. A.*, 1998, **95**, 10100–10105.

70. D. L. Yao, X. Liu, L. D. Hudson and H. D. Webster, Insulin-like growth factor-I given subcutaneously reduces clinical deficits, decreases lesion severity and upregulates synthesis of myelin proteins in experimental autoimmune encephalomyelitis, *Life Sci.*, 1996, **58**, 1301–1306.

71. H. Butzkueven, J. G. Zhang, M. Soilu-Hanninen, H. Hochrein, F. Chionh, K. A. Shipham, B. Emery, A. M. Turnley, S. Petratos, M. Ernst, P. F. Bartlett and T. J. Kilpatrick, LIF receptor signaling limits immune-mediated demyelination by enhancing oligodendrocyte survival, *Nat. Med.*, 2002, **8**, 613–619.

72. R. A. Linker, M. Maurer, S. Gaupp, R. Martini, B. Holtmann, R. Giess, P. Rieckmann, H. Lassmann, K. V. Toyka, M. Sendtner and R. Gold, CNTF is a major protective factor in demyelinating CNS disease: a neurotrophic cytokine as modulator in neuroinflammation, *Nat. Med.*, 2002, **8**, 620–624.

73. A. Flugel, T. Berkowicz, T. Ritter, M. Labeur, D. E. Jenne, Z. Li, J. W. Ellwart, M. Willem, H. Lassmann and H. Wekerle, Migratory activity and functional changes of green fluorescent effector cells before and during experimental autoimmune encephalomyelitis, *Immunity*, 2001, **14**, 547–560.

74. F. Ruffini, R. Furlan, P. L. Poliani, E. Brambilla, P. C. Marconi, A. Bergami, G. Desina, J. C. Glorioso, G. Comi and G. Martino, Fibroblast growth factor-II gene therapy reverts the clinical course and the pathological signs of chronic experimental autoimmune encephalomyelitis in C57BL/6 mice, *Gene Ther.*, 2001, **8**, 1207–1213.

75. P. Villoslada, S. L. Hauser, I. Bartke, J. Unger, N. Heald, D. Rosenberg, S. W. Cheung, W. C. Mobley, S. Fisher and C. P. Genain, Human nerve growth factor protects common marmosets against autoimmune encephalomyelitis by switching the balance of T helper cell type 1 and 2 cytokines within the central nervous system, *J. Exp. Med.*, 2000, **191**, 1799–1806.

76. A. Gritti, E. A. Parati, L. Cova, P. Frolichsthal, R. Galli, E. Wanke, L. Faravelli, D. J. Morassutti, F. Roisen, D. D. Nickel and A. L. Vescovi,

Multipotential stem cells from the adult mouse brain proliferate and self-renew in response to basic fibroblast growth factor, *J. Neurosci.*, 1996, **16**, 1091–1100.

77. H. M. Keyoung, N. S. Roy, A. Benraiss, A. Louissaint Jr, A. Suzuki, M. Hashimoto, W. K. Rashbaum, H. Okano and S. A. Goldman, High-yield selection and extraction of two promoter-defined phenotypes of neural stem cells from the fetal human brain, *Nat. Biotechnol.*, 2001, **19**, 843–850.

78. M. C. Nunes, N. S. Roy, H. M. Keyoung, R. R. Goodman, G. McKhann 2nd, L. Jiang, J. Kang, M. Nedergaard and S. A. Goldman, Identification and isolation of multipotential neural progenitor cells from the sub-cortical white matter of the adult human brain, *Nat. Med.*, 2003, **9**, 439–447.

79. A. L. Vescovi, E. A. Parati, A. Gritti, P. Poulin, M. Ferrario, E. Wanke, P. Frolichsthal-Schoeller, L. Cova, M. Arcellana-Panlilio, A. Colombo and R. Galli, Isolation and cloning of multipotential stem cells from the embryonic human CNS and establishment of transplantable human neural stem cell lines by epigenetic stimulation, *Exp. Neurol.*, 1999, **156**, 71–83.

80. X. J. Li, B. Y. Hu, S. A. Jones, Y. S. Zhang, T. Lavaute, Z. W. Du and S. C. Zhang, Directed differentiation of ventral spinal progenitors and motor neurons from human embryonic stem cells by small molecules, *Stem Cells*, 2008, **26**, 886–893.

81. Y. Yan, D. Yang, E. D. Zarnowska, Z. Du, B. Werbel, C. Valliere, R. A. Pearce, J. A. Thomson and S. C. Zhang, Directed differentiation of dopaminergic neuronal subtypes from human embryonic stem cells, *Stem Cells*, 2005, **23**, 781–790.

82. A. Hicks, T. Schallert and J. Jolkkonen, Cell-based therapies and functional outcome in experimental stroke, *Cell Stem Cell*, 2009, **5**, 139–140.

83. M. Modo, R. P. Stroemer, E. Tang, S. Patel and H. Hodges, Effects of implantation site of stem cell grafts on behavioral recovery from stroke damage, *Stroke*, 2002, **33**, 2270–2278.

84. T. Veizovic, J. S. Beech, R. P. Stroemer, W. P. Watson and H. Hodges, Resolution of stroke deficits following contralateral grafts of con-ditionally immortal neuroepithelial stem cells, *Stroke*, 2001, **32**, 1012–1019.

85. F. Barnabe-Heider and J. Frisen, Stem cells for spinal cord repair, *Cell Stem Cell*, 2008, **3**, 16–24.

86. J. W. McDonald, X. Z. Liu, Y. Qu, S. Liu, S. K. Mickey, D. Turetsky, D. L. Gottlieb and D. W. Choi, Transplanted embryonic stem cells sur-vive, differentiate and promote recovery in injured rat spinal cord, *Nat. Med.*, 1999, **5**, 1410–1412.

87. O. Einstein, O. Ben-Menachem-Tzidon, R. Mizrachi-Kol, E. Reinhartz, N. Grigoriadis and T. Ben-Hur, Survival of neural precursor cells in growth factor-poor environment, Implications for transplantation in chronic disease, *Glia*, 2006, **53**, 449–455.

88. L. Decker, V. Avellana-Adalid, B. Nait-Oumesmar, P. Durbec and A. Baron-Van Evercooren, Oligodendrocyte precursor migration and differentiation: combined effects of PSA residues, growth factors, and substrates, *Mol. Cell. Neurosci.*, 2000, **16**, 422–439.

89. S. Vitry, V. Avellana-Adalid, F. Lachapelle and A. B. Evercooren, Migration and multipotentiality of PSA-NCAM + neural precursors transplanted in the developing brain, *Mol. Cell. Neurosci.*, 2001, **17**, 983–1000.

90. B. D. Yandava, L. L. Billinghurst and E. Y. Snyder, 'Global' cell replacement is feasible *via* neural stem cell transplantation, evidence from the dysmyelinated shiverer mouse brain, *Proc. Natl. Acad. Sci. U. S. A.*, 1999, **96**, 7029–7034.

91. P. N. Anderson, G. Campbell, Y. Zhang and A. R. Lieberman, Cellular and molecular correlates of the regeneration of adult mammalian CNS axons into peripheral nerve grafts, *Prog. Brain Res.*, 1998, **117**, 211–232.

92. J. P. Hammang, D. R. Archer and I. D. Duncan, Myelination following transplantation of EGF-responsive neural stem cells into a myelin-deficient environment, *Exp. Neurol.*, 1997, **147**, 84–95.

93. E. A. Milward, C. G. Lundberg, B. Ge, D. Lipsitz, M. Zhao and I. D. Duncan, Isolation and transplantation of multipotential populations of epidermal growth factor-responsive: neural progenitor cells from the canine brain, *J. Neurosci. Res.*, 1997, **50**, 862–871.

94. T. Ben-Hur, B. Rogister, K. Murray, G. Rougon and M. Dubois-Dalcq, Growth and fate of PSA-NCAM + precursors of the postnatal brain, *J. Neurosci.*, 1998, **18**, 5777–5788.

95. H. S. Keirstead, T. Ben-Hur, B. Rogister, M. T. O'Leary, M. Dubois-Dalcq and W. F. Blakemore, Polysialylated neural cell adhesion molecule-positive CNS precursors generate both oligodendrocytes and Schwann cells to remyelinate the CNS after transplantation, *J. Neurosci.*, 1999, **19**, 7529–7536.

96. T. Ben-Hur, O. Einstein, R. Mizrachi-Kol, O. Ben-Menachem, E. Reinhartz, D. Karussis and O. Abramsky, Transplanted multipotential neural precursor cells migrate into the inflamed white matter in response to experimental autoimmune encephalomyelitis, *Glia*, 2003, **41**, 73–80.

97. Y. Akiyama, O. Honmou, T. Kato, T. Uede, K. Hashi and J. D. Kocsis, Transplantation of clonal neural precursor cells derived from adult human brain establishes functional peripheral myelin in the rat spinal cord, *Exp. Neurol.*, 2001, **167**, 27–39.

98. M. Evans and S. Hunter, Source and nature of embryonic stem cells, *C. R. Biol.*, 2002, **325**, 1003–1007.

99. E. Sasaki, K. Hanazawa, R. Kurita, A. Akatsuka, T. Yoshizaki, H. Ishii, Y. Tanioka, Y. Ohnishi, H. Suemizu, A. Sugawara, N. Tamaoki, K. Izawa, Y. Nakazaki, H. Hamada, H. Suemori, S. Asano, N. Nakatsuji, H. Okano and K. Tani, Establishment of novel embryonic stem cell lines derived from the common marmoset (*Callithrix jacchus*), *Stem Cells*, 2005, **23**, 1304–1313.

100. A. G. Smith, Embryo-derived stem cells, of mice and men, *Annu. Rev. Cell Dev. Biol.*, 2001, **17**, 435–462.
101. H. Wichterle, I. Lieberam, J. A. Porter and T. M. Jessell, Directed differentiation of embryonic stem cells into motor neurons, *Cell*, 2002, **110**, 385–397.
102. M. S. Cho, D. Y. Hwang and D. W. Kim, Efficient derivation of functional dopaminergic neurons from human embryonic stem cells on a large scale, *Nat. Protoc.*, 2008, **3**, 1888–1894.
103. A. Swistowski, J. Peng, Y. Han, A. M. Swistowska, M. S. Rao and X. Zeng, Xeno-free defined conditions for culture of human embryonic stem cells, neural stem cells and dopaminergic neurons derived from them, *PLoS One*, 2009, **4**, e6233.
104. A. Y. Jo, M. Y. Kim, H. S. Lee, Y. H. Rhee, J. E. Lee, K. H. Baek, C. H. Park, H. C. Koh, I. Shin, Y. S. Lee and S. H. Lee, Generation of dopamine neurons with improved cell survival and phenotype maintenance using a degradation-resistant nurr1 mutant, *Stem Cells*, 2009, **27**, 2238–2246.
105. N. Billon, C. Jolicoeur, Q. l. Ying, A. Smith and M. Raff, Normal timing of oligodendrocyte development from genetically engineered, lineage-selectable mouse ES cells, *J. Cell Sci.*, 2002, **115**, 3657–3665.
106. O. Brustle, K. N. Jones, R. D. Learish, K. Karram, K. Choudhary, O. D. Wiestler, I. D. Duncan and R. D. McKay, Embryonic stem cell-derived glial precursors, a source of myelinating transplants, *Science*, 1999, **285**, 754–756.
107. S. Liu, Y. Qu, T. J. Stewart, M. J. Howard, S. Chakrabortty, T. F. Holekamp and J. W. McDonald, Embryonic stem cells differentiate into oligodendrocytes and myelinate in culture and after spinal cord transplantation, *Proc. Natl. Acad. Sci. U. S. A.*, 2000, **97**, 6126–6131.
108. M. Izrael, P. Zhang, R. Kaufman, V. Shinder, R. Ella, M. Amit, J. Itskovitz-Eldor, J. Chebath and M. Revel, Human oligodendrocytes derived from embryonic stem cells, Effect of noggin on phenotypic differentiation *in vitro* and on myelination *in vivo*, *Mol. Cell. Neurosci.*, 2007, **34**, 310–323.
109. G. I. Nistor, M. O. Totoiu, N. Haque, M. K. Carpenter and H. S. Keirstead, Human embryonic stem cells differentiate into oligodendrocytes in high purity and myelinate after spinal cord transplantation, *Glia*, 2005, **49**, 385–396.
110. H. Hentze, R. Graichen and A. Colman, Cell therapy and the safety of embryonic stem cell-derived grafts, *Trends Biotechnol.*, 2007, **25**, 24–32.
111. E. L. Jackson, J. M. Garcia-Verdugo, S. Gil-Perotin, M. Roy, A. Quinones-Hinojosa, S. VandenBerg and A. Alvarez-Buylla, PDGFR alpha-positive B cells are neural stem cells in the adult SVZ that form glioma-like growths in response to increased PDGF signaling, *Neuron*, 2006, **51**, 187–199.
112. K. Takahashi, K. Okita, M. Nakagawa and S. Yamanaka, Induction of pluripotent stem cells from fibroblast cultures, *Nat. Protoc.*, 2007, **2**, 3081–3089.

113. K. Takahashi and S. Yamanaka, Induction of pluripotent stem cells from mouse embryonic and adult fibroblast cultures by defined factors, *Cell*, 2006, **126**, 663–676.

114. J. Yu, M. A. Vodyanik, K. Smuga-Otto, J. Antosiewicz-Bourget, J. L. Frane, S. Tian, J. Nie, G. A. Jonsdottir, V. Ruotti, R. Stewart, I. I. Slukvin and J. A. Thomson, Induced pluripotent stem cell lines derived from human somatic cells, *Science*, 2007, **318**, 1917–1920.

115. S. Yamanaka, Induction of pluripotent stem cells from mouse fibroblasts by four transcription factors, *Cell Prolif.*, 2008, **41**(Suppl. 1), 51–56.

116. J. T. Dimos, K. T. Rodolfa, K. K. Niakan, L. M. Weisenthal, H. Mitsumoto, W. Chung, G. F. Croft, G. Saphier, R. Leibel, R. Goland, H. Wichterle, C. E. Henderson and K. Eggan, Induced pluripotent stem cells generated from patients with ALS can be differentiated into motor neurons, *Science*, 2008, **321**, 1218–1221.

117. S. Karumbayaram, B. G. Novitch, M. Patterson, J. A. Umbach, L. Richter, A. Lindgren, A. E. Conway, A. T. Clark, S. A. Goldman, K. Plath, M. Wiedau-Pazos, H. I. Kornblum and W. E. Lowry, Directed differentiation of human-induced pluripotent stem cells generates active motor neurons, *Stem Cells*, 2009, **27**, 806–811.

118. M. Wernig, J. P. Zhao, J. Pruszak, E. Hedlund, D. Fu, F. Soldner, V. Broccoli, M. Constantine-Paton, O. Isacson and R. Jaenisch, Neurons derived from reprogrammed fibroblasts functionally integrate into the fetal brain and improve symptoms of rats with Parkinson's disease, *Proc. Natl. Acad. Sci. U. S. A.*, 2008, **105**, 5856–5861.

119. C. R. Bjornson, R. L. Rietze, B. A. Reynolds, M. C. Magli and A. L. Vescovi, Turning brain into blood: a hematopoietic fate adopted by adult neural stem cells *in vivo*, *Science*, 1999, **283**, 534–537.

120. A. J. Wagers and I. L. Weissman, Plasticity of adult stem cells, 2004, *Cell*, **116**, 639–648.

121. E. Mezey, K. J. Chandross, G. Harta, R. A. Maki and S. R. McKercher, Turning blood into brain: cells bearing neuronal antigens generated *in vivo* from bone marrow, *Science*, 2000, **290**, 1779–1782.

122. E. Mezey, S. Key, G. Vogelsang, I. Szalayova, G. D. Lange and B. Crain, Transplanted bone marrow generates new neurons in human brains, *Proc. Natl. Acad. Sci. U. S. A.*, 2003, **100**, 1364–1369.

123. S. Wislet-Gendebien, G. Hans, P. Leprince, J. M. Rigo, G. Moonen and B. Rogister, Plasticity of cultured mesenchymal stem cells, switch from nestin-positive to excitable neuron-like phenotype, *Stem Cells*, 2005, **23**, 392–402.

124. M. Alvarez-Dolado, R. Pardal, J. M. Garcia-Verdugo, J. R. Fike, H. O. Lee, K. Pfeffer, C. Lois, S. J. Morrison and A. Alvarez-Buylla, Fusion of bone-marrow-derived cells with Purkinje neurons, cardiomyocytes and hepatocytes, *Nature*, 2003, **425**, 968–973.

125. J. M. Weimann, C. B. Johansson, A. Trejo and H. M. Blau, Stable reprogrammed heterokaryons form spontaneously in Purkinje neurons after bone marrow transplant, *Nat. Cell Biol.*, 2003, **5**, 959–966.

126. P. Dharmasaroja, Bone marrow-derived mesenchymal stem cells for the treatment of ischemic stroke, *J. Clin. Neurosci.*, 2009, **16**, 12–20.
127. Y. Li and M. Chopp, Marrow stromal cell transplantation in stroke and traumatic brain injury, *Neurosci. Lett.*, 2009, **456**, 120–123.
128. L. R. Zhao, W. M. Duan, M. Reyes, C. D. Keene, C. M. Verfaillie and W. C. Low, Human bone marrow stem cells exhibit neural phenotypes and ameliorate neurological deficits after grafting into the ischemic brain of rats, *Exp. Neurol.*, 2002, **174**, 11–20.
129. M. Chopp, X. H. Zhang, Y. Li, L. Wang, J. Chen, D. Lu, M. Lu and M. Rosenblum, Spinal cord injury in rat: treatment with bone marrow stromal cell transplantation, *Neuroreport*, 2000, **11**, 3001–3005.
130. R. D. Tewarie, A. Hurtado, G. J. Ritfeld, S. T. Rahiem, D. F. Wendell, M. M. Barroso, J. A. Grotenhuis and M. Oudega, Bone marrow stromal cells elicit tissue sparing after acute but not delayed transplantation into the contused adult rat thoracic spinal cord, *J. Neurotrauma*, 2009, **26**, 2313–2322.
131. C. Bonilla, M. Zurita, L. Otero, C. Aguayo and J. Vaquero, Delayed intralesional transplantation of bone marrow stromal cells increases endogenous neurogenesis and promotes functional recovery after severe traumatic brain injury, *Brain Inj.*, 2009, **23**, 760–769.
132. R. D. Nandoe Tewarie, A. Hurtado, A. D. Levi, J. A. Grotenhuis and M. Oudega, Bone marrow stromal cells for repair of the spinal cord: towards clinical application, *Cell Transplant.*, 2006, **15**, 563–577.
133. Y. Akiyama, C. Radtke and J. D. Kocsis, Remyelination of the rat spinal cord by transplantation of identified bone marrow stromal cells, *J. Neurosci.*, 2002, **22**, 6623–6630.
134. M. Inoue, O. Honmou, S. Oka, K. Houkin, K. Hashi and J. D. Kocsis, Comparative analysis of remyelinating potential of focal and intravenous administration of autologous bone marrow cells into the rat demyelinated spinal cord, *Glia*, 2003, **44**, 111–118.
135. I. Kassis, N. Grigoriadis, B. Gowda-Kurkalli, R. Mizrachi-Kol, T. Ben-Hur, S. Slavin, O. Abramsky and D. Karussis, Neuroprotection and immunomodulation with mesenchymal stem cells in chronic experimental autoimmune encephalomyelitis, *Arch Neurol.*, 2008, **65**, 753–761.
136. A. M. Clement, M. D. Nguyen, E. A. Roberts, M. L. Garcia, S. Boillee, M. Rule, A. P. McMahon, W. Doucette, D. Siwek, R. J. Ferrante, R. H. Brown Jr, J. P. Julien, L. S. Goldstein and D. W. Cleveland, Wild-type nonneuronal cells extend survival of SOD1 mutant motor neurons in ALS mice, *Science*, 2003, **302**, 113–117.
137. D. H. Hwang, H. J. Lee, I. H. Park, J. I. Seok, B. G. Kim, I. S. Joo and S. U. Kim, Intrathecal transplantation of human neural stem cells overexpressing VEGF provide behavioral improvement, disease onset delay and survival extension in transgenic ALS mice, *Gene Ther.*, 2009, **16**, 1234–1244.
138. K. Yamanaka, S. J. Chun, S. Boillee, N. Fujimori-Tonou, H. Yamashita, D. H. Gutmann, R. Takahashi, H. Misawa and D. W. Cleveland,

Astrocytes as determinants of disease progression in inherited amyotrophic lateral sclerosis, *Nat. Neurosci.*, 2008, **11**, 251–253.

139. S. Kelly, T. M. Bliss, A. K. Shah, G. H. Sun, M. Ma, W. C. Foo, J. Masel, M. A. Yenari, I. L. Weissman, N. Uchida, T. Palmer and G. K. Steinberg, Transplanted human fetal neural stem cells survive, migrate, and differentiate in ischemic rat cerebral cortex, *Proc. Natl. Acad. Sci. U. S. A.*, 2004, **101**, 11839–11844.

140. A. Compston and A. Coles, Multiple sclerosis, *Lancet*, 2002, **359**, 1221–1231.

141. D. A. Dyment and G. C. Ebers, An array of sunshine in multiple sclerosis, *N. Engl. J. Med.*, 2002, **347**, 1445–1447.

142. B. D. Trapp, J. Peterson, R. M. Ransohoff, R. Rudick, S. Mork and L. Bo, Axonal transection in the lesions of multiple sclerosis, *N. Engl. J. Med.*, 1998, **338**, 278–285.

143. D. M. Wingerchuk, C. F. Lucchinetti and J. H. Noseworthy, Multiple sclerosis: current pathophysiological concepts, *Lab. Invest.*, 2001, **81**, 263–281.

144. C. Lucchinetti, W. Bruck, J. Parisi, B. Scheithauer, M. Rodriguez and H. Lassmann, Heterogeneity of multiple sclerosis lesions: implications for the pathogenesis of demyelination, *Ann. Neurol.*, 2000, **47**, 707–717.

145. H. Lassmann, Mechanisms of demyelination and tissue destruction in multiple sclerosis, *Clin. Neurol. Neurosurg.*, 2002, **104**, 168–171.

146. C. Bjartmar, G. Kidd, S. Mork, R. Rudick and B. D. Trapp, Neurological disability correlates with spinal cord axonal loss and reduced N-acetyl aspartate in chronic multiple sclerosis patients, *Ann. Neurol.*, 2000, **48**, 893–901.

147. N. De Stefano, P. M. Matthews, L. Fu, S. Narayanan, J. Stanley, G. S. Francis, J. P. Antel and D. L. Arnold, Axonal damage correlates with disability in patients with relapsing- remitting multiple sclerosis. Results of a longitudinal magnetic resonance spectroscopy study, *Brain*, 1998, **121**, 1469–1477.

148. A. Chang, W. W. Tourtellotte, R. Rudick and B. D. Trapp, Pre-myelinating oligodendrocytes in chronic lesions of multiple sclerosis, *N. Engl. J. Med.*, 2002, **346**, 165–173.

149. A. Compston, Remyelination of the central nervous system, *Mult. Scler.*, 1996, **1**, 388–392.

150. J. W. Prineas, R. O. Barnard, E. E. Kwon, L. R. Sharer and E. S. Cho, Multiple sclerosis: remyelination of nascent lesions, *Ann. Neurol.*, 1993, **33**, 137–151.

151. C. S. Raine and E. Wu, Multiple sclerosis: remyelination in acute lesions, *J. Neuropathol. Exp. Neurol.*, 1993, **52**, 199–204.

152. H. S. Keirstead, J. M. Levine and W. F. Blakemore, Response of the oligodendrocyte progenitor cell population (defined by NG2 labelling) to demyelination of the adult spinal cord, *Glia*, 1998, **22**, 161–170.

153. N. Scolding, R. Franklin, S. Stevens, C. H. Heldin, A. Compston and J. Newcombe, Oligodendrocyte progenitors are present in the normal

adult human CNS and in the lesions of multiple sclerosis, *Brain*, 1998, **121**, 2221–2228.

154. G. Wolswijk, Chronic stage multiple sclerosis lesions contain a relatively quiescent population of oligodendrocyte precursor cells, *J. Neurosci.*, 1998, **18**, 601–609.

155. W. F. Blakemore, J. M. Gilson and A. J. Crang, Transplanted glial cells migrate over a greater distance and remyelinate demyelinated lesions more rapidly than endogenous remyelinating cells, *J. Neurosci. Res.*, 2000, **61**, 288–294.

156. R. A. Rudick, G. R. Cutter, M. Baier, B. Weinstock-Guttman, M. K. Mass, E. Fisher, D. M. Miller and A. W. Sandrock, Estimating long-term effects of disease-modifying drug therapy in multiple sclerosis patients, *Mult. Scler.*, 2005, **11**, 626–634.

157. J. M. Gensert and J. E. Goldman, Endogenous progenitors remyelinate demyelinated axons in the adult CNS, *Neuron*, 1997, **19**, 197–203.

158. O. Brustle, U. Maskos and R. D. McKay, Host-guided migration allows targeted introduction of neurons into the embryonic brain, *Neuron*, 1995, **15**, 1275–1285.

159. J. D. Flax, S. Aurora, C. Yang, C. Simonin, A. M. Wills, L. L. Billinghurst, M. Jendoubi, R. L. Sidman, J. H. Wolfe, S. U. Kim and E. Y. Snyder, Engraftable human neural stem cells respond to developmental cues, replace neurons, and express foreign genes, *Nat. Biotechnol.*, 1998, **16**, 1033–1039.

160. M. T. O'Leary and W. F. Blakemore, Oligodendrocyte precursors survive poorly and do not migrate following transplantation into the normal adult central nervous system, *J. Neurosci. Res.*, 1997, **48**, 159–167.

161. B. Nait-Oumesmar, L. Decker, F. Lachapelle, V. Avellana-Adalid, C. Bachelin and A. B. van Evercooren, Progenitor cells of the adult mouse subventricular zone proliferate, migrate and differentiate into oligodendrocytes after demyelination, *Eur J. Neurosci.*, 1999, **11**, 4357–4366.

162. N. D. Picard-Riera, L. Decker, C. Delarasse, K. Goude, B. Nait-Oumesmar, R. Liblau, D. Pham-Dinh and A. B. van Evercooren, Experimental autoimmune encephalomyelitis mobilizes neural progenitors from the subventricular zone to undergo oligodendrogenesis in adult mice, *Proc. Natl. Acad. Sci. U. S. A.*, 2002, **99**, 13211–13216.

163. A. Tourbah, C. Linnington, C. Bachelin, V. Avellana-Adalid, H. Wekerle and A. B. van Evercooren, Inflammation promotes survival and migration of the CG4 oligodendrocyte progenitors transplanted in the spinal cord of both inflammatory and demyelinated EAE rats, *J. Neurosci. Res.*, 1997, **50**, 853–861.

164. A. Bagri, T. Gurney, Z. He, Y. R. Zou, D. R. Littman, M. Tessier-Lavigne and S. J. Pleasure, The chemokine SDF1 regulates migration of dentate granule cells, *Development*, 2002, **129**, 4249–4260.

165. M. Lu, E. A. Grove and R. J. Miller, Abnormal development of the hippocampal dentate gyrus in mice lacking the CXCR4 chemokine receptor, *Proc. Natl. Acad. Sci. U. S. A.*, 2002, **99**, 7090–7095.

166. A. Belmadani, P. B. Tran, D. Ren, S. Assimacopoulos, E. A. Grove and R. J. Miller, The chemokine stromal cell-derived factor-1 regulates the migration of sensory neuron progenitors, *J. Neurosci.*, 2005, **25**, 3995–4003.

167. R. K. Stumm, C. Zhou, T. Ara, F. Lazarini, M. Dubois-Dalcq, T. Nagasawa, V. Hollt and S. Schulz, CXCR4 regulates interneuron migration in the developing neocortex, *J. Neurosci.*, 2003, **23**, 5123–5130.

168. J. Imitola, K. Raddassi, K. I. Park, F. J. Mueller, M. Nieto, Y. D. Teng, D. Frenkel, J. Li, R. L. Sidman, C. A. Walsh, E. Y. Snyder and S. J. Khoury, Directed migration of neural stem cells to sites of CNS injury by the stromal cell-derived factor 1alpha/CXC chemokine receptor 4 pathway, *Proc. Natl. Acad. Sci. U. S. A.*, 2004, **101**, 18117–18122.

169. Y. P. Yan, K. A. Sailor, B. T. Lang, S. W. Park, R. Vemuganti and R. J. Dempsey, Monocyte chemoattractant protein-1 plays a critical role in neuroblast migration after focal cerebral ischemia, *J. Cereb. Blood Flow Metab.*, 2007, **27**, 1213–1224.

170. A. Belmadani, P. B. Tran, D. Ren and R. J. Miller, Chemokines regulate the migration of neural progenitors to sites of neuroinflammation, *J. Neurosci.*, 2006, **26**, 3182–3191.

171. P. H. Lalive, R. Paglinawan, G. Biollaz, E. A. Kappos, D. P. Leone, U. Malipiero, J. B. Relvas, M. Moransard, T. Suter and A. Fontana, TGF-beta-treated microglia induce oligodendrocyte precursor cell chemotaxis through the HGF-c-Met pathway, *Eur. J. Immunol.*, 2005, **35**, 727–737.

172. J. W. Bulte, T. Ben-Hur, B. R. Miller, R. Mizrachi-Kol, O. Einstein, E. Reinhartz, H. A. Zywicke, T. Douglas and J. A. Frank, MR microscopy of magnetically labeled neurospheres transplanted into the Lewis EAE rat brain, *Magn. Reson. Med.*, 2003, **50**, 201–205.

173. A. S. Arbab, G. T. Yocum, A. M. Rad, A. Y. Khakoo, V. Fellowes, E. J. Read and J. A. Frank, Labeling of cells with ferumoxides-protamine sulfate complexes does not inhibit function or differentiation capacity of hematopoietic or mesenchymal stem cells, *NMR Biomed.*, 2005, **18**, 553–559.

174. M. E. Cohen, N. Muja, N. Fainstein, J. W. Bulte and T. Ben-Hur, Conserved fate and function of ferumoxides-labeled neural precursor cells *in vitro* and *in vivo*, *J. Neurosci. Res.*, 2009.

175. M. Neri, C. Maderna, C. Cavazzin, V. Deidda-Vigoriti, L. S. Politi, G. Scotti, P. Marzola, A. Sbarbati, A. L. Vescovi and A. Gritti, Efficient *in vitro* labeling of human neural precursor cells with superparamagnetic iron oxide particles, relevance for *in vivo* cell tracking, *Stem Cells*, 2008, **26**, 505–516.

176. T. Ben-Hur, R. B. van Heeswijk, O. Einstein, M. Aharonowiz, R. Xue, E. E. Frost, S. Mori, B. E. Reubinoff and J. W. Bulte, Serial *in vivo* MR tracking of magnetically labeled neural spheres transplanted in chronic EAE mice, *Magn. Reson. Med.*, 2007, **57**, 164–171.

177. J. W. Bulte, I. D. Duncan and J. A. Frank, *In vivo* magnetic resonance tracking of magnetically labeled cells after transplantation, *J. Cereb. Blood Flow Metab.*, 2002, **22**, 899–907.

178. A. Mahmood, D. Lu and M. Chopp, Intravenous administration of marrow stromal cells (MSCs) increases the expression of growth factors in rat brain after traumatic brain injury, *J. Neurotrauma*, 2004, **21**, 33–39.
179. L. Coulombel, I. Auffray, M. H. Gaugler and M. Rosemblatt, Expression and function of integrins on hematopoietic progenitor cells, *Acta Haematol.*, 1997, **97**, 13–21.
180. P. B. Tran, D. Ren, T. J. Veldhouse and R. J. Miller, Chemokine receptors are expressed widely by embryonic and adult neural progenitor cells, *J. Neurosci. Res.*, 2004, **76**, 20–34.
181. M. S. Freedman, A. Bar-Or, H. L. Atkins, D. Karussis, F. Frassoni, H. Lazarus, N. Scolding, S. Slavin, K. Le Blanc and A. Uccelli, The therapeutic potential of mesenchymal stem cell transplantation as a treatment for multiple sclerosis: consensus report of the International MSCT Study Group, *Mult. Scler.*, 2010, **16**, 503–510.
182. D. Karussis, C. Karageorgiou, B. Gowda-Kurkalli, A. Vaknin-Dembinsky, J. M. Gomori, I. Kassis, J. W. M. Bulte, T. Ben-Hur and S. Slavin, Safety, phase I/II study with intrathecal and intravenous injection of mesenchymal stem cells in patients with multiple sclerosis and amyotrophic lateral sclerosis, *Mult. Scler.*, 2008, **14**(Suppl. 1), S21.

Mesenchymal Osteogenic Precursors for Bone Repair and Regeneration

NICOLA BALDINI,[a, b] DANTE DALLARI[a] AND FRANCESCA PERUT[b]

[a] Clinica Ortopedica e Traumatologica I; [b] Laboratorio di Fisiopatologia Ortopedica e Medicina Rigenerativa, Istituto Ortopedico Rizzoli, Via di Barbiano 1/10, 40136 Bologna, Italy

11.1 Introduction

Musculoskeletal conditions are major causes of disease around the world. The treatment of congenital skeletal lesions, bone tumors, degenerative osteoarthropathies, metabolic imbalances and fractures often requires extensive bone reconstruction. Bone graft and artificial implants are being extensively used for bone stock replacement, although biological reconstruction would offer several advantages. Fortunately, bone tissue has the ability to regenerate and this peculiar feature can be advantageously exploited. A thorough knowledge of the multiplex factors underlining the bone healing process, including osteogenic cells and growth factors, is mandatory to translate *in vitro* and *in vivo* studies into a clinical setting. This chapter provides an overview of the biology of bone regeneration and some examples of clinical trials where this knowledge has been translated into practice.

Several tools with biological characteristics are available for skeletal regeneration including natural or synthetic biomaterials, homologous bank tissue,

Stem Cell-Based Tissue Repair
Edited by Raphael Gorodetsky and Richard Schäfer
© The Royal Society of Chemistry 2011
Published by the Royal Society of Chemistry, www.rsc.org

engineered tissues, growth factors and adult stem cells. With regard to orthopedic surgery, in recent years several studies have been performed to look for
biocompatible replacements aimed at healing the bone in different conditions.
In the past, the only scaffold resource available was autologous bone, generally
harvested from the iliac crest. However, since the availability of autologous
bone is very limited, an alternative proven to be just as effective as autologous
bone was sought.

The combined use of bone grafts, stem cells and growth factors has the
aim of achieving active healing from a biological point of view while overcoming the disadvantages of limited availability. The use of lyophilized
homologous bone, supplied by musculoskeletal tissue banks, is also well-
established for severe bone defects because it has low antigenic properties
but the same biochemical qualities. Lyophilization destroys both cells and
the organic matrix of the bone, thus eliminating the biological potential of
this material, but makes it an excellent scaffold. Stromal cells harvested
from the bone marrow of the iliac crest may be added to lyophilized bone. In
particular, mesenchymal stem/stromal cells (MSCs) are promising tools
because of their potential to differentiate into cells of osteogenetic and/or
chondrogenetic lineage. These precursor cells may give rise to bone tissue in
the presence of suitable growth factors that are usually released by the
resorption of the bone matrix caused by osteoclasts or platelets. Another
adjuvant is in fact platelet rich plasma (PRP) obtained from autologous blood
and used to release growth factors that play a key role in the regulation of tissue
genesis and development such as platelet-derived growth factor (PDGF),
transforming growth factor β (TGF-β), epidermal growth factor (EGF),
vascular endothelial growth factor (VEGF) and insulin-like growth factor
(IGF). The application of PRP can maintain high concentrations of these
factors where needed.

A main advantage of bone tissue, in terms of healing, lies in its peculiar
ability to remodel and fully regenerate. Bone is a highly vascularized tissue
that undergoes constant remodeling throughout life. Constant remodeling
provides a mechanism for scar-free healing and regeneration of damaged
bone tissue. This property applies not only to the sequence of events that
are initiated in response to injury (*e.g.* fracture), but also to the healing
of bone around and within endosseous implants. Notably, the molecular
mechanisms that regulate skeletal tissue formation during embryological
development are recapitulated during bone repair.[1] The complex series of
integrated cellular events that ultimately lead to bone healing has been extensively investigated in experimental models.[2] The histological phases of bone
healing include:

- an early inflammation phase (hematoma formation, inflammation and
  angiogenesis) with granulation tissue formation;
- a reparative phase, with cartilaginous callus formation and replacement of
  callus by lamellar bone;
- a bone remodeling phase in which the original bone shape is obtained.

Skeletal integrity after bone fracture is usually restored by natural healing mechanisms of bone repair, sometimes with the help of proper orthopedic tools in order to optimize mechanical stability. However, when bone loss is too large as a consequence of trauma, spinal fusion, tumor resection, arthrodesis, metabolic diseases, non- or delayed union, implant failure, as well as old age or poor bone quality, additional strategies are needed to potentiate secondary bone union and bone regeneration; likewise, bone engineering is advocated.[3]

11.2　Mesenchymal Osteogenic Cells

Multipotent precursors in the bone marrow stroma were first described in 1970 by Friedenstein, who demonstrated the osteogenic potential of a minor subpopulation of bone marrow cells. These cells were clonogenic and the term 'colony forming unit fibroblasts' (CFU-Fs) was consequently coined (Figure 11.1).

As a result of their capacity for self-renewal and differentiation, bone marrow-derived stromal cells are also named mesenchymal stem cells.[4] The widely

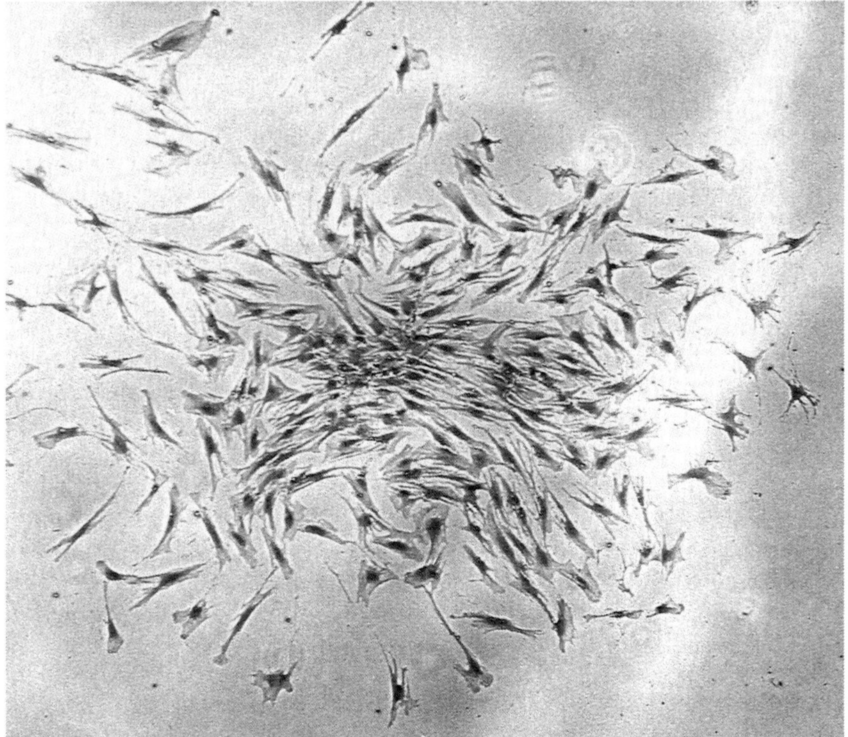

Figure 11.1　Colony-forming unit assay: the presence of potential osteogenic precursors in bone marrow samples is proved by MSC formation of discrete colonies stained with crystal violet.

used term 'MSCs' defines adult cells that are able to differentiate into osteoblasts, adipocytes or chondroblasts *in vitro*, are non-hemopoietic (CD34-), and can be isolated by plastic-adherence[5,6] from a variety of postnatal organs as bone marrow (BM), the cambium layer of the periosteum, adipose tissue, muscle, synovium, dental pulp and amniotic fluid, as well as from fetal tissues such as the umbilical cord blood.

Efforts to enrich MSCs from heterogeneous populations have not been completely successful, as no unique surface antigen is expressed. During the process of differentiation, osteoprogenitor cells acquire specific phenotypes under the control of regulatory factors and transcriptional factors, among which are Runx2, Osterix and a large number of nuclear co-regulators.

Runx2, also known as Cbfa1 (core-binding factor 1), is essential for the commitment of multipotent mesenchymal cells into the osteoblastic lineage.[7] It is expressed in the bipotential precursors of osteoblastic and chondroblastic lineages, but further differentiation to chondrocytes is accompanied by a loss of Cbfa1 expression, whereas its expression is maintained in the osteoblastic lineage. Runx2 induces the expression of bone matrix proteins while keeping the osteoblastic cells at an immature stage. Runx2 interacts with other transcription factors such as Smads 1 and 5, Smad 3 and STAT-1. Runx2 target genes include genes expressed by mature osteoblasts such as osteocalcin, bone sialoprotein, osteopontin and Type I collagen.[8] Osterix (Osx) is an osteoblast-specific transcriptional factor that acts downstream to Runx2/Cbfa1. Osx is a negative regulator of Sox9 and serves as an inhibitor of chondrogenesis and chondrocyte maturation, while promoting osteoblast maturation.[9] β-Catenin, a central component of the cadherin cell adhesion complex, has an essential role in the Wingless/Wnt signaling pathway and plays multiple roles in osteoblasts differentiation. β-Catenin signaling may not be required for the initial commitment of cells into the osteoblastic lineage, based on Runx2 expression, but appears to be essential for their differentiation into functional osteoblasts. Wnts comprise a family of secreted signaling proteins that regulate different developmental processes. Wnt-1-induced secreted protein 1 (WISP-1) expression, during embryonic development, is restricted to osteoblasts and to osteoblastic progenitor cells of the perichondral mesenchyme. WISP-1 is an osteogenic potentiating factor promoting mesenchymal cell proliferation and osteoblastic differentiation while repressing chondrocytic differentiation. Wnt10b inhibits differentiation of pre-adipocytes and promotes osteogenesis by inducing the osteoblastogenic transcription factors Runx2, Dlx5 and osterix and suppressing the adipogenic transcription factors C/EBP-alpha and PPAR-gamma. Many other signaling pathways and factors have been implicated to regulate differentiation of MSCs in bone forming cells such as the factors ATF4, SATB2 and TGF-β, Hedgehog, FGF, ephrin, mitogen-activated protein kinases and sympathetic signaling pathways.[10]

At the molecular level, MSCs exhibit three distinct stages of osteogenic development following osteogenic induction: proliferation, matrix maturation and mineralization (Figure 11.2). The expression levels of osteoblast specific marker genes are finely regulated during these different phases.

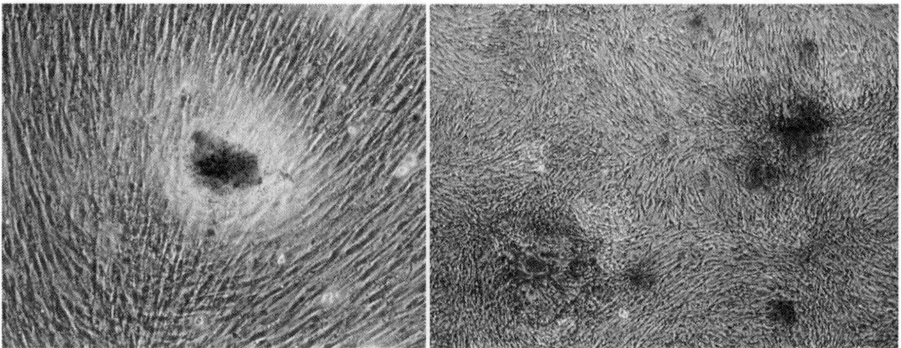

Figure 11.2 Mineralization pattern of MSC cultures stained with alizarin red. Three-dimensional cobblestone nodule ($\times$10, left panel) and less organized Ca–P deposits ($\times$4, right panel).

Alkaline phosphatase expression was found to increase during the proliferation phase and decrease during mineralization. The homeobox gene MSX2, which is implicated in osteoprogenitor cell function, is upregulated during the whole period of differentiation. Collagen 1A1 and osteonectin expression was found to increase from the first stage reaching the maximum during the mineralization stage, while osteocalcin appeared only during matrix formation and the mineralization phases, and is used as a late marker of osteoblast differentiation.[11] Using large-scale expression profiling of clinical samples, gene clusters were associated with the distinct stages of proliferation, matrix maturation and mineralization during the development of the osteoblast phenotype.[11,12]

Different stages of bone repair have been distinguished in the 'fracture healing transcriptome', obtained through the microarray technology, by comparing the gene expression profile of non-fractured with post-fractured rat femurs[13] and in the 'fracture healing proteoma' that describes proteins, which are differentially expressed during bone repair.[14] The bone marrow is filled with a network of stromal cells and hematopoietic cells, with the first group made of both committed cells (*e.g.* adipocytes and osteoblasts) and multipotent precursor cells. The frequency of MSCs in the bone marrow is very low (0.001–0.01%).

Adipose tissue (AT) also contains stem cells similar to BM-derived MSCs. These cells, termed AT-derived stem cells, can be isolated in large numbers from cosmetic liposuction and show multilineage differentiation. The frequency of colony-forming fibroblasts, the actual *in vitro* osteogenic precursor component in the MSC population, has been found to be 1 : 100 in purified AT-derived cells, *i.e.* some 500-fold more than that found in bone marrow.[15]

A perivascular origin for mesenchymal stem cells was supported by some authors.[16,17] Human adult MSC markers were found on pericytes, which provide mechanical support, stability and contractility to small and large vessels, giving rise to adherent, multilineage progenitor cells as they originate osteoblasts, chondroblasts and adipocytes.[18]

The presence of bone precursors in the circulating blood is contentious. Isolation of mesenchymal precursor cells has been obtained with or without prior growth factor mobilization, but the yield is very low and dependent on the isolation method.[19] The number of such precursors may be underestimated owing to the technique of isolation through plastic adherence. Interestingly, their frequency increases after bone fractures, acute burns and myocardial infarction.[20,21]

Osteogenic precursors are commonly isolated from BM-derived adherent cell population. However some authors have demonstrated the ability of non-adherent cells to differentiate into osteoblasts *in vitro* and to form bone when injected into the kidney parenchyma of mice.[22,23] Recently, the ability of non-adherent and adherent bone marrow cells to proliferate and differentiate into osteoblasts was compared. These two populations showed similar characteristics and late adherent cells component is quantitatively relevant, approximately 35% of the BM-derived mononuclear cells.[24] The collection of non-adherent subpopulations increased the total number of BM-derived MSCs up to 36–37%.[25] This finding has to be considered as far as tissue engineering applications are concerned, where non-adherent cells from bone marrow-derived heterogeneous population should represent a complementary source of MSCs. It has been found that this undifferentiated quiescent subpopulation of mesenchymal progenitors may become adherent *in vitro*, begin to proliferate, and differentiate into diverse tissue lineages besides bone, as well as rescue *in vivo* lethally irradiated mice.[26–30]

11.3   Tissue Engineering of Hard Tissues

In addition to the functional properties of the cells, the availability of proper scaffolds able to sustain and/or induce the growth of new, healthy bone tissue is an essential requirement for the repair of large bone defects. Different disciplines (*e.g.* medicine, chemistry, physics, biology and engineering) are involved in improvement and/or integration of three crucial elements as scaffolds (natural or artificial), cells and/or growth factors.[31] Synthetic scaffolds are designed to favor cell attachment, growth and functions. Different strategies have been used to improve the physico-chemical properties of the materials (*i.e.* surface chemistry, wettability, electrical charge, rigidity, micro- and nano-roughness) which directly influence the cellular response.[32] Regenerative medicine approach requires a degradable scaffold, which may be replaced with new tissue. The scaffold degradation rate highly influences the biocompatibility of the biomaterial.[33] BM-derived MSCs are currently used as *in vitro* model to study cell–material interactions.[34] Cell attachment, proliferation and functions are analyzed by multiple techniques to fully understand cell–material integration.

Recently, an extensive characterization of gene expression during MSC *in vitro* differentiation by microarray technology identified genes that are upregulated in MSCs driven to osteogenic differentiation and mineralization.[12]

It was found that the intermediate and late phases of the osteogenic differentiation are characterized by upregulation of bone related genes, among which are ALPL, collagen type X alpha 1, bone morphogenetic protein (BMP) 1, IGF 2, bone sialoprotein, periostin, C-type lectin domain family 3, osteoprotegerin, and osteonectin. Cartilage oligomeric matrix protein was steadily upregulated during MSC differentiation and showed a further peak during mineralization phase, which is characterized by a downregulation of ALPL, COL12A1 and periostin.

Interestingly, a high number of genes related to angiogenesis (through a VEGF-dependent and an angiopoietin-dependent pathway) and to nervous system development (*e.g.* brain-derived neurotrophic factor) are upregulated in MSCs. This suggests their ability to support vascularization and the development of neuronal-like structures. Gene expression analysis may be a useful tool to study *in vitro* cell differentiation into three-dimensional (3D) scaffolds, where morphological assays are difficult to perform and often poorly informative. Confocal microscopy may be a more proper technique to deeply examine adhesion receptors of attachment-dependent cells as osteoblasts.

Extracellular matrix (ECM) proteins including fibronectin, vitronectin, bone sialoprotein and osteopontin[35] contain a specific amino acid sequence, the arginine–glycine–aspartic acid (RGD) motif, which may be recognized by specific integrins expressed by osteoblasts. This physiological interaction has been used in biomaterial development where poly-caprolactone-based polymers (PCL) surfaces, treated by ion-irradiation, were added with RGD oligopeptides. It was found that $\alpha v \beta 3$ integrin, the vitronectin receptor, was strongly expressed on RGD and RGD-ion-irradiated PCL surfaces, while the $\beta 1$ subunit was mainly expressed in osteoblasts cultured on ion-irradiated samples.[36] These findings highlight the delicate relationship between surface structure and cell attachment, which represents a crucial point along the developmental process of tissue regeneration (Figure 11.3).

While cell adhesion is mainly influenced by the material surface, cell growth is influenced by the 3D structure of the scaffold, which partially represents the complexity of bone architecture. The porosity and pore size of the biomaterial and their relationships to the mechanical properties of the scaffolds play a critical role in bone formation, both *in vitro* and *in vivo*. The minimum requirement for pore size is considered to be approximately 100 µm due to cell size, migration requirements and transport. However, pore sizes >300 µm are recommended due to enhanced new bone formation and the formation of capillaries.[37] A proper 3D structure favors nutrient and oxygen transport, and metabolic waste removal.[38] Different strategies are used to reinforce scaffolds without loosening porosity, or to improve the osteoconductivity of a material such as the addition of hydroxyapatite (HA) or α-tricalciumphosphate (TCP) particles to biodegradable polymer, *i.e.* PCL.[39] An oriented migration of human BM-derived MSCs and trabecular osteoblasts along fiber arrangement (known as 'contact guidance') was promoted in a polylactic acid fiber-reinforced polycaprolactone scaffold.[40]

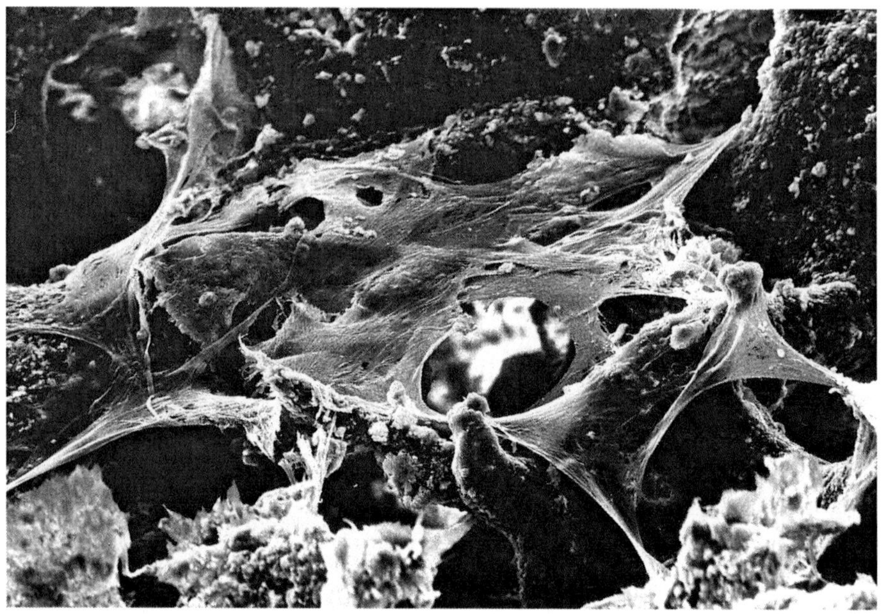

Figure 11.3 Scanning electron microscopy image: adhesion and spreading of human bone marrow stromal cells grown on a 3D bioglass substrate.

11.4   Pre-clinical and Clinical Trials

The value of cell therapy based on BM-derived MSCs should be pre-clinically validated in animal models simulating the clinical setting. The following is an example of such study.

A critical size defect was created in rabbit femurs; histology and histomorphometry were used in the evaluation of healing at two, four and 12 weeks after surgery, showing a progression in the healing rate in critical bone defects treated with BM-derived MSCs plus platelet gel (PG).[41] Based on these encouraging results, a human study was subsequently performed to compare the early osteogenetic potential of lyophilized bone chips alone and chips supplemented with PG and/or BM-derived MSCs in patients undergoing high tibial osteotomy for genu varus. After six and 12 weeks, six months and one year from surgery, patients underwent a clinical and radiographic evaluation. Six weeks after surgery, a computerized tomography (spiral single-slice CT)-guided needle biopsy in the graft insertion site was also performed.

A more active osteogenetic process was observed in group A (lyophilized bone chips/PG) and group B (lyophilized bone chips/PG + BM-derived MSCs) compared with controls. Group B showed a higher bone apposition rate in comparison with controls. At six weeks, the radiographic evaluation was in agreement with the histomorphometric data; it showed that the osteointegration rate was significantly increased by the addition of PG or PG/BM-derived MSCs in comparison with that of controls, and confirmed faster graft integration in group B than in group A. This difference on radiographs remained at three months, but disappeared six and 12 months after surgery. The final clinical outcome of lyophilized bone chips with added PG or PG/BM-derived MSCs, due to the small size of the defects, did not differ from that obtained in the standard surgical procedure. Nevertheless, better healing of the bone gap at the osteotomy site was obtained, while lyophilized bone chips alone in some cases did not achieve complete osteointegration.[42]

The use of these techniques has been extended to the treatment of long bone fracture non-union, osteonecrosis of the femoral head, and in cases of extensive bone loss due to prosthetic revision. In fracture non-union, as defined by the lack of clinical and radiographic evidence of healing after six months from trauma, the bone loses its ability to repair and forms a pathological callus. Thus, it is indispensable to achieve a stable fixation and a good biological support to fix the bone.

The combined use of homologous bone, platelet gel and autologous concentrated stromal cells is a biological stimulus for bone regeneration. In fact, platelets contain several growth factors such as platelet derived growth factor, transforming growth factor β, insulin-like growth factors and vascular endothelial growth factor—all of which positively influence the survival rate, differentiation and proliferation of bone cells. These effects are also exerted on stromal precursor cells, which may transform into osteoblasts, osteocytes and consequently bone tissue. Normally, in fracture treatment the importance of mechanical stability is emphasized as an indispensable factor for healing; however it has been shown that good vascularization of the fracture site carries osteoinductive factors that contribute just as importantly to bone union.[43]

Based on the recognized healing potential of MSCs and PG, the following clinical trial was activated. For the treatment of fracture non-unions, 54 patients (16 of the femur, 19 of the tibia, 10 of the humerus, three of the radius and six of the ulna) were treated with a combination of autologous MSCs, PG and lyophilized homologous bone grafts. Twenty-two non-unions were hypertrophic, and 32 were atrophic. Healing was achieved in 60–90 days in hypertrophic non-unions and 90–180 days in atrophic ones.

Another clinical trial was activated for the conservative treatment of osteonecrosis of the femoral head.[44] This is a condition characterized by a collapse of the skeletal architecture and cartilage surface following necrosis of the bone tissue, resulting in severe pain and functional limitation of the joint. Besides applying cancellous homologous grafts in combination with autologous MSCs and PG, the technique of implanting engineered autologous cartilage and bioceramic cylinders (TruFit) was applied—inspired by knee surgery—to allow the use of 'biological' methods to also advance stage III and IV lesions where subchondral bone collapse had already triggered cartilage degeneration. Currently, in this study 17 patients have been recruited, including 14 with a follow-up of more than six months. The clinical results are encouraging.

Bone engineering is a new exciting frontier for the practicing orthopedic surgeon. Understanding the different factors that recapitulate the complex machinery of bone healing and regeneration must be taken into account to obtain the best clinical results from these powerful techniques. Additional possibilities such as those offered by the use of autologous or homologous MSCs through tissue bank technologies are promising tools, but deserve careful validation in terms of immunogenicity and genomic stability. *Ex vivo* technologies may increase the risk of infection. The effectiveness of recombinant growth factors to modulate cell growth and differentiation can be optimized by a thorough knowledge of the complex cascade of events that, in the specific context of the human setting, regulate the interaction between ligands and receptors. The clinical indications should be probably restricted based on the principles of evidence. In general, the high cost of biologic therapies must be counterbalanced by a careful evaluation of potential benefits.

References

1. C. Ferguson, E. Alpern, T. Miclau and J. A. Helms, Does adult fracture repair recapitulate embryonic skeletal formation?, *Mech. Dev.*, 1999, **87**, 57–66.
2. T. A. Einhorn, C. M. Gundberg, V. J. Devlin and J. Warman, The cell and molecular biology of fracture healing, *Clin. Orthop.*, 1998, **237**, 219–225.
3. K. H. Kraus and C. Kirker-Head, Mesenchymal stem cells and bone regeneration, *Vet. Surg.*, 2006, **35**, 232–242.
4. A. I. Caplan, Mesenchymal stem cells, *J. Orthop. Res.*, 1991, **9**, 641–650.
5. E. M. Horwitz, K. Le Blanc, M. Dominici, I. Mueller, I. Slaper-Cortenbach, F. C. Marini, R. J. Deans, D. S. Krause and A. Keating, Clarification of nomenclature for MSC: The International Society for Cellular Therapy position statement, *Cytotherapy*, 2005, **7**, 393–395.
6. M. Dominici, K. Le Blanc, I. Mueller, I. Slaper-Cortenbach, F. Marini, D. Krause, R. Deans, A. Keating, D. Prockop and E. Horwitz, Minimal criteria for defining multipotent mesenchymal stromal cells. The International Society for Cellular Therapy position statement, *Cytotherapy*, 2006, **8**, 315–317.
7. M. Romero-Prado, C. Blazquez, C. Rodriguez-Navas, J. Munoz, I. Guerriero and E. Delgado-Baeza, Functional characterization of human mesenchymal stem cells that maintain osteochondral fates, *J. Cell. Biochem.*, 2006, **98**, 1457–70.
8. T. Komori, Regulation of bone development and extracellular matrix protein genes by RUNX2, *Cell. Tissue. Res.*, 2010, **339**, 189–195.
9. L. A. Kaback, Y. Soung, A. Naik, N. Smith, E. M. Schwarz, R. J. O'Keefe and H. Drissi, Osterix/Sp7 regulates mesenchymal stem cell mediated endochondral ossification, *J. Cell. Physiol.*, 2008, **214**, 173–182.
10. W. Huang, S. Yang, J. Shao and Y. P. Li, Signaling and transcriptional regulation in osteoblast commitment and differentiation, *Front. Biosci.*, 2007, **12**, 3068–3092.
11. B. Kulterer, G. Friedl, A. Jandrositz, F. Sanchez-Cabo, A. Prokesch, C. Paar, M. Scheideler, R. Windhager, K. H. Preisegger and Z. Trajanoski,

Gene expression profiling of human mesenchymal stem cells derived from bone marrow during expansion and osteoblast differentiation, *BMC Genomics*, 2007, **8**, 70–85.

12. D. Granchi, G. Ochoa, E. Leonardi, V. Devescovi, S. R. Baglio, L. Osaba, N. Baldini and G. Ciapetti, Gene expression patterns related to osteogenic differentiation of bone marrow-derived mesenchymal stem cells during ex vivo expansion, *Tissue Eng. Part C Methods*, 2010, **16**, 511–524.

13. X. Li, R. J. Quigg, J. Zhou, J. T. Ryaby and H. Wang, Early signals for fracture healing, *J. Cell Biochem.*, 2005, **95**, 189–205.

14. X. Li, H. Wang, E. Touma, E. Rousseau, R. J. Quigg and J. T. Ryaby, Genetic network and pathway analysis of differentially expressed proteins during critical cellular events in fracture repair, *J. Cell Biochem.*, 2007, **100**, 527–543.

15. J. K. Fraser, I. Wulur, Z. Alfonso and M. H. Hedrick, Fat tissue: an underappreciated source of stem cells for biotechnology, *Trends Biotechnol.*, 2006, **24**, 150–154.

16. M. Crisan, S. Yap, L. Casteilla, C. W. Chen, M. Corselli, T. S. Park, G. Andriolo, B. Sun, B. Zheng, L. Zhang, C. Norotte, P. N. Teng, J. Traas, R. Schugar, B. M. Deasy, S. Badylak, H. J. Buhring, J. P. Giacobino, L. Lazzari, J. Huard and B. Péault, A perivascular origin for mesenchymal stem cells in multiple human organs, *Cell Stem Cell*, 2008, **3**, 301–313.

17. A. I. Caplan, Why are MSCs therapeutic? New data: new insight, *J. Pathol.*, 2009, **217**, 318–324.

18. G. D. Collett and A. E. Calfield, Angiogenesis and pericytes in the initiation of ectopic calcification, *Circ. Res.*, 2005, **96**, 930–938.

19. C. A. Roufosse, N. C. Direkze, W. R. Otto and N. A. Wright, Circulating mesenchymal stem cells, *Int. J. Biochem. Cell Biol.*, 2004, **36**, 585–597.

20. G. Z. Eghbali-Fatourechi, J. Lamsam, D. Fraser, D. Nagel, B. L. Riggs and S. Khosla, Circulating osteoblast-lineage cells in humans, *New Engl. J. Med.*, 2005, **352**, 1959–1966.

21. J. M. Fox, G. Chamberlain, B. A. Ashton and J. Middleton, Recent advances into the understanding of mesenchymal stem cell trafficking, *Br. J. Haematol.*, 2007, **137**, 491–502.

22. M. W. Long, J. L. Williams and K. G. Mann, Expression of human bone-related proteins in the hematopoietic microenvironment, *J. Clin. Invest.*, 1990, **86**, 1387–1395.

23. K. H. Wlodarski, R. Galus and P. Wodarski, Non-adherent bone marrow cells are a rich source of cells forming bone in vivo, *Folia Biol. (Praha)*, 2004, **50**, 167–173.

24. E. Leonardi, G. Ciapetti, S. R. Baglio, V. Devescovi, N. Baldini and D. Granchi, Osteogenic properties of late adherent subpopulations of human bone marrow stromal cells, *Histochem. Cell Biol.*, 2009, **132**, 547–557.

25. C. Wan, Q. He, M. McCaigue, D. Marsh and G. Li, Nonadherent cell population of human marrow culture is a complementary source of mesenchymal stem cells (MSCs), *J. Orthop. Res.*, 2006, **24**, 21–28.

26. N. Falla, P. Van Vlasselaer, J. Bierkens, B. Borremans, G. Schoeters and U. Van Gorp, Characterization of a 5-fluorouracil-enriched osteoprogenitor population of the murine bone marrow, *Blood.*, 1993, **82**, 3580–3591.
27. P. G. Eipers, S. Kale, R. S. Taichman, G. G. Pipia, N. A. Swords, K. G. Mann and M. W. Long, Bone marrow accessory cells regulate human bone precursor cell development, *Exp. Hematol.*, 2000, **28**, 815–825.
28. M. Dominici, C. Pritchard, J. E. Garlits, T. J. Hofmann, D. A. Persons and E. M. Horwitz, Hematopoietic cells and osteoblasts are derived from a common marrow progenitor after bone marrow transplantation, *Proc. Natl. Acad. Sci. USA.*, 2004, **101**, 11761–11766.
29. U. I. Mödder and S. Khosla, Skeletal stem/osteoprogenitor cells: current concepts, alternate hypotheses, and relationship to the bone remodeling compartment, *J. Cell Biochem.*, 2008, **103**, 393–400.
30. Z. L. Zhang, J. Tong, R. N. Lu, A. M. Scutt, D. Goltzman and D. S. Miao, Therapeutic potential of nonadherent BM-derived mesenchymal stem cells in tissue regeneration, *Bone Marrow Transplant.*, 2009, **43**, 69–81.
31. F. R. Rose and R. O. Oreffo, Bone tissue engineering: hope vs. hype, *Biochem. Biophys. Res. Commun.*, 2002, **292**, 1–7.
32. G. Marletta, G. Ciapetti, C. Satriano, F. Perut, M. Salerno and N. Baldini, Improved osteogenic differentiation of human marrow stromal cells cultured on ion-induced chemically structured poly-epsilon-caprolactone, *Biomaterials*, 2007, **28**, 1132–1140.
33. D. P. Link, J. van den Dolder, J. G. Wolke and J. A. Jansen, The cytocompatibility and early osteogenic characteristics of an injectable calcium phosphate cement, *Tissue Eng.*, 2007, **13**, 493–500.
34. G. Ciapetti, L. Ambrosio, G. Marletta, N. Baldini and A. Giunti, Human bone marrow stromal cells: In vitro expansion and differentiation for bone engineering, *Biomaterials*, 2006, **27**, 6150–6160.
35. K. Anselme, M. Bigerelle, B. Noel, E. Dufresne, D. Judas, A. Iost and P. Hardouin, Qualitative and quantitative study of human osteoblast adhesion on materials with various surface roughnesses, *J. Biomed. Mater. Res.*, 2000, **49**, 155–166.
36. I. Amato, G. Ciapetti, S. Pagani, G. Marletta, C. Satriano, N. Baldini and D. Granchi, Expression of cell adhesion receptors in human osteoblasts cultured on biofunctionalized poly-(epsilon-caprolactone) surfaces, *Biomaterials*, 2007, **28**, 3668–3678.
37. V. Karageorgiou and D. Kaplan, Porosity of 3D biomaterial scaffolds and osteogenesis, *Biomaterials*, 2005, **26**, 5474–5491.
38. K. F. Leong, C. M. Cheah and C. K. Chua, Solid freeform fabrication of three-dimensional scaffolds for engineering replacement tissues and organs, *Biomaterials*, 2003, **24**, 2363–2378.
39. V. Guarino, F. Causa, P. A. Netti, G. Ciapetti, S. Pagani, D. Martini, N. Baldini and L. Ambrosio, The role of hydroxyapatite as solid signal on performance of PCL porous scaffolds for bone tissue regeneration, *J. Biomed. Mater. Res. B Appl. Biomater.*, 2008, **86**, 548–557.

40. V. Guarino, F. Causa, P. Taddei, M. Di Foggia, G. Ciapetti, D. Martini, C. Fagnano, N. Baldini and L. Ambrosio, Polylactic acid fibre-reinforced polycaprolactone scaffolds for bone tissue engineering, *Biomaterials*, 2008, **29**, 3662–3670.
41. D. Dallari, M. Fini, C. Stagni, P. Torricelli, N. Nicoli Aldini, G. Giavaresi, E. Cenni, N. Baldini, A. Cenacchi, A. Bassi, R. Giardino, P. M. Fornasari and A. Giunti, In vivo study on the healing of bone defects treated with bone marrow stromal cells, platelet-rich plasma, and freeze-dried bone allografts, alone and in combination, *J. Orthop. Res.*, 2006, **24**, 877–888.
42. D. Dallari, L. Savarino, C. Stagni, E. Cenni, A. Cenacchi, P. M. Fornasari, U. Albisinni, E. Rimondi, N. Baldini and A. Giunti, Enhanced Tibial Osteotomy Healing with Use of Bone Grafts Supplemented with Platelet Gel or Platelet Gel and Bone Marrow Stromal Cells, *J. Bone Joint. Surg. Am.*, 2007, **89**, 2413–2420.
43. M. Wrotniak, T. Bielecki and T. S. Gazdzik, Current opinion about using the platelet-rich gel in orthopaedics and trauma surgery, *Ortop. Traumatol. Rehabil.*, 2007, **9**, 227–238.
44. V. Gangji, M. Toungouz and J. P. Hauzeur, Stem cell therapy for osteonecrosis of the femoral head, *Expert Opin. Biol. Ther.*, 2005, **5**, 437–442.

CHAPTER 12
Stem Cells and Cartilage Repair

BERND ROLAUFFS,[a, b] ANDREAS BADKE,[b] KUNO
WEISE,[b] ALAN J. GRODZINSKY[c] AND WILHELM K.
AICHER[a, d]

[a] Center for Regenerative Biology and Medicine, University of Tübingen,
Germany; [b] BG Trauma Centre, University of Tübingen Medical Centre,
Tübingen, Germany; [c] Center for Biomedical Engineering, MIT, Cambridge,
USA; [d] Center for Medical Research, University of Tübingen Medical
Faculty, Tübingen, Germany

12.1 MSCs—Cells Suitable for Regeneration of Cartilage Defects?

Mesenchymal stromal cells (MSCs)—sometimes referred to as mesenchymal
stem cells—are capable of differentiating *in vivo* and *in vitro* into mesenchymal
cells including osteoblasts, chondrocytes, adipocytes and possibly others. Thus,
MSCs are important models for the investigation of mechanisms of cellular
differentiation, but also may serve as cellular source for the regeneration of
cartilage and/or bone defects.

For most investigations, MSCs are isolated from bone marrow (BM) aspirates by simple Ficoll® gradient centrifugation to enrich the mononuclear
cells. Others isolate human MSC from adipose tissue,[1,2] muscle, pancreas, or
placenta,[3,4] and even from brain in animals.[5] In diarthrodial joints, MSCs are
derived from the synovial membranes.[6]

For expansion, the MSCs are seeded at low density in cell culture flasks and
expanded in medium containing bovine serum as nutritional supplement.[7]
However, the differentiation capacities of MSCs may alter during *in vitro*

Stem Cell-Based Tissue Repair
Edited by Raphael Gorodetsky and Richard Schäfer
© The Royal Society of Chemistry 2011
Published by the Royal Society of Chemistry, www.rsc.org

expansion. In particular, their chondrogenic potential is reduced after extended periods of *in vitro* expansion.[8,9]

Subsets of MSCs have been described and separated according to their functionally. Among them, the CD271 + CD56 + subset of bone marrow (BM) derived MSCs was isolated, which generated larger cartilaginous micromasses after chondrogenic differentiation.[10] Note that the mineralizing fraction of periosteum-derived progenitor cells expressed CD271, whereas the non-mineralizing cells failed to do so.[11]

MSCs that have been expanded in media containing animal sera cannot be utilized for clinical treatments. Therefore, protocols have been developed to expand human MSCs under appropriate conditions. They include, for instance, the use of human serum to meet the good medical practices (GMP) regulations[12,13] relevant to clinically applicable tissue engineering.

12.2 GMP-compatible Techniques for Stem Cell-based Cartilage Repair

The ABO classification of blood groups was created a 100 years ago,[14] and the basic rules of tolerance or rejection of transplanted tissues were described.[15] Our growing knowledge of immunology together with improved surgical skills in the last century paved the way for both blood transfusions and tissue transplantation to regenerate the complete blood system or replace tissues and organs, respectively. However, the expansion or cloning of individual cells in a laboratory was not required for the transfusion of blood or the application of serum, plasma or extracts. Instead, the cells were isolated and/or purified from whole blood, or blood fractions were prepared by centrifugation, filtration, *etc.* Such preparations were either applied directly to patients or stored for later application.

For transplantation of tissues or whole organs, only a few drops of blood are needed from the donor and recipient to match haplotypes. The tissue is then removed from the donor with the outmost care and implanted in the patient, usually without major manipulations to the recipient. A transplant is not altered in its histological structure; for example, by degradation of the extracellular matrix (ECM) or isolation of cells. In contrast, almost all regimens for tissue engineering with mature cells such as osteoblasts or chondrocytes require disintegration of the ECM of the donor tissue to isolate the desired cells. In addition, *in vitro* manipulations are required for the removal of debris— unfavorable cells or matrix components—and the enrichment of particular cell types for therapeutic use. In addition, their *in vitro* propagation is necessary to generate the required cells numbers for therapy.

Germ-free cultures were impossible to generate prior to the discovery of penicillin in 1928[16] and the development of germ-free work areas for sterile handling of cells. Consequently, the regeneration of damaged tissues, even using somatic cells such as chondrocytes or osteoblasts to regenerate specific functions in the affected areas, was impossible for quite some time both in animal and human studies. About 70 years ago the advent of antibiotics and technical improvements enabling the germ-free isolation and preparation of cells, while sterile culture techniques allowed the development of cell culture protocols from tumors. Later it became possible to use non-transformed cells as well. The first cell line established in a laboratory was the cervical carcinoma HeLa, which was started in 1951–1952. This new technique of culturing cells from mammals *in vitro* under defined conditions immediately raised hopes of being able to utilize healthy cells for curative purposes in both animals and humans.

A major breakthrough in the tissue engineering of human cartilage was made about 15 years ago,[17] when protocols for the isolation and expansion of human chondrocytes to treat circumcised cartilage defects in young patients were developed. Since then they have been further improved. In the meantime, more than 30 000 patients worldwide have received an autologous chondrocyte transplantation (ACT) or a scaffold (= matrix) augmented autologous chondrocyte implantation (MACI).[18–22] The treatment of cartilage defects was the first and, to date, the only cell-based therapy licensed by the US Food and Drug Administration (FDA) and the by European Medicines Agency (EMEA) in the European Union. Nowadays ACT is a standard surgical procedure, and the preparation of autologous cells for the treatment of individual patients is state-of-the-art in industry (Figure 12.1). Regrettably, the success rate is still highly variable.[23–25] It is a well-known fact that the number and quality of chondrocytes found in articular cartilage change with age (Table 12.1).

Chondrocytes from macroscopically intact sites of cartilage of patients suffering from osteoarthritis (OA) express less type II collagen and more of the pro-inflammatory interleukin 1 after cultivation than chondrocytes *ex vivo* (Figure 12.2).

MSCs will possibly provide a cellular basis for cartilage regeneration of the elderly or OA patients in the near future. Extended expansion of chondrocytes even from young healthy donors is accompanied by a dramatic loss of expression of type II collagen, a key component for the function and regeneration of articular cartilage (Figure 12.3). This process is referred to as de-differentiation of the chondrocytes. The loss of expression of type II collagen can be avoided by harvesting larger biopsies, thus reducing the need for cell proliferation *in vitro*. However, this comes at the cost of greater co-morbidity at the site of tissue removal. Optimized cell culture conditions maintain the differentiated phenotype of the cells.

Alternatively, a chondrogenic phenotype may be recovered at least partially by seeding the cells in suitable three-dimensional (3D) carriers, possibly in combination with chondrogenic growth factors such as transforming growth factors (TGF),[26] dexamethasone,[39] or bone morphogenetic proteins (BMP).[27] For experimental purposes the seeding of chondrocytes in alginate beads is the state-of-the-art technique, but alginate is not generally available in GMP quality. Other groups have utilized composites of hyaluronic acid,[28] gelatin,[29] collagens,[30] chitosan,[31] or combinations of such materials.[32] However, the composition of the scaffold influences the metabolism of the chondrocytes and

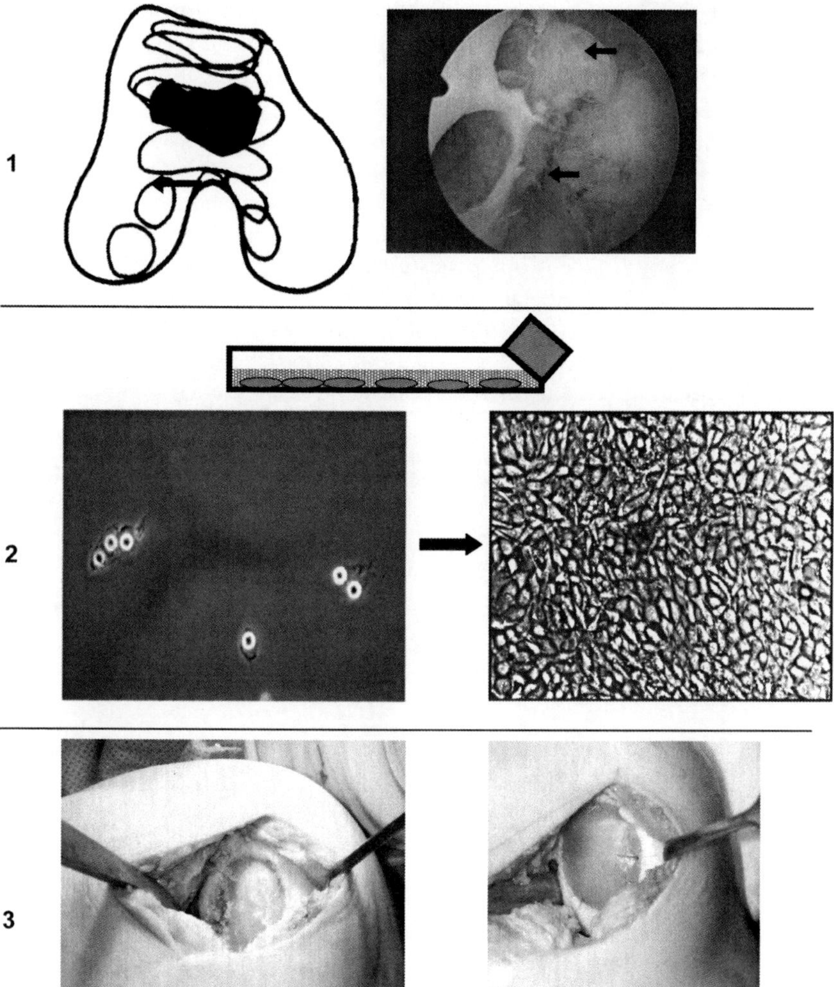

Figure 12.1 Three phases of autologous chondrocyte transplantation. Stage 1 Diagnosis and cell harvest: The cartilage defect is diagnosed to map its localization and size. Typical locations are marked by black circles, the main defect area found in osteoarthritis (OA) patients is marked by the black area (upper left; a black arrow indicates the site for osteochondral biopsy removal for ACI). Biopsies (two 5.0 mm diameter osteochondral plugs; the donor sites after plug removal are indicated by two black arrows) are harvested from the non-weight-bearing articular cartilage of the femoral intercondylar notch of the medial condyle (upper right). Stage 2 Preparation of the implant: the chondrocytes are isolated from the biopsy, expanded *in vitro*, and seeded on a scaffold (middle). Stage 3 Implantation and follow-up: the defect is prepared (lower left) and the construct is implanted in the defect (lower right). Cartilage regeneration by MSCs follows basically this scheme. However, MSCs are not routinely isolated from cartilage but instead from the patient's bone marrow. Furthermore, in stage 2 during *in vitro* expansion of MSCs, chondrogenic or mechanical stimuli must initiate an *in vitro* conditioning of the cells to generate chondrocytes.

Table 12.1 (A) Gross changes of cartilage with age. The number of vital chondrocytes in the tissue, their cytokine-mediated responses as measured for example by TGF-ß1 induced proliferation, and their production of extracellular matrix (ECM). (B) Comparison of chondrocytes harvested from patients diagnosed with osteoarthritis (OA) *versus* patients undergoing autologous chondrocyte transplantation (ACT). Macroscopically healthy appearing cartilage was isolated from the knee cartilage of OA patients undergoing endoprosthesis surgery and from ACT patients in preparation for cartilage tissue engineering after written consent. There were no significant differences in the number of chondrocytes or in cell density, but chondrocytes from OA patients proliferated slower than cells from ACT donors.

A

Age of Patient	*40 y*	*60 y*	*>70 y*
Cells/mm^2	72 000	50 000	40 000
TGF-ß Inducible Proliferation of Cells	+ + + +	+ +	+
Production of ECM	+ + + +	+ +	+
Replicative Capacity	+ + + +	+ + + + +	+ + + +

B

Patient	*OA*	*ACT*
Age	50 + y	Ø35 y
Size of Biopsy	200 mg^{-3} g	200 mg
Chondrocytes/mg cartilage	*ca.* 3000	*ca.* 3000
Proliferation time	*ca.* 18 d	*ca.* 2–20%
Cell Density	*ca.* 6d	*ca.* 0.5–7%

has adverse effects on some materials, including the expression of pro-inflammatory IL-1 (Figure 12.4) or tissue reactions to the scaffold.[33]

Another general problem with the expansion of somatic cells from small biopsies is their aging with each passage of *in vitro* cultivation. In the extreme case, 40–60 population doublings are possible before cells undergo growth arrest or cell death—as reported half a century ago.[34] This natural growth arrest is at least in part associated with cellular aging accompanied by shortening of the telomeres.[35] Consequently, for ACT, the chondrocytes are expanded in a specifically optimized medium to allow the minimum of population doublings while yielding a sufficient number and quality of phenotypically suitable cells. To meet the GMP regulations laid down by the European Medicines Agency[36] and national regulatory authorities, such media must not contain serum from animals such as fetal cattle serum (FCS), but may be enriched by autologous serum preparations from the patient. However, if autologous serum is not available from an individual, pooled AB serum from blood banks can be utilized as surrogate.

Based on experience gathered with mature chondrocytes for ACT[18,20–22] and in combination with the methods developed for the isolation and expansion of

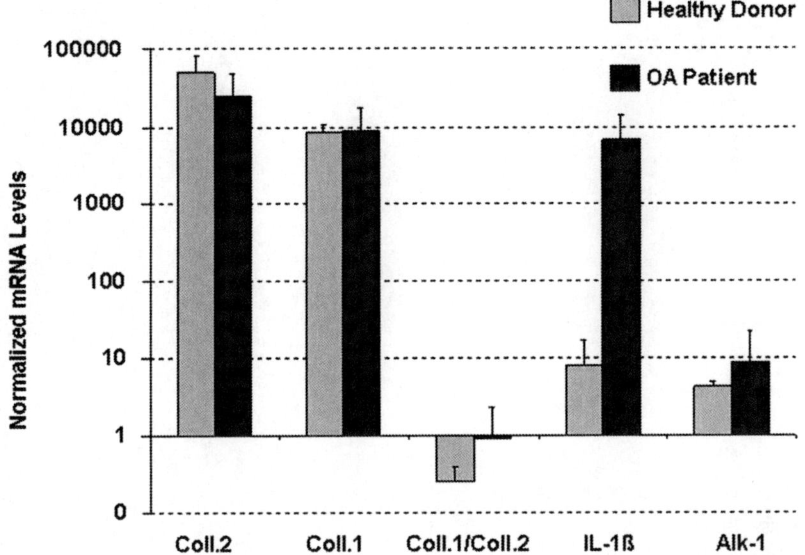

Figure 12.2 Gene expression patterns in OA versus ACT chondrocytes. Chondrocytes were investigated for expression of type I and II collagens, the ratio of the col-1/col-2 expression and transcripts encoding the inflammatory cytokine IL-1, and activin-like-kinase-1 (a marker for irreversible dedifferentiation of chondrocytes). Chondrocytes from OA patients expressed more IL-1 and Alk-1 compared to cells from healthy donors.

MSCs,[13,37] a breakthrough in the use of MSCs to treat cartilaginous defects seems to be within reach. There are several advantages to MSCs: proliferation- and differentiation-competent MSCs can be harvested from various tissues, preferably from bone marrow. They are available in relatively large quantities and in a good quality for various applications. In the last few years several groundbreaking studies have been published focusing on MSCs for various applications,[38] among them repair of cartilage.[39–46]

Some clinical experience with the application of MSCs for regeneration of cartilage or intervertebral disks that was collected recently allowed a first risk evaluation.[47] In one study, a large number of 227 patients was enrolled (duration four years).[48] It was shown that tumors did not occur during the follow-up period (range: three months to two years; mean 11 months). Only 3% of the patients ($n = 7/227$) reported problems that were not related to the MSCs or the carrier implant. Although 1% ($n = 3$) presented with cell-associated problems, all adverse effects were managed by simple interventions.[48]

One challenge for the research and health industry is the task of expanding the cells in strict accordance to EMEA regulations.[36] In the case of MSC expansion *in vitro*, this issue was resolved recently by replacing animal serum or recombinant growth factors not meeting the GMP regulations by human platelet extract and by human plasma.[12,13] Human platelets are a rich source of growth factors including TGF-β1, platelet-derived growth factor (PDGF),

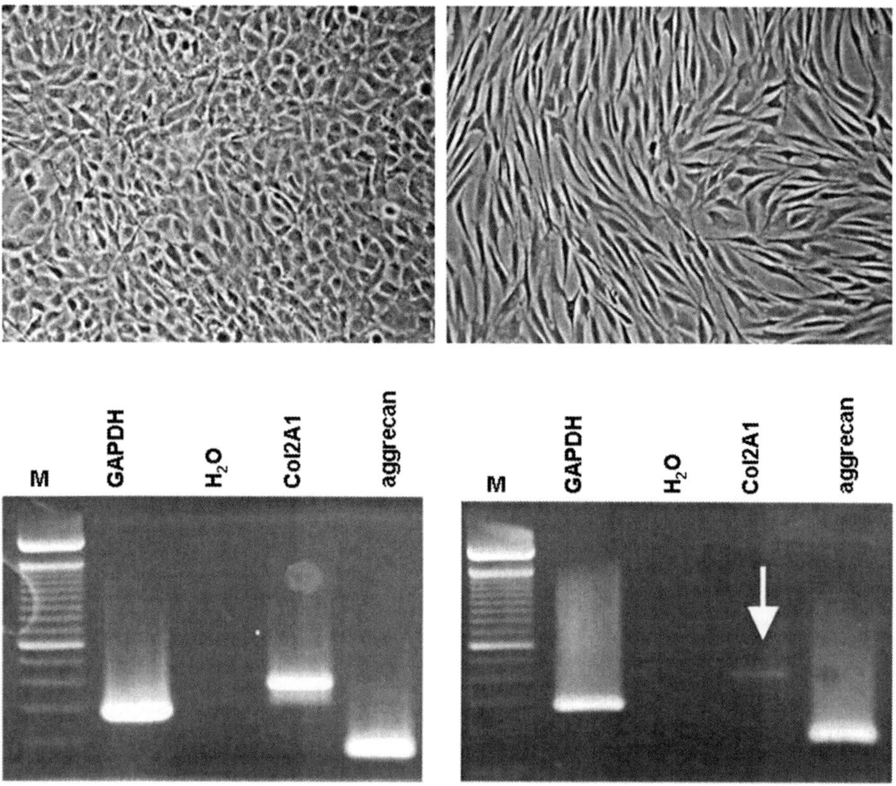

Figure 12.3   Time course of gene expression in chondrocytes from healthy donors. Chondrocytes were expanded in primary culture (approx. 2–3 weeks after inoculation), harvested and spilt for either immediate analysis or for further expansion. Primary culture chondrocytes display a polygonal cellular pattern (top left) and express type II collagen mRNA (bottom left). The chondrocytes de-differentiate during *in vitro* expansion and, after a few passages, display a fibroblast-like cellular pattern (top right) and express only little type II collagen mRNA (bottom right).

endothelial cell growth factor (ECGF) and epidermal growth factor (EGF).[49–52] As the concentrations of growth factors detectable in platelets may vary in a donor-dependent manner, pools of platelets from approximately 4–8 donors are prepared to generate the media supplements needed to expand MSC. The MSCs proliferate well in media enriched with preparations of human fresh frozen platelets (FFP) and plasma (P) from donors of the **AB** serotype, commonly called FFPP medium.[53] The expansion of MSCs in FFPP medium yields significantly higher numbers of cells than the use of commercially available MSC media. Importantly, the osteogenic, chondrogenic and adipogenic differentiation capacities of the MSC are retained.[54] In this way, differentiation-competent MSCs are now available for clinical applications.

Osteogenic and adipogenic differentiation of human MSCs can be induced *in vitro* by stimulating the cells with low molecular weight components: osteoblasts

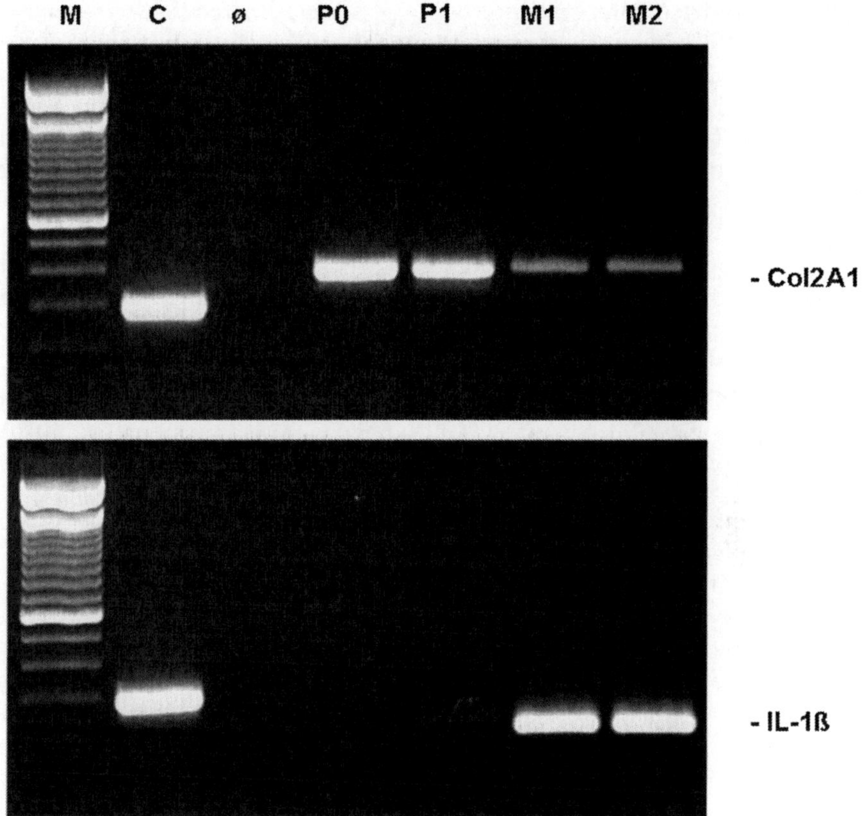

Figure 12.4 Expression of type II collagen and IL-1β by mature chondrocytes in collagenous scaffolds. Primary culture chondrocytes of a healthy donor (P0) were seeded in cell culture flasks or in xenobiotic type I collagen-based matrices and investigated for expression of type II collagen or IL-1β by RT-PCR. After inoculation, the same chondrocytes were investigated in the first passage on cell culture plastic (P1) or in the xenobiotic type I collagen-based matrices (M1, M2). Seeding chondrocytes in the xenobiotic collagen hydrogel induced the expression of the pro-inflammatory cytokine IL-1ß and reduced the expression of type II collagen in comparison to chondrocytes in 2D culture (P1). Amplification of GAPDH (C) served as positive control, omitting the cDNA served as negative control (ø) in the PCR. The size of the products can be estimated from the 100 bp marker ladder (M).

are induced by the addition of β-glycerophosphate, vitamin C and dexamethason to MSCs.[7] All three components are available in clinical-grade purity. Therefore, a GMP-compatible protocol for *in vitro* osteogenic differentiation of MSCs is feasible.[13] It remains to be investigated if MSCs differentiated *in vitro* generate bone after transplantation *in vivo*. Because undifferentiated human MSCs have been applied with success to treat steroid-induced avascular bone necroses,[55] it is hypothesized that estrogenically stimulated MSCs might function for bone repair as well.

Osteogenic differentiation of MSC functions best with BM-derived MSCs, as the repair of bone is one of their major tasks. However, osteoblasts can also be generated from MSCs isolated from tissues other than bone marrow. The pericytes surrounding vessels share many characteristics with the BM-derived MSCs.[3] They, too, can be induced to undergo osteogenic differentiation, albeit with slightly different protocols and outcome (unpublished work).

Adipogenic differentiation of MSCs and pericytes alike is induced *in vitro* by incubating the cells in a high glucose medium and by adding insulin, indometacin, dexamethasone and 3-isobutylxanthine to the cells. With this mixture of factors, the take-up of glucose is facilitated by insulin, whereas indometacin reduces the release of growth-promoting prostaglandins by blocking COX-1. Isobutylxanthine raises the cytoplasmic cAMP concentration, thus further retarding cellular proliferation and promoting differentiation. Here, too, all stimuli are available in qualities suitable for clinical application. Thus, adipogenic differentiation of MSCs following the GMP regulations is also possible *in vitro*.[13]

Several quite different approaches have been taken to achieve efficient chondrogenic differentiation of MSCs *in vitro*. In two-dimensional (2D) cell cultures, chondrogenic differentiation of MSCs is weak even when the cells are incubated in serum-free medium enriched with chondrogenic factors such as glucose, dexamethasone, vitamin C, insulin, proline and TGF-β3. During chondrogenic differentiation the cells mature through several distinct stages in a process closely resembling that of endochondral ossification during embryonal development.[56] Unfortunately, in 2D cultures, the mature chondrocytes do not stop their differentiation at the stage of maturity but spontaneously undergo de-differentiation towards a stage of hypertrophy or acquire a fibroblast-like phenotype characterized by elevated expression of type X or type I collagens, respectively (Figure 12.5). As of today, no specific method has been published which is able to prevent the de-differentiation of MSCs at the stage of mature chondrocytes in 2D cultures.

The fact that chondrocytes reside in a very hydrophilic and slightly acidic ECM has led to the hypothesis that embedding MSCs in hydrogels could facilitate the chondrogenic differentiation of MSCs.[46] Experimentally, this has been achieved with a variety of materials including alginate[57] or other hydrogels enriched with fibrin, hyaluronic acid or chondrogenic factors.[58] At the same time, the hypertrophic de-differentiation of the cells can be avoided in 3D systems.

Chondrogenic differentiation of MSCs can also be achieved in so-called micromass cultures, *i.e.* densely packed clusters of 5×10^5 MSCs in a volume of 20 µL in the presence of the above-mentioned stimuli. Seeding MSCs in micromasses for chondrogenic differentiation has the advantage that the cells are not embedded in a scaffold. Therefore, the scaffold or matrix does not influence the differentiation process (see below). Under optimal conditions, micromasses develop within a few weeks *in vitro* to 1.7–2.1 mm in diameter.

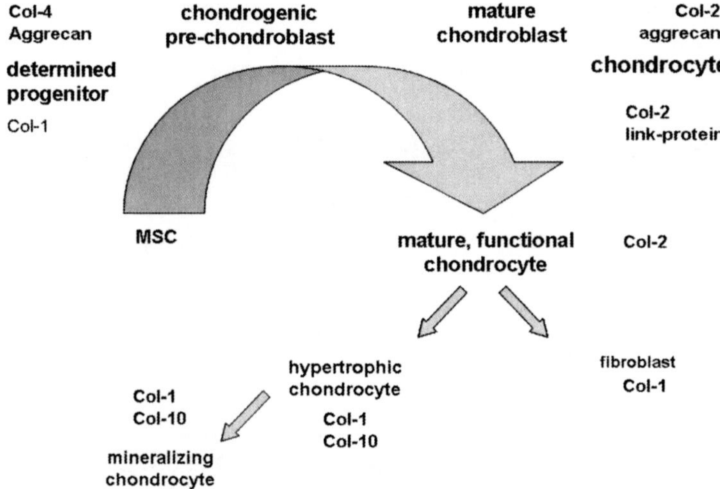

Figure 12.5   Simplified overview of chondrogenic differentiation of MSCs. The stages of chondrogenic differentiation of MSC are labeled in bold letters, the expression of important matrix components is annotated in normal letters. MSCs may differentiate *in vitro* to generate chondrocytes. However, with current techniques the differentiation process does not stop at the stage of mature chondrocytes. *In vitro* the cells de-differentiate to become hypertrophic chondrocytes or cells resembling mineralizing chondrocytes, or de-differentiate to become fibroblasts. The latter are not capable of regenerating a type II collagen-rich extracellular matrix, but instead generate a scar tissue-like, type I collagen-rich matrix.

This would generate a piece of cartilage of sufficient height for many applications, as the cartilage in the weight-bearing zone of an adult knee joint is approximately 2–3 mm thick.

However, micromasses cannot be utilized for implantation into articular sites for several reasons. In micromasses the cells are very densely packed. Therefore, (i) micromasses are far from having the mechanical strength needed to withstand forces in an articulating and weight-bearing joint; and (ii) the nutrition of the cells and waste disposal by diffusion through the cartilaginous matrix cannot support many cells. Consequently, cells in deeper layers of the micromasses tend to undergo necrosis. Furthermore, (iii) due to their round shape, micromasses eventually create a surface reminiscent of cobblestones, but cannot provide the smooth surface needed for articulating joints. Last but not least, (iv) the spherical micromasses would have to be fixed in the cartilage defect with a biocompatible adhesive and filler; because the state-of-the-art fibrin- or gelatin-based tissue glue probably cannot withstand the generated biomechanical forces, such composites from spherical micromasses and filler will probably detach rather quickly and cause further damage in the treated joint, even at remote sites. In summary, the generation of cartilaginous micromasses is a very valuable model for studying pathways of differentiation of MSCs *in vitro*, but their clinical application in weight-bearing joints seems to be unlikely.

12.3 Specific Enrichment of Differentiation-competent MSCs

Ex vivo, MSCs are a rather heterogeneous population of cells and, depending on the tissue of origin, differ not only in their expression of cell surface antigens[59] but also in their differentiation capacities. MSCs isolated from subcutaneous adipose tissue expressed more CD29 and HLA class I antigens than cells from visceral omental adipose tissues.[60] CD29 is the $\beta 1$ component of an integrin receptor. CD29 combines, for instance, with CD49 components to generate receptors for the attachment of MSC to various ECM components including collagens or laminins. Therefore, these differences in receptor expression may influence the adherence of MSCs in the stem cells niches or to the biomaterials utilized as scaffold for tissue engineering. It also influences their migratory potential in the case of wound repair or homing properties.[61]

Recently, experimental evidence was provided that MSCs derived from human bone marrow have a greater tendency to undergo osteogenic differentiation than MSCs isolated from the endometrial (maternal) part of human term placenta.[i] For distinct clinical applications, it would therefore be beneficial to isolate MSCs for bone regeneration from bone marrow, whereas for other applications, placenta-derived MSCs might be beneficial. A chondrogenic differentiation potential was reported in MSCs isolated from the synovial membrane,[6] but the limited numbers of cells available from this membrane require extended expansion of the cells *in vitro*. Therefore, BM- or placenta-derived MSCs will more likely serve as source of cells competent for tissue repair in the future.[62]

Within the BM-derived MSCs, a subpopulation of CD271- and CD56-expressing cells could be defined which was characterized by accelerated proliferation and enhanced chondrogenic differentiation but loss of adipogenic differentiation capacity. The $CD56^{low}$ or $CD56^{negative}$ subset yielded fewer chondrogenic but more adipogenic cells,[10] but there was no difference in osteogenesis in these cells. The latter is not surprising, since BM-derived MSCs are known for their overall good osteogenic differentiation capacity. It remains to be determined whether this is also true for $CD271^{pos}CD56^{pos}$ MSC or pericytes from other sources. Among the MSC-like cells derived from human periosteum, expression of CD271 was found in the osteogenic fraction, whereas the non-mineralizing cells failed to express CD271.[11] This indicates that the stromal cells isolated from distinct tissues may represent cells with distinct characteristics, including the expression of cell surface antigens or differentiation capacities.

---

[i] G. Pilz, Diploma thesis, Faculty of Biology, University of Tübingen, 2008.

Investigations on MSC subsets revealed that the expression of a distinct cell surface antigen seems to be associated with an elevated osteogenic differentiation capacity. In BM-derived MSCs, the expression of this marker is more prominent and correlated with robust osteogenic differentiation. In placenta-derived MSCs (pMSCs), expression of this marker is less prominent and correlated with a reduced osteogenic differentiation capacity. Currently, investigations are ongoing characterizing sorted subsets of MSCs and their osteogenic potential in more detail.[ii]

When MSCs were expanded from single cells, possibly under true clonal conditions, individual clones were shown to be multipotent,[63] suggesting that the individual MSC within the originating bulk population cells did not differ with respect to their differentiation capacities. However, microarray analyses provided evidence that the clonally derived MSCs were an inhomogeneous population.[63] The latter notion was supported by the finding that only 30% of the MSCs were multipotent, but the majority of MSCs presented with various and more restricted differentiation potentials.[64] We possibly owe this conflict of data to the fact that different stages of MSC may exist within a stem cell niche[10] and that different methods for isolation and especially for expansion may enrich different subsets.[65,66] Current knowledge and technology suggest that we will be able to enrich for MSC with a distinct regenerative potential in the near future.

12.4 Effect of Hypoxia on Chondrogenic Differentiation of MSCs

In articular cartilage, the chondrocytes have no access to vascularization and therefore metabolize under bradytrophic conditions. Although the oxygen saturation in cartilage was determined to be rather low, expression of typical genes such as type II collagen is optimal *in vitro* under rather normoxic conditions.[67] Others, however, report that low oxygen tension was beneficial for chondrogenic differentiation of MSCs and for mature chondrocytes.[68–70] We recently provided evidence that the expression of chondrogenic factors such as type II collagen or aggrecan is not enhanced by hypoxia, but that it ameliorated the IL-1-induced catabolic effects.[13] Therefore, hypoxia may be a useful tool for improving implant quality during expansion and chondrogenic differentiation of MSC *in vitro*. Hypoxia may, for instance, compensate for the inflammatory stimuli provided by polymeric scaffolds during hydrolysis.

12.5 Scaffolds for Efficient Chondrogenic Differentiation of MSCs

As outlined above, chondrogenic differentiation of MSCs in micromasses represents a valuable model to investigate this process *in vitro*. But micromasses are not suitable for clinical applications. However, seeding biocompatible or

---

[ii] C. Ulrich, Diploma thesis, Faculty of Biology, University of Tübingen, 2009.

even chondroinductive scaffolds with MSCs and implanting these constructs in articulating joints yielded regeneration of a cartilaginous tissue that allowed athletes to resume their physical activities.[71] As briefly outlined above, suitable biomaterials include components extracted from the ECM such as collagens, gelatin, hyaluronic acid, chondroitin sulfate, fibrin, elastin and alike. Even products from insects (*e.g.* silk or chitosan) or plants (agarose/agar) were investigated. The scaffolds can be produced in different forms as fleeces, membranes, sponges or combinations thereof. One major advantage of these natural components is the fact that they provide natural binding sites for the cellular receptors of the MSCs or chondrocytes. Therefore, both MSCs and chondrocytes alike attach to, and grow on or in, such scaffolds. However, natural components also carry the risk of antigenic or pathogenic effects and, therefore, have to be evaluated carefully.

In contrast, biocompatible and degradable polymers that are suitable for tissue engineering do not usually contain physiological binding sites for cells; for example, an arginine–glycine–aspartic acid (RGD) motif for integrin-mediated binding of cells. Therefore, most cells tend not to bind even to polar surfaces of the polymers in a physiological manner. But there are several advantages to polymers generated in the laboratory:

(i) They can be designed to provide specific elasticity, mechanical strength, degradation kinetics, surface polarity, interconnectivity, shape, *etc.* to meet the needs of the clinical application.

(ii) They can be chemically modified to harbor growth factors, for instance, or to release differentiation factors in a timely order.

(iii) They are very affordable and can be generated with excellent batch-to-batch consistency in their chemical and physical parameters.

(iv) They can be produced easily in GMP quality and stored as dry or wet off-the-shelf products for quite some time.

Furthermore, to improve the attachment of MSCs to such polymers, some manufacturers have developed a coating technology to cover the polymer with a nano-layer of gelatin, providing the peptide (*s.c.*, RGD) motifs for cellular attachment. Gelatin is produced worldwide in huge amounts and in GMP quality for clinical applications. Therefore such polymer–gelatin composites may become the state-of-the art scaffold not only for cartilage tissue engineering but also for many other clinical applications in the near future.

One example of a synthetic scaffold is a self-assembling peptide hydrogel containing unique features for a tissue engineering polymer scaffold such as a nanofiber structure that is almost three orders of magnitude smaller than most polymer microfibers, presenting a unique polymer structure with which cells may interact. In addition, peptide sequences may be designed for specific cell–matrix interactions that influence cell differentiation and tissue formation. Also, the synthetic nature of the peptide minimizes the risk of carrying biological pathogens relative to animal-derived biomaterials.[72]

Besides the attachment of the cells, the degradation of the scaffold material also presents certain problems which remain to be solved. For cartilage regeneration, MSC- or chondrocyte-augmented constructs can be implanted after a short time of incubation in the articular defect and will then mature *in situ*. The scaffold will be degraded and replaced by the ECM produced by the MSCs or the chondrocytic daughter cells.[71] Although biodegradable polymers have several advantages for cartilage tissue engineering, most of them represent esters that split into an acid and alcohol by hydrolytic degradation. Acidification of cartilage may cause adverse effects and the production of proteoglycans may be reduced under lower pH.[73] However, expression of proteoglycans in the evolving new tissue is a prerequisite for the regeneration of a biomechanically functional hyaline cartilage.

The degradation of scaffolds fabricated from protein scaffolds yields amino acids or small peptides. In contrast to the acids generated by degradation of the esters, peptides or amino acids do not change the pH considerably and can be utilized as nutrients by the cells in the constructs. Some adverse effects have been observed with hydrogel matrices generated, for instance, from type I/III-collagen (Figure 12.4). But success of chondrogenic differentiation was achieved by seeding the MSCs in spongiform carriers generated from type I collagen enriched with chondroitin sulfate, TGF-β3 and other chondrogenic factors.[45,74] Under different conditions, however, chondroitin sulfate promoted the osteogenic differentiation of MSCs seeded on textile polymers.[75] Osteogenic differentiation was also promoted by seeding MSC on laminin-5, a component of ectodermal and endodermal tissues.[76]

Furthermore, the stiffness or elasticity of the scaffold modulates the differentiation of MSC as well. Osteogenic differentiation is encouraged on stiff or hard substrates, whereas on soft or elastic substrates myogenic differentiation was observed,[77] indicating that not only the composition of the scaffolds but also structural and physical parameters influence the differentiation of MSCs. The influence of scaffold stiffness may also be responsible in part for the hypertrophic de-differentiation observed in chondrocytes in 2D cultures, as the cell culture vessel mimics a bony environment rather than a hyaline cartilage habitat. The steps of *in vitro* differentiation and de-differentiation of MSCs are thought to emulate the processes of enchondral ossification observed during the development of the bone anlagen and the growth of the extremities during embryonal development (Figure 12.5).

For clinical applications with MSCs it is important to recall that the chondrogenic differentiation potential of MSCs is reduced after expansion of the cells *in vitro*[8] (Figure 12.6). Moreover it is significantly lower in patients with clinical osteoarthritis[9] (Figure 12.7). Note that the adipogenic differentiation of MSCs from OA patients was also reduced in comparison to young donors. Only the osteogenic differentiation capacity of BM-derived MSCs did not seem to be as age-dependent.[9]

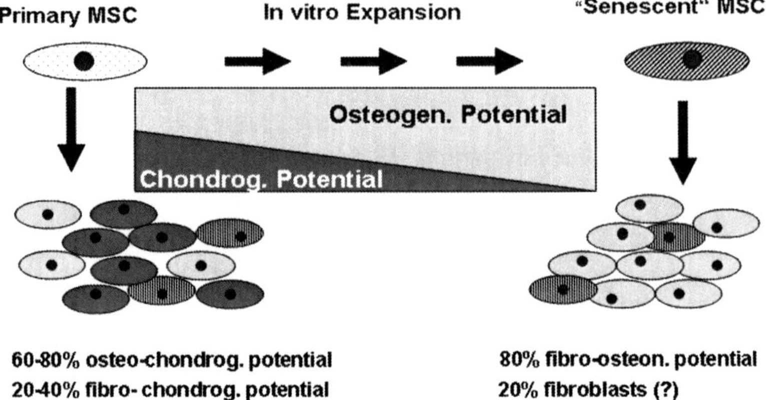

Figure 12.6   Loss of chondrogenic potential in MSCs during expansion *in vitro*. Repeated passaging of MSCs *in vitro* is associated with a gradual loss of chondrocyte-generating clones. The frequency of MSCs generating osteoblasts seems to increase over time. For details see ref. 8.

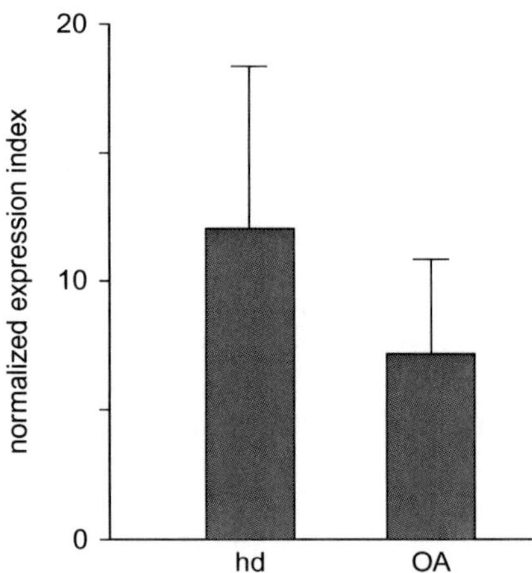

Figure 12.7   Chondrogenic differentiation of BM-derived MSCs from healthy donors and OA patients. MSCs were isolated, expanded and chondrogenic differentiation was induced to compare the chondrogenic potential of the cells. MSCs from OA patients (OA) produced pale micromasses with less type II collagen and proteoglycan, indicating a significantly reduced chondrogenic potential compared to MSC from healthy donors (hd).

12.6   Effects of Biomechanical Stimulation of Chondrocytes and MSCs

Adult articular cartilage functions and responds biomechanically in a depth-dependent fashion due to the depth-dependent nature of the composition,[78,79] structure,[80] and bioelectrical[81] and biomechanical properties[82] of the ECM. The response of articular cartilage to mechanical stimuli is simulated in tissue engineering approaches by the application of exogenous mechanical stimulation leading to greater chondrocyte metabolic activity and ECM production.[83] Increased ECM deposition *in vitro* may increase the mechanical stability of the cell-seeded construct and its ability to withstand loading forces encountered after implantation.[84] Under physiological loading conditions compressive, shear, and tensile forces act upon the cartilage, so those forces were investigated to simulate joint loading and to improve the structural and functional quality of cartilage generated by tissue engineering technology.

Whereas the effects of biomechanical loading on articular chondrocytes have been closely studied, attention is only now turning toward MSCs under loading conditions.[85] However, it remains unclear which loading protocols would be most beneficial, and it is still not possible to engineer cartilage constructs with properties that are comparable to those of the original tissue.[86] However, the engineering of such properties may prove crucial because endogenous repair tissues have inferior biomechanical and biochemical properties,[18,87,88] and poorer wear characteristics[89] which frequently lead to the development of secondary osteoarthritis.[90]

Static compression leads to a reduction of chondrocyte biosynthesis of functionally relevant ECM components[91–96] and may therefore be detrimental in tissue engineering applications. Dynamic compression has stimulatory[72,81,97–102] and both short- and long-term effects on chondrocyte biosynthesis.[81,97,98] Dynamic compression was studied and an optimized alternate day loading protocol was developed which, when applied to chondrocyte-seeded peptide hydrogels, stimulated proteoglycan synthesis up to twice as effectively as in free-swelling cultures. It also resulted in improved mechanical properties, suggesting a large potential of dynamic compression for stimulating proteoglycan synthesis and accumulation for the *in vitro* culture of tissue engineered constructs prior to implantation.[72]

A study has investigated the influence of long-term confined dynamic compression and surface motion under low oxygen tension on tissue-engineered cell–scaffold constructs. It was shown that the combination of loading and low oxygen tension resulted in downregulation of collagen type I mRNA expression, contributing to the stabilization of the chondrocyte phenotype.[103] This suggests that mechanical stimulation combined with low oxygen tension may be an effective tool for modulating the chondrocyte bone anlagen.

However, adverse effects of dynamic compression have also been reported; dynamic compression samples contained 60–80 kDa aggrecan fragments which were detected by both anti-G1 and NITEGE probes, demonstrating the activity of ADAMTS proteases but not MMP-mediated degradation. This suggests that partially mature cartilage tissue engineering constructs may be susceptible to catabolic degradation.[104]

The application of shear stress has also proven to be relevant[105] and possibly a major key player because the brief application of shearing forces applied periodically over a four-week period increased matrix accumulation and improved mechanical performance. Interestingly, the resulting mechanical properties appeared to be associated with the changes in matrix accumulation induced by stimulation, and shear-stimulated constructs contained higher proportions of collagen and proteoglycans and displayed superior mechanical properties compared with the tissues subjected to compressive forces during tissue formation.[86]

It is also important to note that, both under *in vivo* and *in vitro* conditions, the pressurization of the interstitial fluid may support the majority of the applied compressive loads. When cartilage is loaded at ∼1 Hz (*i.e.* walking), only less than 10% of the load remains for a direct compression of the solid ECM phase, suggesting that it is not compression but the indirect application of hydrostatic pressure that is the major contributor to the existing loads in joints during ambulation.[106] At the same time, recent studies using real-time dual-fluoroscopic imaging of patient's knees *in vivo* have shown that, while standing in place for only tens of seconds, cartilage compression as high as 30–35% or more may occur.[107] In such cases, direct compression of the ECM is very important. Investigating the effects of pure hydrostatic pressure application, one study demonstrated increases in the biomechanical tissue properties[108] and also showed synergistic effects between hydrostatic pressure and growth factors on tissue functional properties. Hydrostatic pressure provides a robust method of chondrocyte stimulation because it can be applied both to chondrocytes in monolayer 3D engineered constructs and to explants.[83] However, another study noted that, aside from the beneficial effects intermittent hydrostatic pressure may have on ECM synthesis, its effects on mechanical properties may require longer culture periods to manifest.[109]

When the effects of biomechanical stimulation on MSCs were studied, it was found that stimulation was beneficial for the further differentiation of stem cell tissue engineered constructs.[110] However, while chondrogenesis did occur in MSC-laden hydrogels, the amount of the forming matrix and measures of its mechanical properties were lower than that produced by chondrocytes under the same conditions.[111] Some functionally important tissue characteristics— particularly the glycosaminoglycan content and the resulting mechanical properties—demonstrated that a plateau was reached in MSC-laden constructs, suggesting that diminished capacity was not the result of delayed differentiation.[111] Thus, while MSCs do generate constructs with substantial cartilaginous properties, further studies are needed to investigate their full potential and to design optimized protocols to achieve levels similar to those produced by chondrocytes. However, a suitable periodic application of dynamic compression under 3D environments can encourage BM-derived MSCs to differentiate into chondrocytes and maintain their phenotypes.[112]

12.7 Perspectives

For small, circumcised defects of articular cartilage, tissue engineering utilizing mature chondrocytes from healthy sites of the affected joints seems likely to remain the standard in the near future. However, in cases where autologous chondrocytes are not available (*e.g.* to regenerate a defective intervertebral disk or when osteochondral constructs are needed), it may be possible to employ MSCs in future cell-based therapies. But before we can establish MSC-based therapies for cartilaginous defects, we must first overcome several obstacles such as the selection of optimal chondrogenic MSCs, the efficient differentiation of mature but not hypertrophic chondrocytes, and their adaptation to the mechanical stress in an articulating joint.

The future of cell-based cartilage repair began more than a decade ago.[18] We are confident that MSC-based cartilage regeneration will be an effective tool; our task is now to see that it is put on our research and funding agendas to make it available to meet the needs of tomorrow's patients.

References

1. H. Mizuno, P. A. Zuk and M. Zhu, *et al.*, Myogenic differentiation by human processed lipoaspirate cells, *Plast. Reconstr. Surg.*, 2002, **109**, 199–209, discussion, 210–191.
2. J. B. Mitchell, K. McIntosh and S. Zvonic, *et al.*, Immunophenotype of human adipose-derived cells: temporal changes in stromal-associated and stem cell-associated markers, *Stem Cells*, 2006, **24**, 376–385.
3. M. Crisan, S. Yap and L. Casteilla, *et al.*, A perivascular origin for mesenchymal stem cells in multiple human organs, *Cell Stem Cell*, 2008, **3**, 301–313.
4. S. Barlow, G. Brooke and K. Chatterjee, *et al.*, Comparison of human placenta- and bone marrow-derived multipotent mesenchymal stem cells, *Stem Cells Dev.*, 2008, **17**, 1095–1107.
5. Y. Jiang, B. Vaessen and T. Lenvik, *et al.*, Multipotent progenitor cells can be isolated from postnatal murine bone marrow, muscle, and brain, *Exp. Hematol.*, 2002, **30**, 896–904.
6. C. De Bari, F. Dell'Accio and P. Tylzanowski, *et al.*, Multipotent mesenchymal stem cells from adult synovial membrane, *Arthritis Rheum.*, 2001, **44**, 1928–1942.
7. M. F. Pittenger, J. D. Mosca and K. R. McIntosh, Human mesenchymal stem cells, progenitor cells for cartilage, bone, fat and stroma, *Curr. Topics Microbiol. Immunol.*, 2000, **251**, 3–11.
8. A. Muraglia, R. Cancedda and R. Quarto, Clonal mesenchymal progenitors from human bone marrow differentiate *in vitro* according to a hierarchical model, *J. Cell Sci.*, 2000, **113**, 1161–1166.
9. J. M. Murphy, K. Dixon and S. Beck, *et al.*, Reduced chondrogenic and adipogenic activity of mesenchymal stem cells from patients with advanced osteoarthritis, *Arthritis Rheum.*, 2002, **46**, 704–713.
10. V. L. Battula, S. Treml and P. M. Bareiss, *et al.*, Isolation of functionally distinct mesenchymal stem cell subsets using antibodies against CD56,

CD271 and mesenchymal stem cell antigen-1 (MSCA-1), *Hematologica*, 2009, **94**, 19–30.

11. D. Alexander, F. Schäfer and A. Munz, *et al.*, NGFR: a new osteogenic differentiation marker in mineralizing periosteal cells, *Tissue Eng.*, 2009, **15**, 715.

12. A. Shahdadfar, K. Fronsdal and T. Haug, *et al.*, *In vitro* expansion of human mesenchymal stem cells, choice of serum is a determinant of cell proliferation, differentiation, gene expression, and transcriptome stability, *Stem Cells*, 2005, **23**, 1357–1366.

13. T. Felka, R. Schäfer and P. deZwart, *et al.*, Animal serum-free expansion and differentiation of human mesenchymal stem cells, *Cytotherapy*, 2010, **12**, 143–153.

14. B. Woodward and K. Landsteiner, Over one billion and 38 million lives saved, in *Scientists Greater than Einstein: The Biggest Lifesavers of the Twentieth Century*, B. Woodward, J. Shurkin and D. Gordon, eds. Quill Driver Books, Fresno, CA, 2009, 291–324.

15. G. Schöne, Vergleichende Untersuchungen über die Transplantation von Geschwülsten und von normalen Geweben, *Beitr. klin. Chir.*, 1909, **61**, 1–49.

16. A. Fleming, On the antibacterial action of cultures of a penicillium, with special reference to their use in the isolation of *B. influenzae*, *Br. J. Exp. Pathol.*, 1929, **10**, 226–236.

17. A. Lindahl, M. Brittberg and L. Peterson, Cartilage repair with chondrocytes, clinical and cellular aspects, *Novartis Found. Symp.*, 2003, **249**, 175–186, discussion 186–179.

18. M. Brittberg, L. Peterson and E. Sjogren-Jansson, *et al.*, Articular cartilage engineering with autologous chondrocyte transplantation, A review of recent developments, *J. Bone Joint Surg. Am.*, 2003, **85A**(Suppl 3), 109–115.

19. L. Peterson, T. Minas and M. Brittberg, Treatment of osteochondritis dissecans of the knee with autologous chondrocyte transplantation: results at two to ten years, *J. Bone Joint Surg. Am.*, 2003, **85**, 17–24.

20. W. Bartlett, J. A. Skinner and C. R. Gooding, *et al.*, Autologous chondrocyte implantation versus matrix-induced autologous chondrocyte implantation for osteochondral defects of the knee: a prospective, randomised study, *J. Bone Joint Surg. Br.*, 2005, **87B**, 640–645.

21. P. Behrens, T. Bitter and B. Kurz, *et al.*, Matrix-associated autologous chondrocyte transplantation/implantation (MACT/MACI)—5-year follow-up, *Knee*, 2006, **13**, 194–202.

22. C. Erggelet, P. Kreuz, E. Mrosek, *et al.*, Autologous chondrocyte implantation versus ACI using 3D-bioresorbable graft for the treatment of large full-thickness cartilage lesions of the knee, *Arch. Orthop. Trauma Surg.*, 2009, http://www.springerlink.com/content/p57124034v314777/ [accessed June 2010].

23. U. Horas, R. Schnettler and D. Pelinkovic, *et al.*, Knorpelknochentransplantation versus autogene Chondrocytentransplantation. Eine prospektive vergleichende klinische Studie, *Chirurg*, 2000, **71**, 1090–1097.

24. S. Roberts, I. W. McCall and A. J. Darby, *et al.*, Autologous chondrocyte implantation for cartilage repair, monitoringits success by magnetic resonance imaging and histology, *Arthritis Res. Ther.*, 2003, **5**, R60–R73.

25. G. Knutsen, J. O. Drogset and L. Engebretsen, *et al.*, A randomized trial comparing autologous chondrocyte implantation with microfracture. Findings at five years, *J. Bone Joint Surg. Am.*, 2007, **89**, 2105–2112.

26. P. C. Yaeger, T. L. Masi and J. L. B. de Ortiz, *et al.*, Synergistic action of transforming growth factor-beta and insulin-like growth factor-I induces expression of type II collagen and aggrecan genes in adult human articular chondrocytes, *Exp. Cell Res.*, 1997, **237**, 318–325.

27. T. Gründer, C. Gaissmaier and J. Fritz, *et al.*, BMP-2 enhances the expression of type II collagen and aggrecan in chondrocytes embedded in alginate beads, *Osteoarthritis Cartilage*, 2004, **12**, 559–567.

28. M. Mori, M. Yamaguchi and S. Sumitomo, *et al.*, Hyaluronan-based biomaterials in tissue engineering, *Acta Histochem. Cytochem.*, 2004, **37**, 1–5.

29. H.-W. Kang, Y. Tabata and Y. Ikada, Fabrication of porous gelatin scaffolds for tissue engineering, *Biomaterials*, 1999, **20**, 1339–1344.

30. P. Cherubino, F. A. Grassi and P. Bulgheroni, *et al.*, Autologous chondrocyteimplantation using a bilayer collagen membrane, a preliminary report, *J. Orthop. Surg.*, 2003, **2003**, 10–15.

31. J. S. Mao, L. G. Zhao and Y. J. Yin, *et al.*, Structure andproperties of bilayer chitosan-gelatin scaffolds, *Biomaterials*, 2003, **24**, 1067–1074.

32. C.-H. Chang, H.-C. Liu and C.-C. Lin, *et al.*, Gelatin-chondroitin-hyaluronan tri-copolymer scaffold for cartilage tissue engineering, *Biomaterials*, 2003, **24**, 4853–4858.

33. Y. Fujihara, Y. Asawa and T. Takato, *et al.*, Tissue reactions to engineered cartilage based on poly-L-lactic acid scaffolds, *Tissue Eng. Part A*, 2008, **15**, 1565–1577.

34. L. Hayflick and P. S. Moorhead, The serial cultivation of human diploid cell strains, *Exp. Cell Res.*, 1961, **25**, 585–621.

35. N. E. Sharpless and R. A. DePinho, Telomeres, stem cells, senescence, and cancer, *J. Clin. Invest.*, 2004, **113**, 160–168.

36. European Medicines Agency (EMEA), In-vitro *cultured chondrocyte containing products for cartilage repair of the knee*, EMEA, London, 2009, EMEA/CAT/CPWP/288934/2009, 1–8, www.ema.europa.eu/pdfs/human/cpwp/28893409en.pdf [accessed June 2010].

37. M. F. Pittenger, A. M. Mackay and S. C. Beck, *et al.*, Multilineage potential of adult human mesenchymal stem cells, *Science*, 1999, **284**, 143–147.

38. F. P. Barry and J. M. Murphy, Mesenchymal stem cells, clinical applications and biological characterization, *Int. J. Biochem. Cell Biol.*, 2004, **36**, 568–584.

39. S. Wakitani, T. Goto and S. J. Pineda, *et al.*, Mesenchymal cell-based repair of large, full-thickness defects of articular cartilage, *J. Bone Joint Surg. Am.*, 1994, **76**, 579–592.

40. S. Wakitani, K. Imoto and T. Yamamoto, *et al.*, Human autologous culture expanded bone marrow mesenchymal cell transplantation for repair of cartilage defects in osteoarthritic knees, *Osteoarthritis Cartilage*, 2002, **10**, 199–206.
41. A. I. Caplan, Review: mesenchymal stem cells, cell-based reconstructive therapy in orthopedics, *Tissue Eng.*, 2005, **11**, 1198–1211.
42. K. Uematsu, K. Hattori and Y. Ishimoto, *et al.*, Cartilage regeneration using mesenchymal stem cells and a three-dimensional poly-lactic-glycolic acid (PLGA) scaffold, *Biomaterials*, 2005, **26**, 4273–4279.
43. G. R. Erickson, J. M. Gimble and D. M. Franklin, *et al.*, Chondrogenic potential of adipose tissue-derived stromal cells *in vitro* and *in vivo*, *Biochem. Biophys. Res. Commun.*, 2002, **290**, 763–769.
44. M. R. Pagnotto, Z. Wang and J. C. Karpie, *et al.*, Adeno-associated viral gene transfer of transforming growth factor-β1 to human mesenchymal stem cells improves cartilage repair, *Gene Ther.*, 2007, **14**, 804–813.
45. S. Varghese, N. S. Hwang and A. C. Canver, *et al.*, Chondroitin sulfatebased niches for chondrogenic differentiation of mesenchymal stem cells, *Matrix Biol.*, 2008, **27**, 12–21.
46. N. S. Hwang and J. Elisseeff, Application of stem cells for articular cartilage regeneration, *J. Knee Surg.*, 2009, **22**, 60–71.
47. S. M. Richardson, J. A. Hoyland and R. Mobasheri, *et al.*, Mesenchymal stem cells in regenerative medicine, opportunities and challenges for articular cartilage and intervertebral tissue engineering, *J. Cell Physiol.*, 2010, **222**, 23–32.
48. C. J. Centeno, J. R. Schultz and M. Cheever, *et al.*, Safety and complications reporting on re-implantation of culture-expanded mesenchymal stem cells using autologous platelet lysate technique, *Curr. Stem Cell Res.*, 2010, **5**, 81–93.
49. R. K. Assoian, A. Komoriya and C. A. Meyers, *et al.*, Transforming growth factor-beta in human platelets. Identification of a major storage site, purification, and characterization, *J. Biol. Chem.*, 1983, **258**, 7155–7160.
50. Y. Soma, V. Dvonch and G. R. Grotendorst, Platelet-derived growth factor AA homodimer is the predominant isoform in human platelets and acute human wound fluid, *FASEB J.*, 1992, **6**, 2996–3001.
51. K. Usuki, L. Norberg and E. Larsson, *et al.*, Localization of platelet-derived endothelial cell growth factor in human placenta and purification of an alternatively processed form, *Cell Regul.*, 1990, 577–584.
52. Y. Oka and D. N. Orth, Human plasma epidermal growth factor/beta-urogastrone is associated with blood platelets, *J. Clin. Invest.*, 1983, **72**, 249–259.
53. I. Muller, S. Kordowich and C. Holzwarth, *et al.*, Animal serum-free culture conditions for isolation and expansion of multipotent mesenchymal stromal cells from human BM, *Cytotherapy.*, 2006, **8**, 437–444.
54. T. Felka, R. Schäfer and B. Schewe, *et al.*, Hypoxia reduces the inhibitory effect of IL-1beta on chondrogenic differentiation of FCS-free expanded MSC, *Osteoarthritis Cartilage*, 2009, **17**, 1368–1376.

55. N. Tzaribachev, M. Vaegler and J. Schaefer, *et al.*, Mesenchymal stromal cells, a novel treatment option for steroid-induced avascular osteonecrosis, *Isr. Med. Assoc. J.*, 2008, **10**, 232–234.
56. M. J. Zuscik, M. J. Hilton and X. Zhang, *et al.*, Regulation of chondrogenesis and chondrocyte differentiation by stress, *J. Clin. Invest.*, 2008, **118**, 429–438.
57. K. W. Kavalkovich, R. E. Boynton and J. M. Murphy, *et al.*, Chondrogenic differentiation of human mesenchymal stem cells within an alginate layer culture system, *In vitro Cell. Dev. Biol.*, 2009, **38**, 457–466.
58. S. T. B. Ho, S. M. Cool and J. H. Hui, *et al.*, The influence of fibrin based hydrogels on the chondrogenic differentiation of human bone marrow stromal cells, *Biomaterials*, 2010, **31**, 38–47.
59. D. T. Covas, R. A. Panepucci and A. M. Fontes, *et al.*, Multipotent mesenchymal stromal cells obtained from diverse human tissues share functional properties and gene-expression profile with CD146+ perivascular cells and fibroblasts, *Exp. Hematol.*, 2008, **36**, 642–654.
60. S. Baglioni, M. Francalanci and R. Squecco, *et al.*, Characterization of human adult stem-cell populations isolated from visceral and subcutaneous adipose tissue, *FASEB J.*, 2009, **23**, 3494–3505.
61. C. M. Kolf, E. Cho and R. S. Tuan, Mesenchymal stromal cells. Biology of adult mesenchymal stem cells, regulation of niche, self-renewal and differentiation, *Arthritis Res. Ther.*, 2007, **9**, 204.
62. V. L. Battula, P. M. Bareiss and S. Treml, *et al.*, Human placenta and bone marrow derived MSC cultured in serum-free, b-FGF-containing medium express cell surface frizzled-9 and SSEA-4 and give rise to multilineage differentiation, *Differentiation*, 2007, **75**, 279–291.
63. C. M. DiGirolamo, D. Stokes and D. Colter, *et al.*, Propagation and senescence of human marrow stromal cells in culture, a simple colony-forming assay identifies samples with the greatest potential to propagate and differentiate, *Br. J. Haematol.*, 1999, **107**, 275–281.
64. S. A. Kuznetsov, P. H. Krebsbach and K. Satomura, *et al.*, Single-colony derived strains of human marrow stromal fibroblasts form bone after transplantation *in vivo*, *J. Bone Miner. Res.*, 1997, **12**, 1335–1347.
65. A. Wilson, M. J. Murphy and T. Oskarsson, *et al.*, c-Myc controls the balance between hematopoietic stem cell self-renewal and differentiation, *Genes Dev.*, 2004, **18**, 2747–2763.
66. A. Wilson and A. Trumpp, Bone-marrow hematopoietic stem-cell niches, *Nature Rev. Immunol.*, 2006, **6**, 93–106.
67. S. W. O'Driscoll, J. S. Fitzsimmons and C. N. Commisso, Role of oxygen tension during cartilage formation by periosteum, *J. Orthop. Res.*, 1997, **15**, 682–687.
68. W. Khan, A. Adesida and T. Hardingham, Hypoxic conditions increase hypoxia-inducible transcription factor 2alpha and enhance chondrogenesis in stem cells from the infrapatellar fat pad of osteoarthritis patients, *Arthritis Res. Ther.*, 2007, **9**, R55.

69. C. Murphy, B. Thoms and R. Vaghjiani, *et al.*, Hypoxia. HIF-mediated articular chondrocyte function, prospects for cartilage repair, *Arthritis Res. Ther.*, 2009, **11**, 213.

70. R. J. Egli, J. D. Bastian and R. Ganz, *et al.*, Hypoxic expansion promotes the chondrogenic potential of articular chondrocytes, *J. Orthop. Res.*, 2008, **26**, 977–985.

71. R. Kuroda, K. Ishida and T. Matsumoto, *et al.*, Treatment of a full-thickness articular cartilage defect in the femoral condyle of an athlete with autologous bone-marrow stromal cells, *Osteoarthritis Cartilage*, 2007, **15**, 226–231.

72. J. D. Kisiday, M. Jin and M. A. DiMicco, *et al.*, Effects of dynamic compressiveloading on chondrocyte biosynthesis in self-assembling peptide scaffolds, *J. Biomech.*, 2004, **37**, 595–604.

73. S. D. Waldman, D. C. Couto and S. J. Omelon, *et al.*, Effect of sodium bicarbonate on extracellular pH, matrix accumulation, and morphology of cultured articular chondrocytes, *Tissue Eng.*, 2004, **10**, 1633–1640.

74. M. Hoberg, M. Rudert and A. K. Aicher, Improved chondrogenic differentiation of bone marrow-derived mesenchymal stromal cells in a spongiform collagen scaffold enriched with chondroitin sulfate, unpublished results.

75. M. Wollenweber, H. Domaschke and T. Hanke, *et al.*, Mimicked bioartificial matrix containing chondroitin sulphate on a textile scaffold of poly(3-hydroxybutyrate) alters the differentiation of adult human mesenchymal stem cells, *Tissue Eng.*, 2006, **12**, 345–359.

76. R. F. Klees, R. M. Salasznyk and D. F. Ward, *et al.*, Dissection of the osteogenic effects of laminin-332 utilizing specific LG domains: LG3 induces osteogenic differentiation, but not mineralization, *Exp. Cell Res.*, 2008, **314**, 763–773.

77. A. J. Engler, S. Sen and H. L. Sweeney, *et al.*, Matrix elasticity directs stem cell lineage specification, *Cell*, 2006, **126**, 677–689.

78. A. Maroudas, Physico-chemical properties of articular cartilage, in *Adult Articular Cartilage*, M. A. R. Freeman, ed. Pitman Medical, Tunbridge Wells, UK, 1979, pp. 215–290.

79. M. Venn and A. Maroudas, Chemical composition and swelling of normal and osteoarthrotic femoral head cartilage. I. Chemical composition, *Ann. Rheum. Dis.*, 1977, **36**, 121–129.

80. E. B. Hunziker, Articular cartilage structure in humans and experimental animals, in *Articular Cartilage and Osteoarthritis*, K. E. Kuettner, R. Schleyerbach, J. G. Peyron and V. C. Hascall, ed. Raven Press, New York, 1992, pp. 183–199.

81. M. D. Buschmann, Y. A. Gluzband and A. J. Grodzinsky, *et al.*, Mechanical compression modulates matrix biosynthesis in chondrocyte/agarose culture, *J. Cell. Sci.*, 1995, **108**, 1497–1508.

82. R. M. Schinagl, D. Gurskis and A. C. Chen, *et al.*, Depth-dependent confined compression modulus of full-thickness bovine articular cartilage, *J. Orthop. Res.*, 1997, **15**, 499–506.

83. B. D. Elder and K. A. Athanasiou, Hydrostatic pressure in articular cartilage tissue engineering: from chondrocytes to tissue regeneration, *Tissue Eng. Part B Rev.*, 2009, **15**, 43–53.

84. D. L. Butler, S. A. Goldstein and F. Guilak, Functional tissue engineering: the role of biomechanics, *J. Biomech. Eng.*, 2000, **122**, 570–575.

85. K. E. Wescoe, R. C. Schugar and C. R. Chu, *et al.*, The role of the biochemical and biophysical environment in chondrogenic stem cell differentiation assays and cartilage tissue engineering, *Cell Biochem. Biophys.*, 2008, **52**, 85–102.

86. S. D. Waldman, C. G. Spiteri and M. D. Grynpas, *et al.*, Effect of biomechanical conditioning on cartilaginous tissue formation *in vitro*, *J. Bone Joint Surg. Am.*, 2003, **85A**(Suppl 2), 101–105.

87. T. Furukawa, D. R. Eyre and S. Koide, *et al.*, Biochemical studies on repair cartilage resurfacing experimental defects in the rabbit knee, *J. Bone Joint Surg. Am.*, 1980, **62**, 79–89.

88. H. J. Mankin, The response of articular cartilage to mechanical injury, *J. Bone Joint Surg. Am.*, 1982, **64**, 460–466.

89. S. Nehrer, M. Spector and T. Minas, Histologic analysis of tissue after failed cartilage repair procedures, *Clin. Orthop. Relat. Res.*, 1999, **365**, 149–162.

90. M. A. Davis, W. H. Ettinger and J. M. Neuhaus, *et al.*, The association of knee injury and obesity with unilateral and bilateral osteoarthritis of the knee, *Am. J. Epidemiol.*, 1989, **130**, 278–288.

91. C. R. Lee, A. J. Grodzinsky and M. Spector, Biosynthetic response of passaged chondrocytes in a type II collagen scaffold to mechanical compression, *J. Biomed. Mater. Res. A*, 2003, **64**, 560–569.

92. M. L. Gray, A. M. Pizzanelli and A. J. Grodzinsky, *et al.*, Mechanical and physiochemical determinants of the chondrocyte biosynthetic response, *J. Orthop. Res.*, 1988, **6**, 777–792.

93. M. L. Gray, A. M. Pizzanelli and R. C. Lee, *et al.*, Kinetics of the chondrocytebiosynthetic response to compressive load and release, *Biochim. Biophys. Acta*, 1989, **991**, 415–425.

94. I. L. Jones, A. Klamfeldt and T. Sandstrom, The effect of continuous mechanical pressure upon the turnover of articular cartilage proteoglycans *in vitro*, *Clin. Orthop. Relat. Res.*, 1982, **165**, 283–289.

95. R. Schneiderman, D. Keret and A. Maroudas, Effects of mechanical and osmotic pressure on the rate of glycosaminoglycan synthesis in the human adult femoral head cartilage, an *in vitro* study, *J. Orthop. Res.*, 1986, **4**, 393–408.

96. T. M. Quinn, A. J. Grodzinsky and M. D. Buschmann, *et al.*, Mechanical compression alters proteoglycan deposition and matrix deformation around individual cells in cartilage explants, *J. Cell. Sci.*, 1998, **111**, 573–583.

97. R. L. Sah, Y. J. Kim and J. Y. Doong, *et al.*, Biosynthetic responseof cartilage explants to dynamic compression, *J. Orthop. Res.*, 1989, **7**, 619–636.

98. D. A. Lee and D. L. Bader, Compressive strains at physiological fre-
    quencies influence the metabolism of chondrocytes seeded in agarose,
    J. Orthop. Res., 1997, **15**, 181–188.
99. C. J. Hunter and M. E. Levenston, The influence of repair tissue
    maturation on the response to oscillatory compression in a cartilage
    defect repair model, *Biorheology*, 2002, **39**, 79–88.
100. T. T. Chowdhury, D. L. Bader and J. C. Shelton, *et al.*, Temporal reg-
    ulation of chondrocyte metabolism in agarose constructs subjected to
    dynamic compression, *Arch. Biochem. Biophys.*, 2003, **417**, 105–111.
101. T. Davisson, S. Kunig and A. Chen, *et al.*, Static and dynamic com-
    pression modulate matrix metabolism in tissue engineered cartilage,
    J. Orthop. Res., 2002, **20**, 842–848.
102. R. L. Mauck, S. L. Seyhan and G. A. Ateshian, *et al.*, Influence of seeding
    density and dynamic deformational loading on the developing structure/
    function relationships of chondrocyte-seeded agarose hydrogels, *Ann.
    Biomed. Eng.*, 2002, **30**, 1046–1056.
103. E. Wernike, Z. Li and M. Alini, *et al.*, Effect of reduced oxygen tension
    and long-term mechanical stimulation on chondrocyte-polymer con-
    structs, *Cell Tissue Res.*, 2008, **331**, 473–483.
104. J. D. Kisiday, J. H. Lee and P. N. Siparsky, *et al.*, Catabolic responses
    of chondrocyte-seeded peptide hydrogel to dynamic compression, *Ann.
    Biomed. Eng.*, 2009, **37**, 1368–1375.
105. M. Jin, E. H. Frank and T. M. Quinn,, *et al.*, Tissue shear deformation
    stimulates proteoglycan and protein biosynthesis in bovine cartilage
    explants, *Arch. Biochem. Biophys.*, 2001, **395**, 41–48.
106. M. A. Soltz and G. A. Ateshian, Experimental verification and theoretical
    prediction of cartilage interstitial fluid pressurization at an impermeable
    contact interface in confined compression, *J. Biomech.*, 1998, **31**, 927–934.
107. F. Liu, M. Kozanek and A. Hosseini, *et al.*, *In vivo* tibiofemoral cartilage
    deformation during the stance phase of gait, *J. Biomech.*, 2010, **43**, 658–665.
108. B. D. Elder and K. A. Athanasiou, Synergistic and additive effects of
    hydrostatic pressure and growth factors on tissue formation, *PLoS One*,
    2008, **3**, e2341.
109. J. C. Hu and K. A. Athanasiou, The effects of intermittent hydrostatic
    pressure on self-assembled articular cartilage constructs, *Tissue Eng.*,
    2006, **12**, 1337–1344.
110. V. Terraciano, N. Hwang and L. Moroni, *et al.*, Differential response of
    adult and embryonic mesenchymal progenitor cells to mechanical com-
    pression in hydrogels, *Stem Cells*, 2007, **25**, 2730–2738.
111. R. L. Mauck, X. Yuan and R. S. Tuan, Chondrogenic differentiation and
    functional maturation of bovine mesenchymal stem cells in long-term
    agarose culture, *Osteoarthritis Cartilage*, 2006, **14**, 179–189.
112. Y. Jung, S. H. Kim and Y. H. Kim, *et al.*, The effects of dynamic and
    three-dimensional environments on chondrogenic differentiation of bone
    marrowstromal cells, *Biomed. Mater.*, 2009, **4**, 55009.

Stem Cell-based Replacement Tissue for Heart Repair

AYELET LESMAN AND SHULAMIT LEVENBERG

Bio-Medical Engineering, Technion, Haifa, Israel

13.1 Introduction

The regenerative capacity of human myocardium is insufficient to compensate for the severe loss of cardiomyocytes (CMs) after heart attack (myocardium infarction) or other myocardial diseases. The cell replacement strategy is a promising new approach to treat the infarcted myocardium and is based on transplantation of new pools of healthy cells.

The rationale behind cell transplantation assumes that an increase in the number of functional cells in the affected area can improve the mechanical and conductive properties of the region either directly or *via* bioactive molecules they secrete. To date, many types of stem cells have been proposed to provide the basis for therapeutic management of infarcted myocardium. In this chapter we focus on the derivation of human stem cells for use in clinical applications of this nature. We discuss implementation of therapies based on human embryonic stem cells (hESC) and induced pluripotent stem cells (iPS), and review the potential of adult stem cells, including bone marrow (BM) cells (hematopoietic and mesenchymal stem cells), resident cardiac stem cells and skeletal stem cells (satellite stem cells) in such clinical protocols (Figure 13.1).

Recent breakthroughs in tissue engineering disciplines allowing for the design of biomaterial-based heart tissue constructs have transformed this avenue into a promising approach toward advancing myocardial repair. The various technical strategies and potential biomaterials are discussed, together

Stem Cell-Based Tissue Repair
Edited by Raphael Gorodetsky and Richard Schäfer
© The Royal Society of Chemistry 2011
Published by the Royal Society of Chemistry, www.rsc.org

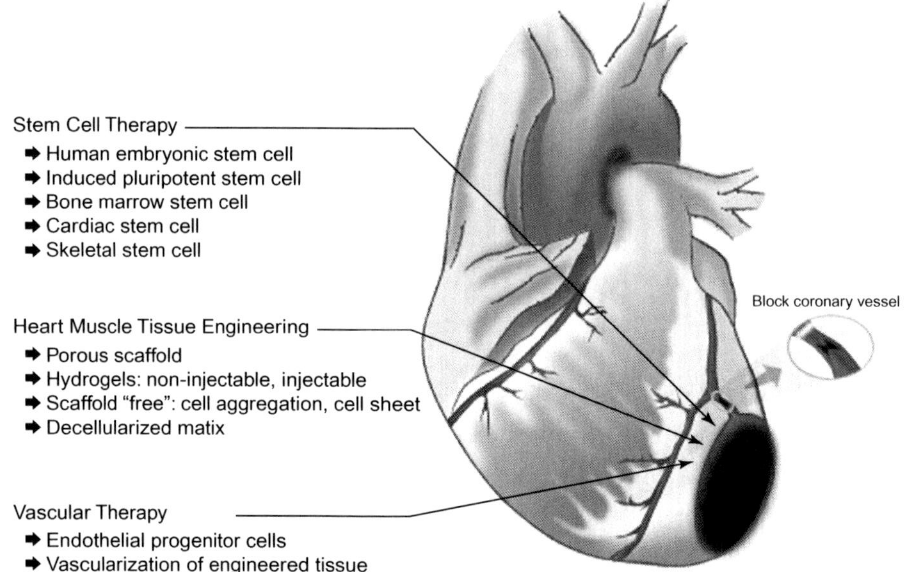

Figure 13.1   Schematic view of the infarcted heart along with current available stem cell therapies discussed in this chapter: various stem cell population; heart muscle tissue engineering; and vascular therapy.

with the prospects of cardiac vascular therapy by means of human endothelial progenitors cells derived from hESC or from the circulation, BM or cord blood. Lastly, as tissue engineering techniques continue to optimize the quality of artificial tissue, vascularization of engineered tissues using hESC derived endothelial cells derived from (hES-ECs) in efforts to augment construct viability, thickness and architecture are introduced (Figure 13.1).

13.2   Cell Therapy in the Heart using Stem Cells

13.2.1   Human Embryonic Stem Cells

hESCs are considered one of the most promising sources of human cells for cardiac repair. Derived from the early embryonic blastocyte stage, these cells bear the capacity to proliferate for prolonged periods *in vitro* and to differentiate to each of the three germ layers (mesoderm, endoderm and ectoderm) when provided with the appropriate cues. In this manner, hESCs can give rise to all cell types of the body.

The first derivation of mouse embryonic stem cells was reported in 1981[1,2] and set the stage for isolation of hESCs almost 20 years later.[3] Since then, extensive research efforts have been invested toward induction and purification of a CM population from hESC pools. Gepstein and colleagues[4] were the first to describe the formation of hESC-derived CMs (hES-CMs) and have been followed by additional groups utilizing different approaches.[5] These *in vitro* derived CMs have been extensively characterized and have been shown to feature unique early-stage cardiac potential. They contain sarcomeric structural patterning, spontaneously pulsate, structurally and functionally integrate with pre-existing cardiac tissue, respond to pharmacological agents and express cardiac-specific genes.[4–8] Moreover, hES-CMs exhibit high proliferative capacity both *in vitro*[9] and *in vivo*,[10] in contrast to the limited proliferative properties reported for mouse ES-CMs. In a study we performed, the number of proliferating hES-CMs was significantly augmented in cocultures with endothelial cells and fibroblasts, seemingly *via* paracrine signaling. Furthermore, upon transplantation into animal hearts, hES-CMs survived, matured and formed functional gap junctions between one another.[11–15]

Derivation of CMs from hESCs is typically accomplished *via* the embryoid body differentiation system. Undifferentiated hESCs are propagated under feeder-free conditions[5] or on a mitotically inactivated fibroblast feeder layer (usually MEF).[3] The undifferentiated hESCs are then removed to suspension to create three-dimensional (3D) aggregates termed embryoid bodies (EBs). Purification of CMs from a mixed population of differentiating cells within the EBs can be achieved through mechanical dissection, Percoll gradient technique, or fluorescence activated cell sorting (FACS). For mechanical dissection, EBs are plated on gelatin-coated culture dishes to give rise to spontaneously beating cells within 4–22 days in 8–10% of the EBs, which can then be separated manually.[4]

However, this technique provides a small CM output and results in an insufficient degree of purity for clinical purposes. Moreover, since the selection is based on beating activity, mature ventricular myocytes—which are more desirable for myocardium repair—are not obtained. FACS-based CM isolation is still limited as appropriate CM-specific surface markers have yet to be identified. Even so, generation of stable transgenic hESC lines expressing a reporter gene (GFP) under the control of a cardiac-specific promoter has been proven successful in effectively isolating pure CM populations using FACS.[16–18]

Two techniques have been designed to stimulate hESC differentiation to CMs. The first is based on the role described for the endoderm in mesodermal differentiation toward the cardiac lineage. Thus, coculturing of hESCs and endodermal-like cells, or cells which generate endodermal signals, results in extensive differentiation toward beating CMs, as reported by Mummery and colleagues.[6] The second approach provides growth factors critical to heart development such as bone morphogenetic proteins (BMP-2 and BMP-4),[19] Activin A[10,20] or basic fibroblast growth factor (bFGF)[21] to promote CM induction from hESCs.[19] Moreover, it was shown that addition of ascorbic acid[22] and 5-azacytidine[23] to culture medium can further induce cardiomyogenesis.

To date, hESCs have not been tested for their potential in repair of human cardiac muscle. However, a number of pre-clinical trials in small rodents have evaluated the contribution of differentiated hESCs in reversal of myocardial infarction.[10–11,20,24,25] Laflamme and coworkers[10] monitored the ability of hES-CMs to form a human myocardium graft after injection to the infarcted heart. They identified a pro-survival protein cocktail that supported CM survival after transplantation, and demonstrated cell proliferation and graft area expansion. Furthermore, the treatment led to attenuation of ventricular dilation and preserved contractile function. Similarly, Caspi *et al.*[11] reported the formation of stable CM grafts of functional benefit after injection of hES-CMs to infarcted rat hearts. In line with these reports, van Laake *et al.*[25] described a temporal but functional improvement of infarcted hearts following hES-CMs transplantation.

13.2.2 Induced Pluripotent Stem Cells

The advent of techniques allowing for generation of iPS cells from adult somatic cells presented a new milestone in developmental research and the study of diseases. iPS cells were first derived from mice in 2006[26] and shortly afterwards in 2007 from human sources.[27,28] Reprogramming was achieved by overexpression of the Oct3/4, Sox2, Klf4 and c-Myc transcriptional factors, yielding cells similar in morphology, gene expression and differentiation potential, both *in vitro* and *in vivo*, to hESCs.[27,28] This advancement enables preparation of patient-specific cell samples for stem cell therapies. However, both low efficiency ($\sim 0.01\%$) and the need for viruses to induce the required gene expression limit the efficacy and practicality of iPS production. Thus, much contemporary research concentrates on stimulating pluripotency repro-gramming without the use of viruses.[29–31]

In vitro differentiation of iPS toward the cardiac lineage has been the focus of much attention in recent years. Mouse iPS-derived CMs exhibited behavior similar to CMs derived from the well-established murine ES cell line.[32–34] More recently, generation of human iPS-derived CMs (iPS-CMs) has also descri-bed[35–40] and reviewed;[41–43] iPS-CMs were shown to express cardiac-specific genes, and markers (*i.e.* Troponin I, sarcomeric α actinin and Connexin 43) to respond to chronotropic agents, and to form functional syncytium *in vitro*. Moreover, the similarity of their response and that of the ESC-CMs to cardio-vascular drugs *in vitro*[37,38] qualify them as a powerful model for *in vitro* cardiac electrophysiological studies and drug screening. To this end, the potential of human cord blood cells as an appropriate immature cell source for repro-gramming has been evaluated, with the reasoning that these cells have not accrued the number of mutations often accumulated in somatic cells over time. These cells demonstrated human iPS cell-generating potential and were driven to form functional CMs *in vitro*.[36]

While their capacity to generate CMs *in vitro* has been proven, with clear advantages over hESC, expected reports of transplantation studies with iPS-CMs will provide further information regarding their true potential toward treatment of infarcted myocardium as well as their safety. Clinical application of iPS will provide for genetically appropriate tissue grafts, remove many political and ethical obstacles, and usher in a new era in stem cell research.

13.2.3 Bone Marrow Stem Cells

The BM is a heterogeneous tissue consisting of stem cell subpopulations which can be characterized into three main groups:

(1) Hematopoietic stem cells (HSCs), which comprise approximately 0.001–0.01% of total BM cell population and are the predecessors of all blood cells in the body;
(2) Mesenchymal stem cells (MSCs), which make up ~0.01% of the total BM cell population; and
(3) Committed progenitor cells, which include endothelial progenitor cells (EPCs).

The next paragraphs focus on HSCs and MSCs. EPCs are discussed in Section 13.4.

The potential of BM cells to differentiate into CMs *in vitro* or *in vivo* has been a major focus of researchers in recent years. Although no consensus currently exists, BM studies have progressed remarkably fast from small animal studies to human clinical trials.

Findings which have described BM cell differentiation toward CMs *in vitro*[44–50] have encouraged researchers around the globe to assess their therapeutic potential toward repair of myocardial infarction. While the true CM-generating potential of BM cells injected into the infarcted myocardium remains controversial,[51–55] they clearly lead to improved cardiac function. Orlic and collaborators were the first to demonstrate that direct injection of BM stem cells into ischemic regions within the heart can lead to a rise in CM levels and enhanced heart function.[54,55] Several other studies have correlated differentiation of transplanted BM cells into functional CMs with improved cardiac function.[56,57] In contrast, several groups have reported limited BM cell plasticity *in vivo,* suggesting a paracrine effect independent of cell therapy outcome.[51–53] In these studies, BM cells were reported to differentiate into mature hematopoietic forming blood cells, while no CMs were detected in the infarcted myocardium. Such varied results can arise from BM population heterogeny, disparities in cell harvesting techniques and yields, and the timing of cell injection.

A review of the clinical status of injected BM stem cells uncovers significantly discrepant results (see ref. 58 for a summary table). More specifically, the REPAIR-AMI[59] and BOOST[60] trials reported improvement of left ventricular ejection fraction, where any differences noted between control and experimental groups became insignificant by the 18-month follow-up examination.[61] A recent meta-analysis[62] summarized improved left ventricular ejection fraction, decreased infarct size and reduced left ventricular end systolic volume upon transplantation of BM cells in a total of 18 eligible studies (999 patients). In contrast, a doubled-blinded, randomized controlled trial[63] demonstrated a reduction in infarct size but no change in left ventricular ejection fraction. These contrasting results can be rooted in disparities in cell harvesting techniques, the timing of cell injection, delivery methods, diversity of patient demographics and the heterogeny of the BM population. Although clinical works have failed to provide conclusive results regarding the advantages of BM-based transplantation toward renewed cardiac function, in all cases the studies indicate that such transplantations are feasible, safe and at least moderately beneficial. Furthermore, the majority of these clinical trials involved autologous BM transplants, bypassing the need to establish donor–recipient compatibility.

MSCs reside in the BM stroma, but have recently been isolated from adipose tissue as well.[64] MSCs are highly adhesive and lack hematopoietic markers, making their isolation fairly simple. Moreover, MSCs are less immunogenic than other cell sources—an additional cost-effective advantage in cell therapy protocols. These cells differentiate into different cell types including osteoblasts, adipocytes, chondrocytes and skeletal muscle cells,[65] and have been shown to give rise to CMs *in vitro*.[48,66] These recent findings have initiated transplantation studies exploring the potential of MSC-derived CMs in treatment of the infarcted heart.

Several pre-clinical studies have described improved ventricular function upon MSC transplantation to the infarcted heart.[67–73] The underlying mechanisms remain unclear but have been suggested to be paracrine-dependent.[67,69,71] A clinical study has already been performed with the use of MSCs.[74] In this trial, 69 patients with acute myocardial infarction underwent intracoronary delivery of autologous BM MSCs and demonstrated improved left ventricular function within six months. Several ongoing trials testing the administration of allogenic and autologous MSCs will influence future planning of MSC applications in heart repair.

13.2.4 Resident Cardiac Stem Cells

The adult heart has traditionally been considered a fully differentiated organ lacking self-regenerative capacities. However, recent findings have discovered a pool of progenitor cells embedded within the heart, which possess proliferative properties and the ability to differentiate into CMs, endothelial cells and smooth muscle cells. The cardiac progenitor cells have been classified into four groups and include:

- cardiac side population (SP) cells;
- c-kit$^+$ cells;
- Islet-1$^+$ cells; and
- stem cell antigen 1 (Sca-1$^+$) cells.

SP cells constitute about 1% of the adult heart[75,76] and are characterized by exclusion of metabolic dyes such as Hoechst and Rhodamine. c-Kit[+] cells express a c-Kit-specific receptor and comprise approximately 0.01% of all myocyte cells.[77] Cardiac stem cells derived from primary cultures of human biopsy patients have been reported to form cardiospheres, where 30% of the cells expressed c-Kit and 20% exhibited full cardiogenic potential.[78] Islet-1[+] cells have been successfully isolated from infantile rat hearts[79,80] and more recently from the human adult hearts as well,[81] where the right atrium was described as the best source for both c-Kit[+] cells and Islet-1[+] cells.[81] Sca-1[+] cells [75] have been isolated from murine hearts, but have not been identified in human hearts. Currently, it is still unclear whether each progenitor cell type is distinct from the next or merely represents progressive stages of the same cell source.

Use of cardiac progenitor populations to generate CMs *in vitro* and to repair myocardium has been the focus of many investigations in recent years. While SP cells have been shown to differentiate into CMs both *in vitro* and *in vivo*,[82–84] their therapeutic potential toward the infarcted heart has not been extensively evaluated. c-Kit[+] cells have been described to regenerate infarcted myocardium and improve cardiac function after transplantation.[81,85] Islet1[+] cells were shown to differentiate *in vitro* to CMs upon coculture with neonatal CMs. Sca-1 + cells isolated from murine hearts differentiated to CMs both *in vitro* and, after cell transplantation, *in vivo*.[86,87] Although unique, resident cardiac stem cells are rare and insufficient to regenerate the infarcted myocardium naturally. Future research will require evaluation of their autologous potential in cell therapy of the heart.

13.2.5 Skeletal Satellite Stem Cells

Satellite cells are adult skeletal stem cells that reside beneath the basal membrane of skeletal muscle tissue. In cases of tissue injury, these cells proliferate and differentiate toward myoblast cells which subsequently fuse to form myotubes and eventually generate mature skeletal muscle fibers. While these cells are further committed than ESCs, they retain a significant degree of multipotency and can give rise to cardiogenic, neurogenic, osteogenic and adipogenic lineages.[88,89]

Skeletal stem cells were among the first stem cell source to be considered for heart repair and were the first to be transplanted in human clinical trials. The ease at which they can be harvested and their notable proliferative capabilities *in vitro* allow for simple expansion and use in autologous transplantation procedures. In addition, skeletal stem cells are less prone to form teratoma and less sensitive to ischemic conditions, making them ideal candidates for cell therapy protocols. However, their inability to couple with host myocardium due to the absence of gap-junctional proteins raises concerns of arrhythmias following transplantation. In addition, these cells do not form CMs *in vivo* and remain committed to the skeletal muscle lineage.[90]

Transplantation studies of skeletal muscle stem cells in animal models demonstrated successful skeletal myoblast grafting which proved beneficial to ventricular contractile functioning.[91,92] Clinical studies applying skeletal stem cells have been well-reviewed by Murry and colleagues,[93] and conclude that hundreds of autologous skeletal stem cells can be generated after expansion *in vitro* and can be effectively engrafted in the scar myocardium area. Conclusions regarding functional heart improvement as well as safety parameters remain to be fully determined. The recent MAGIC clinical trail (2008) reported disappointing results regarding functional heart improvement after injection of autologous skeletal cells to the infarcted heart.[94] This clinical study demonstrated the feasibility of implanting autologous skeletal myoblasts in infarcted hearts, but failed to determine whether this procedure was functionally effective and arrhythmogenic.[94]

13.3   Stem Cell-based Heart Muscle Engineering

Although the CM-differentiating potential and therapeutic capacities of various cell types were presented in the previous sections, cell survival and graft retention rates were not introduced. More than 70% of injected cells die or are washed away within the first days of injection,[95] posing a critical obstacle to direct injection of cells to the hostile infarct environment. However, combination of cells with scaffolds designed to generate engineered heart patches can reduce cell loss and preserve cells in the desired engraftment position *in vivo*. Moreover, tissue engineering principles offer control of tissue properties before transplantation (*i.e.* differentiation, maturation, organization).

This section provides an insight into the main scaffold-based tissue engineering techniques available today. The use of polymeric, biodegradable porous scaffolds on which cells are seeded, adhere and proliferate is described. These scaffolds are expected to degrade with time, leaving space for the tissue to develop.

An introduction to non-injectable and injectable hydrogels is also provided and their use for cell entrapment is summarized. As hydrogels have a high water content and contain natural proteins and/or synthetic polymers, they provide an extracellular matrix (ECM)-like environment appropriate for prolonged support of cell populations.

Scaffold 'free' tissue engineering approaches are introduced including the cell sheets and cell aggregation techniques. In the cell sheet approach, confluent cells can be harvested from a culture dish as an intact layer and serially overlaid to create a 3D tissue construct without any need for biomaterials. Cell aggregation employs cellular self-assembly within rotating shakers to create 3D tissue patches, also without the use of biomaterials. Lastly, an additional approach which isolates natural acellular ECM components from native tissues to serve as scaffolding material is considered.

13.3.1 Porous Scaffolds

Porous matrices are commonly employed as tissue engineering scaffolds. In such applications, cells are seeded directly into the scaffold, filling the micropores, proliferating, adhering to the scaffold walls and assuming the shape of the scaffold support. Particulate leaching methods using microspheres or salt grains can tightly control the scaffold microstructure, and regulate porosity and interconnectivity between pores.[96,97]

Synthetic polymers such as polylactic acid (PLA), polyglycolic acid (PGA) or the co-polymer (PLA-PGA)[98,99] often serve as basic scaffold materials in tissue engineering protocols. Other commonly used porous scaffolds apply the synthetic materials, polyglycerol sebacate (PGS)[100] and polycarprolactone (PCL).[96] In many cases, scaffolds are supplemented with an ECM matrix (*i.e.* matrigel, collagen or fibronectin) to enhance cell adhesion to material surfaces. Synthetic polymers allow for easy tailoring of mechanical properties, morphology definition and control of degradation kinetics. Porous scaffolds can also be derived from natural sources such as collagen, gelatin, fibrin and alginate,[101] all of which facilitate cell adhesion and proliferation. Synthetic and natural polymers are often combined to exploit the advantages of the each, resulting in tight regulation of mechanical scaffold properties coupled with biological cross-talk between cells and their biologically supporting surface.

In a work performed in our laboratory in collaboration with Gepstein and colleagues,[102] hES-CMs, endothelial cells and embryonic fibroblast cells were seeded within a poly-l-lactic acid/poly lactic-co-glycolic acid (PLLA/PLGA) (50/50) scaffold to generate a 3D human, vascularized cardiac muscle construct. The cardiac construct exhibited synchronic beating and was occupied with differentiated CMs arranged in a sarcomeric pattern; tissue construct contractions exhibited physiologically relevant responsiveness to both positive and negative chronotropic agents.[102] Furthermore, addition of endothelial cells and embryonic fibroblast cells to the hES-CM-embedded scaffolds resulted in formation of intense inter-CM vascular networks. Moreover, transplantation of these vascularized human cardiac constructs resulted in tight attachment to the rat myocardium, intense graft vascularization, and formation of functional human blood vessels derived from the pre-transplanted vessels.[14] The hES-CMs were shown to survive and mature to some degree, exhibiting elongation pattern and alignment.[14]

13.3.2 Hydrogels

The use of synthetic hydrogels such as polyethylene-glycol (PEG) or natural hydrogels such as fibrin, collagen, alginate or chitosan have been extensively tested for their cell-supporting properties.[103] In general, hydrogels closely imitate cardiac ECM by providing a soft physiological-like environment. As in the case of porous scaffolds, materials from natural sources provide essential biological cell-scaffold cross-talk, whereas the synthetic materials offer tight control of mechanical properties. Thus, hybrid hydrogels combining both synthetic and natural polymers have been proposed, as in the case of the

PEG-fibrinogen hydrogel.[104] Hydrogels can be polymerized *in vitro*, cultured for the desired time *ex vivo* and then sutured to the infarcted area (non-injectable hydrogels), or injected with cells into the body as a liquid material and then polymerized *in situ via* ultraviolet (UV) light, temperature or pH differences, *etc.*

13.3.2.1 Non-injectable Hydrogels

Zimmermann *et al.*[105] have reported use of the non-injectable hydrogel approach. They describe subjection of an engineered heart tissue composed of neonatal CMs or embryonic chick CMs embedded within a collagen and matrigel matrix to mechanical stretching during *in vitro* cultivation. The engineered heart tissue displayed features comparable with differentiated myocardium, with interconnected, longitudinally oriented cardiac cells.

More recently, a large (thickness/diameter: 1–4 mm/15 mm) force-generating engineered heart tissue prepared from neonatal rat heart cells was described.[106] After its transplantation on rat myocardial infarcts, the engineered heart tissue showed electrical coupling with the native myocardium, with no evidence of arrhythmia. Moreover, the grafted engineered heart tissue prevented further ventricular dilation, induced systolic wall thickening of the infarcted myocardial segments and improved fractional area shortening of infarcted hearts.

On the basis of this report, Guo *et al.*[107] generated cardiac tissue using mESC-derived CMs. mESC-CMs were embedded in a ring shape within a Type I collagen and matrigel matrix, and stretched *in vitro* for seven days; they displayed spontaneous beating movements and response to physical and pharmaceutical stimulation *in vitro*. No teratomas were detected upon subcutaneous transplantation in nude mice for four weeks.

Simpson and associates[108] used a Type I collagen human MSC-embedded matrix to produce a cardiac patch which was transplanted to an infarcted rat heart. While cardiac remodeling was achieved and 23% of the engrafted cells were detectable one week after implantation, they were no longer detectable after four weeks.

Shapira-Schweitzer *et al.*[104] recently described the use of hybrid PEG and fibrinogen hydrogel to support hESC-CMs in 3D tissue formation. The PEG-fibrinogen hydrogel polymerized after exposure to UV light and supported maturation of embedded hES-CM cells within 10–14 days in culture. The matured cells expressed cardiac-specific markers and responded to pharmacological agents.

13.3.2.2 Injectable Hydrogels

Injectable hydrogels facilitate *in vivo* repair of injured tissue by providing a matrix support for cell retention, migration, proliferation and neovascularization. Application of injectable scaffolds is less invasive than transplantation of preformed tissue constructs as they can be delivered *via* catheter and are therefore more clinically appealing.

The pioneering works of the Leor and Cohen groups[109,110] first described injectable alginate hydrogels for repair of infarcted myocardium, where delivery of acellular alginate biomaterial imparted beneficial influence on rat cardiac function.[109] More recently, they demonstrated intracoronary injection of alginate biomaterial to the swine heart, which subsequently reversed left ventricular enlargement and increased scar thickness.[110]

Kofidis and colleagues[111] delivered mESCs in conjunction with injectable matrigel to the infarcted mouse heart. The injectable tissue solidified at body temperature, assumed the geometry of the infarcted zone and led to superior heart functioning when compared to controls animals. A unique, injectable cyclodextrin/poly(ethyleneglycol)-b-polycaprolactone-(dodecanedioic acid)-poly-caprolactone-poly(ethylene glycol) (MPEG-PCL-MPEG) solution designed recently by Wang *et al.* to deliver BM stem cells[112] induced increased cell retention in the graft and vessel density, and prevented scar expansion compared to injection of BM stem cells alone. In parallel, Lu *et al.*[113] examined temperature-sensitive injectable chitosan hydrogel for delivery of mESCs to the infarcted rat heart and reported effective cell transfer correlating with improved cardiac functioning.

13.3.3 Scaffold-free Tissue Engineering

13.3.3.1 Cell Aggregation

As biomaterials can introduce undesirable and sometimes toxic by-products, stimulate unfavorable host responses, or prevent critical cell–cell interactions, scaffold-free tissue engineering techniques have been developed to generate 3D tissues without the use of biomaterials.[114] Such tissues are composed of only cells and the ECM matrix they secrete. Murry *et al.* described a scaffold-free tissue engineering system in which hES-CM cells were forced to aggregate within a rotating shaker to create human cardiac patches.[114] The process allowed for control of patch size and yielded cardiac populations with enriched mature CMs which increased with culture time. In a subsequent publication,[15] they demonstrated that addition of endothelial and fibroblast cells to the original patch components resulted in improved mechanical properties which underwent successful engraftment to the rat heart.

13.3.3.2 Cell Sheets

The cell sheet tissue preparation approach was first described in 2002 by Shimizu *et al.*[115,116] Three-dimensional cardiac tissue can be constructed in this manner by harvesting confluent CM layers from culture dishes and then laying them one over another to form 3D cardiac tissue. Temperature-sensitive culture dishes made of poly(*N*-isopropylacrylamide) (PIPAAm) become hydrophilic and non-adhesive at reduced temperatures, thereby allowing for separation and collection of intact cell layers. Stimulation of protease activity can also induce detachment of whole cell layers. The cell sheet approach preserves cell-to-cell contacts, maintains expression of adhesion proteins, and stimulates secretion of ECM components by cardiac cells, but lacks the mechanical stiffness required for maintenance of physiological conditions. CM sheets have been reported to exhibit synchronously and spontaneously beating activity when partly overlaid.[115] A keynote paper published by Miyahara and his colleagues[70] described the construction of a monolayered cell sheet from adipose tissue-derived MSCs. After transplantation into the infarcted rat heart, the MSC cell sheet gradually grew to form a thick, vascularized graft which included few CMs and undifferentiated MSCs, and improved heart function.

13.3.4 Decellularized Matrix

Cadaveric tissue can be employed to engineer whole organs or tissue segments for transplantation. In this process, samples are first decellularized using detergents, leaving behind all the ECM components and vascular networks. The acellular ECM construct is then recellularized with functional cells such as CMs, endothelial cells or smooth muscle cells and cultured under physiological conditions in efforts to generate tissue patches or whole organs. This approach bears natural blood supply networks, with the potential of enhancing graft viability. Moreover, this approach presents the early stages of organ reconstruction and may serve as an alternative to organ transplants.

Proof of concept studies were performed by Ott and colleagues,[117] demonstrating feasibility of rat heart decellularization while preserving its natural chambers, valves and vasculature. The acellular heart construct was then reseeded with CMs or endothelial cells and sustained in a bioreactor which provided pulsatile flow and pacing. After eight days in culture, the heart construct achieved ~34% recellularization and exhibited pumping function reaching ~2% of adult rat heart potential.

The generation of decellularized tissue construct based on stem cells for heart repair had already been evaluated. Tan *et al.*[118] decellularized the small intestinal submucosa (SIS) to serve as a matrix for MSC cultivation. The SIS–MSC tissue construct was cultivated *in vitro* for 5–7 days and then transplanted into an infarcted rabbit heart and monitored for one month. Rabbit heart function improved after engraftment of acellular SIS and SIS–MSC grafts; however, the SIS–MSC group demonstrated a more profound effect. In addition, MSCs were shown to migrate toward the infarcted area where they differentiated into CMs and smooth muscle cells.

Christman and colleagues proposed combination of decellularized and injectable biomaterial matrices to yield a myocardial matrix more closely mimicking natural specimens.[119] Porcine myocardial tissue was decellularized and processed to form a viscous myocardial matrix with the ability to gel at 37°C. Upon its injection into the rat myocardium, intense neovascularization was detected within the graft area with a significant increase in the number of mature blood vessels at 11 days post-transplantation.

Similarly, Sung and colleagues combined the decellularization and cell sheet engineering approaches[120] to generate thick cardiac patches. Sliced acellular bovine pericardium was used as a scaffold to support MSC cell sheets inserted between the sliced scaffold layers. The cardiac patch was transplanted four weeks after infraction in rats; effective integration of the grafted patch was observed, along with a notable number of neo-vessels and neo-muscles within the graft area. The MSC patch group demonstrated a fourfold increase in blood vessel density compared with the untreated infarct group. In addition, a small fraction of transplanted MSCs expressed mature CM markers.

Although the discipline of valve engineering *via* decellularized matrices is growing rapidly, this is beyond the scope of this chapter.

13.4 Stem Cell-based Vascular Therapy of the Heart

13.4.1 Human Endothelial Progenitor Cells

Human EPCs can be isolated from autologous sources such as circulating peripheral blood or BM, umbilical cord blood or human embryonic stem cells and can give rise to mature endothelial cells. EPCs have been successfully applied to form capillary networks, both *in vitro* and *in vivo*, and to reverse infarction insult of the myocardium.[121–123] Moreover, EPCs are often employed to generate vascular networks within engineered tissues, supporting the formation of thick and complex tissues in the laboratory setting. These engineered vascular networks anastomose with host vasculature and lead to enhanced graft survival.[14,15] This section focuses on transplantation of human EPCs to induce therapeutic angiogenesis of ischemic heart tissues.

EPCs of the peripheral circulation can be characterized and isolated by means of CD34, CD133, KDR and and/or VE cadherin expression levels, where CD34 serves as the most frequently used marker for isolating human EPCs. Pre-clinical studies have already shown that transplantation of EPCs— purified from the peripheral circulation or from the BM—into acute infarcted rodent hearts results in improved vascularization, left ventricular remodeling and contractility.[121–123]

Kawamoto A. *et al.*[122] transplanted EPCs derived from human peripheral blood in infarcted rat hearts and observed their differentiation into mature endothelial cells correlating with improved heart functioning. Similarly, Kocher *et al.*[123] reported that BM-derived EPCs transplanted into ischemic myocardial led to improved cardiac function. Iwasaki *et al.*[121] described a dose-dependent relationship between BM-derived CD34$^+$ cells administered into infarcted rat hearts and preservation of left ventricular functioning. These results instigated a clinical trial evaluating application of circulating EPCs or BM-derived EPCs to acute myocardial infarction patients.[124] In this trial, EPC transplantation showed significant improvement in left ventricular ejection functioning coupled with a decrease in end-systolic volumes. No differences were noted between the two EPC-transplanted groups at the four month follow-up session. Another recent clinical trial provided evidence of feasibility, safety and beneficial effects of heart function after intramyocardial transplantation of autologous CD34[+] stem cells.[125] Similar findings were observed upon transplantation of human umbilical cord blood-derived EPCs into infarcted rat hearts.[126,127]

EPCs derived from hES-ECs represent a promising therapeutic cell source due to their highly proliferative capacities and low immunogenity.[128] Levenberg *et al.* reported successful sorting of PECAM[+] cells from hESC populations.[129] Sorting cells *via* endothelial promoter-regulated green fluorescent protein (GFP) expression provides an additional technique for effective isolation of EPCs from ESC samples.[130] Other researchers have shown that hES-ECs grown on various matrices can self-organize into vascular structures *in vitro*.[131,132] Li *et al.*[133] were the first to transplant ESC-ECs (mouse origin) in the infarcted mouse heart, detecting sustained cell viability for up to eight weeks thereafter and reporting functional improvement of the heart. A recent review summarizing contemporary therapeutic use of hES-EC to reverse ischemic tissues can be seen in ref. 134.

13.4.2 Vascularization of Engineered Tissue using hES-ECs

Vascularization of engineered tissues is a relatively new discipline aimed at generating complex tissue architecture *in vitro* and enhancing tissue viability after transplantation.[135] Such engineered vascular networks serve several functions, including nutrient and oxygen delivery both *in vitro* and *in vivo*. In addition, the endothelial cells secrete paracrine signaling molecules, augmenting differentiation and maturation of the engineered tissue *in vitro*.[136] Upon transplantation, the preformed blood vessels are expected to integrate with host vasculature, enhancing graft perfusion, and to accelerate host neovascularization *via* paracrine signaling pathways.

HES-ECs were shown to successfully form self-assembled 3D vascular networks and to generate vascularized muscle tissue constructs *in vitro*.[15,102,137] The vascularized skeletal muscle tissue constructs were generated by culturing myoblasts, hES-ECs and embryonic fibroblasts within 3D porous scaffolds.[137] Their transplantation into immunodeficient SCID mice led to fusion of hES-EC and host vessels. In a subsequent study, hES-ECs were cultured with hES-CMs and embryonic fibroblasts within 3D porous scaffolds to generate a 3D, vascularized, pulsating cardiac construct.[102] The hES-ECs self-assembled to form endothelial vessel networks, partially covered by a layer of smooth muscle originating from the embryonic fibroblasts. These results are consistent with those of a more recent study[15] in which hES-CMs, hES-ECs and fibroblasts created 3D, vascularized tissue aggregates in the absence of scaffold biomaterial. These vascularized tissue constructs retained the elevated properties of the tissue construct *in vitro* (graft size and cell viability) and improved engraftment potential upon transplantation in the rat heart. These studies highlight the potential of hESC-ECs as a cell source for tissue vascularization processes and the importance of pre-vascularization of implants before transplantation.

13.5 Conclusions

The therapeutic potential of specific human stem cell populations towards repair of cardiac insult has become increasingly evident in recent years. Pre-clinical and clinical studies have determined the feasibility and efficacy of cell transplantation in efforts to enhance myocardial functioning. However, the optimal stem cell candidate has yet to be defined, and probably only simultaneous transplant of various stem cell types will provide the full potential of engraftment procedures. However, significant levels of cell death or loss following injection present a major obstacle facing researchers attempting to design productive engraftment techniques. In this manner, integration of biomaterials in tissue engineering protocols is expected to bear a notable impact on cell delivery efficiency.

In-depth research of fundamental tissue engineering principles will allow for optimization of cell delivery and retention, and also provide techniques for improved tissue formation *in vitro*, yielding enhanced graft survival and functionality after transplantation. Vascularization of the engineered heart muscle has been proven essential in setting the fate of engrafted tissue engraftment constructs. Thus, the optimal stem cell candidate, biomaterial components and vascularization techniques must be appropriately combined to allow for effective cardiac cell therapy.

References

1. M. J. Evans and M. H. Kaufman, *Nature*, 1981, **292**, 154.
2. G. R. Martin, *Proc. Natl. Acad. Sci. U.S.A.*, 1981, **78**, 7634.
3. J. A. Thomson, J. Itskovitz-Eldor, S. S. Shapiro, M. A. Waknitz, J. J. Swiergiel, V. S. Marshall and J. M. Jones, *Science*, 1998, **282**, 1145.
4. I. Kehat, D. Kenyagin-Karsenti, M. Snir, H. Segev, M. Amit, A. Gepstein, E. Livne, O. Binah, J. Itskovitz-Eldor and L. Gepstein, *J. Clin. Invest.*, 2001, **108**, 407.
5. C. Xu, S. Police, N. Rao and M. K. Carpenter, *Circ. Res.*, 2002, **91**, 501.
6. C. Mummery, D. Ward-van Oostwaard, P. Doevendans, R. Spijker, S. van den Brink, R. Hassink, M. van der Heyden, T. Opthof, M. Pera, A. B. de la Riviere, R. Passier and L. Tertoolen, *Circulation*, 2003, **107**, 2733.
7. O. Caspi, I. Itzhaki, G. Arbel, I. Kehat, A. Gepstien, I. Huber, J. Satin and L. Gepstein, *Stem Cells Dev.*, 2009, **18**, 161–172.
8. I. Kehat, A. Gepstein, A. Spira, J. Itskovitz-Eldor and L. Gepstein, *Circ. Res.*, 2002, **91**, 659.
9. M. Snir, I. Kehat, A. Gepstein, R. Coleman, J. Itskovitz-Eldor, E. Livne and L. Gepstein, *Am. J. Physiol. Heart Circ. Physiol.*, 2003, **285**, H2355.

10. M. A. Laflamme, K. Y. Chen, A. V. Naumova, V. Muskheli, J. A. Fugate, S. K. Dupras, H. Reinecke, C. Xu, M. Hassanipour, S. Police, C. O'Sullivan, L. Collins, Y. Chen, E. Minami, E. A. Gill, S. Ueno, C. Yuan, J. Gold and C. E. Murry, *Nat. Biotechnol.*, 2007, **25**, 1015.

11. O. Caspi, I. Huber, I. Kehat, M. Habib, G. Arbel, A. Gepstein, L. Yankelson, D. Aronson, R. Beyar and L. Gepstein, *J. Am. Coll. Cardiol.*, 2007, **50**, 1884.

12. M. A. Laflamme, J. Gold, C. Xu, M. Hassanipour, E. Rosler, S. Police, V. Muskheli and C. E. Murry, *Am. J. Pathol.*, 2005, **167**, 663.

13. I. Kehat, L. Khimovich, O. Caspi, A. Gepstein, R. Shofti, G. Arbel, I. Huber, J. Satin, J. Itskovitz-Eldor and L. Gepstein, *Nat. Biotechnol.*, 2004, **22**, 1282.

14. A. Lesman, M. Habib, O. Caspi, A. Gepstein, G. Arbel, S. Levenberg and L. Gepstein, *Tissue Eng. Part A*, **16**, 115.

15. K. R. Stevens, K. L. Kreutziger, S. K. Dupras, F. S. Korte, M. Regnier, V. Muskheli, M. B. Nourse, K. Bendixen, H. Reinecke and C. E. Murry, *Proc. Natl. Acad. Sci. U.S.A.*, 2009, **106**, 16568.

16. D. Anderson, T. Self, I. R. Mellor, G. Goh, S. J. Hill and C. Denning, *Mol. Ther.*, 2007, **15**, 2027.

17. I. Huber, I. Itzhaki, O. Caspi, G. Arbel, M. Tzukerman, A. Gepstein, M. Habib, L. Yankelson, I. Kehat and L. Gepstein, *FASEB J.*, 2007, **21**, 2551.

18. X. Q. Xu, R. Zweigerdt, S. Y. Soo, Z. X. Ngoh, S. C. Tham, S. T. Wang, R. Graichen, B. Davidson, A. Colman and W. Sun, *Cytotherapy*, 2008, **10**, 376.

19. S. Takei, H. Ichikawa, K. Johkura, A. Mogi, H. No, S. Yoshie, D. Tomotsune and K. Sasaki, *Am. J. Physiol. Heart Circ. Physiol.*, 2009, **296**, H1793.

20. A. Tomescot, J. Leschik, V. Bellamy, G. Dubois, E. Messas, P. Bruneval, M. Desnos, A. A. Hagege, M. Amit, J. Itskovitz, P. Menasche and M. Puceat, *Stem Cells*, 2007, **25**, 2200.

21. P. W. Burridge, D. Anderson, H. Priddle, M. D. Barbadillo Munoz, S. Chamberlain, C. Allegrucci, L. E. Young and C. Denning, *Stem Cells*, 2007, **25**, 929.

22. R. Passier, D. W. Oostwaard, J. Snapper, J. Kloots, R. J. Hassink, E. Kuijk, B. Roelen, A. B. de la Riviere and C. Mummery, *Stem Cells*, 2005, **23**, 772.

23. B. S. Yoon, S. J. Yoo, J. E. Lee, S. You, H. T. Lee and H. S. Yoon, *Differentiation*, 2006, **74**, 149.

24. J. Leor, S. Gerecht, S. Cohen, L. Miller, R. Holbova, A. Ziskind, M. Shachar, M. S. Feinberg, E. Guetta and J. Itskovitz-Eldor, *Heart*, 2007, **93**, 1278.

25. L. W. van Laake, R. Passier, J. Monshouwer-Kloots, A. J. Verkleij, D. J. Lips, C. Freund, K. den Ouden, D. Ward-van Oostwaard, J. Korving, L. G. Tertoolen, C. J. van Echteld, P. A. Doevendans and C. L. Mummery, *Stem Cell Res.*, 2007, **1**, 9.

26. K. Takahashi and S. Yamanaka, *Cell*, 2006, **126**, 663.
27. K. Takahashi, K. Tanabe, M. Ohnuki, M. Narita, T. Ichisaka, K. Tomoda and S. Yamanaka, *Cell*, 2007, **131**, 861.
28. J. Yu, M. A. Vodyanik, K. Smuga-Otto, J. Antosiewicz-Bourget, J. L. Frane, S. Tian, J. Nie, G. A. Jonsdottir, V. Ruotti, R. Stewart, I. L. Slukvin and J. A. Thomson, *Science*, 2007, **318**, 1917.
29. K. Okita, M. Nakagawa, H. Hyenjong, T. Ichisaka and S. Yamanaka, *Science*, 2008, **322**, 949.
30. Y. Shi, C. Desponts, J. T. Do, H. S. Hahm, H. R. Scholer and S. Ding, *Cell Stem Cell*, 2008, **3**, 568.
31. M. Stadtfeld, K. Brennand and K. Hochedlinger, *Curr. Biol.*, 2008, **18**, 890.
32. A. Kuzmenkin, H. Liang, G. Xu, K. Pfannkuche, H. Eichhorn, A. Fatima, H. Luo, T. Saric, M. Wernig, R. Jaenisch and J. Hescheler, *FASEB J.*, 2009, **23**, 4168.
33. G. Narazaki, H. Uosaki, M. Teranishi, K. Okita, B. Kim, S. Matsuoka, S. Yamanaka and J. K. Yamashita, *Circulation*, 2008, **118**, 498.
34. K. Schenke-Layland, K. E. Rhodes, E. Angelis, Y. Butylkova, S. Heydarkhan-Hagvall, C. Gekas, R. Zhang, J. I. Goldhaber, H. K. Mikkola, K. Plath and W. R. MacLellan, *Stem Cells*, 2008, **26**, 1537.
35. H. Gai, E. L. Leung, P. D. Costantino, J. R. Aguila, D. M. Nguyen, L. M. Fink, D. C. Ward and Y. Ma, *Cell Biol. Int.*, 2009, **33**, 1184.
36. A. Haase, R. Olmer, K. Schwanke, S. Wunderlich, S. Merkert, C. Hess, R. Zweigerdt, I. Gruh, J. Meyer, S. Wagner, L. S. Maier, D. W. Han, S. Glage, K. Miller, P. Fischer, H. R. Scholer and U. Martin, *Cell Stem Cell*, 2009, **5**, 434.
37. T. Tanaka, S. Tohyama, M. Murata, F. Nomura, T. Kaneko, H. Chen, F. Hattori, T. Egashira, T. Seki, Y. Ohno, U. Koshimizu, S. Yuasa, S. Ogawa, S. Yamanaka, K. Yasuda and K. Fukuda, *Biochem. Biophys. Res. Commun.*, 2009, **385**, 497.
38. N. Yokoo, S. Baba, S. Kaichi, A. Niwa, T. Mima, H. Doi, S. Yamanaka, T. Nakahata and T. Heike, *Biochem. Biophys. Res. Commun.*, 2009, **387**, 482.
39. J. Zhang, G. F. Wilson, A. G. Soerens, C. H. Koonce, J. Yu, S. P. Palecek, J. A. Thomson and T. J. Kamp, *Circ. Res.*, 2009, **104**, e30.
40. L. Zwi, O. Caspi, G. Arbel, I. Huber, A. Gepstein, I. H. Park and L. Gepstein, *Circulation*, 2009, **120**, 1513.
41. C. Freund and C. L. Mummery, *J. Cell Biochem.*, 2009, **107**, 592.
42. Y. Shiba, K. D. Hauch and M. A. Laflamme, *Curr. Pharm. Des.*, 2009, **15**, 2791.
43. S. Yuasa and K. Fukuda, *Expert Rev. Cardiovasc. Ther.*, 2008, **6**, 803.
44. F. Belema Bedada, A. Technau, H. Ebelt, M. Schulze and T. Braun, *Mol. Cell. Biol.*, 2005, **25**, 9509.
45. M. P. Flaherty, A. Abdel-Latif, Q. Li, G. Hunt, S. Ranjan, Q. Ou, X. L. Tang, R. K. Johnson, R. Bolli and B. Dawn, *Circulation*, 2008, **117**, 2241.
46. R. Koninckx, K. Hensen, A. Daniels, M. Moreels, I. Lambrichts, H. Jongen, C. Clijsters, U. Mees, P. Steels, M. Hendrikx and J. L. Rummens, *Cytotherapy*, 2009, **11**, 778.

47. M. Koyanagi, P. Bushoven, M. Iwasaki, C. Urbich, A. M. Zeiher and S. Dimmeler, *Circ. Res.*, 2007, **101**, 1139.

48. S. Makino, K. Fukuda, S. Miyoshi, F. Konishi, H. Kodama, J. Pan, M. Sano, T. Takahashi, S. Hori, H. Abe, J. Hata, A. Umezawa and S. Ogawa, *J. Clin. Invest.*, 1999, **103**, 697.

49. Y. Wang, C. Feng, J. Xue, A. Sun, J. Li and J. Wu, *J. Physiol. Sci.*, 2009, **59**, 413.

50. J. Yoon, S. C. Choi, C. Y. Park, J. H. Choi, Y. I. Kim, W. J. Shim and D. S. Lim, *Mol. Cells*, 2008, **25**, 216.

51. L. B. Balsam, A. J. Wagers, J. L. Christensen, T. Kofidis, I. L. Weissman and R. C. Robbins, *Nature*, 2004, **428**, 668.

52. C. E. Murry, M. H. Soonpaa, H. Reinecke, H. Nakajima, H. O. Nakajima, M. Rubart, K. B. Pasumarthi, J. I. Virag, S. H. Bartelmez, V. Poppa, G. Bradford, J. D. Dowell, D. A. Williams and L. J. Field, *Nature*, 2004, **428**, 664.

53. J. M. Nygren, S. Jovinge, M. Breitbach, P. Sawen, W. Roll, J. Hescheler, J. Taneera, B. K. Fleischmann and S. E. Jacobsen, *Nat. Med.*, 2004, **10**, 494.

54. D. Orlic, J. Kajstura, S. Chimenti, I. Jakoniuk, S. M. Anderson, B. Li, J. Pickel, R. McKay, B. Nadal-Ginard, D. M. Bodine, A. Leri and P. Anversa, *Nature*, 2001, **410**, 701.

55. D. Orlic, J. Kajstura, S. Chimenti, F. Limana, I. Jakoniuk, F. Quaini, B. Nadal-Ginard, D. M. Bodine, A. Leri and P. Anversa, *Proc. Natl. Acad. Sci. U. S. A.*, 2001, **98**, 10344.

56. J. Kajstura, M. Rota, B. Whang, S. Cascapera, T. Hosoda, C. Bearzi, D. Nurzynska, H. Kasahara, E. Zias, M. Bonafe, B. Nadal-Ginard, D. Torella, A. Nascimbene, F. Quaini, K. Urbanek, A. Leri and P. Anversa, *Circ. Res.*, 2005, **96**, 127.

57. M. Rota, J. Kajstura, T. Hosoda, C. Bearzi, S. Vitale, G. Esposito, G. Iaffaldano, M. E. Padin-Iruegas, A. Gonzalez, R. Rizzi, N. Small, J. Muraski, R. Alvarez, X. Chen, K. Urbanek, R. Bolli, S. R. Houser, A. Leri, M. A. Sussman and P. Anversa, *Proc. Natl. Acad. Sci. U.S.A.*, 2007, **104**, 17783.

58. H. M. Wei, P. Wong, L. F. Hsu and W. Shim, *Singapore Med. J.*, 2009, **50**, 935.

59. V. Schachinger, S. Erbs, A. Elsasser, W. Haberbosch, R. Hambrecht, H. Holschermann, J. Yu, R. Corti, D. G. Mathey, C. W. Hamm, T. Suselbeck, N. Werner, J. Haase, J. Neuzner, A. Germing, B. Mark, B. Assmus, T. Tonn, S. Dimmeler and A. M. Zeiher, *Eur. Heart J.*, 2006, **27**, 2775.

60. K. C. Wollert, G. P. Meyer, J. Lotz, S. Ringes-Lichtenberg, P. Lippolt, C. Breidenbach, S. Fichtner, T. Korte, B. Hornig, D. Messinger, L. Arseniev, B. Hertenstein, A. Ganser and H. Drexler, *Lancet*, 2004, **364**, 141.

61. G. P. Meyer, K. C. Wollert, J. Lotz, J. Steffens, P. Lippolt, S. Fichtner, H. Hecker, A. Schaefer, L. Arseniev, B. Hertenstein, A. Ganser and H. Drexler, *Circulation*, 2006, **113**, 1287.

62. A. Abdel-Latif, R. Bolli, I. M. Tleyjeh, V. M. Montori, E. C. Perin, C. A. Hornung, E. K. Zuba-Surma, M. Al-Mallah and B. Dawn, *Arch. Intern. Med.*, 2007, **167**, 989.

63. S. Janssens, C. Dubois, J. Bogaert, K. Theunissen, C. Deroose, W. Desmet, M. Kalantzi, L. Herbots, P. Sinnaeve, J. Dens, J. Maertens, F. Rademakers, S. Dymarkowski, O. Gheysens, J. Van Cleemput, G. Bormans, J. Nuyts, A. Belmans, L. Mortelmans, M. Boogaerts and F. Van de Werf, *Lancet*, 2006, **367**, 113.

64. P. A. Zuk, M. Zhu, P. Ashjian, D. A. De Ugarte, J. I. Huang, H. Mizuno, Z. C. Alfonso, J. K. Fraser, P. Benhaim and M. H. Hedrick, *Mol. Biol. Cell*, 2002, **13**, 4279.

65. M. F. Pittenger, A. M. Mackay, S. C. Beck, R. K. Jaiswal, R. Douglas, J. D. Mosca, M. A. Moorman, D. W. Simonetti, S. Craig and D. R. Marshak, *Science*, 1999, **284**, 143.

66. M. Shiota, T. Heike, M. Haruyama, S. Baba, A. Tsuchiya, H. Fujino, H. Kobayashi, T. Kato, K. Umeda, M. Yoshimoto and T. Nakahata, *Exp. Cell Res.*, 2007, **313**, 1008.

67. W. Dai, S. L. Hale and R. A. Kloner, *Regen. Med.*, 2007, **2**, 63.

68. W. Dai, S. L. Hale, B. J. Martin, J. Q. Kuang, J. S. Dow, L. E. Wold and R. A. Kloner, *Circulation*, 2005, **112**, 214.

69. M. Gnecchi, H. He, N. Noiseux, O. D. Liang, L. Zhang, F. Morello, H. Mu, L. G. Melo, R. E. Pratt, J. S. Ingwall and V. J. Dzau, *FASEB J.*, 2006, **20**, 661.

70. Y. Miyahara, N. Nagaya, M. Kataoka, B. Yanagawa, K. Tanaka, H. Hao, K. Ishino, H. Ishida, T. Shimizu, K. Kangawa, S. Sano, T. Okano, S. Kitamura and H. Mori, *Nat. Med.*, 2006, **12**, 459.

71. N. Noiseux, M. Gnecchi, M. Lopez-Ilasaca, L. Zhang, S. D. Solomon, A. Deb, V. J. Dzau and R. E. Pratt, *Mol. Ther.*, 2006, **14**, 840.

72. J. G. Shake, P. J. Gruber, W. A. Baumgartner, G. Senechal, J. Meyers, J. M. Redmond, M. F. Pittenger and B. J. Martin, *Ann. Thorac. Surg.*, 2002, **73**, 1919.

73. C. Toma, M. F. Pittenger, K. S. Cahill, B. J. Byrne and P. D. Kessler, *Circulation*, 2002, **105**, 93.

74. S. L. Chen, W. W. Fang, F. Ye, Y. H. Liu, J. Qian, S. J. Shan, J. J. Zhang, R. Z. Chunhua, L. M. Liao, S. Lin and J. P. Sun, *Am. J. Cardiol.*, 2004, **94**, 92.

75. A. M. Hierlihy, P. Seale, C. G. Lobe, M. A. Rudnicki and L. A. Megeney, *FEBS Lett*, 2002, **530**, 239.

76. C. M. Martin, A. P. Meeson, S. M. Robertson, T. J. Hawke, J. A. Richardson, S. Bates, S. C. Goetsch, T. D. Gallardo and D. J. Garry, *Dev. Biol.*, 2004, **265**, 262.

77. A. P. Beltrami, L. Barlucchi, D. Torella, M. Baker, F. Limana, S. Chimenti, H. Kasahara, M. Rota, E. Musso, K. Urbanek, A. Leri, J. Kajstura, B. Nadal-Ginard and P. Anversa, *Cell*, 2003, **114**, 763.

78. E. Messina, L. De Angelis, G. Frati, S. Morrone, S. Chimenti, F. Fiordaliso, M. Salio, M. Battaglia, M. V. Latronico, M. Coletta, E. Vivarelli, L. Frati, G. Cossu and A. Giacomello, *Circ. Res.*, 2004, **95**, 911.

79. K. L. Laugwitz, A. Moretti, J. Lam, P. Gruber, Y. Chen, S. Woodard, L. Z. Lin, C. L. Cai, M. M. Lu, M. Reth, O. Platoshyn, J. X. Yuan, S. Evans and K. R. Chien, *Nature*, 2005, **433**, 647.

80. A. Moretti, L. Caron, A. Nakano, J. T. Lam, A. Bernshausen, Y. Chen, Y. Qyang, L. Bu, M. Sasaki, S. Martin-Puig, Y. Sun, S. M. Evans, K. L. Laugwitz and K. R. Chien, *Cell*, 2006, **127**, 1151.

81. A. Itzhaki-Alfia, J. Leor, E. Raanani, L. Sternik, D. Spiegelstein, S. Netser, R. Holbova, M. Pevsner-Fischer, J. Lavee and I. M. Barbash, *Circulation*, 2009, **120**, 2559.

82. S. X. Liang, T. Y. Tan, L. Gaudry and B. Chong, *Int. J. Cardiol.*, **138**, 40.

83. T. Oyama, T. Nagai, H. Wada, A. T. Naito, K. Matsuura, K. Iwanaga, T. Takahashi, M. Goto, Y. Mikami, N. Yasuda, H. Akazawa, A. Uezumi, S. Takeda and I. Komuro, *J. Cell Biol.*, 2007, **176**, 329.

84. O. Pfister, F. Mouquet, M. Jain, R. Summer, M. Helmes, A. Fine, W. S. Colucci and R. Liao, *Circ. Res.*, 2005, **97**, 52.

85. M. Rota, M. E. Padin-Iruegas, Y. Misao, A. De Angelis, S. Maestroni, J. Ferreira-Martins, E. Fiumana, R. Rastaldo, M. L. Arcarese, T. S. Mitchell, A. Boni, R. Bolli, K. Urbanek, T. Hosoda, P. Anversa, A. Leri and J. Kajstura, *Circ. Res.*, 2008, **103**, 107.

86. K. Matsuura, T. Nagai, N. Nishigaki, T. Oyama, J. Nishi, H. Wada, M. Sano, H. Toko, H. Akazawa, T. Sato, H. Nakaya, H. Kasanuki and I. Komuro, *J. Biol. Chem.*, 2004, **279**, 11384.

87. H. Oh, S. B. Bradfute, T. D. Gallardo, T. Nakamura, V. Gaussin, Y. Mishina, J. Pocius, L. H. Michael, D. R. Behringer, D. J. Garry, M. L. Entman and M. D. Schneider, *Proc. Natl. Acad. Sci. U. S. A.*, 2003, **100**, 12313.

88. N. Arsic, D. Mamaeva, N. J. Lamb and A. Fernandez, *Exp. Cell Res.*, 2008, **314**, 1266.

89. A. Asakura, M. Komaki and M. Rudnicki, *Differentiation*, 2001, **68**, 245.

90. H. Reinecke, V. Poppa and C. E. Murry, *J. Mol. Cell. Cardiol.*, 2002, **34**, 241.

91. C. E. Murry, R. W. Wiseman, S. M. Schwartz and S. D. Hauschka, *J. Clin. Invest.*, 1996, **98**, 2512.

92. D. A. Taylor, B. Z. Atkins, P. Hungspreugs, T. R. Jones, M. C. Reedy, K. A. Hutcheson, D. D. Glower and W. E. Kraus, *Nat. Med.*, 1998, **4**, 929.

93. C. E. Murry, L. J. Field and P. Menasche, *Circulation*, 2005, **112**, 3174.

94. P. Menasche, O. Alfieri, S. Janssens, W. McKenna, H. Reichenspurner, L. Trinquart, J. T. Vilquin, J. P. Marolleau, B. Seymour, J. Larghero, S. Lake, G. Chatellier, S. Solomon, M. Desnos and A. A. Hagege, *Circulation*, 2008, **117**, 1189.

95. J. Muller-Ehmsen, P. Whittaker, R. A. Kloner, J. S. Dow, T. Sakoda, T. I. Long, P. W. Laird and L. Kedes, *J. Mol. Cell. Cardiol.*, 2002, **34**, 107.

96. S. J. Hollister, *Nat. Mater.*, 2005, **4**, 518.

97. G. Antonios and J. S. T. Mikos, *J. Biotechnol., ISSN*, 2000, **3**, 114.

98. S. Levenberg, N. F. Huang, E. Lavik, A. B. Rogers, J. Itskovitz-Eldor and R. Langer, *Proc. Natl. Acad. Sci. U. S. A.*, 2003, **100**, 12741.
99. M. Levy-Mishali, J. Zoldan and S. Levenberg, *Tissue Eng. Part A*, 2009, **15**, 935.
100. J. Gao, P. M. Crapo and Y. Wang, *Tissue Eng.*, 2006, **12**, 917.
101. J. Leor, S. Aboulafia-Etzion, A. Dar, L. Shapiro, I. M. Barbash, A. Battler, Y. Granot and S. Cohen, *Circulation*, 2000, **102**, III56.
102. O. Caspi, A. Lesman, Y. Basevitch, A. Gepstein, G. Arbel, I. H. Habib, L. Gepstein and S. Levenberg, *Circ. Res.*, 2007, **100**, 263.
103. M. W. Tibbitt and K. S. Anseth, *Biotechnol. Bioeng.*, 2009, **103**, 655.
104. K. Shapira-Schweitzer, M. Habib, I. Gepstein and D. Seliktar, *J. Mol. Cell. Cardiol.*, 2009, **46**, 213.
105. W. H. Zimmermann, K. Schneiderbanger, P. Schubert, M. Didie, F. Munzel, J. F. Heubach, S. Kostin, W. L. Neuhuber and T. Eschenhagen, *Circ. Res.*, 2002, **90**, 223.
106. W. H. Zimmermann, I. Melnychenko, G. Wasmeier, M. Didie, H. Naito, U. Nixdorff, A. Hess, L. Budinsky, K. Brune, B. Michaelis, S. Dhein, A. Schwoerer, H. Ehmke and T. Eschenhagen, *Nat. Med.*, 2006, **12**, 452.
107. X.-M. Guo, Y.-S. Zhao, H.-X. Chang, C.-Y. Wang, L. L. E. X.-A. Zhang, C.-M. Duan, L.-Z. Dong, H. Jiang, J. Li, Y. Song and X. Yang, *Circulation*, 2006, **113**, 2229.
108. D. Simpson, H. Liu, T. H. Fan, R. Nerem and S. C. Dudley Jr, *Stem Cells*, 2007, **25**, 2350.
109. N. Landa, L. Miller, M. S. Feinberg, R. Holbova, M. Shachar, I. Freeman, S. Cohen and J. Leor, *Circulation*, 2008, **117**, 1388.
110. J. Leor, S. Tuvia, V. Guetta, F. Manczur, D. Castel, U. Willenz, O. Petnehazy, N. Landa, M. S. Feinberg, E. Konen, O. Goitein, O. Tsur-Gang, M. Shaul, L. Klapper and S. Cohen, *J. Am. Coll. Cardiol.*, 2009, **54**, 1014.
111. T. Kofidis, D. R. Lebl, E. C. Martinez, G. Hoyt, M. Tanaka and R. C. Robbins, *Circulation*, 2005, **112**, I173.
112. T. Wang, X. J. Jiang, Q. Z. Tang, X. Y. Li, T. Lin, D. Q. Wu, X. Z. Zhang and E. Okello, *Acta Biomater.*, 2009, **5**, 2939.
113. W. N. Lu, S. H. Lu, H. B. Wang, D. X. Li, C. M. Duan, Z. Q. Liu, T. Hao, W. J. He, B. Xu, Q. Fu, Y. C. Song, X. H. Xie and C. Y. Wang, *Tissue Eng. Part A*, 2009, **15**, 1437.
114. K. R. Stevens, L. Pabon, V. Muskheli and C. E. Murry, *Tissue Eng. Part A*, 2009, **15**, 1211.
115. T. Shimizu, M. Yamato, Y. Isoi, T. Akutsu, T. Setomaru, K. Abe, A. Kikuchi, M. Umezu and T. Okano, *Circ. Res.*, 2002, **90**, e40.
116. T. Shimizu, M. Yamato, A. Kikuchi and T. Okano, *Biomaterials*, 2003, **24**, 2309.
117. H. C. Ott, T. S. Matthiesen, S. K. Goh, L. D. Black, S. M. Kren, T. I. Netoff and D. A. Taylor, *Nat. Med.*, 2008, **14**, 213.

118. M. Y. Tan, W. Zhi, R. Q. Wei, Y. C. Huang, K. P. Zhou, B. Tan, L. Deng, J. C. Luo, X. Q. Li, H. Q. Xie and Z. M. Yang, *Biomaterials*, 2009, **30**, 3234.

119. J. M. Singelyn, J. A. DeQuach, S. B. Seif-Naraghi, R. B. Littlefield, P. J. Schup-Magoffin and K. L. Christman, *Biomaterials*, 2009, **30**, 5409.

120. H. J. Wei, C. H. Chen, W. Y. Lee, I. Chiu, S. M. Hwang, W. W. Lin, C. C. Huang, Y. C. Yeh, Y. Chang and H. W. Sung, *Biomaterials*, 2008, **29**, 3547.

121. H. Iwasaki, A. Kawamoto, M. Ishikawa, A. Oyamada, S. Nakamori, H. Nishimura, K. Sadamoto, M. Horii, T. Matsumoto, S. Murasawa, T. Shibata, S. Suehiro and T. Asahara, *Circulation*, 2006, **113**, 1311.

122. A. Kawamoto, H. C. Gwon, H. Iwaguro, J. I. Yamaguchi, S. Uchida, H. Masuda, M. Silver, H. Ma, M. Kearney, J. M. Isner and T. Asahara, *Circulation*, 2001, **103**, 634.

123. A. A. Kocher, M. D. Schuster, M. J. Szabolcs, S. Takuma, D. Burkhoff, J. Wang, S. Homma, N. M. Edwards and S. Itescu, *Nat. Med.*, 2001, **7**, 430.

124. V. Schachinger, B. Assmus, M. B. Britten, J. Honold, R. Lehmann, C. Teupe, N. D. Abolmaali, T. J. Vogl, W. K. Hofmann, H. Martin, S. Dimmeler and A. M. Zeiher, *J. Am. Coll. Cardiol.*, 2004, **44**, 1690.

125. D. W. Losordo, R. A. Schatz, C. J. White, J. E. Udelson, V. Veereshwarayya, M. Durgin, K. K. Poh, R. Weinstein, M. Kearney, M. Chaudhry, A. Burg, L. Eaton, L. Heyd, T. Thorne, L. Shturman, P. Hoffmeister, K. Story, V. Zak, D. Dowling, J. H. Traverse, R. E. Olson, J. Flanagan, D. Sodano, T. Murayama, A. Kawamoto, K. F. Kusano, J. Wollins, F. Welt, P. Shah, P. Soukas, T. Asahara and T. D. Henry, *Circulation*, 2007, **115**, 3165.

126. T. Murohara, H. Ikeda, J. Duan, S. Shintani, K. Sasaki, H. Eguchi, I. Onitsuka, K. Matsui and T. Imaizumi, *J. Clin. Invest.*, 2000, **105**, 1527.

127. I. Ott, U. Keller, M. Knoedler, K. S. Gotze, K. Doss, P. Fischer, K. Urlbauer, G. Debus, N. von Bubnoff, M. Rudelius, A. Schomig, C. Peschel and R. A. Oostendorp, *FASEB J.*, 2005, **19**, 992.

128. S. Levenberg, J. Zoldan, Y. Basevitch and R. Langer, *Blood*, 2007, **110**, 806.

129. S. Levenberg, J. S. Golub, M. Amit, J. Itskovitz-Eldor and R. Langer, *Proc. Natl. Acad. Sci. U. S. A.*, 2002, **99**, 4391.

130. S. Kim and H. A. von Recum, *Tissue Eng. Part A*, 2009, **15**, 3709.

131. S. Gerecht-Nir, A. Ziskind, S. Cohen and J. Itskovitz-Eldor, *Lab. Invest.*, 2003, **83**, 1811.

132. K. E. McCloskey, M. E. Gilroy and R. M. Nerem, *Tissue Eng.*, 2005, **11**, 497.

133. Z. Li, J. C. Wu, A. Y. Sheikh, D. Kraft, F. Cao, X. Xie, M. Patel, S. S. Gambhir, R. C. Robbins, J. P. Cooke and J. C. Wu, *Circulation*, 2007, **116**, I46.

134. Z. Li, Z. Han and J. C. Wu, *J. Cell. Biochem.*, 2009, **106**, 194.
135. T. Kaully, K. Kaufman-Francis, A. Lesman and S. Levenberg, *Tissue Eng. Part B Rev.*, 2009, **15**, 159.
136. D. L. Brutsaert, *Physiol. Rev.*, 2003, **83**, 59.
137. S. Levenberg, J. Rouwkema, M. Macdonald, E. S. Garfein, D. S. Kohane, D. C. Darland, R. Marini, C. A. van Blitterswijk, R. C. Mulligan, P. A. D'Amore and R. Langer, *Nat. Biotechnol.*, 2005, **23**, 879.

CHAPTER 14

Regeneration of the Vascular System

M. SCHLEICHER,[a] A. J. HUBER,[a] H. P. WENDEL[b] AND
U. A. STOCK[a]

[a] Department of Thoracic and Cardiovascular Surgery, University Hospital
Tübingen, Tübingen, Germany; [b] Surgery of Congenital Heart Defects And
Pediatric Cardio-surgery, Children's University Hospital Tübingen,
Tübingen, Germany

14.1 Introduction

Current blood vessel replacement concepts using either prosthetic or biological
conduit devices are, despite excellent mid-term results, associated with major
limitations.[1] Prosthetic blood vessel substitutes such as polytetrafluoroethylene
(PTFE) less than 5 mm in diameter require life-long anticoagulation and
are susceptible to infections. Biological conduits (*e.g.* autologous greater
saphenous veins) have better hemodynamic characteristics and avoid long-term
anticoagulation, but are limited in availability and were originally designed for
low-pressure conditions. In the long term they fail due to intimal hyperplasia
and fibrosis. Allografts such as superficial femoral arteries have the best
hemodynamic properties, avoid anticoagulation completely and are resistant to
infections to a certain extent. However, due to organ scarcity, their availability is
limited. None of the currently available blood vessel replacement devices possesses any regenerative or growth potential. This shortcoming is crucial especially for treatment of pediatric patients and grafts with repeated injury (*e.g.*
hemodialysis shunts). The multidisciplinary approach of tissue engineering[2]

Stem Cell-Based Tissue Repair
Edited by Raphael Gorodetsky and Richard Schäfer

Published by the Royal Society of Chemistry, www.rsc.org

might offer an attractive way of overcoming these shortcomings and developing a blood vessel substitute identical to natural human veins and arteries.

Tissue engineering is an evolving multidisciplinary field of research which combines basic research such as chemistry, material processing and cell biology with clinical disciplines such as surgery.[3] The principal concept uses scaffolds made out of acellular bioresorbable or biodegradable or decellularized xenogenous material, formed in case of blood vessels in tubes of the shape of the organ structures to be replaced. These scaffolds are seeded with autologous cells such as vascular wall cells (smooth muscle cells and endothelial cells), circulating stem cells or progenitor cells. Once cells are attached to the three-dimensional (3D) scaffold, this newly generated cell–polymer construct can be used to replace diseased or deformed body parts (*e.g.* blood vessels). As cellular structures and matrix develop, the scaffold degrades gradually. After complete degradation and absorption or resorption, the engineered tissue vessel remains without any artificial material.

In 2001, Shinoka and colleagues began a human trial evaluating the use of polyglycolic acid and ε-caprolactone or L-lactide grafts in patients with single ventricle physiology. They reported late clinical and radiological surveillance of a cohort of 25 patients who underwent implantation of tissue-engineered vascular grafts as extracardiac cavopulmonary conduits. The pressure conditions of such grafts are extremely low (10–15 mmHg). They observed no aneurysm formation but graft stenosis.[4]

In recent years simple seeding of the starter matrices has turned out to be insufficient.[5,6] Studies have shown that the application of physical signals (*e.g.* flow and pressure) remain crucial for a homogenous cell seeding on surface and deeper tissue.[7] In order to apply these signals a dynamic tissue culture of up to six weeks is required.[8] Although early *in vivo* experiments proved the feasibility of this concept,[9] it imposes a variety of demands for successful transfer from pre-clinical large animal set-ups to clinical arena. Due to the imminent risk of bacterial and fungal infection throughout the entire *in vitro* culture (it takes up to six weeks from cell harvest to implantation of the engineered product), any potential production facility will need to meet good manufacturing practice (GMP) standard. Furthermore, the extremely time-consuming process involves a challenging infrastructure and enormous costs.

In 1987 Peter Zilla *et al.* described the first clinical surface engineering application of vascular PTFE grafts.[10] This approach was based on seeding autologous endothelial cells on PTFE grafts for distal femoro-popliteal bypass surgery. Even though this surface engineering approach resulted in an improved long-term function[11] during the following 20 years, only a couple of hundred clinical applications have been described. Again, the necessary infrastructure and resulting costs resulted in reluctant application.[12]

A recent study by our group on tissue-engineered aortic tissue in a large animal model indicates that sufficient collagenous tissue formation occurs, in particular after implantation.[13] However, elastic fiber formation was reduced significantly. This observed lack in mature elastin structures in tissue engineered constructs had been described before[14,15] and seemed a serious

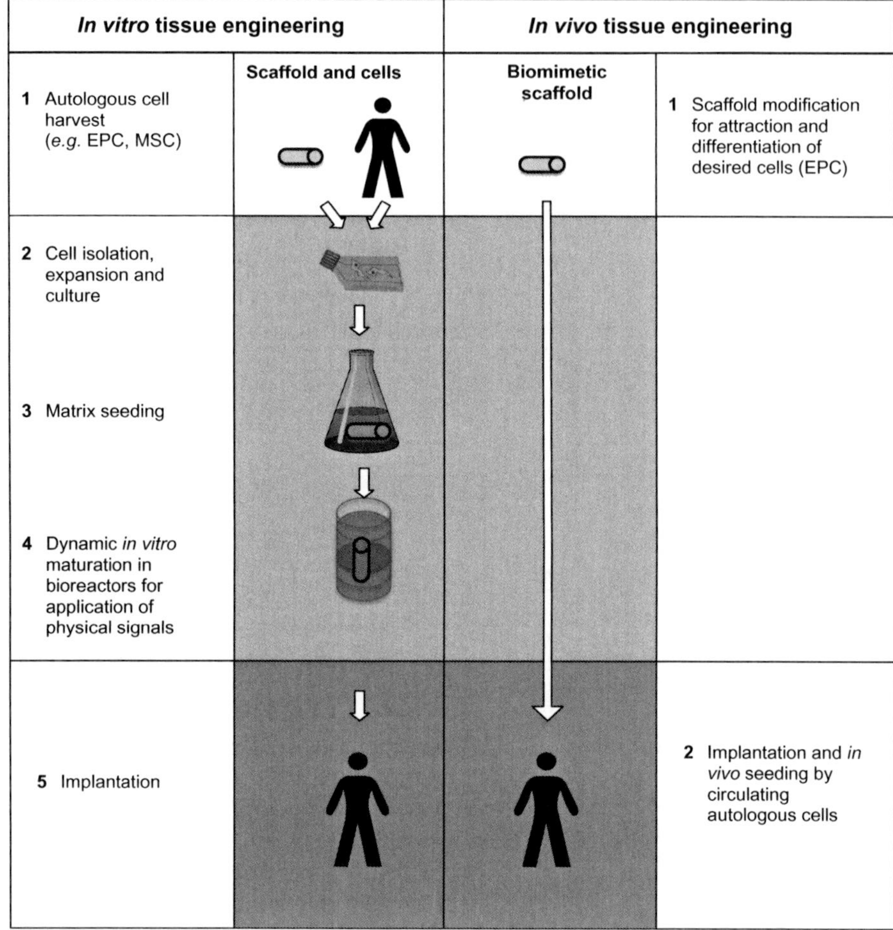

Figure 14.1   Concept of *in vitro* and *in vivo* blood vessel tissue engineering.

limitation for achieving a functional tissue-engineered blood vessel (TEBV) for high-pressure circulation in general.[16]

This chapter presents a novel concept to pursue the regeneration of the vascular system while eliminating the shortcomings outlined above.[17] This *in vivo* tissue engineering concept utilizes the natural regenerative potential of humans to enable either surface or complete tissue engineering of blood vessels. Figure 14.1 shows a comparison of *in vitro* and *in vivo* blood vessel tissue engineering.

14.2   The Concept of *In vivo* Tissue Engineering

The concept of *in vivo* tissue engineering avoids all cell and tissue culture steps and encourages the regenerative potential of the patient's body to populate a starter matrix.[18] Accordingly the objective is to implant off-the-shelf blood vessel scaffolds without any pre-implantation cell seeding. Autologous cells with regenerative potential circulating in the blood stream are supposed to recognize this scaffold, adhere to it, and differentiate to functional vascular cells. Depending on the scaffold, endothelialization of the surface or even complete repopulation of the scaffold will be achieved. To provide stimulating conditions for this process, the matrices provided undergo specific modifications.

The obvious advantage of this concept is that these enhanced scaffolds are produced in stock and are implanted without the need for further manipulation. *In vivo* tissue engineering does not require any cell and tissue culturing and, in consequence, eliminates the need for the costly and sophisticated infrastructures required by GMP and the US Food and Drug Administration (FDA) for the conformed cell and tissue culturing necessary for *in vitro* tissue engineering. The risk of bacterial and fungal infection during cell and tissue culture is eliminated. Moreover patients will undergo only one medical intervention, as there is no need to harvest autologous cells prior to implantation.

14.3   Starter Matrices

Currently preferred matrices for *in vivo* tissue engineering consist of biopolymeric materials such as:

- polyhydroxyoctanoate (PHO);[19]
- poly-4-hydroxybutyrate (P4HB);[20]
- fibrin–polylactide;[21]
- hyaluronan;[22] and
- collagen-based[23] or decellularized xenogenous tissue (*e.g.* porcine aorta)[24].

Porcine blood vessels are very similar to human arteries and offer excellent hemodynamic properties. A variety of decellularization concepts are pursued,[25–27] which all aim to remove all potentially immunological active cells while preserving crucial extracellular matrix architecture. Recent studies using multiphoton laser scanning microscopy (LSM)[28] identified previously not visible alterations of collagenous and elastic structures.[29]

Implantation of decellularized grafts in animals resulted in spontaneous re-endothelialization in most cases. As seen in PTFE grafts with and without endothelial cell seeding and allograft implantations, this recellularization is limited in humans to a couple of millimeters whereas they work in all kinds of animal models.[30]

An alternative to porcine aortic tissue is porcine small intestine submucosa (SIS). Badylak *et al.*[31] excised both carotid arteries in a dog and replaced them with either a saphenous vein graft or a SIS graft. They observed progressive replacement with fibrous connective tissue and microvasculature similar to mature host tissue, and progression of endothelialization with time. Again the potential transferability from animals to humans is highly speculative and the design of man-made SIS blood vessels remains less sophisticated compared with natural human or porcine blood vessels.

Another starter matrix material was clinically introduced in 2002. The graft material consisted of decellularized bovine ureter (SGVG 100, CryoLife Inc., GA, USA) for hemodialysis[32] and infrainguinal bypass surgery.[33] Uncomplicated segments of post-implantation specimens in hemodialysis patients showed myofibroblastic ingrowth, but no luminal endothelialization and no vascularization of the wall other than at sites of needle puncture; 50% showed a severe adventitial host inflammatory response with dominant granulomatous and eosinophil-rich infiltrates. Inflammation was present in grafts with various complications such as stenosis, thrombosis and aneurysm formation. In a small series of 12 patients receiving an infrainguinal bypass, ten needed early explantation due to massive aneurysmatic dilatation between seven days and 18 months after implantation.

14.4 Attracting Host Cells and Masking Inflammatory Structures

Because decellularized xenograft blood vessels in humans show practically no repopulation and endothelialization but inflammation and degeneration, it is crucial to enable and propagate a directed repopulation and, in parallel, eliminate the inflammatory and degenerative potential. Consistent enhancement of the *in vivo* concept is based on a directed recruitment of cell populations with the potential to differentiate into functional heart valve cells by highly specific attraction structures. Masking of surface structures is required in order to eliminate, or at least mitigate, inflammation and thrombogenicity of the exposed extracellular matrix (ECM) proteins.

Different approaches have been described for immobilization of circulating cells on expanded polytetrafluoroethylene (ePTFE) grafts. Rotmans *et al.* used anti-CD34 human antibodies to attract circulating endothelial progenitor cells (EPCs).[34] *In vivo* cell seeding with anti-CD34 antibodies enabled an accelerated endothelialization but stimulated intimal hyperplasia in porcine arteriovenous ePTFE grafts. Walluscheck *et al.* modified ePTFE grafts with immobilized synthetic peptides containing arginine–glycine–aspartic acid (RGD).[35,36] Graft coating with a RGD peptide significantly improved endothelial cell attachment and retention in their *in vitro* perfusion experiments. But the RGD peptide sequence is a recognition sequence for cell adhesion on extracellular matrix for various cell types and therefore not specific for heart valve progenitor cells. So far none of the applied attracting structures have been employed with heart valve tissues.

Wojciechowski *et al.* studied the potential of hematopoietic stem and progenitor cells (HSPCs) to be captured directly from circulating blood *in vivo*. Vascular shunt prototypes coated internally with P-selectin were inserted into the femoral artery of rats. After one hour, blood perfusion immunofluorescence microscopy and flow cytometric analysis revealed successful capture of mononuclear cells positive for the HSPC surface marker CD34.[37]

A variety of approaches are being pursued for masking the immunogenic potential of decellularized scaffolds. Wu *et al.* (2007) tested decellularized porcine aortic valve leaflets coated with poly(3-hydroxybutyrate-co-3-hydroxyhexanoate) (PHBHHx) in sheep.[38] The *in vivo* results indicated that coating with PHBHHx reduced calcification and promoted repopulation of hybrid valves with the recipient's cells resembling native valve tissue. Stamm *et al.* also showed improved biological and biomechanical stability using aortic porcine valves coated with biodegradable poly(hydroxybutyrate) (PHB).[39] Tedder *et al.* conducted experiments showing that decellularized porcine pericardium coated with penta-galloyl glucose (PGG) (a collagen-binding polyphenol) exhibited excellent biaxial mechanical properties, did not calcify *in vivo* (rats) and supported infiltration by host fibroblasts and subsequent matrix remodeling.[23]

An alternative approach utilizes a co-immobilization of growth factors to direct the differentiation of adhering cell populations into functional vascular wall cells. A potential candidate is vascular endothelial growth factor (VEGF). Crombez *et al.* presented experiments with VEGF immobilized on ePTFE grafts; VEGF affected cell migration behavior but did not effect endothelial cell adhesion.[40]

Therefore, a specific triggering for cell homing and immobilization remains a compulsive basic requirement for successful *in vivo* engineering approaches.

14.5   Cell Populations for *In vivo* Tissue Engineering

The majority of approaches focus on attracting endothelial progenitor cells (EPC) circulating in the patient's blood stream to the surface of blood vessel implants.[41] These cells posses the ability to differentiate into mature endothelial cells.[42] Further studies addressing the integration of EPCs into mature endothelium even found that a small fraction of these cells can also trans-differentiate into smooth muscle cells *in vitro*.[43] These cells can establish an autologous endothelial cell layer on the surface of blood vessels, achieving a non-thrombogenic blood-tissue interface. They might also regulate immune and inflammatory reaction and possess a unique way to protect the implanted graft from degeneration. For complete repopulation of the interstitial area of the blood vessel matrix, appropriate circulating cell populations with the capacity to synthesize and mediate remodeling of extracellular matrix remain to be identified.

14.6   Surface Engineering of Heart Valves using Aptamer Technology

A promising technique for homing of EPCs to graft surfaces is the use of a star-polyethyleneglycol coating (starPEG, SusTech GmbH, Darmstadt, Germany)[44] combined with immobilized aptamers. Aptamers are single-stranded DNA or RNA molecules that can fold into a 3D structure. Within this

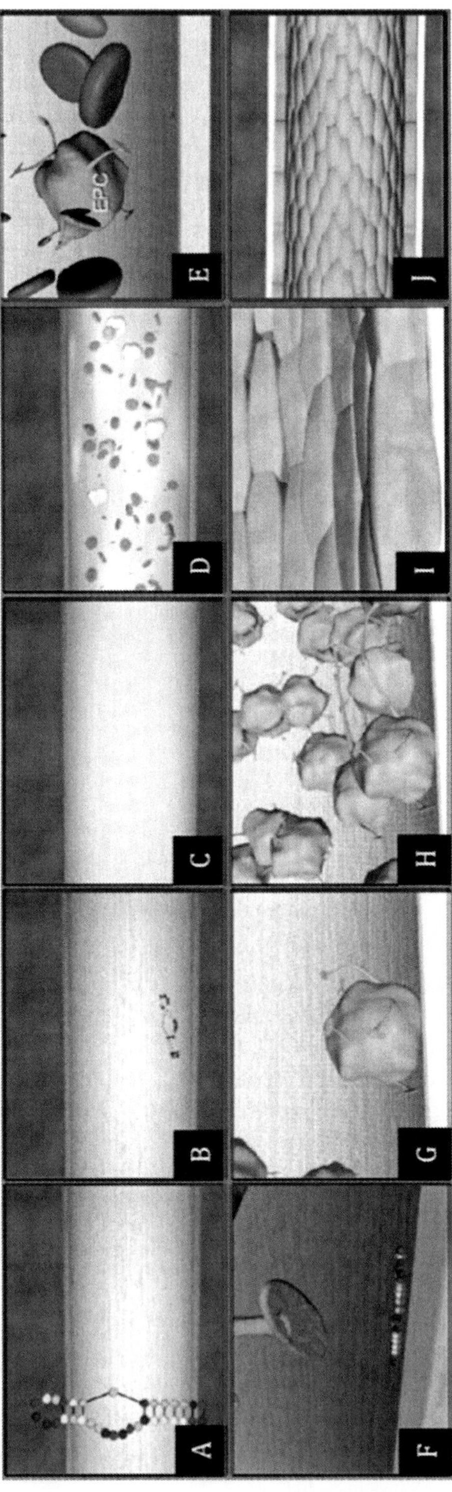

Figure 14.2 *In vivo* engineering concept by immobilization of EPCs on biological surface: (A–C) Aptamer coupling on blood vessel graft; (D–F) EPC immobilization by adhesion to aptamers; (G) EPC differentiation to endothelial cells (EC); (H) EC adhesion; (I) EC proliferation; and (J) EC migration.

structure they bind with ultra-high affinity and specificity to their target molecules, which include proteins, peptides, enzymes, antibodies and cell surface receptors.[45–47] Aptamers are generated from a combinatorial library in a cascade process called systematic evolution of ligands by exponential enrichment (SELEX).[48]

Aptamers have significant advantages compared with other targeting agents. They are very small, bind their target molecules with very high affinity and specificity, cause no immunogenic reactions, and are nontoxic.[46,47] In addition, they are very easily synthesized *in vitro*.[48]

In case of *in vivo* tissue engineering, aptamers can be generated for various cell populations possessing regenerative potential. Aptamers can be covalently immobilized directly onto the graft matrices or onto masking coatings such as starPEG, and bind to circulating target cells. Aptamers targeting porcine CD31-positive cells have already been successfully developed and covalently bound to starPEG-coated small PTFE discs and starPEG-coated small polydimethylsiloxane (PDMS) discs, which were incubated with whole anticoagulated porcine blood in a modified chandler loop model.[49] This study showed that aptamers against porcine CD31-positive cells are able to attach these progenitor cells selectively on synthetic devices.[50] Currently used matrices for bioprosthetic blood vessel substitutes have various limitations. On glutaraldehyde-fixed scaffolds, cells are unable to grow due to glutaraldehyde toxicity.[51] Decellularized xenogeneic tissues still offer a high immunogenic and thrombogenic potential.[52,53]

Coating of scaffolds with starPEG might overcome these limitations. StarPEG is a six-arm star-shaped polyethylenglycol polymer, easily coated to irregular surfaces. StarPEG avoids unspecific protein adsorption and cell adhesion by formation of an ultra-thin polymer network of unique density.[54] Therefore, it is highly qualified to serve as masking substrate for toxic or immunogenic and thrombogenic blood vessel graft surfaces. After polymerization starPEG has free active groups available for use for coupling cell attracting molecules such as aptamers. The coupling of aptamers enhances the adhesion of their target cells to blood vessel surface. Following adhesion, and depending on the scaffold, cells can differentiate, proliferate and migrate on the surface (Figure 14.2). The ultimate goal is to achieve a confluent repopulation and endothelialization of the scaffold. Aptamers are excellently qualified as attractors for circulating progenitor cells. We believe that the application of aptamers will enable a unique target-oriented *in vivo* tissue engineering of scaffolds.

References

1. T. Huynh, G. Abraham, J. Murray, K. Brockbank, P. O. Hagen and S. Sullivan, Remodeling of an acellular collagen graft into a physiologically responsive neovessel, *Nat. Biotechnol.*, 1999, **17**, 1083–1086.
2. R. Langer and J. P. Vacanti, Tissue engineering, *Science*, 1993, **260**, 920–926.

3. R. Skalak andC. F. Fox,*Tissue Engineering*, Liss, New York, 1988 [cited in: R. M. Nerem, *Ann. Biomed. Eng.*, 1991, **19**, 529–533].

4. N. Hibino, E. McGillicuddy, G. Matsumura, Y. Ichihara, Y. Naito, C. Breuer and T. Shinoka, Late-term results of tissue-engineered vascular grafts in humans, *J. Thorac. Cardiovasc. Surg.*, 2010, **139**, 431–436.

5. T. Shinoka, D. Shum-Tim, P. X. Ma, R. E. Tanel, N. Isogai, R. Langer, J. P. Vacanti and J. E. Hayer Jr., Creation of viable pulmonary artery autografts through tissue engineering, *J. Thorac. Cardiovasc. Surg.*, 1998, **115**, 536–545.

6. S. L. Mitchell and L. E. Niklason, Requirements for growing tissue-engineered vascular grafts, *Cardiovasc. Pathol.*, 2003, **12**, 59–64.

7. F. Opitz, K. Schenke-Layland, W. Richter, D. P. Martin, I. Degenkolbe, T. Wahlers and U. A. Stock, Tissue engineering of ovine aortic blood vessel substitutes using applied shear stress and enzymatically derived vascular smooth muscle cells, *Ann. Biomed. Eng.*, 2004, **32**, 212–222.

8. L. E. Niklason, J. Gao, W. M. Abbott, K. K. Hirschi, S. Houser, R. Marini and R. Langer, Functional arteries grown *in vitro*, *Science*, 1999, **284**, 489–493.

9. S. P. Hoerstrup, I. Cummings Mrcs, M. Lachat, F. J. Schoen, R. Jenni, S. Leschka, S. Nevenschwander, D. Schmidt, A. Hol, C. Guenter, M. Goessi, M. Genoni and G. Zund, Functional growth in tissue-engineered living, vascular grafts: follow-up at 100 weeks in a large animal model, *Circulation*, 2006, **114**, I159–166.

10. P. Zilla, R. Fasol, M. Deutsch, T. Fischlein and E. Minar, Endothelial cell seeding of polytetrafluoroethylene vascular grafts in humans: a preliminary report, *J. Vasc. Surg.*, 1987, **6**, 535–541.

11. L. Bordenave, P. Fernandez, M. Rémy-Zolghadri, S. Villars, R. Daculsi and D. Hidy, *In vitro* endothelialized ePTFE prostheses: clinical update 20 years after the first realization, *Clin. Hemorheol. Microcirc.*, 2005, **33**, 227–234.

12. S. F. Williams, D. P. Martin, D. M. Horowitz and O. P. Peoples, PHA applications: addressing the price performance issue. I. Tissue engineering, *Int. J. Biol. Macromol.*, 1999, **25**, 111–121.

13. F. Opitz, K. Schenke-Layland, T. U. Cohnert, B. Starcher, K. J. Halbhuber, D. P. Martin and U. A. Stock, Tissue engineering of aortic tissue: dire consequence of suboptimal elastic fiber synthesis *in vivo*, *Cardiovasc. Res.*, 2004, **63**, 719–730.

14. J. H. Campbell, J. L. Efendy and G. R. Campbell, Novel vascular graft grown within recipient's own peritoneal cavity, *Circ. Res.*, 1999, **85**, 1173–1178.

15. L. E. Niklason, W. Abbott, J. Gao, B. Klagges, K. K. Hirschi, K. Ulubayram, N. Conroy, R. Jones, A. Vasanawala, S. Sanzgin and R. Langer, Morphologic and mechanical characteristics of engineered bovine arteries, *J. Vasc. Surg.*, 2001, **33**, 628–638.

16. A. Patel, B. Fine, M. Sandig and K. Mequanint, Elastin biosynthesis: the missing link in tissue-engineered blood vessels, *Cardiovasc. Res.*, 2006, **71**, 40–49.

17. M. Schleicher, H. P. Wendel, O. Fritze and U. A. Stock, *In vivo* tissue engineering of heart valves: evolution of a novel concept, *Regen. Med.*, 2009, **4**, 613–619.
18. A. de Mel, G. Jell, M. M. Stevens and A. M. Seifalian, Biofunctionalization of biomaterials for accelerated *in situ* endothelialization: a review, *Biomacromolecules*, 2008, **9**, 2969–2979.
19. D. Shum-Tim, U. Stock, J. Hrkach, T. Shinoka, J. Lien, M. A. Moses, A. Stamp, G. Taylor, A. H. Moran, W. Landis, R. Langer, J. P. Vacanti and J. E. Mayer Jr, Tissue engineering of autologous aorta using a new biodegradable polymer, *Ann. Thorac. Surg.*, 1999, **68**, 2298–2304.
20. U. A. Stock, T. Sakamoto, S. Hatsuoka, D. P. Martin, M. Nagashima, A. M. Moran, M. A. Moses, P. N. Khalil, F. J. Schoen, J. P. Vacanti and J. E. Mayer Jr, Patch augmentation of the pulmonary artery with bioabsorbable polymers and autologous cell seeding, *J. Thorac. Cardiovasc. Surg.*, 2000, **120**, 1158–1167.
21. B. Tschoeke, T. C. Flanagan, S. Koch, M. S. Harwoko, T. Deichmann, V. Ellå, J. S. Sachweh, H. Kellomåki, T. Gries, T. Schmitz-Rode and S. Jokenhoevel, Tissue-engineered small-caliber vascular graft based on a novel biodegradable composite fibrin-polylactide scaffold, *Tissue Eng. Part A*, 2009, **15**, 1909–1918.
22. B. Zavan, V. Vindigni, S. Lepidi, I. Iacopetti, G. Avruscio, G. Abatangelo and R. Cortivo, Neoarteries grown *in vivo* using a tissue-engineered hyaluronan-based scaffold, *FASEB J.*, 2008, **22**, 2853–2861.
23. M. E. Tedder, J. Liao, B. Weed, C. Stabler, H. Zhang, A. Simionescu and D. T. Simionescu, Stabilized collagen scaffolds for heart valve tissue engineering, *Tissue Eng. Part A*, 2009, **15**, 1257–1268.
24. G. F. Liu, Z. J. He, D. P. Yang, X. F. Han, T. F. Guo, C. G. Hao, H. Ha and C. L. Nie, Decellularized aorta of fetal pigs as a potential scaffold for small diameter tissue engineered vascular graft, *Chin. Med. J.*, 2008, **121**, 1398–1406.
25. P. M. Dohmen, F. Costa, S. V. Lopes, S. Yoshi, F. P. Souza, R. Vilani, H. B. Costa and W. Konertz, Results of a decellularized porcine heart valve implanted into the juvenile sheep model, *Heart Surg. Forum*, 2005, **8**, E100–104.
26. A. Bader, T. Schilling, O. E. Teebken, G. Brandes, T. Herden, G. Steinhoff and A. Haverich, Tissue engineering of heart valves–human endothelial cell seeding of detergent acellularized porcine valves, *Eur. J. Cardiothorac. Surg.*, 1998, **14**, 279–284.
27. G. J. Wilson, D. W. Courtman, P. Klement, J. M. Lee and H. Yeger, Acellular matrix: a biomaterials approach for coronary artery bypass and heart valve replacement, *Ann. Thorac. Surg.*, 1995, **60**, S353–358.
28. O. Fritze, M. Schleicher, K. König, K. Schenke-Layland, U. Stock and C. Harasztosi, Facilitated non-invasive visualization of collagen and elastin in blood vessels, *Tissue Eng. Part C Methods*, 2010 [Epub ahead of print].
29. J. Zhou, O. Fritze, M. Schleicher, H. P. Wendel, K. Schenke-Layland, C. Harasztosi, S. Hu and U. A. Stock, Impact of heart valve decellularization

on 3-D ultrastructure, immunogenicity and thrombogenicity, *Biomaterials*, 2010, **31**, 2549–2554.

30. L. Bordenave, M. Rémy-Zolghadri, P. Fernandez, R. Bareille and D. Midy, Clinical performance of vascular grafts lined with endothelial cells, *Endothelium*, 1999, **6**, 267–275.

31. S. F. Badylak, R. Record, K. Lindberg, J. Hodde and K. Park, Small intestinal submucosa: a substrate for *in vitro* cell growth, *J. Biomater. Sci. Polym.*, 1998, **9**, 863–878.

32. C. R. Darby, D. Roy, D. Deardon and A. Cornall, Depopulated bovine ureteric xenograft for complex haemodialysis vascular access, *Eur. J. Vasc. Endovasc. Surg.*, 2006, **31**, 181–186.

33. V. Tolva, G. B. Bertoni, S. Trimarchi, V. Grassi, M. Fusari and V. Rampoldi, Unreliability of depopulated bovine ureteric xenograft for infra inguinal bypass surgery: mid-term results from two vascular centres, *Eur. J. Vasc. Endovasc. Surg.*, 2007, **33**, 214–216.

34. J. I. Rotmans, J. M. Heyligers and H. J. Verhagen, *In vivo* cell seeding with anti-CD34 antibodies successfully accelerates endothelialization but stimulates intimal hyperplasia in porcine arteriovenous expanded polytetrafluoroethylene grafts, *Circulation*, 2005, **112**, 12–18.

35. K. P. Walluscheck, G. Steinhoff. S. Kelm and A. Haverich, Improved endothelial cell attachment on ePTFE vascular grafts pretreated with synthetic RGD-containing peptides, *Eur. J. Vasc. Endovasc. Surg.*, 1996, **12**, 321–330.

36. K. P. Walluscheck, G. Steinhoff and A. Haverich, Endothelial cell seeding of deendothelialised human arteries: improvement by adhesion molecule induction and flow-seeding technology, *Eur. J. Vasc. Endovasc. Surg.*, 1996, **12**, 46–53.

37. J. C. Wojciechowski, S. D. Narasipura, N. Charles, D. Mickelsen, K. Rana, M. L. Blair and M. R. King, Capture and enrichment of CD34-positive haematopoietic stem and progenitor cells from blood circulation using P-selectin in an implantable device, *Br. J. Haematol.*, 2008, **140**, 673–681.

38. S. Wu, Y. L. Liu, B. Cui, X. H. Qu and G. Q. Chen, Study on decellularized porcine aortic valve/poly (3-hydroxybutyrate-co-3-hydroxyhexanoate) hybrid heart valve in sheep model, *Artif. Organs.*, 2007, **31**, 689–697.

39. C. Stamm, A. Khosravi, N. Grabow, K. Schmohl, N. Treckmann, A. Drechsel, M. Nan, K. P. Schmitz, A. Haubold and G. Steinhoff, Biomatrix/polymer composite material for heart valve tissue engineering, *Ann. Thorac. Surg.*, 2004, **78**, 2084–2092.

40. M. Crombez, P. Chevallier and R. C. Gaudreault, Improving arterial prosthesis neo-endothelialization: application of a proactive VEGF construct onto PTFE surfaces, *Biomaterials*, 2005, **26**, 7402–7409.

41. S. Kaushal, G. E. Amiel, K. J. Guleserian, O. M. Shapira, T. Perry, F. W. Sutherland, E. Robkin, A. H. Horan, F. J. Schoen, A. Atala, S. Soker, J. Bischoff and J. E. Mayor Jr, Functional small-diameter neovessels created

using endothelial progenitor cells expanded *ex vivo*, *Nat. Med.*, 2001, **7**, 1035–1040.

42. M. G. Frid, V. A. Kale and K. R. Stenmark, Mature vascular endothelium can give rise to smooth muscle cells *via* endothelial-mesenchymal trans-differentiation: *in vitro* analysis, *Circ. Res.*, 2002, **90**, 1189–1196.

43. M. Hristov, W. Erl and P. C. Weber, Endothelial progenitor cells: mobilization, differentiation, and homing, *Arterioscler. Thromb. Vasc. Biol.*, 2003, **23**, 1185–1189.

44. J. Hoffmann, J. Groll, J. Heuts, H. Rong, D. Klee, G. Ziemer, M. Moeller and H. P. Wendel, Blood cell and plasma protein repellent properties of star-PEG-modified surfaces, *J. Biomater. Sci. Polym. Ed.*, 2006, **17**, 985–996.

45. D. S. Wilson and J. W. Szostak, *In vitro* selection of functional nucleic acids, *Annu. Rev. Biochem.*, 1999, **68**, 611–647.

46. S. M. Nimjee, C. P. Rusconi and B. A. Sullenger, Aptamers: an emerging class of therapeutics, *Annu. Rev. Med.*, 2005, **56**, 555–583.

47. E. N. Brody and L. Gold, Aptamers as therapeutic and diagnostic agents, *J. Biotechnol.*, 2000, **74**, 5–13.

48. C. Tuerk and L. Gold, Systematic evolution of ligands by exponential enrichment: RNA ligands to bacteriophage T4 DNA polymerase, *Science*, 1990, **249**, 505–510.

49. A. B. Chandler, *In vitro* thrombotic coagulation of the blood: a method for producing a thrombus, *Lab. Invest.*, 1958, **7**, 110–114.

50. J. Hoffmann, A. Paul, M. Harwardt, J. Groll, T. Reeswinkel, D. Klee, M. Moeller, H. Fischer, T. Walker, T. Greiner, G. Ziemer and H. P. Wendel, Immobilized DNA aptamers used as potent attractors for porcine endothelial precursor cells, *J. Biomed. Mater. Res. A*, 2008, **84**, 614–621.

51. M. J. Maizato, O. Z. Higa, M. B. Mathor, M. A. Camillo and P. J. Spencer, Glutaraldehyde-treated bovine pericardium: effects of lyophilization on cytotoxicity and residual aldehydes, *Artif. Organs.*, 2003, **27**, 692–694.

52. M. T. Kasimir, E. Rieder, G. Seebacher, A. Nigisch, B. Dekan, E. Wolner, G. Weigel and P. Simon, Decellularization does not eliminate thrombogenicity and inflammatory stimulation in tissue-engineered porcine heart valves, *J. Heart Valve Dis.*, 2006, **15**, 278–286.

53. M. T. Kasimir, G. Weigel, J. Sharma, E. Rieder, G. Seebacher, E. Wolner and P. Simon, The decellularized porcine heart valve matrix in tissue engineering: platelet adhesion and activation, *Thromb. Haemost.*, 2005, **94**, 562–567.

54. J. Groll, W. Haubensak, T. Ameringer and M. Moeller, Ultrathin coatings from isocyanate terminated star PEG prepolymers: patterning of proteins on the layers, *Langmuir*, 2005, **21**, 3076–3083.

Regenerative Strategies for the Endocrine Pancreas: From Islets to Stem Cells and Tissue Reprogramming

JUAN DOMÍNGUEZ-BENDALA AND CAMILLO RICORDI

Diabetes Research Institute, University of Miami, 1450 NW 10th Avenue, Miami, FL 33136, USA

15.1 Current Cell Therapies for Diabetes: Limitations

Beta cells cluster with other endocrine cells in discrete structures termed the islets of Langerhans. The cause of type I diabetes (also known as juvenile diabetes, even if it is not exclusively diagnosed in young patients) is an auto-immune process that eventually destroys these beta cells. This is manifested in a dysregulation of glucose metabolism, which can only be treated by either exogenous insulin administration or transplantation. The former cannot replicate the complex metabolic control exerted by the islets in healthy subjects and prolonged insulin use is often associated with the onset of life-threatening complications.[1–12] As for the latter, solid organ transplantation is highly effective but risky.[13–18] Islet transplantation, in contrast, was saluted from very early on as a safer and less invasive alternative.[19–24] The process involves the enzymatic/mechanical dissociation of the islets from the rest of the tissues of the pancreas,[25] followed by a centrifugation that enriches for the fractions with the higher proportion of islets. These are infused by gravity in the portal vein of the patient and become entrapped as a result of size restrictions as the blood vessels decrease in diameter.[26] This embolization is followed by revascularization within a few weeks.[27–29] In the early times of islet transplantation, rapid loss of graft function was a common consequence of the use of steroid-based immunosuppression. This changed in 2000 with the introduction of a glucocorticoid-free regime that extended very substantially the survival and function of the graft.[30]

For all the success of islet transplantation, complete insulin independence may not be achieved until a critical mass of islets has been infused. Often, this may require more than one donor.[30–32] The compounded problem of the scarce availability of organs for processing and the need for immunosuppression are clear indications that, in its current form, this therapy will not become the treatment of choice for type I diabetes anytime soon.

15.2   Regeneration and Adult Stem Cells

The premise—and promise—of stem cells for the treatment of diabetes is the same for both adult and embryonic types: they could represent a potentially unlimited, self-renewable reservoir of beta cells. An ideal situation would be one in which stem cells from the affected organ could be either (a) primed to regenerate it *in vivo* or (b) extracted, propagated *in vitro*, differentiated and re-implanted in the patient. An ongoing controversy surrounds the field, as the evidence for stem cells in the adult pancreas with the ability to repopulate the islets is often contradictory. On the one hand, proponents of the 'ductal stem cell' hypothesis[33] argue that the repeated observation that single beta cells often appear to bud out from the ductal tissue is not coincidental. Additional evidence has been gathered over the years from numerous *in vitro* studies in which cultured ductal cells could be led to express endocrine cell markers after certain treatments.[34–36] Increases in the expression of the master beta cell regulator Pdx1,[37] insulin[38] and overall 'beta cell sprouting' in ductal tissue samples have been observed following either pathological or experimental injury of the pancreas.[39–41]

On the other hand, Dor and collaborators,[42] using sophisticated lineage-tracing techniques, established in 2004 that normal turnover and regeneration of beta cells occurs mainly by replication and not from stem cells. This seemed to settle the debate for a while, until additional lineage-tracing experiments indicated that the ducts may indeed contribute to the endocrine component of the pancreas after all.[43,44] To confuse things even more, an elegant series of experiments conducted by Xu and colleagues[45] determined that the fetal endocrine developmental program can be re-ignited in the adult pancreas. Indeed, a specific injury model (pancreatic duct ligation) induced the reappearance in the pancreas of *bona fide* Neurogenin-3$^+$ endocrine stem cells, largely absent from the adult organ under any other circumstance. Of note, these cells were not necessarily associated either with the ducts or with the islets.

While the evidence is therefore inconsistent with a universal explanation of islet regeneration, to date nobody has been able to claim the isolation and expansion of any putative endocrine stem cell from the adult pancreas. Therefore, researchers have been looking elsewhere to find adult stem cell types that could be used for regenerative purposes. The workhorse of all adult stem cells remains the mesenchymal stem cell (MSC), a plastic-adherent, rapidly dividing cell type that can be isolated from many tissues, including the pancreas.[46] The induced differentiation *in vitro* along the pancreatic cell lineage and subsequent transplantation of MSCs has been only partially successful in animals.[47–58] However, the direct administration of undifferentiated MSCs to diabetic animals has yielded some intriguing results,[59–62] perhaps as an indirect consequence of the well-known immunomodulatory, anti-inflammatory, pro-angiogenic and trophic properties of these cells.[63–70]

Another adult stem cell used in a similar setting is the hematopoietic stem cell (HSC), which can be easily banked for wide human leukocyte antigen (HLA) representation (chiefly from cord blood)[71] or harvested from the very same patient. In addition to their use in novel interventions aimed at resetting the immunological clock of diabetes,[72,73] bone marrow-derived cells are also being directly administered into the pancreas of the patient through interventional radiology techniques. While there is still a void on the cellular/molecular mechanisms by which HSCs may exert their action when delivered in such fashion, this approach has already generated highly promising results on type II diabetic patients when used in combination with hyperbaric oxygen therapy.[74] Several Phase I/II trials looking at the local delivery of MSCs and bone marrow cells are presently underway (www.clinicaltrials.gov).

15.3   Embryonic Stem Cells

The only self-renewable cells with the unequivocal ability to generate beta cells are the embryonic stem (ES) cells obtained from the inner cell mass of the early embryo.[75] However, progress at inducing this process *in vitro* has been a rather tortuous road, plagued by wrong turns,[76–78] safety concerns[79–81] and overall low efficiency.[81,82] It is now generally accepted that the method described by D'Amour *et al.*,[81] with its subsequent modifications,[83,84] is the best representation of the state-of-the-art.

This protocol is based on a first critical step that had eluded many researchers before, namely the effective generation of definitive endoderm from human embryonic stem (hES) cells.[85] Still, even in its most advanced shape, it still yields less than 10% of beta cells—which, in many instances, fail to respond to glucose stimulation *in vitro*. Attempts at improving both parameters (yield and function) led to the development of an alternative approach in which the cells were allowed to complete their maturation upon transplantation in animals. However, the solution of the previous problems[83] came at the expense of a very high (>20%) incidence of teratogenic lesions, caused by the *in vivo* expansion of carryover undifferentiated cells. The much-awaited use of hES cells for type I diabetes clinical trials will not happen until these concerns have been completely addressed. Some of the strategies presented thus far to prevent the formation of tumors (and/or eliminate them once formed) include:

- extensive screening for undifferentiated cellular by-products;[86]
- the genetic engineering of hES cells with specific suicide genes that would be activated only in the cells that remained in a proliferative state after differentiation has been induced;[87–91] and
- the use of encapsulation techniques to physically contain the spread of the tumor.[92–95]

15.4   Reprogramming

Reprogramming is a newly coined term for an old concept, trans-differentiation. The general idea is to convert a terminally differentiated tissue into another, which, if applied to the generation of functional beta cells, may have therapeutic applications. Abundant cases of pancreas-to-liver[96–99] and liver-to-pancreas[100,101] interconversion have been extensively described in the literature—a phenomenon undoubtedly linked to the shared ancestry of both organs.[102–110]

Among the first attempts at inducing liver trans-differentiation was a report by Ferber *et al.* earlier this decade.[111] Their strategy was to ectopically express in the liver parenchyma the gene *Pdx1*, a key regulator of pancreatic development[112] and adult beta cell homeostasis.[113] In order to accomplish this, they injected diabetic mice with an adenovirus harboring a constitutively active *Pdx1* cassette. The expression of the foreign gene was preferentially observed in the liver, where it activated several beta cell-specific genes. A substantial reduction in blood sugar levels was reported, and insulin expression persisted in the liver long after the adenovirus had been cleared from the system.[114] This was suggestive of an irreversible activation of a pancreatic program in at least some of the hepatic cells that expressed the gene transiently. Additional research into the molecular mechanism of this liver-to-pancreas transdifferentiation indicated that these cells may need to rapidly 'de-differentiate' before adopting the pancreatic program.[115]

Several other groups reported similar results with slight variations over the main theme, including the simultaneous use of several 'reprogramming' genes.[116–123] However, with the possible exception of the full liver-to-pancreas conversion reported in transgenic frogs by Horb *et al.*,[124] most of these approaches have yielded at best a 'hybrid' between hepatocytes and true beta cells.

Success at achieving full reprogramming towards beta cells might require the concerted choice of an even closer starting material and the appropriate combination of genes. This was the reasoning behind the experiments recently conducted by Zhou *et al.*,[125] who screened 20 pancreatic transcription factors to come up with a combination of three (Pdx1, Ngn3 and MafA) whose viral-mediated delivery to the pancreas of recipient mice led to the trans-differentiation of exocrine tissue into beta cells. These newly created cells could be detected as early as three days after the adenoviral injection, and their numbers kept increasing for up to three months. Consistent with the results previously reported by Ber and collaborators,[114] this observation seems to confirm that an initial 'push' is sufficient to induce the reprogramming, which then becomes permanent despite the clearance of the virus from the system.

As encouraging as this may be from a translational perspective, few would argue that direct *in vivo* reprogramming would be a risky choice, especially because of the potential side-effects of adenoviruses. Nevertheless, these experiments represent a valid proof of principle that could open the door to the utilization of the plentiful leftover tissue that is currently discarded after islet isolation procedures. In yet another clinical setting, candidate tissues could be

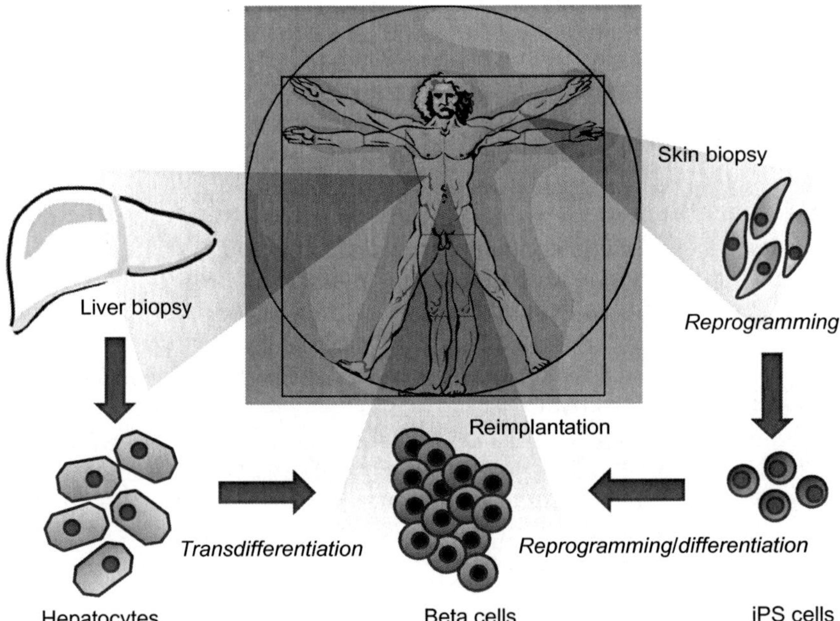

Figure 15.1 Theoretical *ex vivo* reprogramming approaches for the autologous treatment of type I diabetes. A liver biopsy from the patient should provide enough cells for direct *in vitro* reprogramming of hepatocytes into beta cells, which could be subsequently re-implanted in the patient. Alternatively, cells from the skin or other accessible tissues of the patient could be reprogrammed (de-differentiated) into iPS cells and later reprogrammed and/or differentiated into beta cells. While de-differentiation adds another step to the direct trans-differentiation approach, it may actually result in a better and more controlled recapitulation of beta cell ontogenesis. Non-autologous interventions may also be developed from the *in vitro* trans-differentiation of the acinar byproducts of regular islet isolations.

biopsied from the very patient, expanded and reprogrammed *ex vivo* and then implanted back. While autoimmunity would still have to be addressed separately, such approach would get around the issue of allorejection. In this context, another potential use of reprogramming would be to generate patient-matched induced pluripotent (iPS) cells. These cells, which are functionally indistinguishable from hES cells, can be derived from a variety of mature somatic cells following the forced expression of a minimal number of pluripotency genes, chiefly *Oct3/4* and *Sox2*.[126–128]

The development of this revolutionary technique has indeed jump-started the field of personalized regenerative medicine, entirely bypassing the need for the much more unwieldy and ethically troublesome somatic cell nuclear transfer (SCNT). We have already witnessed the generation of iPS cells from type I diabetic patients[129] and other disease models are likely to follow suit soon. The approach here might require a double round of reprogramming experiments: one to obtain self-renewable iPS cells genetically matched to the patient and another to differentiate them along the beta cell lineage (Figure 15.1). While the use of viruses as reprogramming agents is not indicated for clinical purposes, progress at finding non-viral alternatives has been nothing short of frantic over the past couple of years. These include the use of protein transduction,[130–132] episomal vectors[133] and small molecules,[134–137] as well as the use of adult stem cells (more predisposed to be reprogrammed) as substrates for the generation of iPS cells.[138]

15.5 Conclusions

Although still in its infancy, the field of pancreatic regeneration has already spawned multiple avenues of research that have proven their worth in animal models. The clinical translation of these findings is always challenging, but we are just starting to unveil the possibilities. A prediction about which one of these methods or cell types will eventually lead to an effective regeneration therapy would be futile at this point. In fact, many of these approaches, rather than being mutually exclusive, are likely to end up employed in concert. For instance, we could envision a strategy wherein patient-tailored iPS cells are differentiated into beta cells *in vitro* and then co-transplanted with MSCs (also isolated from the patient) for enhanced engraftment and function. Or we could expand the latter *ex vivo* and then reprogram them in one single step along the pancreatic endocrine lineage. Either of the above could be additionally aided by HSC-based interventions designed to reboot the immune system. The options are almost endless, and the groundbreaking success in the field of islet transplantation will unquestionably facilitate their path to the clinic.

Acknowledgements

J. D-B. and C. R. are funded by the Diabetes Research Institute Foundation (DRIF), the National Institutes of Health (NIH) and the Juvenile Diabetes Research Foundation (JDRF).

References

1. S. B. Chaudhary, F.A. Liporace, A. Gandhi, B. G. Donley, M. S. Pinzur and S. S. Lin, Complications of ankle fracture in patients with diabetes, *J. Am. Acad. Orthop. Surg.*, 2008, **16**, 159–170.
2. DCCTRG, The effect of intensive treatment of diabetes on the development and progression of long-term complications in insulin-dependent diabetes mellitus. The Diabetes Control and Complications Trial Research Group, *N. Engl. J. Med.*,1993, **329**, 977–986.
3. A. D. Deshpande, M. Harris-Hayes and M. Schootman, Epidemiology of diabetes and diabetes-related complications, *Phys. Ther.*, 2008, **88**, 1254–1264.
4. A. Kaparianos, E. Argyropoulou, F. Sampsonas, K. Karkoulias, M. Tsiamita and K. Spiropoulos, Pulmonary complications in diabetes mellitus, *Chron. Respir. Dis.*, 2008, **5**, 101–108.
5. P. Kar and R. I. Holt, The effect of sulphonylureas on the microvascular and macrovascular complications of diabetes, *Cardiovasc. Drugs Ther.*, 2008, **22**, 207–213.
6. S. H. Meeuwisse-Pasterkamp, M. M. van der Klauw and B. H. Wolffenbuttel, Type 2 diabetes mellitus: prevention of macrovascular complications, *Expert Rev. Cardiovasc. Ther.*, 2008, **6**, 323–341.
7. B. M. Nathan and A. Moran, Metabolic complications of obesity in childhood and adolescence: more than just diabetes, *Curr. Opin. Endocrinol. Diabetes Obes.*, 2008, **15**, 21–29.
8. B. Rosenn, Obesity and diabetes: a recipe for obstetric complications, *J. Matern. Fetal Neonatal Med.*, 2008, **21**, 159–164.
9. A. Shakil, R. J. Church and S. S. Rao, Gastrointestinal complications of diabetes, *Am. Fam. Physician*, 2008, **77**, 1697–1702.
10. A. O. Stirban and D. Tschoepe, Cardiovascular complications in diabetes: targets and interventions, *Diabetes Care*, 2008, **31**(Suppl 2), S215–221.
11. G. W. Taylor and W. S. Borgnakke, Periodontal disease: associations with diabetes, glycemic control and complications, *Oral Dis.*, 2008, **14**, 191–203.
12. S. Watkinson and R. Seewoodhary, Ocular complications associated with diabetes mellitus, *Nurs. Stand.*, 2008, **22**, 51–57, quiz 58, 60.
13. G. W. Burke, G. Ciancio and H. W. Sollinger, Advances in pancreas transplantation, *Transplantation*, 2004, **77**, S62–67.
14. A. J. Fabrega, P. A. Rivas and R. Pollak, Pancreas-kidney transplantation for intensivists: perioperative care and complications, *J. Intensive Care Med.*, 1994, **9**, 281–289.
15. C. E. Freise, S. Narumi, P. G. Stock and J. S. Melzer, Simultaneous pancreas-kidney transplantation: an overview of indications, complications, and outcomes, *West J. Med.*, 1999, **170**, 11–18.
16. R. Landgraf, Impact of pancreas transplantation on diabetic secondary complications and quality of life, *Diabetologia*, 1996, **39**, 1415–1424.
17. L. B. Melton, Pancreas transplantation, *Semin. Nephrol.*, 1992, **12**, 256–266.

18. J. D. Pirsch, C. Andrews, D. E. Hricik, M. A. Josephson, A. B. Leichtman, C. Y. Lu, L. B. Melton, V. K. Rao, R. R. Riggio, R. J. Stratta and M. R. Weir, Pancreas transplantation for diabetes mellitus, *Am. J. Kidney Dis.*, 1996, **27**, 444–450.
19. L. Inverardi, N. S. Kenyon and C. Ricordi, Islet transplantation: immunological perspectives, *Curr. Opin. Immunol.*, 2003, **15**, 507–511.
20. S. Merani and A. M. Shapiro, Current status of pancreatic islet transplantation, *Clin. Sci. (Lond.)*, 2006, **110**, 611–625.
21. T. B. Murdoch, D. McGhee-Wilson, A. M. Shapiro and J. R. Lakey, Methods of human islet culture for transplantation, *Cell Transplant.*, 2004, **13**, 605–617.
22. A. Pileggi, R. Alejandro and C. Ricordi, Clinical islet transplantation, *Minerva Endocrinol.*, 2006, **31**, 219–232.
23. C. Ricordi and T. B. Strom, Clinical islet transplantation: advances and immunological challenges, *Nat. Rev. Immunol.*, 2004, **4**, 259–268.
24. E. Seung, J. P. Mordes, D. L. Greiner and A. A. Rossini, Induction of tolerance for islet transplantation for type 1 diabetes, *Curr. Diab Rep.*, 2003, **3**, 329–335.
25. C. Ricordi, P. E. Lacy, E. H. Finke, B. J. Olack and D. W. Scharp, Automated method for isolation of human pancreatic islets, *Diabetes*, 1988, **37**, 413–420.
26. O. Korsgren, T. Lundgren, M. Felldin, A. Foss, B. Isaksson, J. Permert, N. H. Persson, E. Rafael, M. Ryden, K. Salmela, A. Tibell, G. Tufveson and B. Nilsson, Optimising islet engraftment is critical for successful clinical islet transplantation, *Diabetologia*, 2008, **51**, 227–232.
27. P. O. Carlsson and G. Mattsson, Oxygen tension and blood flow in relation to revascularization in transplanted adult and fetal rat pancreatic islets, *Cell Transplant.*, 2002, **11**, 813–820.
28. P. O. Carlsson, F. Palm and G. Mattsson, Low revascularization of experimentally transplanted human pancreatic islets, *J. Clin. Endocrinol. Metab.*, 2002, **87**, 5418–5423.
29. E. Lammert, G. Gu, M. McLaughlin, D. Brown, R. Brekken, L. C. Murtaugh, H. P. Gerber, N. Ferrara and D. A. Melton, Role of VEGF-A in vascularization of pancreatic islets, *Curr. Biol.*, 2003, **13**, 1070–1074.
30. A. M. Shapiro, J. R. Lakey, E. A. Ryan, G. S. Korbutt, E. Toth, G. L. Warnock, N. M. Kneteman and R. V. Rajotte, Islet transplantation in seven patients with type 1 diabetes mellitus using a glucocorticoid-free immunosuppressive regimen, *N. Engl. J. Med.*, 2000, **343**, 230–238.
31. T. Froud, C. Ricordi, D. A. Baidal, M. M. Hafiz, G. Ponte, P. Cure, A. Pileggi, R. Poggioli, H. Ichii, A. Khan, J. V. Ferreira, A. Pugliese, V. V. Esquenazi, N. S. Kenyon and R. Alejandro, Islet transplantation in type 1 diabetes mellitus using cultured islets and steroid-free immunosuppression: Miami experience, *Am. J. Transplant.*, 2005, **5**, 2037–2046.
32. J. F. Markmann, S. Deng, X. Huang, N. M. Desai, E. H. Velidedeoglu, C. Lui, A. Frank, E. Markmann, M. Palanjian, K. Brayman, B. Wolf, E. Bell, M. Vitamaniuk, N. Doliba, F. Matschinsky, C. F. Barker and

A. Naji, Insulin independence following isolated islet transplantation and single islet infusions, *Ann. Surg.*, 2003, **237**, 741–749: , discussion 749–750.

33. S. Bonner-Weir, L. A. Baxter, G. T. Schuppin and F. E. Smith, A second pathway for regeneration of adult exocrine and endocrine pancreas. A possible recapitulation of embryonic development, *Diabetes*, 1993, **42**, 1715–1720.

34. S. Bonner-Weir, M. Taneja, G. C. Weir, K. Tatarkiewicz, K. H. Song, A. Sharma and J. J. O'Neil, *In vitro* cultivation of human islets from expanded ductal tissue, *Proc. Natl. Acad. Sci. U. S. A.*, 2000, **97**, 7999–8004.

35. H. Noguchi, G. Xu, S. Matsumoto, H. Kaneto, N. Kobayashi, S. Bonner-Weir and S. Hayashi, Induction of pancreatic stem/progenitor cells into insulin-producing cells by adenoviral-mediated gene transfer technology, *Cell Transplant.*, 2006, **15**, 929–938.

36. S. Yatoh, R. Dodge, T. Akashi, A. Omer, A. Sharma, G. C. Weir and S. Bonner-Weir, Differentiation of affinity-purified human pancreatic duct cells to beta-cells, *Diabetes*, 2007, **56**, 1802–1809.

37. A. Sharma, D. H. Zangen, P. Reitz, M. Taneja, M. E. Lissauer, C. P. Miller, G. C. Weir, J. F. Habener and S. Bonner-Weir, The homeodomain protein IDX-1 increases after an early burst of proliferation during pancreatic regeneration, *Diabetes*, 1999, **48**, 507–513.

38. A. Martin-Pagola, G. Sisino, G. Allende, J. Dominguez-Bendala, R. Gianani, H. Reijonen, G. T. Nepom, C. Ricordi, P. Ruiz, J. Sageshima, G. Ciancio, G. W. Burke and A. Pugliese, Insulin protein and proliferation in ductal cells in the transplanted pancreas of patients with type 1 diabetes and recurrence of autoimmunity, *Diabetologia*, 2008, **51**, 1803–1813.

39. L. Rosenberg and A. I. Vinik, Trophic stimulation of the ductular-islet cell axis: a new approach to the treatment of diabetes, *Adv. Exp. Med. Biol.*, 1992, **321**, 95–104, discussion, 105–109.

40. W. Gepts, Pathologic anatomy of the pancreas in juvenile diabetes mellitus, *Diabetes*, 1965, **14**, 619–633.

41. C. V. Weaver, R. L. Sorenson and H. C. Kaung, Immunocytochemical localization of insulin-immunoreactive cells in the pancreatic ducts of rats treated with trypsin inhibitor, *Diabetologia*, 1985, **28**, 781–785.

42. Y. Dor, J. Brown, O. I. Martinez and D. A. Melton, Adult pancreatic beta-cells are formed by self-duplication rather than stem-cell differentiation, *Nature*, 2004, **429**, 41–46.

43. S. Bonner-Weir, A. Inada, S. Yatoh, W. C. Li, T. Aye, E. Toschi and A. Sharma, Transdifferentiation of pancreatic ductal cells to endocrine beta-cells, *Biochem. Soc. Trans.*, 2008, **36**, 353–356.

44. A. Inada, C. Nienaber, H. Katsuta, Y. Fujitani, J. Levine, R. Morita, A. Sharma and S. Bonner-Weir, Carbonic anhydrase II-positive pancreatic cells are progenitors for both endocrine and exocrine pancreas after birth, *Proc. Natl. Acad. Sci. U. S. A.*, 2008, **105**, 19915–19919.

45. X. Xu, J. D'Hoker, G. Stange, S. Bonne, N. De Leu, X. Xiao, M. Van de Casteele, G. Mellitzer, Z. Ling, D. Pipeleers, L. Bouwens, R. Scharfmann, G. Gradwohl and H. Heimberg, Beta cells can be generated from endogenous progenitors in injured adult mouse pancreas, *Cell*, 2008, **132**, 197–207.
46. L. da Silva Meirelles, P. C. Chagastelles and N. B. Nardi, Mesenchymal stem cells reside in virtually all post-natal organs and tissues, *J. Cell. Sci.*, 2006, **119**, 2204–2213.
47. L. B. Chen, X. B. Jiang and L. Yang, Differentiation of rat marrow mesenchymal stem cells into pancreatic islet beta-cells, *World J. Gastroenterol.*, 2004, **10**, 3016–3020.
48. K. S. Choi, J. S. Shin, J. J. Lee, Y. S. Kim, S. B. Kim and C. W. Kim, *In vitro* trans-differentiation of rat mesenchymal cells into insulin-producing cells by rat pancreatic extract, *Biochem. Biophys. Res. Commun.*, 2005, **330**, 1299–1305.
49. J. Sun, Y. Yang, X. Wang, J. Song and Y. Jia, Expression of Pdx-1 in bone marrow mesenchymal stem cells promotes differentiation of islet-like cells *in vitro*, *Sci. China C, Life Sci.*, 2006, **49**, 480–489.
50. Y. Li, R. Zhang, H. Qiao, H. Zhang, Y. Wang, H. Yuan, Q. Liu, D. Liu, L. Chen and X. Pei, Generation of insulin-producing cells from PDX-1 gene-modified human mesenchymal stem cells, *J. Cell. Physiol.*, 2007, **211**, 36–44.
51. O. Karnieli, Y. Izhar-Prato, S. Bulvik and S. Efrat, Generation of insulin-producing cells from human bone marrow mesenchymal stem cells by genetic manipulation, *Stem Cells*, 2007, **25**, 2837–2844.
52. J. Xu, Y. Lu, F. Ding, X. Zhan, M. Zhu and Z. Wang, Reversal of diabetes in mice by intrahepatic injection of bone-derived GFP-murine mesenchymal stem cells infected with the recombinant retrovirus-carrying human insulin gene, *World J. Surg.*, 2007, **31**, 1872–1882.
53. T. Masaka, M. Miyazaki, G. Du, M. Hardjo, M. Sakaguchi, M. Takaishi, K. Kataoka, K. Yamamoto and N. H. Huh, Derivation of hepato-pancreatic intermediate progenitor cells from a clonal mesenchymal stem cell line of rat bone marrow origin, *Int. J. Mol. Med.*, 2008, **22**, 447–452.
54. C. Moriscot, F. de Fraipont, M. J. Richard, M. Marchand, P. Savatier, D. Bosco, M. Favrot and P. Y. Benhamou, Human bone marrow mesenchymal stem cells can express insulin and key transcription factors of the endocrine pancreas developmental pathway upon genetic and/or microenvironmental manipulation *in vitro*, *Stem Cells*, 2005, **23**, 594–603.
55. X. H. Wu, C. P. Liu, K. F. Xu, X. D. Mao, J. Zhu, J. J. Jiang, D. Cui, M. Zhang, Y. Xu and C. Liu, Reversal of hyperglycemia in diabetic rats by portal vein transplantation of islet-like cells generated from bone marrow mesenchymal stem cells, *World J. Gastroenterol.*, 2007, **13**, 3342–3349.
56. E. Hisanaga, K. Y. Park, S. Yamada, H. Hashimoto, T. Takeuchi, M. Mori, M. Seno, K. Umezawa, I. Takei and I. Kojima, A simple method to induce differentiation of murine bone marrow mesenchymal

cells to insulin-producing cells using conophylline and betacellulin-delta4, *Endocrin. J.*, 2008, **55**, 535–543.

57. C. Chang, D. Niu, H. Zhou, F. Li and F. Gong, Mesenchymal stem cells contribute to insulin-producing cells upon microenvironmental manipulation *in vitro*, *Transplant. Proc.*, 2007, **39**, 3363–3368.

58. R. M. Baertschiger, D. Bosco, P. Morel, V. Serre-Beinier, T. Berney, L. H. Buhler and C. Gonelle-Gispert, Mesenchymal stem cells derived from human exocrine pancreas express transcription factors implicated in beta-cell development, *Pancreas*, 2008, **37**, 75–84.

59. Q. Y. Dong, L. Chen, G. Q. Gao, L. Wang, J. Song, B. Chen, Y. Z. Xu and L. Sun, Allogeneic diabetic mesenchymal stem cells transplantation in streptozotocin-induced diabetic rat, *Clin. Invest Med.*, 2008, **31**, E328–337.

60. C. Chang, X. Wang, D. Niu, Z. Zhang, H. Zhao and F. Gong, Mesenchymal stem cells adopt beta-cell fate upon diabetic pancreatic microenvironment, *Pancreas*, 2009, **38**, 275–281.

61. C. F. Chang, K. H. Hsu, S. H. Chiou, L. L. Ho, Y. S. Fu and S. C. Hung, Fibronectin and pellet suspension culture promote differentiation of human mesenchymal stem cells into insulin producing cells, *J. Biomed. Mater. Res. A*, 2008, **86**, 1097–1105.

62. F. E. Ezquer, M. E. Ezquer, D. B. Parrau, D. Carpio, A. J. Yanez and P. A. Conget, Systemic administration of multipotent mesenchymal stromal cells reverts hyperglycemia and prevents nephropathy in type 1 diabetic mice, *Biol. Blood Marrow Transplant.*, 2008, **14**, 631–640.

63. R. Abdi, P. Fiorina, C. N. Adra, M. Atkinson and M. H. Sayegh, Immunomodulation by mesenchymal stem cells: a potential therapeutic strategy for type 1 diabetes, *Diabetes*, 2008, **57**, 1759–1767.

64. P. K. Mishra, Bone marrow-derived mesenchymal stem cells for treatment of heart failure: is it all paracrine actions and immunomodulation?, *J. Cardiovasc. Med. (Hagerstown)*, 2008, **9**, 122–128.

65. K. Le Blanc and O. Ringden, Immunomodulation by mesenchymal stem cells and clinical experience, *J. Intern Med.*, 2007, **262**, 509–525.

66. K. Ozaki, K. Sato, I. Oh, A. Meguro, R. Tatara, K. Muroi and K. Ozawa, Mechanisms of immunomodulation by mesenchymal stem cells, *Int. J. Hematol.*, 2007, **86**, 5–7.

67. Y. X. Xu, L. Chen, R. Wang, W. K. Hou, P. Lin, L. Sun, Y. Sun and Q.Y. Dong, Mesenchymal stem cell therapy for diabetes through paracrine mechanisms, *Med. Hypotheses*, 2008, **71**, 390–393.

68. S. Sadat, S. Gehmert, Y. H. Song, Y. Yen, X. Bai, S. Gaiser, H. Klein and E. Alt, The cardioprotective effect of mesenchymal stem cells is mediated by IGF-I and VEGF, *Biochem. Biophys. Res. Commun.*, 2007, **363**, 674–679.

69. Y. Wu, L. Chen, P. G. Scott and E. E. Tredget, Mesenchymal stem cells enhance wound healing through differentiation and angiogenesis, *Stem Cells*, 2007, **25**, 2648–2659.

70. S. G. Ball, C. A. Shuttleworth and C. M. Kielty, Mesenchymal stem cells and neovascularization: role of platelet-derived growth factor receptors, *J. Cell. Mol. Med.*, 2007, **11**, 1012–1030.

71. D. S. Hong and H. J. Deeg, Hemopoietic stem cells: sources and applications, *Med. Oncol.*, 1994, **11**, 63–68.
72. J. C. Voltarelli, C. E. Couri, A. B. Stracieri, M. C. Oliveira, D. A. Moraes, F. Pieroni, M. Coutinho, K. C. Malmegrim, M. C. Foss-Freitas, B. P. Simoes, M. C. Foss, E. Squiers and R. K. Burt, Autologous non-myeloablative hematopoietic stem cell transplantation in newly diagnosed type 1 diabetes mellitus, *JAMA*, 2007, **297**, 1568–1576.
73. C. E. Couri and J. C. Voltarelli, Autologous stem cell transplantation for early type 1 diabetes mellitus, *Autoimmunity*, 2008, **41**, 666–672.
74. E. Estrada, F. Valacchi, E. Nicora, S. Brieva, C. Esteve, L. Echevarria, T. Froud, K. Bernetti, S. Messinger Cayetano, O. Velazquez, R. Alejandro and C. Ricordi, Combined treatment of intrapancreatic autologous bone marrow stem cells and hyperbaric oxygen in Type 2 diabetes mellitus, *Cell Transplant.*, 2008, **17**, 1295–1304.
75. J. A. Thomson, J. Itskovitz-Eldor, S. S. Shapiro, M. A. Waknitz, J. J. Swiergiel, V. S. Marshall and J. M. Jones, Embryonic stem cell lines derived from human blastocysts, *Science*, 1998, **282**, 1145–1147.
76. N. Lumelsky, O. Blondel, P. Laeng, I. Velasco, R. Ravin and R. McKay, Differentiation of embryonic stem cells to insulin-secreting structures similar to pancreatic islets, *Science*, 2001, **292**, 1389–1394.
77. H. Baharvand, H. Jafary, M. Massumi and S. K. Ashtiani, Generation of insulin-secreting cells from human embryonic stem cells, *Dev. Growth Differ.*, 2006, **48**, 323–332.
78. M. Hansson, A. Tonning, U. Frandsen, A. Petri, J. Rajagopal, M. C. Englund, R. S. Heller, J. Hakansson, J. Fleckner, H. N. Skold, D. Melton, H. Semb and P. Serup, Artifactual insulin release from differentiated embryonic stem cells, *Diabetes*, 2004, **53**, 2603–2609.
79. T. Fujikawa, S. H. Oh, L. Pi, H. M. Hatch, T. Shupe and B. E. Petersen, Teratoma formation leads to failure of treatment for type I diabetes using embryonic stem cell-derived insulin-producing cells, *Am. J. Pathol.*, 2005, **166**, 1781–1791.
80. P. Blyszczuk, J. Czyz, G. Kania, M. Wagner, U. Roll, L. St-Onge and A. M. Wobus, Expression of Pax4 in embryonic stem cells promotes differentiation of nestin-positive progenitor and insulin-producing cells, *Proc. Natl. Acad. Sci. U. S. A.*, 2003, **100**, 998–1003.
81. K. A. D'Amour, A. G. Bang, S. Eliazer, O. G. Kelly, A. D. Agulnick, N. G. Smart, M. A. Moorman, E. Kroon, M. K. Carpenter and E. E. Baetge, Production of pancreatic hormone-expressing endocrine cells from human embryonic stem cells, *Nat. Biotechnol.*, 2006, **24**, 1392–1401.
82. J. Jiang, M. Au, K. Lu, A. Eshpeter, G. Korbutt, G. Fisk and A. S. Majumdar, Generation of insulin-producing islet-like clusters from human embryonic stem cells, *Stem Cells*, 2007, **25**, 1940–1953.
83. E. Kroon, L. A. Martinson, K. Kadoya, A. G. Bang, O. G. Kelly, S. Eliazer, H. Young, M. Richardson, N. G. Smart, J. Cunningham, A. D. Agulnick, K. A. D'Amour, M. K. Carpenter and E. E. Baetge, Pancreatic

endoderm derived from human embryonic stem cells generates glucose-responsive insulin-secreting cells *in vivo*, *Nat. Biotechnol.*, 2008, **26**, 443–452.

84. A. B. McLean, K. A. D'Amour, K. L. Jones, M. Krishnamoorthy, M. J. Kulik, D. M. Reynolds, A. M. Sheppard, H. Liu, Y. Xu, E. E. Baetge and S. Dalton, Activin a efficiently specifies definitive endoderm from human embryonic stem cells only when phosphatidylinositol 3-kinase signaling is suppressed, *Stem Cells*, 2007, **25**, 29–38.

85. K. A. D'Amour, A. D. Agulnick, S. Eliazer, O. G. Kelly, E. Kroon and E. E. Baetge, Efficient differentiation of human embryonic stem cells to definitive endoderm, *Nat. Biotechnol.*, 2005, **23**, 1534–1541.

86. G. Vogel, Cell biology. Ready or not? Human ES cells head toward the clinic, *Science*, 2005, **308**, 1534–1538.

87. K. Tamada, X. P. and F. C. Brunicardi, Molecular targeting of pancreatic disorders, *World J. Surg.*, 2005, **29**, 325–333.

88. X. P. Wang, K. Yazawa, J. Yang, D. Kohn, W. E. Fisher and F. C. Brunicardi, Specific gene expression and therapy for pancreatic cancer using the cytosine deaminase gene directed by the rat insulin promoter, *J. Gastrointest. Surg.*, 2004, **8**, 98–108, discussion 106–108.

89. K. Yazawa, W. E. Fisher and F. C. Brunicardi, Current progress in suicide gene therapy for cancer, *World J. Surg.*, 2002, **26**, 783–789.

90. M. Schuldiner, J. Itskovitz-Eldor and N. Benvenisty, Selective ablation of human embryonic stem cells expressing a 'suicide' gene, *Stem Cells*, 2003, **21**, 257–265.

91. M. U. Fareed and F. L. Moolten, Suicide gene transduction sensitizes murine embryonic and human mesenchymal stem cells to ablation on demand—a fail-safe protection against cellular misbehavior, *Gene Ther.*, 2002, **9**, 955–962.

92. J. Beck, R. Angus, B. Madsen, D. Britt, B. Vernon and K. T. Nguyen, Islet encapsulation: strategies to enhance islet cell functions, *Tissue Eng.*, 2007, **13**, 589–599.

93. S. M. Dang, S. Gerecht-Nir, J. Chen, J. Itskovitz-Eldor and P. W. Zandstra, Controlled, scalable embryonic stem cell differentiation culture, *Stem Cells*, 2004, **22**, 275–282.

94. A. Fort, N. Fort, C. Ricordi and C. L. Stabler, Biohybrid devices and encapsulation technologies for engineering a bioartificial pancreas, *Cell Transplant.*, 2008, **17**, 997–1003.

95. G. Orive, R. M. Hernandez, A. Rodriguez Gascon, R. Calafiore, T. M. Chang, P. de Vos, G. Hortelano, D. Hunkeler, I. Lacik and J. L. Pedraz, History, challenges and perspectives of cell microencapsulation, *Trends Biotechnol.*, 2004, **22**, 87–92.

96. M. S. Rao, R. S. Dwivedi, V. Subbarao, M. I. Usman, D. G. Scarpelli, M. R. Nemali, A. Yeldandi, S. Thangada, S. Kumar and J. K. Reddy, Almost total conversion of pancreas to liver in the adult rat: a reliable model to study transdifferentiation, *Biochem. Biophys. Res. Commun.*, 1988, **156**, 131–136.

97. M. S. Rao and J. K. Reddy, Hepatic transdifferentiation in the pancreas, *Semin. Cell Biol.*, 1995, **6**, 151–156.

98. M. S. Rao, V. Subbarao and J. K. Reddy, Induction of hepatocytes in the pancreas of copper-depleted rats following copper repletion, *Cell Differ.*, 1986, **18**, 109–117.

99. C. N. Shen, J. M. Slack and D. Tosh, Molecular basis of transdifferentiation of pancreas to liver, *Nat. Cell Biol.*, 2000, **2**, 879–887.

100. H. K. Wolf, J. L. Burchette Jr., J. A. Garcia and G. Michalopoulos, Exocrine pancreatic tissue in human liver: a metaplastic process?, *Am. J. Surg. Pathol.*, 1990, **14**, 590–595.

101. B. C. Lee, J. D. Hendricks and G. S. Bailey, Metaplastic pancreatic cells in liver tumors induced by diethylnitrosamine, *Exp. Mol. Pathol.*, 1989, **50**, 104–113.

102. D. Melton, Signals for tissue induction and organ formation in vertebrate embryos, *Harvey Lect.*, 1997, **93**, 49–64.

103. G. Deutsch, J. Jung, M. Zheng, J. Lora and K. S. Zaret, A bipotential precursor population for pancreas and liver within the embryonic endoderm, *Development*, 2001, **128**, 871–881.

104. J. Jung, M. Zheng, M. Goldfarb and K. S. Zaret, Initiation of mammalian liver development from endoderm by fibroblast growth factors, *Science*, 1999, **284**, 1998–2003.

105. F. Lemaigre and K. S. Zaret, Liver development update: new embryo models, cell lineage control, and morphogenesis, *Curr. Opin. Genet. Dev.*, 2004, **14**, 582–590.

106. K. D. Tremblay and K. S. Zaret, Distinct populations of endoderm cells converge to generate the embryonic liver bud and ventral foregut tissues, *Dev. Biol.*, 2005, **280**, 87–99.

107. H. Yoshitomi and K. S. Zaret, Endothelial cell interactions initiate dorsal pancreas development by selectively inducing the transcription factor Ptf1a, *Development*, 2004, **131**, 807–817.

108. K. S. Zaret, Hepatocyte differentiation, from the endoderm and beyond, *Curr. Opin. Genet. Dev.*, 2001, **11**, 568–574.

109. K. S. Zaret, Liver specification and early morphogenesis, *Mech. Dev.*, 2000, **92**, 83–88.

110. J. M. Wells and D. A. Melton, Vertebrate endoderm development, *Annu. Rev. Cell Dev. Biol.*, 1999, **15**, 393–410.

111. S. Ferber, A. Halkin, H. Cohen, I. Ber, Y. Einav, I. Goldberg, I. Barshack, R. Seijffers, J. Kopolovic, N. Kaiser and A. Karasik, Pancreatic and duodenal homeobox gene 1 induces expression of insulin genes in liver and ameliorates streptozotocin-induced hyperglycemia, *Nat. Med.*, 2000, **6**, 568–572.

112. J. Jonsson, L. Carlsson, T. Edlund and H. Edlund, Insulin-promoter-factor 1 is required for pancreas development in mice, *Nature*, 1994, **371**, 606–609.

113. U. Ahlgren, J. Jonsson, L. Jonsson, K. Simu and H. Edlund, Beta-cell-specific inactivation of the mouse Ipf1/Pdx1 gene results in loss of the beta-cell phenotype and maturity onset diabetes, *Genes Dev.*, 1998, **12**, 1763–1768.

114. I. Ber, K. Shternhall, S. Perl, Z. Ohanuna, I. Goldberg, I. Barshack, L. Benvenisti-Zarum, I. Meivar-Levy and S. Ferber, Functional, persistent, and extended liver to pancreas transdifferentiation, *J. Biol. Chem.*, 2003, **278**, 31950–31957.

115. I. Meivar-Levy, T. Sapir, S. Gefen-Halevi, V. Aviv, I. Barshack, N. Onaca, E. Mor and S. Ferber, Pancreatic and duodenal homeobox gene 1 induces hepatic dedifferentiation by suppressing the expression of CCAAT/enhancer-binding protein beta, *Hepatology*, 2007, **46**, 898–905.

116. D. Q. Tang, S. Lu, Y. P. Sun, E. Rodrigues, W. Chou, C. Yang, L. Z. Cao, J. J. Chang and L. J. Yang, Reprogramming liver-stem WB cells into functional insulin-producing cells by persistent expression of Pdx1- and Pdx1-VP16 mediated by lentiviral vectors, *Lab. Invest.*, 2006, **86**, 83–93.

117. A. Y. Wang, A. Ehrhardt, H. Xu and M. A. Kay, Adenovirus transduction is required for the correction of diabetes using Pdx-1 or Neurogenin-3 in the liver, *Mol. Ther.*, 2007, **15**, 255–263.

118. H. Kaneto, T. A. Matsuoka, Y. Nakatani, T. Miyatsuka, M. Matsuhisa, M. Hori and Y. Yamasaki, A crucial role of MafA as a novel therapeutic target for diabetes, *J. Biol. Chem.*, 2005, **280**, 15047–15052.

119. H. Kaneto, T. Miyatsuka, Y. Fujitani, H. Noguchi, K. H. Song, K. H. Yoon and T. A. Matsuoka, Role of PDX-1 and MafA as a potential therapeutic target for diabetes, *Diabetes Res. Clin. Pract.*, 2007, **77**(Suppl 1), S127–137.

120. H. Kaneto, T. Miyatsuka, T. Shiraiwa, K. Yamamoto, K. Kato, Y. Fujitani and T. A. Matsuoka, Crucial role of PDX-1 in pancreas development, beta-cell differentiation, and induction of surrogate beta-cells, *Curr. Med. Chem.*, 2007, **14**, 1745–1752.

121. T. A. Matsuoka, H. Kaneto, R. Stein, T. Miyatsuka, D. Kawamori, E. Henderson, I. Kojima, M. Matsuhisa, M. Hori and Y. Yamasaki, MafA regulates expression of genes important to islet beta-cell function, *Mol. Endocrinol.*, 2007, **21**, 2764–2774.

122. T. Miyatsuka, H. Kaneto, Y. Kajimoto, S. Hirota, Y. Arakawa, Y. Fujitani, Y. Umayahara, H. Watada, Y. Yamasaki, M. A. Magnuson, J. Miyazaki and M. Hori, Ectopically expressed PDX-1 in liver initiates endocrine and exocrine pancreas differentiation but causes dysmorphogenesis, *Biochem. Biophys. Res. Commun.*, 2003, **310**, 1017–1025.

123. H. Kojima, M. Fujimiya, K. Matsumura, P. Younan, H. Imaeda, M. Maeda and L. Chan, NeuroD-betacellulin gene therapy induces islet neogenesis in the liver and reverses diabetes in mice, *Nat. Med.*, 2003, **9**, 596–603.

124. M. E. Horb, C. N. Shen, D. Tosh and J. M. Slack, Experimental conversion of liver to pancreas, *Curr. Biol.*, 2003, **13**, 105–115.

125. Q. Zhou, J. Brown, A. Kanarek, J. Rajagopal and D. A. Melton, *In vivo* reprogramming of adult pancreatic exocrine cells to beta-cells, *Nature*, 2008, **455**, 627–632.

126. K. Takahashi and S. Yamanaka, Induction of pluripotent stem cells from mouse embryonic and adult fibroblast cultures by defined factors, *Cell*, 2006, **126**, 663–676.

127. K. Takahashi, K. Tanabe, M. Ohnuki, M. Narita, T. Ichisaka, K. Tomoda and S. Yamanaka, Induction of pluripotent stem cells from adult human fibroblasts by defined factors, *Cell*, 2007, **131**, 861–872.
128. J. Yu, M. A. Vodyanik, K. Smuga-Otto, J. Antosiewicz-Bourget, J. L. Frane, S. Tian, J. Nie, G. A. Jonsdottir, V. Ruotti, R. Stewart, I. I. Slukvin and J. A. Thomson, Induced pluripotent stem cell lines derived from human somatic cells, *Science*, 2007, **318**, 1917–1920.
129. R. Maehr, S. Chen, M. Snitow, T. Ludwig, L. Yagasaki, R. Goland, R. L. Leibel and D. A. Melton, Generation of pluripotent stem cells from patients with type 1 diabetes, *Proc. Natl. Acad. Sci. U. S. A.*, 2009, **106**, 15768–15773.
130. H. Zhou, S. Wu, J. Young Joo, S. Zhu, D. Wook Han, T. Lin, S. Trauger, G. Bien, S. Yao, Y. Zhu, G. Siuzdak, H. Scholer, L. Duan and S. Ding, Generation of induced pluripotent stem cells using recombinant proteins, *Cell Stem Cell*, 2009, **4**, 381–384.
131. M. Bosnali and F. Edenhofer, Generation of transducible versions of transcription factors Oct4 and Sox2, *Biol. Chem.*, 2008, **389**, 851–861.
132. D. Kim, C. H. Kim, J. I. Moon, Y. G. Chung, M. Y. Chang, B. S. Han, S. Ko, E. Yang, K. Y. Cha, R. Lanza and K. S. Kim, Generation of human induced pluripotent stem cells by direct delivery of reprogramming proteins, *Cell Stem Cell*, 2009, **4**, 472–476.
133. J. Yu, K. Hu, K. Smuga-Otto, S. Tian, R. Stewart, I. I. Slukvin and J. A. Thomson, Human induced pluripotent stem cells free of vector and transgene sequences, *Science*, 2009, **324**, 797–801.
134. D. Huangfu, K. Osafune, R. Maehr, W. Guo, A. Eijkelenboom, S. Chen, W. Muhlestein and D. A. Melton, Induction of pluripotent stem cells from primary human fibroblasts with only Oct4 and Sox2, *Nat. Biotechnol.*, 2008, **26**, 1269–1275.
135. D. Huangfu, R. Maehr, W. Guo, A. Eijkelenboom, M. Snitow, A. E. Chen and D. A. Melton, Induction of pluripotent stem cells by defined factors is greatly improved by small-molecule compounds, *Nat. Biotechnol.*, 2008, **26**, 795–797.
136. Y. Shi, C. Desponts, J. T. Do, H. S. Hahm, H. R. Scholer and S. Ding, Induction of pluripotent stem cells from mouse embryonic fibroblasts by Oct4 and Klf4 with small-molecule compounds, *Cell Stem Cell*, 2008, **3**, 568–574.
137. Y. Shi, J. T. Do, C. Desponts, H. S. Hahm, H. R. Scholer and S. Ding, A combined chemical and genetic approach for the generation of induced pluripotent stem cells, *Cell Stem Cell*, 2008, **2**, 525–528.
138. J. B. Kim, V. Sebastiano, G. Wu, M. J. Arauzo-Bravo, P. Sasse, L. Gentile, K. Ko, D. Ruau, M. Ehrich, D. van den Boom, J. Meyer, K. Hubner, C. Bernemann, C. Ortmeier, M. Zenke, B. K. Fleischmann, H. Zaehres and H. R. Scholer, Oct4-induced pluripotency in adult neural stem cells, *Cell*, 2009, **136**, 411–419.

Regeneration of the Lower Urinary Tract: Clinical Applications and Future Outlook

MARKUS RENNINGER, BASTIAN AMEND, JÖRG SEIBOLD, GERHARD FEIL, ARNULF STENZL AND KARL-DIETRICH SIEVERT

Department of Urology, University of Tuebingen, Germany

16.1 Introduction

The ethical concerns about the use of human embryonic stem cells (hESCs), the limited source of available organs for transplantation, the immunological difficulties of heterologous tissue transplantation and organ restrictions due to limited blood supply or complicated functional structure have called medical disciplines into action. Research at a translation level is taking place exploring different sources for cell-based therapies of human tissue and to evaluate reconstructive urological approaches. Today, urological tissue engineering is a rapidly evolving field; however, the potential of pluripotent and/or adult stem cells in basic research and transfer into clinical practice has yet to be realized.

16.1.1 First Applications in Urology

The treatment of urinary stress incontinence is the only one introduced into clinical practice and is not without its limitations as only certain institutions have been permitted to initiate clinical trials (*e.g.* University of Essen,

Stem Cell-Based Tissue Repair
Edited by Raphael Gorodetsky and Richard Schäfer
© The Royal Society of Chemistry 2011
Published by the Royal Society of Chemistry, www.rsc.org

Germany; Innovacell Biotechnology AG, Innsbruck, Austria; Nagoya University Hospital, Japan; and, Royal Oak, MI, USA).

16.1.2 Legal Requirements

In recent years the European Medicines Agency (EMEA) and the Committee for Advanced Therapy (CAT) have instigated European Union guidelines for country-specific institutions. More enhanced guidelines that clearly articulate 'bench-to-bed' protocols must be tested prior to their approval in order to be incorporated into daily clinical reconstructive urology. These protocols can then be used as a basis for reconstructive Good Manufacturing Practice (GMP). GMP development is essential as cell culture conditions must be regimented to minimize and even exclude risk so that cell changes or cancer cannot occur.[1]

The arduous process that defines the correct source of stem cells, identifies suitable adult progenitor cells with regenerative potential in the respective recipient's tissue and incorporates the different GMP requirements necessary for stem cell culture and *in vitro* differentiation conditions, while leading to the generation of functional and transplantable *in vivo* tissue grafts, has yet to be undertaken.

Previous ambitious and astonishing predictions need to be adjusted to ensure a realistic and accurate viewpoint. Urological tissue engineering should be positioned as 'a young field' that promises to influence urological treatment in the 'near future'.[2] Once its feasibility is proven, the ethical and legal points of view considered and addressed, and the scientific consistency ensured, clinical trials should be initiated to ensure patient safety and benefit. On the one hand, patients should be well-informed so as not to choose the latest experimental approach as the first therapeutic step, but on the other hand they have to be selected cautiously for randomized clinical trials to exclude a bias in always offering a technique based on tissue engineering as a salvage therapy or last option.

Promising reports of tissue engineering with a urological background were urethral reconstructions which successfully used seeded und unseeded acellular matrices in clinical trials.[3] Furthermore, clinical reports about bladder wall reconstruction have been published.[4] Treatment options using cell therapy for incontinence[5,6] and infertility[7] will probably soon be clinically approved. Despite its histological complexity,[8] some progress has been made even in renal tissue formation by nuclear transfer, three-dimensional (3D) culture and transplantation in an animal model.[9]

16.2 Graft Generation

Tissue engineering covers the whole field of stem cells and seeding of matrices to generate functioning tissue or even a whole organ. Successful tissue engineering requires the following aspects to be considered.

16.2.1   Stem Cell Sources

Current sources for cell harvesting appear to be unlimited. The transdifferentiation of adult-sourced stem cells such as bone marrow, hair follicle and adipose tissue has been proven.[10] Adult stem cells are still esteemed to play the central role in creating truly biomimetic tissue. Their favorable characteristics are their potency, their autologous approach, their relative abundance and their primarily noninvasive extraction. Conversely, their limiting factor may be the potentially disease-affected tissue from which they are derived.

Other outstanding human adult stem cell sources successfully reprogrammed *in vitro* into a pluripotent state using selected criteria and culture conditions, were called 'pluripotent human adult germline stem cells' (haGSC).[11] In case of mice, germline-derived pluripotent stem cells (gPS[12]) could also be generated, and were further phenotypically and functionally characterized by all hallmarks of pluripotency. Stem cells derived from other sources have shown a pluripotent character expressing both embryonic and adult stem cell markers with the ability to differentiate into derivates of each germ layer. Amniotic fluid-derived stem cells[13–16] and placenta-derived stem cells[17] are under investigation in the urological environment, while novel techniques for the characterization and isolation of these stem cells are being probed.

16.2.2   Alternatives in Pluripotent Stem Cell Research

Due to ethical concerns about embryonic stem cells and their non-autologous nature in clinical application, alternative cell resources such as inducible pluripotent stem cells (iPS)[18] might play a more suitable role for autologous tissue engineering in the near future, especially using induced pluripotent cells (iPS), generated by the Yamanaka factors in the absence of potential oncogene c-Myc, as recent publications have demonstrated.[19,20]

A human retroviral-induced pluripotent dermal fibroblast derived stem cell has seemingly opened the possibility of overcoming stem cell availability and shortages; however, it remains unclear if those 'created stem cells' fulfill requirements or have a potential increased malignancy risk, even though it is now possible to induce those cells without the previously mentioned viral vectors.[19]

iPSs are a controversial but indispensable tool in pluripotency and differentiation research. At the present time, their use for tissue engineering approaches is complicated by their pluripotent state, their low abundance during the generation process, and the need to resolve the issue of their functional integration without tumorigenicity. Human embryonic stem cells (hESCs) remain the gold standard in stem cell research and influence GMP-conform handling of pluripotency (*e.g.* the feeder-free stem cell culture in the absence of viruses and murine).

Perhaps other cell biological techniques such as somatic cell nuclear transfer[9,21] or new non-stable, transient reprogramming techniques to induce multipotent behavior without tumorigenicity might give rise to new developments in tissue engineering applications; however, further investigations are necessary. Continued basic research in this field is fundamental and absolutely crucial for a fundamental knowledge of self-renewal, differentiation, tissue formation, cell–cell interaction, de-differentiation and the induction of malignancy. In the future, there may be a way to organize a stem cell collection to enhance treatment possibilities for individuals using an autologous approach or even a genetic match for heterologous transplantation.

16.2.3 From Bench to Bed

A further critical step forward would be the establishment of independent rodent animal models with more preclinical characteristics following GMP guidelines, which can then be converted into Good Clinical Practice (GCP) conform clinical trials to ensure patient safety and to verify therapeutic clinical superiority in reconstructive urological surgery or other cell-based therapies.

The risks and benefits of tissue engineering-based treatments must also be evaluated and compared to already established treatment options. Currently, there is only limited literature suggesting any superior tissue engineering-based treatments for patients as an alternative to conventional therapies.[22] Therapeutic tissue regeneration can only be promoted if it leads to physiological graft integration in the host organism with re-established function.

16.2.4 Culture Media for Tissue Engineering and 3D Matrices

Tissue engineering advances will be taken a step further with the development of dynamic and multi-dimensional cell culture conditions and cell seeding on different pre-fabricated natural or synthetic biomaterials as scaffold components. These developments are required, combined with more effective surgical techniques, to promote definitive vascularization and neutralization resulting in a completely regenerated and functioning urological organ. Even the importance of improvements in cell culture technologies, such as the development of selective media or further identification of suitable markers for isolation, will lead to a more extensive characterization of specific progenitors and pre-differentiated or further differentiated cell types used for future tissue engineering processes.

16.2.5 Host Requirements

Physiological tissue formation or the replacement of multi-layered and multi-dimensional complex tissue grafts is not only impaired by the difficulties of the cell culture and tissue engineering techniques, but also by the disease-affected surrounding tissue in the recipient or by the reduced general health condition of the patient. Taken together, these criteria seem to be a virtually insurmountable obstacle for complete graft regeneration.

16.2.6 Developmental Biology: The Key to Understanding Tissue Engineering?

A formidable knowledge of developmental biology is crucial for the development of an advanced fundamental understanding of the molecular and pathophysiological tissue regeneration process,[23,24] and translational research is essential in order to achieve the paramount goal of a functional organ or implantable tissue regeneration.

16.2.7 Current Tissue Engineering: Realistic Assessment

Previous ambitious and astonishing predictions must be adjusted to ensure a realistic and accurate viewpoint. Urological tissue engineering should be positioned as 'a young field' that promises to influence urological treatment in the 'near future'.[2] Once its feasibility is proven, the ethical and legal points of view considered and addressed, and scientific consistency ensured, clinical trials can be initiated to verify the benefits of tissue engineering for specific needs of urology and patient safety.

16.2.8 First Clinical Trials in Urology for Seeded Matrices and Injectable Cell Suspensions

After Atala and colleagues published the implementation of a seeded bladder wall reconstruction,[4] El-Kassaby *et al.* investigated the use of seeded and unseeded similar acellular matrices for patients in urethral reconstruction.[3] The cell therapy research groups closest to receiving clinical approval for this procedure are:

- Innovacell Biotechnology AG, Innsbruck, Austria,[5,6] for urinary stress incontinence; and
- Schlatt's group (University of Pittsburg, USA) for infertility.[7]

16.2.9 Successful Tissue Engineering Components

Successful tissue engineering must comprise an end-to-end process of stem cell and matrix seeding to generate functioning tissue or even a whole organ. The criteria for deciding which tissue engineering approach is appropriate depends primarily on the extent of the tissue, organ or organ part that needs to be replaced, which in turn, determines the timeline. These different graft preparations or tissue transplantation strategies should be evaluated and the best path determined. Some strategies might include the following.

16.2.9.1 Cell Injections

One option might be to re-establish the dysfunctional stromal or organ environment with an injection of suspended undifferentiated, pre-differentiated or differentiated stem cells with the result that those cells achieve improved function over time.[25] Examples are the investigation of treatments for urinary stress incontinence or reflux into the upper urinary tract (see Section 16.3.3).

16.2.9.2 Graft Generation

To avoid any matrix or other foreign material, options to resurface organs with mono- or multi-layered grafts (*e.g.* urothelium[26]) need to be evaluated. If an extensive, fully functioning organ wall is needed, a more mature and 3D graft construction may be seeded with various types of differentiated cell types to ensure the immediate occurrence of the desired function. These graft models reflect some of the individual strategies in urological tissue regeneration which need to be evaluated.

16.2.9.3 Generate Complete Organs

Despite its histological complexity,[8] there has been a huge progression in the development of renal tissue formation using nuclear transfer, 3D culture and transplantation in an animal model.[9] Prior to these publications, Atala's group generated the loop of Henle as a functional 3D construct.[9,27,28]

16.2.10 Requirements for GMP, Grafts and Hosts

Another challenge faced by researchers is to determine how the graft and the recipient's organ should be prepared. Stem cell cultures and differentiating media must conform to GMP guidelines and scaffolds should possess the same physical and biological properties as the recipient's connective tissue. The cells should be seeded on scaffolds with a molecular surface structure as close as possible to the human extracellular matrix (ECM) and with a causal histogenesis background; in some cases, they may have to be pre-conditioned by mechanical simulation in a bioreactor.[29] The graft's cellular components should have biological, immunological and physiological functions similar to those of the native and healthy tissue. The recipient's organism has to be treated using minimally invasive surgical techniques to harvest the original autologous cells. The surrounding host tissue should be prepared to avoid scar tissue transformation, optimize vascularization and diminish physical pressure or stress on the graft during the time of integration and regeneration. The pathology affecting the original tissue should be removed or changed to recreate a healthy micro-environment for successful graft integration.

16.2.11 Scaffolds and Structural Support of Cells

In addition to the option of bringing cells into the host environment to re-establish function, the field of urology could require the replacement of more complex parts of tissue or even a complete organ composed of different tissues itself (*e.g.* kidney, bladder wall). Therefore, the development of biomaterials that join the biological and mechanical characteristics of the host extracellular matrix and which can be seeded with the different tissue cells types being replaced is expected to be inevitable.

Prior to the inner-corporal regeneration in the host, an important aspect of tissue engineering is appropriate surgical handling that does not risk the implant's survival. The biomaterials used should be degradable and absorbable in the host tissue without causing inflammation or a foreign body response that would prevent the healing process or even destroy the graft.

At the time of implantation, scaffolds or matrices slowly degenerate during the healing process and are replaced by secreted extracellular proteins of the newly developed cells. The physical characteristics of the extracellular scaffold must be the same or at least as similar as possible to those of the recipient's tissue. The microstructure architecture should offer the best possible option to ensure that these cells have the required nutrition to regenerate physiological function.

16.2.11.1 *Commonly Used Extracellular Scaffolds*

Manufactured artificial scaffolds are common synthetic polymers such as polyglycolic acid (PGA), polylactic acid (PLA) and poly(lactic-co-glycolic acid) (PLGA),[30] and are FDA-approved. The physical properties and architectural structures of synthetic scaffolds—as well as their biocompatibility and immunotolerance—might be well-understood and sufficiently manufactured, but the more complex aspects of the ultra structural, molecular requirements of adhesion molecules, cell–cell and cell–surface interaction and niche formatting process is not yet generally known or sufficiently reproducible.

Due to their artificial nature, synthetic scaffolds are not directly degradable with enzymes but by chemical hydrolysis. The biological behavior of these materials can be influenced by altering molecular weight, crystallinity and co-polymer content which allows 3D manufacturing into different emerging forms.[31–33] Factors such as biocompatibility,[34] fiber diameter, orientation[35] and elasticity[36] influence the tissue formatting process, while serum absorption onto biomaterial surfaces should also be taken into account.[37] The distribution of the different cell types into the scaffold surface, as well as their lifelike proliferation, remains a challenge.

Organ tissue is transformed into an acellular matrix through the removal of cellular components by means of chemical and mechanical manipulation [*e.g.* bladder submucosa[3] and small intestine submucosa (SIS®),[38] organ-specific acellular matrix[39–44]]. This acellular matrix may offer the possibility of providing the remaining specific growth factors and required proteins for the process of regeneration after processing. Examples of decellularization of natural and orthotropic tissues such as full thickness bladder are supposed to retain the micro-architectural and tensile properties of a healthy bladder wall, the SIS[45,49] and the amnion.[46–48]

Although artificial manufactured scaffolds have the advantage of a standardized and reproducible manufacturing process that easily complies with GMP requirements, the seeded cells often do not find an ideal situation for regeneration. Opposite to synthetic matrices, the advantages of collagen are:

- the natural abundance of this molecule in the human body;
- its easy and natural degradation and absorption regulated by cross-linking of the fibers and the expression of adhesion sequences for different cell types regulation; and
- the phenotype and proliferation properties of ingrowing fibroblasts[50] and chondrocytes.[51]

The polysaccharide alginate has also been easily used as a delivery vehicle for single cells[52] due to its ability to gel and immobilize single cells.

During the healing process, factors within the host influence the regeneration outcome as much as that of the pathology of the recipient's organ. In particular, an organ-specific acellular matrix seems to promote growth due to its collagen and elastin structures while providing the proper micro-environment for the required cell types. Acellular tissue influences the differentiation status of the seeding cells and supports cell growth and integrity.[53]

In an optimal setting using bladder submucosa, collagen matrices for urethral replacement, a success rate of 85% was reported;[54] whereas in a randomized clinical trial, acellularized matrices were less effective than buccal mucosa in the treatment of urethral strictures when the underlying tissue bed was affected by fibrosis.[3] It is worthwhile mentioning that in these trials the matrix used was produced using radiation, amongst others, which might have destroyed the previously mentioned growth factors.

Other biomaterial supports that would generate 3D cell growth are naturally derived collagen and alginate. The preservation of a 3D tissue structure is the primary scaffold function to ensure proper orientation in regenerative tissue growth. An artificial extracellular matrix can even deliver critical bioactive molecules such as adhesion peptides[55] and growth factors[56] to optimize cell growth and the integration process through cell adhesion, proliferation, migration and differentiation.[57]

Two previously complete separate developed approaches (cellular organ tissue without any cells *vs.* previously seeded cells on a manufactured scaffold) seem to join in the use of organ specific matrix pre-seeded with tissue specific cells.

16.2.12   Cell Engineering Limitations (Aging)

In addition to aged or disease-affected primary autologous tissues, a further limitation of cell-based tissue engineering can be the *in vitro* expansion of the cells derived from organs with low regenerative capacity and the requirement to keep these cells in a desired differentiation state during the period of expansion. These aspects could be incorporated by extended biopsies and improved growth media.[58–61]

16.2.13   Cell Injectables/Single Cells

Single cells can be implanted with or without hydrogels as carriers.[62] In addition, minimally invasive techniques using endoscopic devices (*e.g.* for external urethral sphincter repair by muscular regeneration[6,63]) were developed and showed significant sphincter function improvement. Using autologous cloned porcine myoblasts, the generated muscle fibers were demonstrated histologically. Reliable multi-center clinical trials have still to be performed, as does confirmation of the one-year follow-up data in both genders. Further improvements could be achieved using human adult testicles as a stem cell source for different reconstructive strategies in urology.

With the isolation, characterization and differentiation towards derivates of all three human germ layers, adult human germline stem cells were introduced into regenerative medicine as pluripotent and adult stem cells.[11] The first clinical approach is the introduction of a porcine animal model using an injectable solution of myogenic pre-differentiated adult human germline stem cells for reinforcement of the external urinary sphincter.[64]

16.2.14   Hosting and Nutrification

Due to the lack of macroscopic and microscopic vascularization *in vitro*, the problem of oxygen diffusion and nutrient molecules in the development of 3D tissue constructs had to be solved and led to the development of the bioreactor technology.[65,66] The goal of tissue engineering is not only to provide a vivid piece of tissue, but to re-establish lost host functionality. After surgical implantation, the problem of tissue nutrification initially depends on diffusion and, in the process of integration and regeneration, on the vascularisation from the surrounding and affected tissue.[67,68] Mertsching and colleagues reported an approach using acellular gut and re-epithelizing the vessels prior to implantation to be able to re-anatomize them to the blood supply.[68]

For example, a seeded construct wrapped in a flap of the omentum was proposed in ureter replacement to serve as a bioreactor for the multilayer graft after transplantation *in vivo*,[69,70] resulting in improved vascularization and integration.[71] The technique using an omentum flap was also propagated for autologous neobladder construction.[4]

Newer enhancement strategies of scaffold tissue engineering incorporate advanced knowledge of intracellular and extracellular signal transduction and cell–cell interaction which also make use of the modulating function of key molecules such as growth factors and adhesion molecules. With these biofunctionalized scaffolds, cell seeding and proliferation,[72,73] enhanced vascularization[74] may be further promoted.

Different techniques have been applied, for example, using growth factors such as vascular endothelial growth factor (VEGF) as a supplement in the cell media, as adhesive molecules on scaffold products or in using genetically modified cells.[73,74] Despite these benefits, the scientists must ensure strict adherence to GMP protocols and, in the case of VEGF, the promotion of tumor development must be excluded.

16.2.15 Functional Control of Cells

The functional control of differentiation of cells *in vitro* is primarily dependent on two conditions, *i.e.* lineage derivation and differentiation-inducing factors. The functional control of cells in the definite graft is achieved by:

- its niche;
- cell–cell interactions;
- cell–ECM interactions;
- auto- and paracrine- cytokine stimulation;
- interactions with the immune system; and
- the best possible supply of blood vessels and nervous system.

For example, in the urinary bladder as a complex functional organ, volume compliance depends on the sensory and motor abilities of specialized cells of the muscular layer with cell–cell interaction and communication, which should promote functional control over the tissue-engineered neo-organ. On the one hand, due to the lack of specialized ultra-structures promoting neural attraction and guidance, only a small amount of neural ingrowth has been reported at the periphery of free grafts without any functional implications.[75] However, organ-specific acellular matrices seem to be capable of promoting organ-specific regeneration as demonstrated for the bladder in the animal model.[41]

The strongest factor for contractile function of muscular tissue-engineered grafts seems to the dynamic environment in the cell culture,[76] in contrast to this under cardiac muscle differentiating conditions of ES-cells commonly spontaneously contracting element are shown.[77] Perhaps combining various pre-differentiating conditions, stable or transient induction of muscle lineage specific promotors and further cultivation with supply of differentiating factors might assist in the production of the ideal tissue-engineered organ that provides sensory perceptions and contractile mechanisms for even a further improved functional approach in reconstructive urology.[78] Further cultivation with a supply of differentiating factors might help to obtain the ideal engineered tissue, providing sensory perceptions and contractile mechanisms before solving the next step of vessel and nerve supply or functional contact with other specialized tissue such as urothelium for an even further improved functional approach in reconstructive urology.

In addition to the optimization of the *in vivo* setting for the cells, other factors might influence the hosting and development of those cells such as additional electrical peripheral nerve stimulation, leading towards the restoration of urological voiding functions in order to influence grafted muscular organs until a full innervation and physiological function is reached.

16.2.16 Histological Evidence

A new technique is intended to replace standard surgery or at least obtain a better outcome than standard practice. To increase the storage of a urinary bladder, tissue engineering faces the tremendous challenge of ensuring that all layers of a functional bladder wall, which can expand (storage) and contract (in a coordinated function with the sphincter to evacuate), have sensibility and are non-absorbable for urine. A fully functional bladder wall—including muscles, innervation and vascularization—covered by a sensory but non absorbable surface was realized for the first time in the porcine model by Fraser and colleagues using *in vitro* tissue-engineered autologous bladder urothelium and smooth muscle cells derived from the uterus for bladder augmentation.[79] *In vitro* human and porcine urothelial cells can be cultivated under certain conditions, inducing a differentiated urinary barrier.[24,80]

Tissue engineering does not end with graft implantation. It is then that the integration process begins in the recipient and the regeneration of the organism to re-establish function occurs.

Harvesting any tissue or cells, which can be used for tissue engineering or a stem cell approach, should to be performed as minimally invasive as possible. A minimally invasive approach to harvest cells from the urinary bladder for urothelial cell cultures was performed by Nagele *et al.* through bladder washings and avoiding cystoscopic biopsies.[81] Although progenitor cells are more likely to be found at the basement membrane region in rat urothelium, highly proliferative cells with good differentiation potentials were found to be ideal for seeding grafts.[82] The approach to expand these harvested urothelial cells and stratify those with calcium to induce the development of urothelium does not necessarily demand progenitor cells.[81] Harvesting requires exclusion of a malignant process in the urothelium, although benign processes might also influence proliferation growth.[22,37]

16.3 From Urological Tissue Engineering to Functional Urological Organ Replacement

16.3.1 Urethra

Urethral regeneration is probably the most achievable project in regenerative urology today. The organ often demonstrates the result of the disease (commonly non-malignant), a stricture, where the traditional endosurgical techniques do not always lead to the best therapeutic long-term outcome. As opposed to the endosurgical technique, the open buccal mucosa transplantation approach is a feasible technique and has advanced the field of reconstructive urological surgery.[83] In developing an approach that includes tissue engineering possibilities, the urologist can schedule the reconstruction to a time when the graft is available. Primary vascularization or even innervation is not immediately necessary because of the permeability, which allows for nitrification from the beginning due to the thickness of the graft.

Several synthetic and collagen-based biomaterials have been investigated in acellular and cell-seeded animal models with different outcomes.[38,43,44,53] In different approaches the neo-urethra demonstrated a normal urothelial cellular surface with organized muscle bundles. These results were confirmed clinically with human bladder acellular collagen matrices[84] and transferred into the clinic.[3] Despite the fact that initial urethral repair with this tubularized but acellular material was not feasible, it caused graft contracture and resulted in stricture formation.[85] Furthermore, the transplantation of a seeded matrix with autologous cells resulted in the formation of new tissue similar to native tissue.[86] Further enhancements might involve the cell selection and adding a supplement of growth factor such as VEGF.[72–73,87]

Related to the use of buccal mucosa is the use of tissue-engineered buccal mucosa from a small biopsy. A small clinical study using this tissue-engineered buccal mucosa followed patients with a diagnosis of lichen sclerosus (LS, also known as balanitis xerotica obliterans, or BXO) and reported clinical difficulties.[88] The use of SIS for the treatment of urethral stricture demonstrated a success rate of about 85%, with a poorer outcome for the distal penile urethra.[89,90]

But to establish a new standard, the new tissue engineering approach must prove superior to conventional treatment. When transforming laboratory results it must be borne in mind that, in most cases, the region of investigation in the model does not include a pathological process. This has been changed recently by the use of an existing pathological model to demonstrate the outcome of urethral reconstruction by tissue-engineered urothelium.[91]

16.3.2 Bladder

The urinary bladder represents a significant challenge for the reconstructive surgeon compared with urethral reconstruction. Due to the lack of adequate vascularization, graft survival in early stages after *in vivo* implantation is a common problem. Reconstructive tissue engineering issues, especially in this case, are performed with a focus on tissue expansion, matrix-based seromuscular graft generation and transplantation techniques.

In one pre-clinical trial, bladder augmentation was performed with allogenic acellular bladder matrices seeded with autologous urothelial and muscular cells from biopsies.[75] The compliance of the cell-seeded tissue-engineered bladders showed almost no deviation from pre-operative values and a histologically normal cellular structure consisting of a tri-layer of urothelium, submucosa and muscle cells. Homologous bladder acellular matrix in a dog model showed a more complete regeneration, which was explained by the concentration of collagen components in the matrix influencing smooth muscle regeneration.[42]

Tissue-engineered constructs of cocultured functional cellular components seeded on matrices with complex physiological properties are today's concept. Smooth muscle grafts lined with urothelium and nutrified by a pedicle have been used as an alternative,[79] but not yet brought to a clinical application.

The additional wrapping of the construct with omentum seems to optimize nutrition, demonstrating an improvement in multilayered urothelial seeding,[69,70] capacity and compliance.[4]

Cell culture of detrusor muscle derived myocytes histologically and immunohistochemically showed[75] a phenotype similar to the original smooth muscle cells and some further physiological aspects (*e.g.* contraction upon biochemical stimulation). In electrophysiological experiments in cell cultures, a contraction-like reaction could be measured[92] suggesting that, after performing a single cell culture followed by new syncytial growth, muscle-like physiological cell function can be obtained.

It is vital to support the development of tissue-engineered muscular organs because urological function, storage and voiding always depend on contractile elements. Dynamic cell culture might be an important step[76] in the further development of a muscular phenotype of myogenic pre-differentiated cells. Three-dimensional scaffolds supplemented with bioactive factors[37] are inevitable for further tissue-like culture of human smooth muscle cells.

After all, the graft depends on a neurovascular supply to gain full physiological function (low pressure storage to protect the upper urinary tract, evacuation connected to the urethral sphincter to provide continence).[4] Thus, the completely tissue-engineered bladder is a very complex undertaking combining different issues such as a low-pressure storage, excluding reflux into the upper urinary tract, and arbitrary voiding.

16.3.4   External Urethral Sphincter and Vesicoureteral Reflux

The main tissue engineering principles of external urethral sphincter repair or better reinforcement are the most common cell-based approaches with single cells that function as an enhancer between the whole amounts of potentially degenerative and/or injury-affected cells in the receiving tissue itself.

Bulking agents are used in daily urological routine and are endoscopically administered in a minimally invasive way. Different approaches have been investigated such as injection of synthetic or biological bulking agents, cell suspensions, implantation or encapsulating techniques, or a combination of those.[93] The requirements for the injectables are non-antigenicity, a stable localization and stable volume in the tissue, and safety for the patient. The only suitable urological indication is for the treatment of the pediatric vesicoureteral reflux,[94] whereas for urinary incontinence, it has been proven not to last in the long term.[95] However, Cochrane analysis of those therapies revealed unsatisfactory results due to the small trials and their moderate quality.[96]

As a result, injectable autologous chondrocytes have demonstrated a therapeutical benefit of vesicoureteral reflux in a porcine model.[97] In combination with the biodegradable carrier alginate, the transplanted cells improved vesicoureteral reflux without any evidence of obstruction. Clinical trials showed a similar success rate in comparison with other soluble injectables.[98]

Myoblasts have been investigated for the therapy of stress urinary incontinence since 2000;[99] after injection in an open procedure, labeled myoblasts were detected in regenerative myofibers. For the regeneration of the external urethral sphincter, derived sphincteric muscle cells were tested and showed their regenerative potential in mice.[100] Spincter-derived muscle cells were also tested in a porcine animal model with similar results,[111] but a more comprehensive approach with a superior regenerative potential is the derivation of quiescent myoblasts, that play a crucial role in sceletal muscle regeneration out of muscle biopsies and the injection of these cells in the urinary sphincter. Yiou *et al.*[101] demonstrated integration of the injected myoblasts and their innervation leading to enforcement of the external urethral sphincter. The approach has been performed several times in pre-clinical[63] and clinical settings in case of urinary incontinence, leading to a significant improvement even after one-year follow-up in both sexes[5,6] and showing a clear superiority to injections with collagen.[102]

16.3.5 Penis

Penile reconstructive surgery is an emerging concern after urogenital trauma, penile carcinoma, severe erectile dysfunction, ambiguous genitalia, hypospadias, epispadias and transsexualism. The operative cure for Peyronie's disease can cause large and serious functional tissue defect of the corpora cavernosa in addition to a possible injury of the innervation and function of the smooth muscle inside the corpora.

Numerous materials for the reconstruction of the corporal bodies have been investigated such as SIS, bovine pericardium, and saphenous vein and alloderm. Reconstructive issues in patients with SIS and alloderm showed consistent clinical results.[103] Tissue engineering may be a suitable option to solve the problem of functionality (erectile function), but there are limitations in terms of the suitable tissue for transplantation or reconstruction of the penis mostly consisting of endothelial cells and smooth muscle cells. Grafts generated from scrotal dermis have been used to improve the penile dimension. The caused vascular defect related to this approach might be addressed with cultured endothelial cells for the penile defects.[104]

Yet approaches in rat animal models have been established for smooth muscle derived stem cells replacing corporal smooth muscle with purpose of improving the erectile function.[105] A further animal model was established in rabbits by using autologous cavernosal smooth muscle and endothelial cells, seeding them on a acellular collagen matrix[106,107] and leading to an engineered functional penile tissue. Even the feasibility of engineering the entire pendular penile corporal bodies from cavernosal collagen matrices seeded with autologous cells in a rabbit model using a multistep static/dynamic procedure was been shown; the male animals with reconstructed penises successfully impregnated females.[108]

16.3.6 Kidney

Kidney replacement has been the domain of whole organ transplantation and heterogenic organs are transplanted, with the need for immunosuppression, every day in clinical practice. However, available organs for transplantation are limited and most of the patients are young and have to undergo frequent dialysis with an increased risk of morbidity and mortality.

Kidneys are considered a complex anatomical and functional structure[8] due to their complex function bound to the nephrone as the smallest functional unit in the kidney. Tissue engineering is complex in this field, but the first advances have been made in the successful construction of an extracorporeal renal

support system, tissue-engineered out of synthetic and biological components.[109,110] A further approach in kidney tissue engineering was achieved by
the therapeutical cloning (somatic nuclear transfer of a skin cell into an enu-
cleated oocyte) in a bovine animal model.[9] These 'cloned' cells were seeded onto
a biodegradable scaffold (three collagen-coated cylindrical polycarbonate
membranes connected to catheters leading into a reservoir) and implanted
subcutaneously in the animal from which the nucleus was derived. Histological
evidence of vascularization and self-organization into glomeruli- and tubuli-
like structures was confirmed and the collected fluid revealed renal specific
molecules without any immune response.

16.4   Conclusions

Tissue engineering and stem cell research is ongoing in all different fields of
reconstructive medicine which aims to restore organ functions lost by degenerative and even oncological diseases, trauma or (genetic) handicap and will
change the picture of reconstructive urology.

   Pre-clinical animal models leading to well-designed clinical trials seek to gain
proof that tissue engineering based approaches might be superior to established
therapies in the near future. Acquiring good clinical follow-up data will take
some time, but these data must be obtained in order to establish a tissue
engineering approach in daily clinical routine. At the same time the debate
about the use of stem cell continues and ethical expectations are set in most
countries with a good medical standard.

   In basic research, the search for new sources of (pluri-)potent adult stem cells
will continue. The encouraging news is that there are new cell sources, for which
isolation strategies need to be specified and a precise characterization for the
precise cell location are required. Apart from the possibility of differentiating
cells, it will be important to gain knowledge of how the host can be set up to use
those cells or tissue engineering tissues to lead to the restoration of a full
functioning organ or organ part. The first results can be reported in urology as
the use of progenitor cells to re-establish external urethral sphincter function.
Others like the use of tissue engineering urothelium are close to being approved
for urethral reconstruction or augmentation of the urinary bladder.

References

1. K. D. Sievert, J. Hennenlotter and I. Laible, *et al.*, The periprostatic
   autonomic nerves—bundle or layer?, *Eur. Urol.*, 2008, **54**, 1109–1116.
2. K. D. Sievert, B. Amend and A. Stenzl, Tissue engineering for the lower
   urinary tract: a review of a state of the art approach, *Eur. Urol.*, 2007, **52**,
   1580–1589.
3. A. el-Kassaby, T. AbouShwareb and A. Atala, Randomized comparative
   study between buccal mucosal and acellular bladder matrix grafts in
   complex anterior urethral strictures, *J. Urol.*, 2008, **179**, 1432–1436.

  4. A. Atala, S. B. Bauer, S. Soker, J. J. Yoo and A. B. Retik, Tissue-engineered autologous bladders for patients needing cystoplasty, *Lancet*, 2006, **367**, 1241–1246.
  5. M. Mitterberger, R. Marksteiner and E. Margreiter E, *et al.*, Autologous myoblasts and fibroblasts for female stress incontinence: a 1-year follow-up in 123 patients, *BJU Int.*, 2007, **100**, 1081–1085.
  6. M. Mitterberger, G. M. Pinggera and R. Marksteiner, *et al.*, Adult stem cell therapy of female stress urinary incontinence, *Eur. Urol.*, 2008, **53**, 169–175.
  7. S. Schlatt, J. Ehmcke and K. Jahnukainen, Testicular stem cells for fertility preservation: preclinical studies on male germ cell transplantation and testicular grafting, *Pediatr. Blood Cancer*, 2009, **53**, 274–80.
  8. H. Auchincloss and J. V. Bonventre, Transplanting cloned cells into therapeutic promise, *Nat. Biotechnol.*, 2002, **20**, 665–666.
  9. R. P. Lanza, H. Y. Chung and J. J. Yoo, *et al.*, Generation of histocompatible tissues using nuclear transplantation, *Nat. Biotechnol.*, 2002, **20**, 689–696.
 10. T. Hodgkinson, X. F. Yuan and A. Bayat, Adult stem cells in tissue engineering, *Expert Rev. Med. Devices*, 2009, **6**, 621–640.
 11. S. Conrad, M. Renninger and J. Hennenlotter, *et al.*, Generation of pluripotent stem cells from adult human testis, *Nature*, 2008, **456**, 344–349.
 12. K. Ko, N. Tapia and G. Wu, *et al.*, Induction of pluripotency in adult unipotent germline stem cells, *Cell Stem Cell*, 2009, **5**, 87–96.
 13. M. Cananzi, A. Atala and P. De Coppi, Stem cells derived from amniotic fluid: new potentials in regenerative medicine, *Reprod. Biomed., Online*, 2009, **18**(Suppl 1), 17–27.
 14. P. De Coppi, G. Bartsch Jr., and M. M. Siddiqui, *et al.*, Isolation of amniotic stem cell lines with potential for therapy, *Nat. Biotechnol.*, 2007, **25**, 100–106.
 15. T. Aboushwareb and A. Atala, Stem cells in urology, *Nat. Clin. Pract. Urol.*, 2008, **5**, 621–631.
 16. A. Toda, M. Okabe, T. Yoshida and T. Nikaido, The potential of amniotic membrane/amnion-derived cells for regeneration of various tissues, *J. Pharmacol. Sci.*, 2007, **105**, 215–228.
 17. D. M. Delo, P. De Coppi, G. Bartsch Jr. and A. Atala, Amniotic fluid and placental stem cells, *Methods Enzymol.*, 2006, **419**, 426–438.
 18. K. Takahashi, K. Tanabe and M. Ohnuki, *et al.*, Induction of pluripotent stem cells from adult human fibroblasts by defined factors, *Cell*, 2007, **131**, 861–872.
 19. K. Okita, M. Nakagawa, H. Hyenjong, T. Ichisaka and S. Yamanaka, Generation of mouse induced pluripotent stem cells without viral vectors, *Science*, 2008, **322**, 949–953.
 20. M. Wernig, A. Meissner and R. Foreman, *et al.*, *In vitro* reprogramming of fibroblasts into a pluripotent ES-cell-like state, *Nature*, 2007, **448**, 318–324.

21. K. Hochedlinger and R. Jaenisch, Nuclear transplantation, embryonic stem cells, and the potential for cell therapy, *N. Engl. J. Med.*, 2003, **349**, 275–286.
22. D. Wood and J. Southgate, Current status of tissue engineering in urology, *Curr. Opin. Urol.*, 2008, **18**, 564–569.
23. S. D. Silva-Barbosa, G. S. Butler-Browne, J. P. Di Santo and V. Mouly, Comparative analysis of genetically engineered immunodeficient mouse strains as recipients for human myoblast transplantation, *Cell Tranplant.*, 2005, **14**, 457–467.
24. A. M. Turner, R. Subramaniam, D. F. Thomas and J. Southgate, Generation of a functional, differentiated porcine urothelial tissue *in vitro*, *Eur. Urol.*, 2008, **54**, 1423–1432.
25. H. Strasser, R. Marksteiner and E. Margreiter, *et al.*, [Stem cell therapy for urinary incontinence], *Urologe A*, 2004, **43**, 1237–1241 [in German].
26. S. Maurer, G. Feil and A. Stenzl, [*In vitro* stratified urothelium and its relevance in reconstructive urology], *Urologe A*, 2005, **44**, 738–742 [in German].
27. A. Joraku, K. A. Stern, A. Atala and J. J. Yoo, *In vitro* generation of three-dimensional renal structures, *Methods*, 2009, **47**, 129–133.
28. C. Becker and G. Jakse, Stem cells for regeneration of urological structures, *Eur. Urol.*, 2007, **51**, 1217–1228.
29. S. Korossis, F. Bolland, E. Ingham, J. Fisher, J. Kearney and J. Southgate, Review: tissue engineering of the urinary bladder: considering structure-function relationships and the role of mechanotransduction, *Tissue Eng.*, 2006, **12**, 635–644.
30. J. L. Pariente, B. S. Kim and A. Atala, *In vitro* biocompatibility evaluation of naturally derived and synthetic biomaterials using normal human bladder smooth muscle cells, *J. Urol.*, 2002, **167**, 1867–1871.
31. L. E. Freed, G. Vunjak-Novakovic and R. J. Biron, *et al.*, Biodegradable polymer scaffolds for tissue engineering, *Biotechnology (N.Y.)*, 1994, **12**, 689–693.
32. R. R. Harris, D. Wilcox, R. L. Bell and G. W. Carter, The role of tissue mast cells in polyacrylamide gel-induced inflammation in mice, *Inflamm. Res.*, 1998, **47**, 104–108.
33. A. G. Mikos, M. D. Lyman, L. E. Freed and R. Langer, Wetting of poly(L-lactic acid) and poly(DL-lactic-co-glycolic acid) foams for tissue culture, *Biomaterials*, 1994, **15**, 55–58.
34. G. C. Bartsch Jr, V. Malinova, B. E. Volkmer, R. E. Hautmann and B. Rieger, CO-alkene polymers are biocompatible scaffolds for primary urothelial cells *in vitro* and *in vivo*, *BJU Int.*, 2007, **99**, 447–453.
35. S. C. Baker, N. Atkin and P. A. Gunning, *et al.*, Characterisation of electrospun polystyrene scaffolds for three-dimensional *in vitro* biological studies, *Biomaterials*, 2006, **27**, 3136–3146.
36. G. Rohman, J. J. Pettit, F. Isaure, N. R. Cameron and J. Southgate, Influence of the physical properties of two-dimensional polyester

substrates on the growth of normal human urothelial and urinary smooth muscle cells *in vitro*, *Biomaterials*, 2007, **28**, 2264–2274.

37. S. C. Baker and J. Southgate, Towards control of smooth muscle cell differentiation in synthetic 3D scaffolds, *Biomaterials*, 2008, **29**, 3357–3366.

38. F. Chen, J. J. Yoo and A. Atala, Acellular collagen matrix as a possible 'off the shelf' biomaterial for urethral repair, *Urology*, 1999, **54**, 407–410.

39. J. J. Yoo, J. Meng, F. Oberpenning and A. Atala, Bladder augmentation using allogenic bladder submucosa seeded with cells, *Urology*, 1998, **51**, 221–225.

40. S. E. Dahms, H. J. Piechota, R. Dahiya, T. F. Lue and E. A. Tanagho, Composition and biomechanical properties of the bladder acellular matrix graft: comparative analysis in rat, pig and human, *Br. J. Urol.*, 1998, **82**, 411–419.

41. H. J. Piechota, C. A. Gleason and S. E. Dahms, *et al.*, Bladder acellular matrix graft: *in vivo* functional properties of the regenerated rat bladder, *Urol. Res.*, 1999, **27**, 206–213.

42. K. D. Sievert, T. Fandel and J. Wefer, *et al.*, Collagen I, III ratio in canine heterologous bladder acellular matrix grafts, *World J. Urol.*, 2006, **24**, 101–109.

43. B. P. Kropp, B. L. Eppley and C. D. Prevel, *et al.*, Experimental assessment of small intestinal submucosa as a bladder wall substitute, *Urology*, 1995, **46**, 396–400.

44. B. P. Kropp, S. Badylak and K. B. Thor, Regenerative bladder augmentation: a review of the initial preclinical studies with porcine small intestinal submucosa, *Adv. Exp. Med. Biol.*, 1995, **385**, 229–235.

45. F. Bolland, S. Korossis and S. P. Wilshaw, *et al.*, Development and characterisation of a full-thickness acellular porcine bladder matrix for tissue engineering, *Biomaterials*, 2007, **28**, 1061–1070.

46. S. P. Wilshaw, J. Kearney, J. Fisher and E. Ingham, Biocompatibility and potential of acellular human amniotic membrane to support the attachment and proliferation of allogeneic cells, *Tissue Eng. Part A*, 2008, **14**, 463–472.

47. Y. Zhang, D. Frimberger, E. Y. Cheng, H. K. Lin and K. P. Kropp, Challenges in a larger bladder replacement with cell-seeded and unseeded small intestinal submucosa grafts in a subtotal cystectomy model, *BJU Int.*, 2006, **98**, 1100–1105.

48. Y. Zhang, B. P. Kropp, H. K. Lin, R. Cowan and E. Y. Cheng, Bladder regeneration with cell-seeded small intestinal submucosa, *Tissue Eng.*, 2004, **10**, 181–187.

49. A. M. Kajbafzadeh, S. Payabvash and A. H. Salmasi, *et al.*, Time-dependent neovasculogenesis and regeneration of different bladder wall components in the bladder acellular matrix graft in rats, *J. Surg. Res.*, 2007, **139**, 189–202.

50. F. H. Silver and G. Pins, Cell growth on collagen: a review of tissue engineering using scaffolds containing extracellular matrix, *J. Long Term Eff. Med. Implants*, 1992, **2**, 67–80.
51. A. E. Sams and A. J. Nixon, Chondrocyte-laden collagen scaffolds for resurfacing extensive articular cartilage defects, *Osteoarthr. Cartil.*, 1995, **3**, 47–59.
52. O. Smidsrod and G. Skjak-Braek, Alginate as immobilization matrix for cells, *Trends Biotechnol.*, 1990, **8**, 71–78.
53. K. D. Sievert, M. E. Bakircioglu, L. Nunes, R. Tu, R. Dahiya and E. A. Tanagho, Homologous acellular matrix graft for urethral reconstruction in the rabbit: histological and functional evaluation, *J. Urol.*, 2000, **163**, 1958–1965.
54. A. W. El-Kassaby, A. B. Retik, J. J. Yoo and A. Atala, Urethral stricture repair with an off-the-shelf collagen matrix, *J. Urol.*, 2003, **169**, 170–173, discussion 173.
55. R. O. Hynes, Integrins: versatility, modulation, and signaling in cell adhesion, *Cell*, 1992, **69**, 11–25.
56. T. F. Deuel and N. Zhang, Growth factors. In: R. P. Lanza, R. Langer and J. Vacanti (eds.), *Principles of Tissue Engineering*, Academic Press, NewYork, 1997, pp. 129–142.
57. B. S. Kim and D. J. Mooney, Engineering smooth muscle tissue with a predefined structure, *J. Biomed. Mater Res.*, 1998, **41**, 322–332.
58. B. G. Cilento, M. R. Freeman, F. X. Schneck, A. B. Retik and A. Atala, Phenotypic and cytogenetic characterization of human bladder urothelia expanded *in vitro*, *J. Urol.*, 1994, **152**, 665–670.
59. M. Liebert, A. Hubbel and M. Chung *et al.*, Expression of mal is associated with urothelial differentiation *in vitro*: identification by differential display reverse-transcriptase polymerase chain reaction, *Differentiation*, 1997, **61**, 177–185.
60. J. A. Puthenveettil, M. S. Burger and C. A. Reznikoff, Replicative senescence in human uroepithelial cells, *Adv. Exp. Med. Biol.*, 1999, **462**, 83–91.
61. D. Delo, D. Eberli, A. Atala and S. Soker, Engineering of urinary sphincter muscle for the ageing patient population, *J. Urol.*, 2008, **179**, 515.
62. J. J. Yoo, I. Lee and A. Atala, Cartilage rods as a potential material for penile reconstruction, *J. Urol.*, 1998, **160**, 1164–1168, discussion 1178.
63. M. Mitterberger, G. M. Pinggera and R. Marksteiner, *et al.*, Functional and histological changes after myoblast injections in the porcine rhabdosphincter, *Eur. Urol.*, 2007, **52**, 1736–1743.
64. M. Renninger, C. Selent and G. Feil, *et al.*, Successful transfer of human adult germ line stem cells in the urinary sphincter of mini-pigs, *Eur. Urol. Suppl.*, 2009, **8**, 348.
65. M. Lovett, K. Lee, A. Edwards and D. L. Kaplan, Vascularization strategies for tissue engineering, *Tissue Eng. Part B Rev.*, 2009, **15**, 353–370.
66. M. R. Ladd, S. J. Lee, A. Atala and J. J. Yoo, Bioreactor maintained living skin matrix, *Tissue Eng. Part A*, 2009, **15**, 861–868.
67. S. Suh, J. Kim, J. Shin, K. Kil, K. Kim and H. Kim, Use of omentum as an *in vivo* cell culture system in tissue engineering, *Asaio J.*, 2004, **50**, 464–467.

68. H. Mertsching, J. Schanz and V. Steger, *et al.*, Generation and transplantation of an autologous vascularized bioartificial human tissue, *Transplantation*, 2009, **88**, 203–210.
69. H. Baumert, D. Mansouri and G. Fromont, *et al.*, Terminal urothelium differentiation of engineered neoureter after *in vivo* maturation in the 'omental bioreactor', *Eur. Urol.*, 2007, **52**, 1492–1498.
70. H. Baumert, P. Simon and M. Hekmati, *et al.*, Development of a seeded scaffold in the great omentum: feasibility of an *in vivo* bioreactor for bladder tissue engineering, *Eur. Urol.*, 2007, **52**, 884–890.
71. K. Hattori, A. Joraku, T. Miyagawa, K. Kawai, R. Oyasu and H. Akaza, Bladder reconstruction using a collagen patch prefabricated within the omentum, *Int. J. Urol.*, 2006, **13**, 529–537.
72. B. Burgu, L. S. McCarthy, V. Shah, D. A. Long, D. T. Wilcox and A. S. Woolf, Vascular endothelial growth factor stimulates embryonic urinary bladder development in organ culture, *BJU Int.*, 2006, **98**, 217–225.
73. Y. Guan, L. Ou and G. Hu, *et al.*, Tissue engineering of urethra using human vascular endothelial growth factor gene-modified bladder urothelial cells, *Artif. Organs*, 2008, **32**, 91–99.
74. S. Soker, M. Machado and A. Atala, Systems for therapeutic angiogenesis in tissue engineering, *World J. Urol.*, 2000, **18**, 10–18.
75. F. Oberpenning, J. Meng, J. J. Yoo and A. Atala, De novo reconstitution of a functional mammalian urinary bladder by tissue engineering, *Nat. Biotechnol.*, 1999, **17**, 149–155.
76. G. du Moon, G. Christ, J. D. Stitzel, A. Atala and J. J. Yoo, Cyclic mechanical preconditioning improves engineered muscle contraction, *Tissue Eng. Part A*, 2008, **14**, 473–482.
77. K. Guan, J. Rohwedel and A. M. Wobus, Embryonic stem cell differentiation models: cardiogenesis, myogenesis, neurogenesis, epithelial and vascular smooth muscle cell differentiation *in vitro*, *Cytotechnology*, 1999, **30**, 211–226.
78. S. Goudenege, D. F. Pisani and B. Wdziekonski, *et al.*, Enhancement of myogenic and muscle repair capacities of human adipose-derived stem cells with forced expression of MyoD, *Mol. Ther.*, 2009, **17**, 1064–1072.
79. M. Fraser, D. F. Thomas, E. Pitt, P. Harnden, L. K. Trejdosiewicz and J. Southgate, A surgical model of composite cystoplasty with cultured urothelial cells: a controlled study of gross outcome and urothelial phenotype, *BJU Int.*, 2004, **93**, 609–616.
80. W. R. Cross, I. Eardley, H. J. Leese and J. Southgate, A biomimetic tissue from cultured normal human urothelial cells: analysis of physiological function, *Am. J. Physiol. Renal Physiol.*, 2005, **289**, F459–468.
81. U. Nagele, S. Maurer and G. Feil, *et al.*, *In vitro* investigations of tissue-engineered multilayered urothelium established from bladder washings, *Eur. Urol.*, 2008, **54**, 1414–1422.
82. E. A. Kurzrock, D. K. Lieu, L. A. Degraffenried, C. W. Chan and R. R. Isseroff, Label-retaining cells of the bladder: candidate urothelial stem cells, *Am. J. Physiol. Renal Physiol.*, 2008, **294**, F1415–1421.

83. M. R. Markiewicz, M. A. Lukose, J. E. Margarone 3rd, G. Barbagli, K. S. Miller and S. K. Chuang, The oral mucosa graft: a systematic review, *J. Urol.*, 2007, **178**, 387–394.
84. A. Atala, L. Guzman and A. B. Retik, A novel inert collagen matrix for hypospadias repair, *J. Urol.*, 1999, **162**, 1148–1151.
85. P. J. le Roux, Endoscopic urethroplasty with unseeded small intestinal submucosa collagen matrix grafts: a pilot study, *J. Urol.*, 2005, **173**, 140–143.
86. R. E. De Filippo, H. G. Pohl, J. J. Yoo and A. Atala, Total penile urethra replacement with autologous cell-seeded collagen matrices, *J. Urol.*, 2002, **167**, 152–153.
87. M. Youssif, H. Shiina and S. Urakami, *et al.*, Effect of vascular endothelial growth factor on regeneration of bladder acellular matrix graft: histologic and functional evaluation, *Urology*, 2005, **66**, 201–207.
88. S. Bhargava, J. M. Patterson, R. D. Inman, S. MacNeil and C. R. Chapple, Tissue-engineered buccal mucosa urethroplasty-clinical outcomes, *Eur. Urol.*, 2008, **53**, 1263–1269.
89. R. Fiala, A. Vidlar, R. Vrtal, K. Belej and V. Student, Porcine small intestinal submucosa graft for repair of anterior urethral strictures, *Eur. Urol.*, 2007, **51**, 1702–1708, discussion 1708..
90. E. Palminteri, E. Berdondini, F. Colombo and E. Austoni, Small intestinal submucosa (SIS) graft urethroplasty: short-term results, *Eur. Urol.*, 2007, **51**, 1695–1701, discussion 1701.
91. J. Seibold, C. Selent and E. Gustafsson, *et al.*, Investigations on iatrogenic induced stricture development in a large animal model, *Eur. Urol. Suppl.*, 2009, **8**, 189.
92. D. N. Wood, R. A. Brown and C. H. Fry, Characterization of the control of intracellular [Ca2 +] and the contractile phenotype of cultured human detrusor smooth muscle cells, *J. Urol.*, 2004, **172**, 753–757.
93. A. F. Kotb, L. Campeau and J. Corcos, Urethral bulking agents: techniques and outcomes, *Curr. Urol. Rep.*, 2009, **10**, 396–400.
94. J. Seibold, M. Werther, K. D. Sievert and A. Stenzl, [Long-term results after endoscopic subureteral injection for VUR using dextranomer/ hyaluronic acid copolymer: techniques and outcomes: a five years experience.], *Urologe A*, 2010, **49**, 536–539 [in German].
95. J. Seibold, M. Werther, B. Amend, A. Stenzl and K. D. Sievert, Stress urinary incontinence after radical prostatectomy: Long term effects of endoscopic injection with dextranomer/hyaloronic acid copolymer, *Eur. Urol. Suppl.*, 2009, **8**, 338.
96. P. E. Keegan, K. Atiemo, J. Cody, S. McClinton and R. Pickard, Periurethral injection therapy for urinary incontinence in women, *Cochrane Database Syst. Rev.*, 2007, **CD003881**.
97. A. Atala, W. Kim, K. T. Paige, C. A. Vacanti and A. B. Retik, Endoscopic treatment of vesicoureteral reflux with a chondrocyte-alginate suspension, *J. Urol.*, 1994, **152**, 641–643, discussion 644.

98. D. A. Diamond and A. A. Caldamone, Endoscopic correction of vesicoureteral reflux in children using autologous chondrocytes: preliminary results, *J. Urol.*, 1999, **162**, 1185–1188.
99. M. B. Chancellor, T. Yokoyama and S. Tirney, *et al.*, Preliminary results of myoblast injection into the urethra and bladder wall: a possible method for the treatment of stress urinary incontinence and impaired detrusor contractility, *Neurourol. Urodyn.*, 2000, **19**, 279–287.
100. R. Yiou, J. P. Lefaucheur and A. Atala, The regeneration process of the striated urethral sphincter involves activation of intrinsic satellite cells, *Anat. Embryol. (Berl.)*, 2003, **206**, 429–435.
101. R. Yiou, J. J. Yoo and A. Atala, Restoration of functional motor units in a rat model of sphincter injury by muscle precursor cell autografts, *Transplantation*, 2003, **76**, 1053–1060.
102. H. Strasser, R. Marksteiner and E. Margreiter, *et al.*, Autologous myoblasts and fibroblasts versus collagen for treatment of stress urinary incontinence in women: a randomised controlled trial, *Lancet*, 2007, **369**, 2179–2186.
103. L. D. Knoll, Use of small intestinal submucosa graft for the surgical management of Peyronie's disease, *J. Urol.*, 2007, **178**, 2474–2478, discussion 2478.
104. A. Pilatz, D. Schultheiss and A. I. Gabouev, *et al.*, Isolation of primary endothelial and stromal cell cultures of the corpus cavernosum penis for basic research and tissue engineering, *Eur. Urol.*, 2005, **47**, 710–718, discussion 718–719.
105. G. Nolazco, I. Kovanecz and D. Vernet, *et al.*, Effect of muscle-derived stem cells on the restoration of corpora cavernosa smooth muscle and erectile function in the aged rat, *BJU Int.*, 2008, **101**, 1156–1164.
106. J. Wefer, N. Schlote and N. Sekido, *et al.*, Tunica albuginea acellular matrix graft for penile reconstruction in the rabbit: a model for treating Peyronie's disease, *BJU Int.*, 2002, **90**, 326–331.
107. T. G. Kwon, J. J. Yoo and A. Atala, Autologous penile corpora cavernosa replacement using tissue engineering techniques, *J. Urol.*, 2002, **168**, 1754–1758.
108. K. L. Chen, D. Eberli, J. J. Yoo and A. Atala, Bioengineered corporal tissue for structural and functional restoration of the penis, *Proc. Natl. Acad. Sci. U. S. A.*, 2010, **107**, 3346–3350.
109. P. Aebischer, T. K. Ip, G. Panol and P. M. Galletti, The bioartificial kidney: progress towards an ultrafiltration device with renal epithelial cells processing, *Life Support Syst.*, 1987, **5**, 159–168.
110. R. P. Lanza, J. L. Hayes and W. L. Chick, Encapsulated cell technology, *Nat. Biotechnol.*, 1996, **14**, 1107–1111.
111. M. Mitterberger, G. M. Pinggera, R. Marksteiner, E. Margreiter, R. Plattner, G. Klima and H. Strasser, *Eur. Urol.*, 2007, **52**, 1736–1743.

Effective Tissue Repair and Immunomodulation by Mesenchymal Stem Cells within a Milieu of Cytokines

PHILIP LIM,[a] SHYAM A. PATEL[a, b] AND PRANELA RAMESHWAR*[,a]

[a] New Jersey Medical School-UMDNJ, Department of Medicine, Division of Hematology/Oncology, Newark, NJ, USA; [b] Graduate School of Biomedical Sciences-UMDNJ, Newark, NJ, USA

17.1 Overview

17.1.1 Introduction

Mesenchymal stem cells (MSCs) are non-hematopoietic stem cells found primarily in adult bone marrow and are relatively easy to isolate and culture. They are also capable of self-renewal and, given proper stimulation, have the ability to differentiate into various mature cell types. Furthermore, recent data suggest that MSCs may not be limited to differentiation into cells of mesodermal cell lineages, but may also capable of trans-differentiation into cell types of endodermal or ectodermal origin. In addition to their multipotential ability, MSCs exhibit immunomodulatory effects—both immunosuppressive and immunostimulatory—on their local tissue environment in the presence of cytokines such as interferon (IFN)-γ or interleukin (IL)-6. This ability to modulate the

Stem Cell-Based Tissue Repair
Edited by Raphael Gorodetsky and Richard Schäfer
© The Royal Society of Chemistry 2011
Published by the Royal Society of Chemistry, www.rsc.org

immune response allows for the potential transplantation of MSCs across allogeneic barriers. The combination of these characteristics makes MSCs a novel therapeutic option for tissue repair. More specifically, MSCs may prove to be a viable therapeutic option in the fields of cardiac, neuronal and osteogenic tissue repair. This chapter discusses in detail the application of MSCs in tissue repair and the potential drawbacks of MSC therapy in the context of a milieu of cytokines.

Stem cells are undifferentiated and can be found in different organs. They have two important characteristics that distinguish them from other cell types, *i.e.* the abilities to self-renew and to differentiate along multiple lineages to generate specialized cells. More recently, stem cells have been shown to exhibit plasticity by being able to cross germ layers; to be specific, a stem cell of mesodermal origin could form cells that are traditionally ectodermal or endodermal.

Despite various cell types, this chapter focuses on MSCs and their potential for tissue repair. The International Society for Cellular Therapy suggests the following basic requirements for stem cells to be designated as MSCs:[1]

- plastic adherence;
- expression of CD73, CD90 and CD105;
- negativity for various hematopoietic markers; and
- multipotency.

Despite this suggestion, several studies have shown differences in the marker expression profile. For example, MHC class II expression has been demonstrated by various investigators;[2] the significance of this characteristic is discussed in detail later.

The multipotent differentiation capability of MSCs is further demonstrated by their ability to differentiate into other mesodermal cells such as fibroblasts, smooth muscle and stromal cells.[3] There is also evidence that MSCs exhibit plasticity through trans-differentiation to form ectodermal dopamine neurons.[4,5] The ability to differentiate MSCs into a variety of mature cell types may lead to the implantation of MSCs at sites of tissue injury and ultimately result in restoration of the damaged tissue infrastructure.

In order to elicit proper differentiation, specific cellular pathways must be triggered through the interaction of cytokines and MSCs. In addition, cytokines are involved with controlling the proliferation rate and immunomodulatory properties of MSCs. A cytokine is a soluble protein produced by cells to elicit or inhibit an immune response, and can be subdivided into various groups such as interleukins, lymphokines and chemokines.[6] Functionally, cytokines are involved in normal homeostatic regulation such as host immunity and tolerance.

The immunological control exhibited by cytokines is important because MSCs are also considered to be gatekeeper cells in the bone marrow and function both as antigen presenting cells and immune suppressors.[7] For example, MSCs will revert to an immunosuppressive phenotype in the presence of IFN-γ in order to prevent prolonged inflammation, thus maintaining homeostasis.[8] Of further importance is the effect MSCs can have on their surrounding environment. Studies suggest that MSCs have the capacity to release soluble factors that can alter the tissue microenvironment during repair.[9] Therefore, cytokines have a bimodal function in tissue repair through their interaction with MSCs and also neighboring cells within the microenvironment.

In addition to their multipotency and immune regulatory ability, MSCs are being extensively studied due to their ease of isolation from various tissue sources and expansion ability *in vitro*. MSCs are primarily isolated from adult bone marrow, but have also been found in the placenta, umbilical cord, and fetal blood, bone marrow and liver.[10–13] The bone marrow cavity is resident to MSCs and hematopoietic stem cells (HSCs).[14] MSCs are found surrounding the higher oxygen content areas around the blood vessels, while HSCs are located mainly in the low oxygen, endosteal areas of the bone marrow.[15,16] Once harvested from the bone marrow, MSCs are plastic-adherent and proliferate as fibroblastic, spindle-shaped cells. The ability of MSCs to expand readily in culture has led to the investigation of potentially culturing large quantities of MSCs *via* an automated cell culture process for therapeutic use.[17]

In summary, the multipotential ability of MSCs—in addition to their immunomodulatory properties and their high *in vitro* expansion potential—makes them an attractive therapeutic tool for a variety of clinical applications in the field of tissue repair. However, further investigations are needed to expand MSCs on a large scale as these cells appear to be one of the leading candidates for therapy.

17.1.2   Cytokines and Immune Properties of MSCs

A discussion of the biology of MSCs within a milieu of cytokines would be incomplete without considering the immunomodulatory properties of these stem cells. The concept of immunosuppression by MSCs has been investigated by several laboratories due to the cells' potential therapeutic application across allogeneic barriers. The interaction between cytokines and MSCs plays an important role in the bimodal immunological properties of MSCs, as immune responses can be enhanced in some conditions and suppressed by others. A fundamental principle that should be considered before elaborating on the precise mechanisms by which MSCs mediate immune modulation is the expression of low levels of MHC class II and reduced expression of the co-stimulatory molecules, B7-1 (CD80) and B7-2 (CD86).[18] Perhaps reduced expression of co-stimulatory molecules might explain a mechanism by which MSCs evade immune detection upon transplantation in an allogeneic host.[18]

The immunosuppressive properties of MSCs are vast and are currently more supported than the immunostimulatory properties.[14] The mechanisms by which MSC-mediated immunosuppression occur are diverse with regard to soluble mediators. These mechanisms include, but are not limited to, production of prostaglandin (PG) E_2, induction of the regulatory T cell (T_{reg}) subset, and response to pre-treatment with IFN-γ.[18–21]

PGE_2 has been suggested to promote the reduction in T cell proliferation upon MSC engraftment.[19] MSCs from allogeneic lung grafts, for example, have been shown to decrease CD4+ and CD8+ T cell proliferation upon activation by allogeneic stimuli.[18,19] Of particular interest is that the immunosuppressive effect was due to a soluble factor rather than physical cell contact.[19] Blockade of PGE_2 synthesis by indomethacin, an inhibitor of cyclooxygenase 1 and cyclooxygenase 2, eliminated these effects.[19]

Induction of T_{reg} proliferation has been suggested in a study on the usage of MSCs during cardiac transplant.[18] The T_{reg} population comprises a subset of tolerance-promoting lymphocytes that express CD4, CD25 and the forkhead-box (Fox) transcription factor P3.[22] Polarization towards T_{reg} production is facilitated by transforming growth factor (TGF)-γ and inhibited by IL-6.[23] IL-6 signaling is extensive in chronic inflammation since it is a pro-inflammatory cytokine that induces the expression of acute phase reactants.[24] Since T_{reg} cells hinder the exacerbation of colitis and allergic reactions while promoting self-tolerance, the application of MSCs to disease states involving inflammation or autoimmunity may be beneficial if MSCs induce expansion of the T_{reg} population.[23,25] Although the mechanism by which MSCs may induce T_{reg} proliferation is unclear, a study has suggested that expression of leukemia inhibitory factor (LIF) may be involved.[20]

A major player likely to be involved in MSC-mediated immunosuppression is IFN-γ, secreted by T-helper (T_H) 1 cells. IFN-γ treatment of MSCs leads to induction of the co-inhibitory molecule B7-H1 (CD274) on MSCs.[21,26] T cells from IFN-γ knockout mice fail to allow MSCs to significantly suppress T cell proliferation, indicating a role for this cytokine in the immune responses of MSCs.[21] Thus, both soluble mediators and cellular contact are components of MSC-mediated immunosuppression, with the cytokine IFN-γ exhibiting a key role.

The milieu of other soluble mediators that have been reported to play a role in MSC-mediated immune inhibition includes hepatocyte growth factor (HGF) and the T_H2-associated cytokine, IL-10. MSCs transduced to express IL-10 have been shown to decrease the severity of inflammatory arthritis *in vivo*.[27] Upon administration of IL-10-expressing MSCs, serum IL-6 levels were decreased, suggesting downregulation of the inflammatory response.[27] Further mechanistic studies are warranted before MSCs can be used in the treatment of rheumatoid arthritis.

While the majority of evidence supports a role for MSCs in damage control and immunosuppression, these conclusions are not definitive. One study demonstrated that tacrolimus, an inhibitor of IL-2 transcription and therefore T cell signaling, was necessary for allogeneic MSCs to survive and differentiate into osteogenic cells.[28] These studies suggest that MSCs may not be intrinsically immunoprivileged and that the immunosuppressive effects may be due to MSC-independent inhibition of the local milieu of cytokines such as IL-2.

The immune-enhancing properties of MSCs are not as well-documented as the immunosuppressive properties and thus have not received as much attention. In B cells of the spleen and blood, MSCs have been shown to induce IgG production.[29] The underlying mechanism of this phenomenon may involve pro-inflammatory cytokines such as IL-6 since this cytokine is important for immunoglobulin secretion.[29] MSCs can also induce IFN-γ production in lymphocytes.[29] The immunomodulatory effect of MSCs may be dose-dependent, as low numbers of MSCs can stimulate mixed lymphocyte cultures whereas high numbers are inhibitory.[30] To date, the data on immune enhancement are just beginning to be unraveled compared with studies on the immune suppressive effects of MSCs.

The dual immune properties of MSCs can be advantageous for clinical applications. However, effective use of stem cell therapy requires a comprehensive understanding of interactions among cytokines. The current therapeutic potential of MSCs spans numerous inflammatory and autoimmune disease states including osteoarthritis, graft-versus-host disease and Charcot–Marie–Tooth disease.[31–33] As stated above, the immune mechanisms of MSCs within various types of injuries remain unclear. Until further research into MSC-mediated immunological regulation is performed, our knowledge of MSC behavior within a milieu of cytokines is limited. These investigations need to be completed to allow for appropriate application of these cells and other forms of cell-based therapy to diverse clinical situations.

17.2   MSCs in Regenerative Medicine

17.2.1   MSCs in Cardiac Tissue Repair

Heart disease has been the leading cause of death for the last 80 years and a major cause of disability in the United States.[34] Recent advances in pharmaceuticals and medical treatment are resulting in more individuals surviving initial cardiovascular incidents and, therefore, morbidity from vascular disease continues to rise.[35] Following an atherosclerosis-related myocardial infarction, cardiomyocytes undergo ischemia-induced death which eventually leads to scar formation and reduced contractility.[36] The reduced contractility of the heart suggests that there is a lack of effective intrinsic mechanisms for myocardial repair and regeneration.[37] Therefore, an exogenous source of stem cells, such as MSCs, could potentially be used as a therapeutic approach to treat patients who have had a myocardial infarction.

Some studies have suggested that MSCs may be capable of differentiating into muscle tissue, including cardiac myocytes, when exposed to proper stimulation.[38] For example, in a recent study, MSCs that were co-cultured with rat cardiac myocytes showed increased expression cardiac myocyte-specific genes.[39] Furthermore, the addition of HGF and insulin-like growth factor-1 (IGF-1) to co-cultures of MSCs and cardiomyocytes enhances the expression of cardiac transcription factor GATA-4.[40] As a part of the GATA family of transcription factors, GATA-4 regulates the expression of genes necessary for cardiac contraction.[40] In addition, another group demonstrated that cargiogenic induction can occur by treatment of the MSCs with the DNA demethylating agent, 5-azacytidine.[38] The mechanisms may involve demethylation of the glycogen synthase kinase-3β (GSK-3β) promoter.[38] Some scientists have suggested that, by placing MSCs in an appropriate microenvironment, MSCs may be able differentiate into myocardial tissue and serve as a potential source for cardiac tissue repair. However, it is important to realize that there is no clear consensus on whether MSCs can differentiate into functional cardiomyocytes. Currently, the most obvious application of MSCs for cardiac purposes is that MSCs exert paracrine effects that lead to suppression of detrimental inflammation induced by hypoxic damage. As research in this area progresses, the future may see broader applications for MSCs for treatment of cardiac injury.

The delivery method of MSCs into the site of injury is also a critical factor to consider when discussing MSCs as a therapy for cardiac repair. One of the major routes of delivery under investigation is an intravenous system.[41] Intravenous administration is an attractive strategy because it is not invasive and also allows for repeated administration.[41] However, one of its major limitations is the low percentage of MSCs that ultimately migrate to the injured heart.[41] In an attempt to improve the efficiency of MSCs entering the area of injury, post-myocardial infarction rats were given an intravenous injection of MSCs plus granulocyte colony-stimulating factor (G-CSF). This combined method of treatment demonstrated increased MSC mobility from the bone marrow and into the ischemic myocardium.[41] Although the exact mechanism behind this migration is unknown, it is hypothesized that G-CSF may promote the chemotaxis of MSCs toward cardiac stromal cell derived factor-1 (SDF-1) by increasing the surface expression on MSCs of CXCR4 (CD184), the chemokine receptor for SDF-1.[41] However, further research is needed to improve prospects of stem cell-based therapy for myocardial infarction as there was minimal overall improvement in function despite the increased migration of MSCs to ischemic cardiac tissue.[41,42]

One possible reason for the marginal improvement in function could be a significant cell death rate of implanted MSCs. To improve the survival of MSCs in cardiac repair, the cells were transduced with a member of the PI(3)K signaling pathway Akt.[42] The data show that Akt-transduced MSCs hampered ventricular remodeling and established functional cardiac improvement within a few days.[42] Akt-transduced MSCs were found to exhibit paracrine actions on the surrounding tissues by upregulating the release of soluble factors such as vascular endothelial growth factor (VEGF), fibroblast growth factor-2 (FGF-2), HGF and IGF-1 after exposure to hypoxic conditions.[42] Of particular importance is the expression of VEGF and IGF-1, which are angiogenic and pro-survival factors, respectively, as they have been shown to protect cardiac myocytes from apoptotic stimuli and promote tube formation of endothelial cells.[43] Through these studies, it is evident that growth factors play a crucial role in the ability of MSCs to illicit proper cardiac tissue repair.

The application of MSCs following a myocardial infarction is quite plausible due to their ability to migrate to cardiac tissue and promote cardiac tissue repair *via* paracrine actions. Possible mechanisms of paracrine suppression include the release of TGF-β, which has been shown to induce the production of T$_{regs}$.[18] These are highly significant findings because, in conjunction with previous studies on MSC-mediated facilitation of cardiac allograft survival, MSCs may be able to promote cardiac repair through recruitment of immune-inhibitory subsets of T lymphocytes.[18,44] It is important, however, to further explore the effects of cytokines and MSCs within the cardiac tissue micro-environment to obtain a complete understanding of the benefits and limitations of MSC administration.

17.2.2   MSCs in Application for Central Nervous System Disorders

Despite intense research efforts, stroke is the one of the most frequent causes of adult onset neurological disability in the United States.[34] In contrast to axons in the peripheral nervous system (PNS), those in the central nervous system (CNS) have a very limited ability for self-repair following a stroke or injury.[45] The only current available treatment for stroke is administered during the first few hours of onset, and many patients are not able to receive and benefit from this treatment in such a short time window.[46] Without treatment, significant neuronal damage can occur and result in long-term or permanent loss in sensory, language, visual and motor systems. However, MSCs are known to have therapeutic benefits after cerebral ischemia and may prove to be a valid therapeutic treatment option for neuronal injury.[47]

The potential use of MSCs for neuronal repair would require an understanding of surrounding injured tissues and how the factors found within these regions could establish a cross-talk with the implanted cells. Human MSCs cultured in supernatant derived from ischemic brain extracts increased production of brain-derived neurotrophic factor (BDNF), nerve growth factor (NGF), VEGF and HGF, lending further credibility to the hypothesis that increased expression of trophic factors is responsible for the beneficial effects of MSCs.[48] In addition, MSCs are capable of modulating the immune and inflammatory response following ischemia by upregulating the expression of MHC class I and a series of anti-inflammatory and anti-apoptotic-related factors such as latent TGF-β binding protein-2 and other TGF-β family members.[49] By modulating both the inflammatory and immune responses and decreasing apoptosis, MSCs are capable of limited neuronal rescue following an ischemic event.

Although MSCs alone seem to be an effective approach in the treatment of brain injuries, evidence shows that their therapeutic potency may be increased when combined with molecular agonists and genetic modification. For example, the beneficial effect of MSCs on cerebral ischemia was enhanced with the additional administration of erythropoietin (EPO) and adrenomedullin.[50,51] EPO has been shown to promote a cholinergic neurogenesis from MSCs while causing a reduction in toxic amyloid-β *via* neprilysin synthesis within these neurons.[52] EPO appears to be successful in this regard because it ameliorates the effects of hypoxia, which can result in metabolic derangements and decreased ability of MSCs to handle cellular stress, including glutamate toxicity.[52] In addition to its neuroprotective effect, EPO also plays a role in the proliferation and differentiation of stem cells.[50] The combination treatment of MSCs and EPO improved cell proliferation and neurogenesis, and may account for an increase in long-term memory observed after ischemic events in the brain.[50] Thus MSC therapy for neurodegeneration and stroke might be able to be optimized by concurrent EPO administration.[52]

In addition to these findings, MSCs genetically modified to express BDNF, glial cell line-derived neurotrophic factor (GDNF) and FGF were therapeutically more efficient than control MSCs.[51–54] MSCs transfected with the BDNF or GDNF genes resulted in improved function and less ischemic injury in a rat model.[53–55] Through either a combination treatment or genetic modification, the use of MSCs as an approach for the treatment of cerebrovascular accidents is promising.

In addition to their ability to differentiate into osteoblasts, chondroblasts, adipocytes and stromal cells, MSCs have the potential to trans-differentiate into non-mesenchymal tissues such as neurons. When grown in neuronal progenitor basal medium augmented with cyclic AMP (cAMP), iso-butylmethylxanthine, NGF and insulin, MSCs were induced to differentiate into neuronal-like cells and expressed a neuronal profile.[56] MSCs have also been induced to differentiate into specific neuronal cells such as dopamine neurons.[4,5] By means of *in vitro* manipulation with sonic hedgehog, FGF-8, and basic FGF, MSCs trans-differentiated into cells that not only expressed dopamine neuron-specific markers, but also constitutively secreted dopamine.[4] This ability to trans-differentiate across different cell types provides an avenue through which MSCs can be used to treat neurological pathologies.

Furthermore, some of the most promising therapeutic uses for MSCs involve the treatment of neurodegenerative diseases such as Alzheimer's disease, Parkinson's disease and Huntington's disease. As the name implies, neurodegenerative diseases involve the death and atrophy of neurons in the brain. Thus far, no cure exists for neurodegenerative diseases and treatment is limited to the alleviation of only some of the symptoms.[57] The previously described dopamine neurons derived from MSCs could potentially become a therapy for Parkinson's disease, but they would not be suitable for the many other neurodegenerative diseases. In order to treat the variety of neural cells involved in such diseases, neural stem cells (NSCs) are the optimum choice for therapeutic treatment as they can differentiate into all major neural cell types.[58] Unfortunately, endogenous NSCs exist only in very limited quantities and, therefore, may not suitable for autologous transplantation.[58] However, culturing MSCs in NSC media supplemented with epidermal growth factor (EGF) and bFGF results in the formation of NSCs that develop first as neurosphere-like structures and express characteristic NSC antigens.[58] When these MSC-derived neurospheres are co-cultured with primary astrocytes, they are able to differentiate into mature neurons with synaptic transmission that had tetrodotoxin-sensitive action potentials.[58] Although the differentiation potential of MSCs into neurons *in vitro* is dependent on the induction conditions, their plasticity is one of the key characteristics that make MSCs a potential therapy for a multitude of neurological injuries.

17.2.3   MSCs in Bone Repair

The prospect of using MSCs in disease of the bone and other connective tissues has received much attention. The sequence of events that occurs following bone fracture are quite complex at the molecular level and, in several cases, may account for ineffective healing.[59] The release of cytokines from the site of injury and the migration of MSCs are vital to the healing process.[59] MSC therapy has also been proposed for repair of cartilaginous structures like the intervertebral disc.[60] The methods that are currently being investigated include MSC induction of nucleus pulposus and annulus fibrosis cells.[60] Nonetheless, these therapies have not entered the clinic and much research remains to be done if MSCs are to regenerate vertebral disc in humans.

The induction of bone and cartilage repair appears to depend on cytokines and other soluble mediators.[61] A cooperative role between bone morphogenetic protein (BMP)-2 and the immunosuppressive cytokine, TGF-β, has been demonstrated for polarization of MSCs towards the chondrocyte lineage.[61] Although TGF-β alone can promote chondrocyte differentiation from MSCs, the combination of these proteins is more effective as bone morphogenetic protein (BMP)-2 enhances the effects of TGF-β.[61] Intracellular signal transduction pathways that may mediate the effects of these cytokines include the mitogen-activated protein kinase and SMAD pathways.[61,62] Modulators of these pathways can thereby alter the differentiation and fate of multipotent MSCs. Adenoviral systems have been used to introduce TGF-β and BMP-2 for the purposes of cartilage differentiation, as confirmed by synthesis of proteoglycans, glycosaminoglucans and collagen.[63] By using a combination of induction factors rather than single factors such as TGF-β alone, the dose of adenovirus can be reduced while maintaining the targeted effect.[63] These clinical strategies may spare patients the negative effects of high dose adenoviral therapy in the future.

Regarding bone repair, MSCs can undergo osteogenic differentiation when exposed to induction media containing dexamethasone, ascorbic acid and other factors.[62] Enhancement of osteogenic differentiation occurs with FGF-2.[62] Although numerous factors can enhance osteogenic differentiation, equally important are factors that prevent this effect. The Hedgehog signaling pathway appears to inhibit osteocyte formation from MSCs.[64] The cytokines TGF-β and BMP-2, which facilitate the development of a chondrogenic phenotype, also inhibit differentiation into the osteogenic lineage.[62] Confirmation of the osteogenic phenotype can be demonstrated by the production of type I collagen and alkaline phosphatase.[65] Thus far, a limited amount of clinical evidence has accrued for the use of MSCs in bone repair such as orthopedic regeneration of femoral head necrosis.[66] Autologous MSC transplantation to the site of osteonecrosis in mongrel dogs led to successful viability and osteogenic differentiation of MSCs at the damaged site.[66] Studies such as these offer hope for future MSC therapy in humans once clinical challenges are met.

The sources of MSCs in mammals are quite diverse, but the potential for bone repair is evident in most sources. MSCs derived from adipose tissue have been studied for orthopedic regeneration.[62] The mechanisms by which MSCs from particular sources polarize towards osteogenic cells are gradually being uncovered. Recently, non-coding 19–23 nucleotide regulatory molecules known as microRNAs have been implicated in these mechanisms.[67] For example, miR-196a has been shown to enhance differentiation of adipose tissue-derived MSCs into osteogenic cells *via* degradation of the endogenous homeobox protein, HOXC8, message.[67] The cAMP/protein kinase A signal transduction pathway leads to secretion of osteogenic cytokines such as IGF-1, BMP-2 and IL-11.[68] Stem cells from adipose tissue may offer great potential since ethical considerations are bypassed and could replace the invasive procedure of obtaining bone marrow aspirates. Multipotential precursor cells of the skin, which are found within hair follicles, have the ability to develop into osteocytes and chondrocytes in traumatic environments.[65] They can aid in repair when transplanted into a bone fracture.[65]

Although the outlook for bone repair by MSCs is promising, the delivery method of MSCs and their practical application in bone repair poses a few challenges. Nonetheless, scientists are overcoming these problems as research into the field continues. Thus far, intra-articular injections of autologous MSCs have been shown to successfully regenerate knee joints.[69] MSCs have been used to assess the compatibility of various biomaterials and membranes since they differentiate into a wide range of cell types.[70] Polycarbonates containing poly(ethylene)glycol (PEG) promote induction of MSCs into osteocytes.[71] Scaffolds for tissue engineering are thus important if MSCs can be delivered successfully, since gene expression appears to depend on the scaffold.[52,70] Scaffolds also function to maintain cellular support and differentiation, which are variables that must be optimized prior to the implementation of MSCs for bone repair.[62]

17.3 Challenges for MSC Therapy

It is becoming clear that, as the field of stem cell research progresses, stem cell therapy hold promise for future application in regenerative medicine. Thus far, this chapter has discussed numerous potential applications for MSCs in the repair of various pathological conditions. However, the picture is incomplete without considering the challenges to stem cell therapy. Challenges to the use of stem cells are most evident for embryonic stem (ES) cells, as these cells can lead to teratoma formation and pose major ethical concerns regarding their acquisition.[71] On the other hand, the prospects for MSC therapy in particular seem promising for numerous pathological states and appear to pose far fewer problems that ES cell therapy. Despite the vast therapeutic potential for MSCs, there exist a few limitations to their use, as is also true for ES cells. Among the major obstacles to the use of MSCs in the clinic are malignant transformation, senescence-induced loss of function, immune-mediated rejection of MSCs and limited survivability of engrafted cells.[71–75] Each of these challenges is discussed below.

17.3.1   Oncogenesis

The prospect of transformation of MSCs has been a major qualm among scientists in recent years. In a murine model, for example, MSCs from the bone marrow have the ability to undergo malignant transformation, resulting in fibrosarcoma.[72,76] Although the mechanisms by which MSC transformation occurs are not completely understood, it has been suggested that increased telomerase activity and increased expression of the oncogene *c-myc* play an important role.[76] In dermatofibrosarcoma protuberans (a slow-cycling malignancy of the skin), derivatives of MSCs have been suspected as the neoplastic cells.[77] In addition to these tumors, a recent study demonstrated that MSC-like cells were found in 65% of gastric adenocarcinomas.[78] Lipomas, or benign tumors of adipose tissue, have also been found to contain a population of MSCs.[79] It has been suggested that these stem cells may function in initiating or maintaining certain connective tissue tumors, and thus the enthusiasm for translating MSCs to the bedside should be slightly cautioned on these grounds.[78]

17.3.2   Senescence and Genetic Instability

Another notable drawback to using MSCs in the clinic involves the consequences of cellular aging and genetic instability.[73,74] MSC senescence increases the likelihood of transformation, presumably due to the accumulation of genetic damage with repeated cell division cycles.[73] During MSC senescence, MSCs lose their ability to differentiate, divide and mobilize as efficiently as young MSCs.[74,80] The accumulation of advanced glycation end-products and reactive oxygen and nitrogen species may be the reason underlying the loss of MSC function as these cells age. As for genetic instability, a subpopulation of cells with groups of MSCs—known as multipotent adult progenitor cells (MAPCs)—have been shown to be susceptible to insults of genomic integrity due to the requirement for many cell division cycles to occur before MAPCs arise within a group of MSCs.[81] To date, it is unclear as to whether MSCs age because of their genetic profile or the environment in which they are administered.[80] It is possible that both of these factors play a role in combination with one another. Understanding the mechanisms of aging in MSCs can shed light on potential methods to counteract these obstacles to administration of MSCs in the clinic.

17.3.3   Immunological Considerations

Aside from these issues, another challenge that MSC therapy must face is immune-mediated rejection.[2,75] MHC class II expression from MSCs makes these cells inherently immunogenic and confers them with the potential for reaction against host tissue.[2] Alongside B lymphocytes, macrophages and dendritic cells, MSCs are considered to be antigen-presenting cells based on MHC class II expression.[2,82] A putative method by which to diminish the immunogenicity of MSCs may be *via* pre-treatment with high levels IFN-γ, which has been shown to downregulate MHC class II expression in MSCs.[2] For the prevention of immune incompatibility, it has been hypothesized that knockout of MHC class II genes may prevent allogeneic graft rejection and promote tolerance.[83]

Regarding the challenge to survival of transplanted MSCs, these stem cells have been shown to undergo apoptosis upon transplantation site.[84] In MSC therapy for cardiac repair, for instance, MSCs transplanted into the heart die throughout the first four days after transplantation, demonstrating the practical challenges to MSC therapy for tissue repair.[84] Cytokines appear to play a vital role in determining the viability of MSCs at transplantation sites.[75] MSCs release a variety of cytokines and growth factors into their local microenvironment including FGF-2, IGF-1, and VEGF, most of which promote cell survival or confer anti-apoptotic function. Since programmed cell death appears to hinder MSC applications, methods to combat apoptosis in stem cells have been suggested as a means of evading this problem.[75] Some biochemical methods have attempted to bypass these challenges. For example, over-expression of transglutaminase has been shown to increase adhesion of MSCs to the extracellular matrix.[85] Thus, *ex vivo* experimental manipulations of MSCs may have the ability to restore MSC viability and promote successful engraftment.[85]

In addition to these challenges, a wide variety of factors must be considered prior to the translation of MSCs into the clinic. These include, but are not limited to, the localization and homing of MSCs upon transplantation.[86] The interaction between MSCs and the cells of the target microenvironment pose limitations to their effectiveness in therapy.[86] The use of MSCs as a vehicle for gene delivery is hampered by the development of auto-antibodies on the grounds of MHC mismatch, as shown in the case of EPO delivery for treatment of anemia.[87] Before the field of stem cell biology can see the benefits of clinical application (*e.g.* repair of the degenerating nervous system, cardiovascular system and bone), numerous challenges must be met, and therefore, much further research is warranted to address these challenges.

17.4 Conclusions

This chapter discusses the promises and challenges regarding stem cell therapies. They are difficult to generalize since, at any site of tissue injury, there would be variations in inflammatory mediators such as immunostimulatory and immunosuppressive cytokines. While these factors might be beneficial in the repair process, they could also reduce the efficacy of stem cell therapy in the microenvironment. Further complications arise when considering that stem

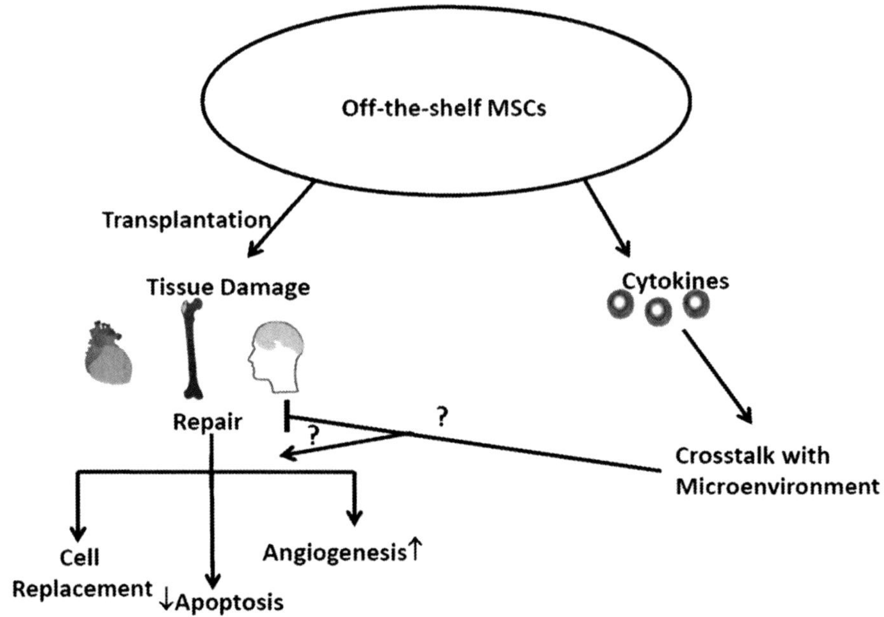

Figure 17.1   Summary of the crosstalk between stem cells and cytokines. Left: an ideal case where injuries are treated with 'off-the-shelf' stem cells for the repair of tissue damage. Right: changes in the microenvironment with inflammatory mediators such as cytokines. These cytokines interact with the stem cells, in this case MSCs, resulting in a bidirectional cross-talk. The influence of these responses on tissue repair depends on the cross-talk, for which the mechanisms are undetermined.

cells produce inflammatory mediators themselves, thereby developing a cross-talk with the immune microenvironment.[9,88] Effective stem cell therapies can only be expected if the mechanisms of interaction between stem cells and microenvironmental factors are unraveled through basic science discoveries (Figure 17.1).

References

1. M. Dominici, K. Le Blanc, I. Mueller, I. Slaper-Cortenbach, F. Marini, D. Krause, R. Deans, A. Keating, D. J. Prockop and E. Horwitz, Minimal criteria for defining multipotent mesenchymal stromal cells. The International Society for Cellular Therapy position statement, *Cytotherapy*, 2006, **8**, 315–317.
2. T. C. Tang, K. A. Trzaska, S. V. Smirnov, S. V. Kotenko, S. K. Schwander, J. J. Ellner and P. Rameshwar, Down-regulation of MHC II in mesenchymal stem cells at high IFN-gamma can be partly explained by cytoplasmic retention of CIITA, *J. Immunol.*, 2008, **180**, 1826–1833.

3. P. A. Zuk, M. Zhu. H. Mizuno, J. Huang, J. W. Futrell, A. J. Katz, P. Benhaim, H. P. Lorenz and M. H. Hedrick, Multilineage cells from human adipose tissue: implications for cell-based therapies, *Tissue Eng.*, 2001, **7**, 211–228.

4. K. A. Trzaska, E. V. Kuzhikandathil and P. Rameshwar, Specification of a dopaminergic phenotype from adult human mesenchymal stem cells, *Stem Cells*, 2007, **25**, 2797–2808.

5. R. Barzilay, I. Kan, T. Ben-Zur, S. Bulvik, E. Melamed and D. Offen, Induction of human mesenchymal stem cells into dopamine-producing cells with different differentiation protocols, *Stem Cells Dev.*, 2008, **17**, 547–554.

6. J. G. Cannon, Inflammatory cytokines in nonpathological states, *News Physiol. Sci.*, 2000, **15**, 298–303.

7. J. L. Chan, K. C. Tang, A. P. Patel, L. M. Bonilla, N. Pierobon, N. M. Ponzio and P. Rameshwar, Antigen-presenting property of mesenchymal stem cells occurs during a narrow window at low levels of interferon-gamma, *Blood*, 2006, **107**, 4817–4824.

8. M. Castillo, K. Liu, L. Bonilla and P. Rameshwar, The immune properties of mesenchymal stem cells, *Int. J. Biomed. Sci.*, 2007, **3**, 100–104.

9. D. J. Prockop, 'Stemness' does not explain the repair of many tissues by mesenchymal stem/multipotent stromal cells (MSCs), *Clin. Pharmacol. Ther.*, 2007, **82**, 241–243.

10. P. Bianco, M. Riminucci, S. Gronthos and P. G. Robey, Bone marrow stromal cells: nature, biology, and potential applications, *Stem Cells*, 2001, **19**, 180–192.

11. Y. Zhang, C. Li, X. Jiang, S. Zhang, Y. Wu, B. Liu, P. Tang and N. Mao, Human placenta-derived mesenchymal progenitor cells support culture expansion of long-term culture-initiating cells from cord blood CD34+ cells, *Exp. Hematol.*, 2004, **32**, 657–664.

12. E. J. Gang, J. A. Jeong, S. H. Hong, S. H. Hwang, S. W. Kim, I. H. Yang, C. Ahn, H. Han and H. Kim, Skeletal myogenic differentiation of mesenchymal stem cells isolated from human umbilical cord blood, *Stem Cell*, 2004, **22**, 617–624.

13. C. Campagnoli, I. A. Roberts, S. Kumar, P. R. Bennett, I. Bellantuono and N. M. Fisk, Identification of mesenchymal stem/progenitor cells in human first-trimester fetal blood, liver, and bone marrow, *Blood*, 2001, **98**, 2396–2402.

14. S. A. Patel, L. Sherman, J. Munoz and P. Rameshwar, Immunological properties of mesenchymal stem cells and clinical implications, *Arch. Immunol. Ther. Exp.*, 2008, **56**, 1–8.

15. S. K. Nilsson, H. M. Johnston and J. A. Coverdale, Spatial localization of transplanted hemopoietic stem cells: inferences for the localization of stem cell niches, *Blood*, 2001, **97**, 2293–2299.

16. J. J. Minguell, A. Erices and P. Conget, Mesenchymal stem cells, *Exp. Biol. Med.*, 2001, **226**, 507–520.

17. R. J. Thomas, A. Chandra, Y. Liu, P. C. Hourd, P. P. Conway and D. J. Williams, Manufacture of a human mesenchymal stem cell population using an automated cell culture platform, *Cytotechnology*, 2007, **55**, 31–39.

18. F. Casiraghi, N. Azzollini, P. Cassis, B. Imberti, M. Morigi, D. Cugini, R. A. Cavinato, M. Todeschini, S. Solini, A. Sonzogni, N. Perico, G. Remuzzi and M. Noris, Pretransplant infusion of mesenchymal stem cells prolongs the survival of a semiallogeneic heart transplant through the generation of regulatory T cells, *J. Immunol.*, 2008, **181**, 3933–3946.

19. L. Jarvinen, L. Badri, S. Wettlaufer, T. Ohtsuka, T. J. Standiford, G. B. Toews, D. J. Pinsky, M. Peters-Golden and V. N. Lama, Lung resident mesenchymal stem cells isolated from human lung allografts inhibit T cell proliferation *via* a soluble mediator, *J. Immunol.*, 2008, **181**, 4389–4396.

20. A. Nasef, C. Mazurier, S. Bouchet, S. François, A. Chapel, D. Thierry, N. C. Gorin and L. Fouillard, Leukemia inhibitory factor: role in human mesenchymal stem cells mediated immunosuppression, *Cell. Immunol.*, 2008, **253**, 16–22.

21. H. Sheng, Y. Wang, Y. Jin, Q. Zhang, Y. Zhang, L. Wang, B. Shen, S. Yin, W. Liu, L. Cui and N. Li, A critical role of IFNgamma in priming MSC-mediated suppression of T cell proliferation through up-regulation of B7-H1, *Cell Res.*, 2008, **18**, 846–857.

22. A. Dolganiuc, E. Paek, K. Kodys, J. Thomas and G. Szabo, Myeloid dendritic cells of patients with chronic HCV infection induce proliferation of regulatory T lymphocytes, *Gastroenterology*, 2008, **135**, 2119–2127.

23. S. Dominitzki, M. C. Fantini, C. Neufert, A. Nikolaev, P. R. Galle, J. Scheller, G. Monteleone, S. Rose-John, M. F. Neurath and C. Becker, Cutting edge: trans-signaling *via* the soluble IL-6R abrogates the induction of FoxP3 in naive CD4 + CD25 T cells, *J. Immunol.*, 2007, **179**, 2041–2045.

24. S. Tenhumberg, G. H. Waetzig, A. Chalaris, B. Rabe, D. Seegert, J. Scheller, S. Rose-John and J. Grötzinger, Structure-guided optimization of the interleukin-6 trans-signaling antagonist sgp130, *J. Biol. Chem.*, 2008, **283**, 27200–27207.

25. F. Meiler, S. Klunker, M. Zimmermann, C. A. Akdis and M. Akdis, Distinct regulation of IgE, IgG4 and IgA by T regulatory cells and toll-like receptors, *Allergy*, 2008, **63**, 1455–1463.

26. P. Rameshwar, IFNgamma and B7-H1 in the immunology of mesenchymal stem cells, *Cell Res.*, 2008, **18**, 805–806.

27. J. J. Choi, S. A. Yoo, S. J. Park, Y. J. Kang, W. U. Kim, I. H. Oh and C. S. Cho, Mesenchymal stem cells overexpressing interleukin-10 attenuate collagen-induced arthritis in mice, *Clin. Exp. Immunol.*, 2008, **153**, 269–276.

28. N. Kotobuki, Y. Katsube, Y. Katou, M. Tadokoro, M. Hirose and H. Ohgushi, *In vivo* survival and osteogenic differentiation of allogeneic rat bone marrow mesenchymal stem cells (MSCs), *Cell Transplant.*, 2008, **17**, 705–712.

29. I. Rasmusson, K. Le Blanc, B. Sundberg and O. Ringdén, Mesenchymal stem cells stimulate antibody secretion in human B cells, *Scand. J. Immunol.*, 2007, **65**, 336–343.

30. K. Le Blanc, L. Tammik, B. Sundberg, S. E. Haynesworth and O. Ringdén, Mesenchymal stem cells inhibit and stimulate mixed lymphocyte cultures

and mitogenic responses independently of the major histocompatibility complex, *Scand. J. Immunol.*, 2003, **57**, 11–20.

31. F. H. Chen and R. S. Tuan, Mesenchymal stem cells in arthritic diseases, *Arthritis Res. Ther.*, 2008, **10**, 223.

32. H. Li, Z. Guo, X. Jiang, H. Zhu, X. Li and N. Mao, Mesenchymal stem cells alter migratory property of T and dendritic cells to delay the development of murine lethal acute graft-versus-host disease, *Stem Cells*, 2008, **26**, 2531–2541.

33. A. Leal, T. E. Ichim, A. M. Marleau, F. Lara, S. Kaushal and N. H. Riordan, Immune effects of mesenchymal stem cells: implications for Charcot-Marie-Tooth disease, *Cell. Immunol.*, 2008, **253**, 11–15.

34. W. Rosamond, K. Flegal, K. Furie, A. Go, K. Greenlund, N. Haase, S. M. Hailpern, M. Ho, V. Howard, B. Kissela, S. Kittner, D. Lloyd-Jones, M. McDermott, J. Meigs, C. Moy, G. Nichol, C. O'Donnell, V. Roger, P. Sorlie, J. Steinberger, T. Thom, M. Wilson and Y. Hong, American Heart Association Statistics Committee and Stroke Statistics Subcommittee, Heart disease and stroke statistics—2008 update: a report from the American Heart Association Statistics Committee and Stroke Statistics Subcommittee, *Circulation*, 2008, **117**, e25–146.

35. F. B. Hu, M. J. Stampfer, J. E. Manson, F. Grodstein, G. A. Colditz, F. E. Speizer and W. C. Willett, Trends in the incidence of coronary heart disease and changes in diet and lifestyle in women, *N. Engl. J. Med.*, 2000, **343**, 530–537.

36. Y. T. Sia, T. G. Parker, J. N. Tsoporis, P. Liu, A. Adam and J. L. Rouleau, Long-term effects of carvedilol on left ventricular function, remodeling, and expression of cardiac cytokines after large myocardial infarction in the rat, *J. Cardiovasc. Pharmacol.*, 2002, **39**, 73–87.

37. S. Ohnishi, H. Ohgushi, S. Kitamura and N. Nagaya, Mesenchymal stem cells for the treatment of heart failure, *Int. J. Hematol.*, 2007, **86**, 17–21.

38. J. Cho, P. Rameshwar and J. Sadoshima, Distinct roles of glycogen synthase kinase (GSK)-3alpha and GSK-3beta in mediating cardiomyocyte differentiation in murine bone marrow-derived mesenchymal stem cells, *J. Biol. Chem.*, 2009, **284**, 36647–36658.

39. T. Wang, Z. Xu, W. Jiang and A. Ma, Cell-to-cell contact induces mesenchymal stem cell to differentiate into cardiomyocyte and smooth muscle cell, *Int. J. Cardiol.*, 2006, **109**, 74–81.

40. Z. Li, T. X. Gu and Y. H. Zhang, Hepatocyte growth factor combined with insulin like growth factor-1 improves expression of GATA-4 in mesenchymal stem cells cocultured with cardiomyocytes, *Chin. Med. J. (Engl.)*, 2008, **121**, 336–340.

41. Z. Cheng, X. Liu, L. Ou, X. Zhou, Y. Liu, X. Jia, J. Zhang, Y. Li and D. Kong, Mobilization of mesenchymal stem cells by granulocyte colony-stimulating factor in rats with acute myocardial infarction, *Cardiovasc. Drugs Ther.*, 2008, **22**, 363–371.

42. M. Gnecchi, H. He, N. Noiseux, O. D. Liang, L. Zhang, F. Morello, H. Mu, L. G. Melo, R. E. Pratt, J. S. Ingwall and V. J. Dzau, Evidence

supporting paracrine hypothesis for Akt-modified mesenchymal stem cell-mediated cardiac protection and functional improvement, *FASEB J.*, 2006, **20**, 661–669.

43. S. Sadat, S. Gehmert, Y. H. Song, Y. Yen, X. Bai, S. Gaiser, H. Klein and E. Alt, The cardioprotective effect of mesenchymal stem cells is mediated by IGF-I and VEGF, *Biochem. Biophys. Res. Commun.*, 2007, **363**, 674–679.

44. S. A. Patel, J. R. Meyer, S. J. Greco, K. E. Corcoran, M. Bryan and P. Rameshwar, Mesenchymal stem cells protect breast cancer cells through regulatory T cells: role of mesenchymal stem cell-derived TGF-β, *J. Immunol.*, 2010, **185**, 5885–5894.

45. R. Zietlow, E. L. Lane, S. B. Dunnett and A. E. Rosser, Human stem cells for CNS repair, *Cell Tissue Res.*, 2008, **331**, 301–322.

46. C. A. Cronin, C. J. Weisman and R. H. Llinas, Stroke treatment: beyond the three-hour window and in the pregnant patient, *Ann. N.Y. Acad. Sci.*, 2008, **1142**, 159–178.

47. D. G. Phinney and I. Isakova, Plasticity and therapeutic potential of mesenchymal stem cells in the nervous system, *Curr. Pharm. Des.*, 2005, **11**, 1255–1265.

48. X. Chen, Y. Li, L. Wang, M. Katakowski, L. Zhang, J. Chen, Y. Xu, S. C. Gautam and M. Chopp, Ischemic rat brain extracts induce human marrow stromal cell growth factor production, *Neuropathology*, 2002, **22**, 275–279.

49. H. Ohtaki, J. H. Ylostalo, J. E. Foraker, A. P. Robinson, R. L. Reger, S. Shioda and D. J. Prockop, Stem/progenitor cells from bone marrow decrease neuronal death in global ischemia by modulation of inflammatory/immune responses, *Proc. Natl. Acad. Sci. U.S.A.*, 2008, **105**, 14638–14643.

50. E. Esneault, E. Pacary, D. Eddi, T. Freret, E. Tixier, J. Toutain, O. Touzani, P. Schumann-Bard, E. Petit, S. Roussel and M. Bernaudin, Combined therapeutic strategy using erythropoietin and mesenchymal stem cells potentiates neurogenesis after transient focal cerebral ischemia in rats, *J. Cereb. Blood Flow Metab.*, 2008, **28**, 1552–1563.

51. K. Hanabusa, N. Nagaya, T. Iwase, T. Itoh, S. Murakami, Y. Shimizu, W. Taki, K. Miyatake and K. Kangawa, Adrenomedullin enhances therapeutic potency of mesenchymal stem cells after experimental stroke in rats, *Stroke*, 2005, **36**, 853–858.

52. L. Danielyan, R. Schäfer, A. Schulz, T. Ladewig, A. Lourhmati, M. Buadze, A. L. Schmitt, S. Verleysdonk, D. Kabisch, K. Koeppen, G. Siegel, B. Proksch, T. Kluba, A. Eckert, C. Köhle, T. Schöneberg, H. Northoff, M. Schwab and C. H. Gleiter, Survival, neuron-like differentiation and functionality of mesenchymal stem cells in neurotoxic environment: the critical role of erythropoietin, *Cell Death Differ.*, 2009, **16**, 1599–1614.

53. K. Kurozumi, K. Nakamura, T. Tamiya, Y. Kawano, M. Kobune, S. Hirai, H. Uchida, K. Sasaki, Y. Ito, K. Kato, O. Honmou, K. Houkin,

I. Date and H. Hamada, BDNF gene-modified mesenchymal stem cells promote functional recovery and reduce infarct size in the rat middle cerebral artery occlusion model, *Mol. Ther.*, 2004, **9**, 189–197.

54. K. Kurozumi, K. Nakamura, T. Tamiya, Y. Kawano, K. Ishii, M. Kobune, S. Hirai, H. Uchida, K. Sasaki, Y. Ito, K. Kato, O. Honmou, K. Houkin, I. Date and H. Hamada, Mesenchymal stem cells that produce neurotrophic factors reduce ischemic damage in the rat middle cerebral artery occlusion model, *Mol. Ther.*, 2005, **11**, 96–104.

55. N. Ikeda, N. Nonoguchi, M. Z. Zhao, T. Watanabe, Y. Kajimoto, D. Furutama, F. Kimura, M. Dezawa, R. S. Coffin, Y. Otsuki, T. Kuroiwa and S. Miyatake, Bone marrow stromal cells that enhanced fibroblast growth factor-2 secretion by herpes simplex virus vector improve neurological outcome after transient focal cerebral ischemia in rats, *Stroke*, 2005, **36**, 2725–2730.

56. T. Tondreau, M. Dejeneffe, N. Meuleman, B. Stamatopoulos, A. Delforge, P. Martiat, D. Bron and L. Lagneaux, Gene expression pattern of functional neuronal cells derived from human bone marrow mesenchymal stromal cells, *BMC Genomics*, 2008, **9**, 166.

57. M. S. Forman, J. Q. Trojanowski and V. M. Lee, Neurodegenerative diseases: a decade of discoveries paves the way for therapeutic breakthroughs, *Nat. Med.*, 2004, **10**, 1055–1063.

58. L. Fu, L. Zhu, Y. Huang, T. D. Lee, S. J. Forman and C. Shih, Derivation of neural stem cells from mesenchymal stem cells: evidence for a bipotential stem cell population, *Stem Cells Dev.*, 2008, **17**, 1109–1122.

59. F. N. Kwong and M. B. Harris, Recent developments in the biology of fracture repair, *J., Am. Acad. Orthop. Surg.*, 2008, **16**, 619–625.

60. A. Hiyama, J. Mochida and D. Sakai, Stem cell applications in intervertebral disc repair, *Cell. Mol. Biol.*, 2008, **54**, 24–32.

61. B. Shen, A. Wei, H. Tao, A. D. Diwan and D. D. Ma, BMP-2 enhances TGF-beta3-mediated chondrogenic differentiation of human bone marrow multipotent mesenchymal stromal cells in alginate bead culture, *Tissue Eng. Part A.*, 2008, **15**, 1311–1320.

62. H. Tapp, E. N. Hanley, J. C. Patt and H. E. Gruber, Adipose-derived stem cells: characterization and current application in orthopaedic tissue repair, *Exp. Biol. Med.*, 2008, **234**, 1–9.

63. A. F. Steinert, G. D. Palmer, C. Pilapil, N. Ulrich, C. H. Evans and S. C. Ghivizzani, Enhanced *in vitro* chondrogenesis of primary mesenchymal stem cells by combined gene transfer, *Tissue Eng. Part A.*, 2008, **15**, 1127–1139.

64. M. Plaisant, C. Fontaine, W. Cousin, N. Rochet, C. Dani and P. Peraldi, Activation of hedgehog signaling inhibits osteoblast differentiation of human mesenchymal stem cells, *Stem Cells*, 2008, **27**, 703–713.

65. J. F. Lavoie, J. A. Biernaskie, Y. Chen, D. Bagli, B. Alman, D. R. Kaplan and F. D. Miller, Skin-derived precursors differentiate into skeletogenic cell types and contribute to bone repair, *Stem Cells Dev.*, 2008, **18**, 893–906.

66. Z. Yan, D. Hang, C. Guo and Z. Chen, Fate of mesenchymal stem cells transplanted to osteonecrosis of femoral head, *J. Orthop. Res.*, 2009, **27**, 442–446.

67. Y. J. Kim, S. W. Bae, S. S. Yu, Y. C. Bae and J. S. Jung, miR-196a regulates proliferation and osteogenic differentiation in mesenchymal stem cells derived from human adipose tissue, *J. Bone Miner. Res.*, 2009, **24**, 816–825.

68. R. Siddappa, A. Martens, J. Doorn, A. Leusink, C. Olivo, R. Licht, L. van Rijn, C. Gaspar, R. Fodde, F. Janssen, C. van Blitterswijk and J. de Boer, cAMP/PKA pathway activation in human mesenchymal stem cells *in vitro* results in robust bone formation *in vivo*, *Proc. Natl. Acad. Sci. U.S.A.*, 2008, **105**, 7281–7286.

69. H. Alfaqeh, M. Y. Norhamdan, K. H. Chua, H. C. Chen, B. S. Aminuddin and B. H. Ruszymah, Cell based therapy for osteoarthritis in a sheep model: gross and histological assessment, *Med. J. Malaysia.*, 2008, **63**(Suppl A), 37–38.

70. K. Madhumathi, N. S. Binulal, H. Nagahama, H. Tamura, K. T. Shalumon, N. Selvamurugan, S. V. Nair and R. Jayakumar, Preparation and characterization of novel beta-chitin-hydroxyapatite composite membranes for tissue engineering applications, *Int. J. Biol. Macromol.*, 2008, **44**, 1–5.

71. T. Briggs, M. D. Treiser, P. F. Holmes, J. Kohn, P. V. Moghe and T. L. Arinzeh, Osteogenic differentiation of human mesenchymal stem cells on poly(ethylene glycol)-variant biomaterials, *J. Biomed. Mater. Res. A.*, 2009, **91**, 975–984.

72. A. S. Correia, S. V. Anisimov, J. Y. Li and P. Brundin, Growth factors and feeder cells promote differentiation of human embryonic stem cells into dopaminergic neurons: a novel role for fibroblast growth factor-20, *Front. Neurosci.*, 2008, **2**, 26–34.

73. H. Li, X. Fan, R. C. Kovi, Y. Jo, B. Moquin, R. Konz, C. Stoicov, E. Kurt-Jones, S. R. Grossman, S. Lyle, A. B. Rogers, M. Montrose and J. Houghton, Spontaneous expression of embryonic factors and p53 point mutations in aged mesenchymal stem cells: a model of age-related tumorigenesis in mice, *Cancer Res.*, 2007, **67**, 10889–10898.

74. A. Stolzing, E. Jones, D. McGonagle and A. Scutt, Age-related changes in human bone marrow-derived mesenchymal stem cells: consequences for cell therapies, *Mech. Ageing Dev.*, 2008, **129**, 163–173.

75. R. Z. Shi and Q. P. Li, Improving outcome of transplanted mesenchymal stem cells for ischemic heart disease, *Biochem. Biophys. Res Commun.*, 2008, **376**, 247–250.

76. M. Miura, Y. Miura, H. M. Padilla-Nash, A. A. Molinolo, B. Fu, V. Patel, B. M. Seo, W. Sonoyama, J. J. Zheng, C. C. Baker, W. Chen, T. Ried and S. Shi, Accumulated chromosomal instability in murine bone marrow mesenchymal stem cells leads to malignant transformation, *Stem Cells*, 2006, **24**, 1095–1103.

77. D. S. Behroozan, A. Glaich and L. H. Goldberg, Dermatofibrosarcoma protuberans following tanning bed use, *J. Drugs Dermatol.*, 2005, **4**, 751–754.
78. H. Cao, W. Xu, H. Qian, W. Zhu, Y. Yan, H. Zhou, X. Zhang, X. Xu, J. Li, Z. Chen and X. Xu, Mesenchymal stem cell-like cells derived from human gastric cancer tissues, *Cancer Lett.*, 2009, **274**, 61–71.
79. T. M. Lin, H. W. Chang, K. H. Wang, A. P. Kao, C. C. Chang, C. H. Wen, C. S. Lai and S. D. Lin, Isolation and identification of mesenchymal stem cells from human lipoma tissue, *Biochem. Biophys. Res. Commun.*, 2007, **361**, 883–889.
80. S. Sethe, A. Scutt and A. Stolzing, Aging of mesenchymal stem cells, *Ageing Res. Rev.*, 2006, **5**, 91–116.
81. N. A. Habib and M. Y. Gordon, Clinical applications of stem cell therapy—the pros and cons of stem cell sources, *Regenerative Med.*, 2006, **1**, 301–302.
82. F. Morandi, L. Raffaghello, G. Bianchi, F. Meloni, A. Salis, E. Millo, S. Ferrone, V. Barnaba and V. Pistoia, Immunogenicity of human mesenchymal stem cells in HLA-class I-restricted T-cell responses against viral or tumor-associated antigens, *Stem Cells*, 2008, **26**, 1275–1287.
83. M. Yang and L. Liu, MHC II gene knockout in tissue engineering may prevent immune rejection of transplants, *Med. Hypotheses*, 2008, **70**, 798–801.
84. M. Zhang, D. Methot, V. Poppa, Y. Fujio, K. Walsh and C. E. Murry, Cardiomyocyte grafting for cardiac repair: graft cell death and anti-death strategies, *J. Mol. Cell. Cardiol.*, 2001, **33**, 907–921.
85. H. Song, W. Chang, S. Lim, H. S. Seo, C. Y. Shim, S. Park, K. J. Yoo, B. S. Kim, B. H. Min, H. Lee, Y. Jang, N. Chung and K. C. Hwang, Tissue transglutaminase is essential for integrin-mediated survival of bone marrow-derived mesenchymal stem cells, *Stem Cells*, 2007, **25**, 1431–1438.
86. R. Summer and A. Fine, Mesenchymal progenitor cell research limitations and recommendations, *Proc. Am. Thorac. Soc.*, 2008, **5**, 707–710.
87. P. M. Campeau, M. Rafei, M. François, E. Birman, K. A. Forner and J. Galipeau, Mesenchymal stromal cells engineered to express erythropoietin induce anti-erythropoietin antibodies and anemia in allorecipients, *Mol. Ther.*, 2009, **17**, 369–372.
88. S. A. Patel, A. C. Heinrich, B. Y. Reddy and P. Rameshwar, Inflammatory mediators: parallels between cancer biology and stem cell therapy, *J. Inflamm. Res.*, 2009, **2**, 13–19.

Homing of Mesenchymal Stromal Cells

REINHARD HENSCHLER,[a] ERIKA DEAK[a] AND RICHARD SCHÄFER[b, c]

[a] DRK Institute of Transfusion Medicine and Immune Hematology, Sandhofstrasse 1, 60528, Frankfurt, Germany; [b] Harvard Stem Cell Institute, Harvard University, 7 Divinity Avenue, Cambridge, MA 02138, USA; [c] Department of Stem Cell and Regenerative Biology, Harvard University, 7 Divinity Avenue, Cambridge, MA 02138, USA

18.1 The Story Begins with Infusing Cells: Definition of Homing End-points

The term 'homing' has been generally used to characterize the way in which cells present in the circulation localize into specific tissues. However, there is no internationally agreed consensus defining the respective tissues. In a strict way, this would be the tissue of origin of the transplanted cells, *e.g.* mesenchymal stromal cells (MSCs) derived from the bone marrow (BM) would home preferentially to bone marrow. However, most groups nowadays refer to 'homing' as the entry of a circulating cell into any type of tissue, generally assuming that there are preferred target tissues such as heart tissue when analyzing a myocardial infarct model. In addition, consensus is lacking as to whether the term 'homing' applies merely to any egress of circulating cells from the vessel into an interstitial space within a tissue, or whether 'homed' cells have to reach specific environments as part of the homing pathway. For example, anchoring in a 'stem cell niche' would be the prerequisite to allow a stem cell to survive and to

Stem Cell-Based Tissue Repair
Edited by Raphael Gorodetsky and Richard Schäfer
© The Royal Society of Chemistry 2011
Published by the Royal Society of Chemistry, www.rsc.org

undergo any further developmental decision and initiate daughter cells to develop along differentiation events (see Chapter 2). Still, there is a general agreement that a 'homed' cell is a cell that has left the bloodstream and has entered a tissue environment. This definition is applied in this context in this chapter.

18.2 Distribution of Intravenous Infused MSCs into Different Tissues: Human and Primate Studies Led the Way

A study in 2000 examined the intravenous (i.v.) administration of MSCs to patients with breast cancer[1] given as an adjuvant to hematopoietic stem cells (HSCs). It became clear that MSCs are well-tolerated at the dose of $1–5\times10^{6}\,kg^{-1}$ and that they can be tracked until total disappearing from the circulation.[2] In the following years, a number of pre-clinical studies examined in more detail in which tissues i.v. infused MSCs accumulate. Devine *et al.*[3] found in non-human primates that MSCs survived well for weeks up to several months in gastrointestinal tissue, as well as in kidney, lung, liver, thymus and skin. Fouillard *et al.*[4] and Chapel *et al.*[5] found MSCs beneficial after homing to the bone marrow in patients with aplastic anemia, and to multiple organs after a radiation-induced multiorgan failure, respectively.[4,6] MSCs could also be detected at low frequencies in several tissues when i.v. infused into a non-diseased organism in animal models. In these studies, the MSCs were tracked by *ex vivo* investigations using histology molecular biology methods such as polymerase chain reaction (PCR).[3,7,8] The latter study performed a more systematic kinetic assessment in a non-injury murine model using enhanced green fluorescent protein (eGFP) transfected murine MSCs in mice, and demonstrated that MSCs were also found to accumulate in regions of damaged tissue such as myocardial infarct. Moreover, homing of MSCs can also be assessed by *in vivo* imaging.[9]

The evidence for accumulation of MSCs in different tissues, including a local pathological situation, is summarized in this chapter and presented in Table 18.1.

18.3 Interactions of MSCs with Cells in the Lung

Over the last five years, the majority of pre-clinical, as well as clinical studies, have analyzed MSC homing after systemic application. After intravenous delivery of MSCs the lungs are the first capillary barrier. Do MSCs remain in the lungs? Lee *et al.*[2] recently provided a quantification of human MSCs accumulation in the lung over time. Nearly all infused MSCs could be traced in the lung of mice in the first 15 minutes after infusion. Over the course of four days, however, the human specific signal in the lungs decreased in an exponential way to 0.01%. Clinical studies with human MSCs in humans did not report significant symptoms such as dyspnoea or chest pain.[1] However it is

Table 18.1 Molecules shown to be involved in homing of MSCs to target tissues.

Class of molecules	Molecule and endothelial ligand(s)	Evidence	Reference
Selectins/selectin ligands	CD44–E-selectin	Homing to bone marrow after glycan engineering of MSCs	21
	unknown ligand –P-selectin	Rolling in ear vessels	48
	L-selectin–unknown ligand	*None as yet*	-
Integrins	alpha4/beta1–VCAM1	Cardiac homing blocked by anti-VCAM-1	25
		Cardiac homing blocked by anti-beta1 integrin	26
	Beta 2 integrins	*None as yet*	25
Chemokine receptors	CCR2–MCP-1 or MCP-3	Transgenic expression in heart of CCR2 ligand attracts i.v. applied MSCs	27
		Injection of MCP-3 into heart recruits MSCs	28
		Migration to breast tumors	38
	CXCR4–SDF1	Overexpressed CXCR4 leads to cardiac homing of MSCs	29
	CCR7–unknown ligand	Homing of i.v. MSCs into wounded skin	37
Matrix metallo-proteinases (MMPs)	Membrane-type MMPs–extra-cellular matrix	Migration through base-ment membranes, partially after culture in hypoxia, or regulating chemokine receptor responsiveness	41, 42, 43

possible that murine MSCs (mMSCs) behave differently with regard to their ability to induce lung damage, most likely also reflecting lung retention. We recently showed in a C57BL/6 clonogenic murine MSCs transfusion model that murine MSCs, in contrast to human MSCs, clog in murine lungs.[10] Interestingly, accumulation of MSCs in the lung did not compromise their beneficial effects in a cardiac infarct model in mice using human MSCs.[2]

Andrews *et al.*[8] described the integration of mMSCs into lung airway epithelia. In a bleomycin injury model, Ortiz *et al.*[11] have also demonstrated increased homing of MSCs in the lungs. MSCs were shown to integrate into lung epithelial tissue. Together, these data demonstrate not only the possibility that MSCs can pass through the lung barrier, but also an inherent tendency that at least some MSCs may remain in lung tissue and integrate. Ortiz *et al.*[11] also investigated the engraftment of MSCs in a lung injury model after bleomycin exposure. Their data suggest an enhanced engraftment of MSCs to the lung, including integration of the homed MSCs.

18.4   Homing of MSCs to the Bone Marrow

The bone marrow is the most common source of MSCs. Many investigations have confirmed the positive effect of MSCs to stimulate hematopoietic regeneration. This has been clinically exploited for example in cord blood transplantation, where addition of third-party MSCs was shown to downregulate the clonal predominance of one of two concomitantly transplanted cord blood units.[12] Our group[13] and later Erices *et al.*[14] have used the NOD/SCID mouse model to demonstrate the presence of MSCs in human G-CSF mobilized blood or cord blood. It is this possible that preparations of $CD34^+$ progenitors may contain functional MSC (precursor) cells. Moreover, Wynn *et al.*[15] have suggested that only subpopulations of MSCs may be capable of homing to bone marrow—at least in the NOD/SCID mouse model they used.

Children with the devastating disease osteogenesis imperfecta (OI) suffer from an enzyme defect in the collagen synthesis pathway. Horwitz *et al.*[16] showed that this defect can be corrected by cultured MSCs. Similarly, LeBlanc *et al.*[17] treated an OI child pre-birth with BM-derived osteogenic cells. In a murine model, the engraftment of the osteoprogenitors has been systematically characterized using a GFP transgene.[18,19]

Rombouts *et al.*[20] have elegantly demonstrated that propagation of MSCs in culture induced a severe (almost total) engraftment failure of murine MSCs in the bone marrow. This engraftment defect of MSCs in bone marrow was further investigated, and after systematic work, Sackstein *et al.*[21] came up with a protocol to glycol-conjugate a CD44 epitope, thus rendering cultured, non BM-homing MSCs capable of engrafting into bone marrow. Moreover, Mendez-Ferrer *et al.*[22] employed two-photon intravital microscopy, thus tracking an MSC-like cell which keeps contact to the most primitive hematopoietic cells close to the putative osteoblast niche in murine bone marrow.

18.5   Homing into Cardiac Tissue

Cardiac repair has been an area of intense interest for a variety of cellular therapeutics, including MSCs. Stasis or very reduced blood flow, and the possibility to interfere with catheters to release progenitor or other effector cells types in specific areas have led to the generation of a number of promising data on the mechanisms by which MSCs can reach damaged myocardium. Using a swine model, Freyman *et al.*[23] showed that delivery into heart vessels results in superior homing of MSCs compared with intramyocardial injection and intravenous infusion. It has been difficult to demonstrate differentiation of MSCs into specific cell types in damaged myocardium, but Martens *et al.*[24] have observed homed MSCs to stabilize new vessel formation during healing of myocardial ischemic events. Segers *et al.*[25] found in a myocardial infarct model that pretreatment of microvascular endothelial cells with proinflammatory stimuli was required to induce MSC extravasation, and that this could be inhibited by an anti-VCAM1 but not an anti-ICAM1 antibody. Their data point to the involvement of beta1 integrins in MSC homing to cardiac tissue. In a similar murine model, Ip *et al.*[26] investigated factors which are upregulated in ischemic myocardium. Although the chemokine SDF-1 and E-selectin were among the most highly upregulated homing molecules, only integrin beta1 but not the SDF-1 receptor CXCR4 could be confirmed as involved in cardiac homing of MSCs.

However, different approaches led to other results: Belema-Bedada *et al.*[27] and Schenk *et al.*[28] demonstrated the involvement of CCR2 receptor chemokine ligands in cardiac homing of MSCs. Cheng *et al.*[29] transplanted MSCs which overexpressed CXCR4 receptor in a rat myocardial infarct model. However, all these studies did not investigate the concomitant involvement of integrins by chemokines, which is a major pathway for the homing of both leukocytes and hematopoietic stem and progenitor cells.[30,31] Thus, taking the published data together, it is likely that chemokine-mediated activation of integrin may be operative in myocardial extravasation of MSCs.

18.6   Homing of MSCs to Other Tissues

MSCs have been demonstrated to be beneficial in acute kidney failure. In line with the tubular tissue damage, intra-arterially administered MSCs accumulated in peritubular areas *via* CD44-hyaluronicacid interaction.[32] It has not been determined whether this is occurring at a blood vessel wall side, or rather reflecting interstitial retention. Morigi *et al.*[33] showed in an acute renal injury model that homed MSCs limited capillary alterations and neutrophil infiltration, indicating a local action of MSCs.

Ponte *et al.*[34] reported that multiple chemokines act on MSCs to stimulate migration *in vitro*. As reported at the beginning of this chapter, a number of studies have shown an increased intestinal homing of MSCs after irradiation. Zhang *et al.*[35] showed a CXCR4 dependent engraftment of MSCs into murine intestinal tissue, pointing to a more general role of the SDF-1–CXCR4 pathway in homing of MSCs to tissues. Sordi *et al.*[36] demonstrated that a set of distinct chemokine receptors is expressed on pancreas-derived MSCs, acting on their migration to inflamed pancreatic tissue. Skin could be a target for MSC-mediated tissue regeneration as well. Sasaki *et al.*[37] reported that GFP-positive MSCs migrate into wounded skin using the CCR7 receptor, are locally retained and acquired epithelial antigens. IN addition, tumor tissue seems to provide chemokines which attract MSCs.[38] Thus, there is increasing accumulating evidence that chemokine receptors are operative in migration of systemically applied MSCs to tissues. Still, the tissue specificity which may be achieved by selective action of chemokines or other homing receptors, or the involvement of selectin and in many cases integrin-type receptors, and the potentially synergistic interplay between these pathways remain to be elucidated.

It is furthermore possible that MSCs can be preconditioned towards activating their migration and, subsequently, homing. This may be a general principle, as shown by Rosova *et al.*[39] who demonstrated that MSCs kept in 1% oxygen show increased AKT phosphorylation, along with increased motility and upregulation of the c-met receptor. Stimulation with, among other cytokines, hepatocyte growth factor (HGF) can render MSCs susceptible to upregulation of the CXCR4 pathway.[40]

Further to adhesion on endothelial cells, MSCs would transmigrate and start an interstitial migration path. As a mechanism, matrix metalloproteinases (MMPs) have been found operational in endothelial adhesion and transmigration of MSCs. Annabi *et al.*[41] discovered that membrane-type MMP is upregulated by hypoxia and stimulates matrix invasion by MSCs; this was confirmed in a similar study by de Becker *et al.*[42] who also studied transendothelial migration through BM microvessels and basement membranes. Linking these observations back to chemokine signals, Ries *et al.*[43] showed that chemokine SDF and proinflammatory cytokines are major regulators of MMP activation modulation in MSCs.

18.7 Is Homing Required at all for MSCs to Initiate Treatment Effects?

An indirect, but homing-regulated pathway for MSC action has been proposed by Husnain *et al.*[44] who showed in a murine model that transplanted MSCs overexpressing IGF-1 homing to bone marrow are crucially involved in the mobilization of hematopoietic cells into the circulation, in turn stimulating cardiac repair. Lee *et al.*[2] went a step further by elegantly demonstrating that i.v. administered human MSCs, in a murine myocardial infarct model, produce a tumor necrosis factor (TNF) related protein GAS6 in the lungs. The authors demonstrated that recombinant GAS6 could mediate the cardioprotection in the same way as MSCs.

Little is known so far as to where the majority of systemically applied MSCs end up after delivery. It is clear that only a minority of the MSCs can be traced after a few days post-transplantation.[2] In line with these results, Francois *et al.*[45] observed that irradiation induces a generally increased engraftment of MSCs in all tissues except the lungs. It is thus likely that only the total amount of surviving MSCs is increased in irradiated tissues, allowing their improved survival. Whether specific initial homing signals or MSC-endothelial cell interactions already govern the later fate of the transplanted MSCs is also unknown to date.

18.8 Immunosuppressive Functions of MSCs: A Function of Specific Homing Signals?

Casiraghi *et al.*[46] made an interesting observation that, in a murine allotolerance model, infusion of MSCs ameliorates outcome. However, this effect is lost if HSCs are co-administered. This indicates that, among other reasons, HSC co-transplantation might alter homing of MSCs away from lymphatic tissue. Inflammatory stimuli have already been indicated to have the potential to direct MSC homing. Wu *et al.*[47] found that inflamed tissue in heart allografts attracts MSCs which engraft. So far it has been difficult to demonstrate that MSCs home to lymphatic tissues. This may in part be due to their absent expression of L-selectin and beta2 integrins, which are critical homing signals towards secondary lymphatic organs.[31,48]

In a murine sepsis model, Németh *et al.*[49] found that MSCs are likely act on macrophages to modulate release of anti-inflammatory stimuli. It is not clear whether this process occurs in the vascular bed or in an extravascular space. So far, areas where MSCs can exert specific immunosuppressive effects remain elusive.[50]

18.9 Possible Future Directions in the Exploration of Homing Mechanisms of MSCs

Most importantly, the search for homing of MSCs has been both time- and energy-consuming. It is predicted that novel techniques such as iron oxide nanoparticles, in combinations with non-radioactive and non-invasive imaging techniques, may improve possibilities to trace homing MSCs.

Hsiao *et al.*[51] demonstrated that MSCs can principally be detected at a single cell level using nuclear magnetic resonance (NMR) imaging. Kraitchman *et al.*[52] followed allogeneic MSCs in a dog model using a similar approach, detecting MSCs in ischemic heart tissue after systemic i.v. infusion. Dual modality imaging has been found helpful to trace MSCs labeled by supraparamagnetic iron particles.[53] Findings that specific culture conditions may influence this behavior will become of increasing importance.[40,54] In addition, to study the endogenous behavior such as events which induce mobilization of MSCs[55,56] may lead to new conclusions on how to apply MSCs and which type of MSCs may be appropriate. It is clear that all these studies imply more sound knowledge and understanding of the mechanisms which act on MSCs and which decide on their homing behavior.

References

1. O. N. Koc and H. M. Lazarus, Mesenchymal stem cells: heading into the clinic, *Bone Marrow Transplant.*, 2001, **27**, 235–239.
2. R. H. Lee, A. A. Pulin, M. J. Seo, D. J. Kota, J. Ylostalo, B. L. Larson, L. Semprun-Prieto, P. Delafontaine and D. J. Prockop, Intravenous

hMSCs improve myocardial infarction in mice because cells embolized in lung are activated to secrete the anti-inflammatory protein, TSG-6, *Cell Stem Cell*, 2009, **5**, 54–63.

3. S. M. Devine, C. Cobbs, M. Jennings, A. Bartholomew and R. Hoffman, Mesenchymal stem cells distribute to a wide range of tissues following systemic infusion into nonhuman primates, *Blood*, 2003, **101**, 2999–3001.

4. L. Fouillard, M. Bensidhoum, D. Bories, H. Bonte, M. Lopez, A. M. Moseley, A. Smith, S. Lesage, F. Beaujean, D. Thierry, P. Gourmelon, A. Najman and N. C. Gorin, Engraftment of allogeneic mesenchymal stem cells in the bone marrow of a patient with severe idiopathic aplastic anemia improves stroma, *Leukemia*, 2003, **2**, 474–476.

5. A. Chapel, J. M. Bertho, M. Bensidhoum, L. Fouillard, R. G. Young, J. Frick, C. Demarquay, F. Cuvelier, E. Mathieu, F. Trompier, N. Dudoignon, C. Germain, C. Mazurier, J. Aigueperse, J. Borneman, N. C. Gorin, P. Gourmelon and D. Thierry, Mesenchymal stem cells home to injured tissues when co-infused with hematopoietic cells to treat a radiation-induced multi-organ failure syndrome, *J. Gene Med.*, 2003, **5**, 1028–1038.

6. M. Bensidhoum, A. Chapel, S. Francois, C. Demarquay, C. Mazurier, L. Fouillard, S. Bouchet, J. M. Bertho, P. Gourmelon and J. Aigueperse, *et al.*, Homing of *in vitro* expanded Stro-1_ or Stro-1+ human mesenchymal stem cells into the NOD/SCID mouse and their role in supporting human CD34 cell engraftment, *Blood*, 2004, **103**, 3313–3319.

7. J. Gao, J. E. Dennis, R. F. Muzic, M. Lundberg and A. I. Caplan, The dynamic *in vivo* distribution of bone marrow-derived mesenchymal stem cells after infusion, *Cells Tissues Organs*, 2001, **169**, 12–20.

8. F. Anjos-Afonso, E. K. Siapati and D. Bonnet, In vivo contribution of murine mesenchymal stem cells into multiple cell-types under minimal damage conditions, *J. Cell Sci.*, 2004, **117**, 5655–5664.

9. J. M. Hill, A. J. Dick, V. K. Raman, R. B. Thompson, Z. X. Yu, K. A. Hinds, B. S. Pessanha, M. A. Guttman, T. R. Varney, B. J. Martin, C. E. Dunbar, E. R. McVeigh and R. J. Lederman, Serial cardiac magnetic resonance imaging of injected mesenchymal stem cells, *Circulation*, 2003, 1009–1014.

10. E. Deak, B. Rüster, L. Keller, K. Eckert, I. Fichtner, E. Seifried and R. Henschler, Suspension medium influences interaction of mesenchymal stromal cells with endothelium and pulmonary toxicity after transplantation in mice, *Cytotherapy*, 2010, **12**, 260–264.

11. L. A. Ortiz, F. Gambelli, C. McBride, D. Gaupp, M. Baddoo, N. Kaminski and D. G. Phinney, Mesenchymal stem cell engraftment in lung is enhanced in response to bleomycin exposure and ameliorates its fibrotic effects, *Proc. Natl. Acad. Sci. U.S.A.*, 2003, **100**, 8407–8411.

12. D. W. Kim, Y. J. Chung, T. G. Kim, Y. L. Kim and I. H. Oh, Cotransplantation of third-party mesenchymal stromal cells can alleviate single-donor predominance and increase engraftment from double cord transplantation, *Blood*, 2004, **103**, 1941–1948.

13. S. R. Goan, I. Junghahn, M. Wissler, M. Becker, J. Aumann, U. Just, G. Martiny-Baron, I. Fichtner and R. Henschler, Donor stromal cells from human blood engraft in NOD/SCID mice, *Blood*, 2000, **96**, 3971–3978.
14. A. A. Erices, C. I. Allers, P. A. Conget, C. V. Rojas and J. J. Minguell, Human cord blood-derived mesenchymal stem cells home and survive in the marrow of immunodeficient mice after systemic infusion, *Cell Transplant.*, 2003, **12**, 555–561.
15. R. F. Wynn, C. A. Hart, C. Corradi-Perini, L. O'Neill, C. A. Evans, J. E. Wraith, L. J. Fairbairn and I. Bellantuono, A small proportion of mesenchymal stem cells strongly expresses functionally active CXCR4 receptor capable of promoting migration to bone marrow, *Blood*, 2004, **104**, 2643–2645.
16. E. M. Horwitz, K. Le Blanc, M. Dominici, I. Mueller, I. Slaper-Cortenbach, F. C. Marini, R. J. Deans, D. S. Krause and A. Keating, Clarification of the nomenclature for MSC: The International Society for Cellular Therapy position statement, *Cytotherapy*, 2005, **7**, 393–395.
17. K. Le Blanc, C. Götherström, O. Ringdén, M. Hassan, R. McMahon, E. Horwitz, G. Anneren, O. Axelsson, J. Nunn, U. Ewald, S. Nordén-Lindeberg, M. Jansson, A. Dalton, E. Aström and M. Westgren, Fetal mesenchymal stem-cell engraftment in bone after *in utero* transplantation in a patient with severe osteogenesis imperfecta, *Transplantation*, 2005, **79**, 1607–1614.
18. R. F. Pereira, M. D. O'Hara, A. V. Laptev, K. W. Halford, M. D. Pollard, R. Class, D. Simon, K. Livezey and D. J. Prockop, Marrow stomal cells as a source of progenitor cells for nonhemtapoietic tissues in transgenic mice with a phenotype of osteogenesis imperfecta, *Proc. Natl. Acad. Sci. U.S.A.*, 1998, **95**, 1142–1147.
19. R. Marino, C. Martinez, K. Boyd, M. Dominici, T. J. Hofmann and E. M. Horwitz, Transplantable marrow osteoprogenitors engraft in discrete saturable sites in the marrow microenvironment, *Exp. Hematol.*, 2008, **36**, 360–368.
20. W. J. Rombouts and R. E. Ploemacher, Primary murine MSC show highly efficient homing to the bone marrow but lose homing ability following culture, *Leukemia*, 2003, **17**, 160–170.
21. R. Sackstein, J. S. Merzaban, D. W. Cain, N. M. Dagia, J. A. Spencer, C. P. Lin and R. Wohlgemuth, *Ex vivo* glycan engineering of CD44 programs human multipotent mesenchymal stromal cell trafficking to bone, *Nat. Med.*, 2008, **14**, 181–187.
22. S. Mendez-Ferrer, T. V. Mitchurina, F. Ferraro, A. Mazloom, B. MacArthur, S. Lira, D. T. Scadden, A. Maáyan, G. N. Enikolopov and P. S. Frenette, Coordinated regulation of hematopoietic and mesenchymal stem cells in a bone marrow niche, *Blood*, 2009, **114**(22)Abstract 2.
23. T. Freyman, G. Polin, H. Osman, J. Crary, M. Lu, L. Cheng, M. Palasis and R. L. Wilensky, A quantitative, randomized study evaluating three methods of mesenchymal stem cell delivery following myocardial infarction, *Eur. Heart J.*, 2006, **27**, 1114–1122.

24. T. P. Martens, F. See, M. D. Schuster, H. P. Sondermeijer, M. M. Hefti, A. Zannettino, S. Gronthos, T. Seki and S. Itescu, Mesenchymal lineage precursor cells induce vascular network formation in ischemic myocardium, *Nat. Clin. Pract. Cardiovasc. Med.*, 2006, **3**, S18–S22.
25. V. F. M. Segers, I. Van Riet, L. J. Andries, K. Lemmens, M. J. Demolder, A. J. M. L. De Becker, M. M. Kockx and G. W. De Keulenaer, Mesenchymal stem cell adhesion to cardiac microvascular endothelium: activators and mechanisms, *Am. J. Physiol. Heart Circ. Physiol.*, 2006, **290**, H1370–H1377.
26. J. E. Ip, Y. Wu, J. Huang, L. Zhang, R. E. Pratt and V. J. Dzau, Mesenchymal stem cells use integrin b1 not CXC chemokine receptor 4 for myocardial migration and engraftment, *Mol. Biol. Cell*, 2007, **18**, 2873–2882.
27. F. Belema-Bedada, S. Uchida, A. Martire, S. Kostin and T. Braun, Efficient homing of multipotent adult MSCs depends on FROUNT-mediated clustering of CCR2, *Cell Stem Cell*, 2008, **2**, 566–575.
28. S. Schenk, N. Mal, A. Finan, M. Zhang, M. Kiedrowski, Z. Popovic, P. M. McCarthy and M. S. Penn, Monocyte chemotactic protein-3 is a myocardial MSC homing factor, *Stem Cells*, 2007, **25**, 245–251.
29. Z. Cheng, L. Ou, X. Zhou, F. Li, X. Jia, Y. Zhang, X. Liu, Y. Li, C. A. Ward, L. G. Melo and D. Kong, Targeted migration of mesenchymal stem cells modified with CXCR4 gene to infarcted myocardium improves cardiac performance, *Mol. Ther.*, 2008, **16**, 571–579.
30. D. D. Wagner and P. S. Frenette, The vessel wall and its interactions, *Blood*, 2008, **111**, 5271–5281.
31. R. Förster, A. C. Davalos-Misslitz and A. Rot, CCR7 and its ligands: balancing immunity and tolerance, *Nat. Rev. Immunol.*, 2008, **5**, 362–371.
32. M. B. Herrera, B. Bussolati, S. Bruno, L. Morando, G. Mauriello-Romanazzi, F. Sanavio, I. Stamenkovic, L. Biancone and G. Camussi, Exogenous mesenchymal stem cells localize to the kidney by means of CD44 following acute tubular injury, *Kidney Int.*, 2007, **72**, 430–441.
33. M. Morigi, C. Rota, T. Montemurro, E. Montelatici, V. Lo Cicero, B. Imberti, M. Abbate, C. Zoja, P. Cassis, L. Longaretti, P. Rebulla, M. Introna, C. Capelli, A. Benigni, G. Remuzzi and L. Lazzari, Life-sparing effect of human cord-blood mesenchymal stem cells in experimental acute kidney injury, *Stem Cells*, 2010, **28**, 513–522.
34. A. L. Ponte, E. Marais, N. Gallay, A. Langonne, B. Delorme, O. Herault, P. Charbord and J. Domenech, The *in vitro* migration capacity of human bone marrow mesenchymal stem cells: comparison of chemokine and growth factor chemotactic activities, *Stem Cells*, 2007, **25**, 1737–1745.
35. J. Zhang, J.-F. Gong, W. Zhang, W.-M. Zhu and J.-S. Li, Effects of transplanted bone marrow mesenchymal stem cells on the irradiated intestine of mice, *J. Biomed. Sci.*, 2008, **15**, 585–594.
36. V. Sordi, M. L. Malosio, F. Marchesi, A. Mercalli, R. Melzi, T. Giordano, N. Belmonte, G. Ferrari, B. E. Leone, F. Bertuzzi, G. Zerbini, P. Allavena, E. Bonifacio and L. Piemonti, Bone marrow mesenchymal stem cells

express a restricted set of functionally active chemokine receptors capable of promoting migration to pancreatic islets, *Blood*, 2005, **106**, 419–427.

37. M. Sasaki, R. Abe, Y. Fujita, S. Ando, D. Inokuma and H. Shimizu, Mesenchymal stem cells are recruited into wounded skin and contribute to wound repair by transdifferentiation into multiple skin cell type, *J. Immunol.*, 2008, **180**, 2581–2587.

38. R. M. Dwyer, S. M. Potter-Beirne, K. A. Harrington, A. J. Lowery, E. Hennessy, J. M. Murphy, F. P. Barry, T. O'Brien and M. J. Kerin, Monocyte chemotactic protein-1 secreted by primary breast tumors stimulates migration of mesenchymal stem cells, *Clin. Cancer Res.*, 2007, **13**, 5020–5027.

39. I. Rosova, M. Dao, B. Capoccia, D. Link and J. A. Nolta, Hypoxic preconditioning results in increased motility and improved therapeutic potential of human mesenchymal stem cells, *Stem Cells*, 2008, **26**, 2173–2182.

40. M. Shi, J. Li, L. Liao, B. Chen, B. Li, L. Chen, H. Jia and R. C. Zhao, Regulation of CXCR4 expression in human mesenchymal stem cells by cytokine treatment: role in homing efficiency in NOD/SCID mice, *Haematologica*, 2007, **92**, 897–904.

41. B. Annabi, Y.-T. Lee, S. Turcotte, E. Naud, R. R. Desrosiers, M. Champagne, N. Eliopoulos, J. Galipeau and R. Beliveau, Hypoxia promotes murine bone-marrow-derived stromal cell migration and tube formation, *Stem Cells*, 2003, **21**, 337–347.

42. A. De Becker, P. Van Hummelen, M. Bakkus, I. Vande Broek, J. De Wever, M. De Waele and I. Van Riet, Migration of culture-expanded human mesenchymal stem cells through bone marrow endothelium is regulated by matrix metalloproteinase-2 and tissue inhibitor of metalloproteinase-3, *Haematologica*, 2007, **92**, 440–449.

43. C. Ries, V. Egea, M. Karow, H. Kolb, M. Jochum and P. Neth, MMP-2, MT1-MMP, and TIMP-2 are essential for the invasive capacity of human mesenchymal stem cells: differential regulation by inflammatory cytokines, *Blood*, 2007, **109**, 4055–4063.

44. K. H. Husnain, J. Shujia, M. I. Niagara and A. Muhammad, IGF-1–Overexpressing Mesenchymal stem cells accelerate bone marrow stem cell mobilization *via* paracrine activation of SDF-1_/CXCR4 signaling to promote myocardial repair, *Circ. Res.*, 2008, **103**, 1300–1308.

45. S. Francois, M. Bensidhoum, M. Mouiseddine, C. Mazurier, B. Allenet, A. Semont, J. Frick, A. Sache, S. Bouchet, D. Thierry, P. Gourmelon, N.-C. Gorin and A. Chapel, Local irradiation not only induces homing of human mesenchymal stem cells at exposed sites but promotes their widespread engraftment to multiple organs: a study of their quantitative distribution after irradiation damage, *Stem Cells*, 2006, **24**, 1020–1029.

46. F. Casiraghi, N. Azzollini, P. Cassis, B. Imberti, M. Morigi, D. Cugini, R. A. Cavinato, M. Todeschini, S. Solini, A. Sonzogni, N. Perico, G. Remuzzi and M. Noris, Pretransplant infusion of mesenchymal stem cells prolongs the survival of a semiallogeneic heart transplant through the generation of regulatory T cells, *J. Immunol.*, 2008, **181**, 3933–3946.

47. G. D. Wu, J. A. Nolta, Y.-S. Jin, M. L. Barr, H. Yu, V. A. Starnes and D. V. Cramer, Migration of mesenchymal stem cells to heart allografts during chronic injection, *Immunobiology*, 2003, **75**, 679–685.

48. B. Rüster, S. Gottig, R. J. Ludwig, R. Bistrian, S. Muller, E. Seifried, J. Gille and R. Henschler, Mesenchymal stem cells display coordinated rolling and adhesion behavior on endothelial cells, *Blood*, 2006, **108**, 3938–3944.

49. K. Németh, A. Leelahavanichkul, P. S. Yuen, B. Mayer, A. Parmelee, K. Doi, P. G. Robey, K. Leelahavanichkul, B. H. Koller, J. M. Brown, X. Hu, I. Jelinek, R. A. Star and E. Mezey, Bone marrow stromal cells attenuate sepsis *via* prostaglandin E(2)-dependent reprogramming of host macrophages to increase their interleukin-10 production, *Nat. Med.*, 2009, **15**, 42–49.

50. A. J. Nauta and W. E. Fibbe, Immunomodulatory properties of mesenchymal stromal cells, *Blood*, 2007, **110**, 3499–3506.

51. J. K. Hsiao, M. F. Tai, H. H. Chu, S. T. Chen, H. Li, D. M. Lai, S. T. Hsieh, J. L. Wang and H. M. Liu, Magnetic nanoparticle labeling of mesenchymal stem cells without transfection agent: cellular behavior and capability of detection with clinical 1.5 T magnetic resonance at the single cell level, *Magn. Reson. Med.*, 2007, **58**, 717–724.

52. D. L. Kraitchman, M. Tatsumi, W. D. Gilson, T. Ishimori, D. Kedziorek, P. Walczak, W. P. Segars, H. H. Chen, D. Fritzges, I. Izbudak, R. G. Young, M. Marcelino, M. F. Pittenger, M. Solaiyappan, R. C. Boston, B. M. W. Tsui, R. L. Wahl and J. L. Bulte, Dynamic imaging of allogeneic mesenchymal stem cells trafficking to myocardial infarction, *Circulation*, 2005, **112**, 1451–1461.

53. P. Walczak, J. Zhang, A. A. Gilad, D. A. Kedziorek, J. Ruiz-Cabello, R. G. Young, M. F. Pittenger, P. C. M. van Zijl, J. Huang and J. W. M. Bulte, Dual-modality monitoring of targeted intraarterial delivery of mesenchymal stem cells after transient ischemia, *Stroke*, 2008, **39**, 1569–1574.

54. I. A. Potapova, P. R. Brink, I. S. Cohen and S. V. Doronin, Culturing of human mesenchymal stem cells as three-dimensional aggregates induces functional expression of CXCR4 that regulates adhesion to endothelial cells, *J. Biol. Chem.*, 2008, **283**, 13100–13107.

55. G. Y. Rochefort, B. Delorme, A. Lopez, O. Herault, P. Bonnet, P. Charbord, V. Eder and J. Domenech, Multipotential mesenchymal stem cells are mobilized into peripheral blood by hypoxia, *Stem Cells*, 2006, **24**, 2202–2208.

56. I. M. Barbash, P. Chouraqui, J. Baron, M. S. Feinberg, S. Etzion, A. Tessone, L. Miller, E. Guetta, D. Zipori, L. H. Kedes, R. A. Kloner and J. Leon, Systemic delivery of bone marrow-derived mesenchymal stem cells to the infarcted myocardium: feasibility, cell migration, and body distribution, *Circulation*, 2003, **108**, 863–868.